Digital Television Fundamentals

McGraw-Hill Video & Audio Books

Digital Television Fundamentals

Design and Installation of Video and Audio Systems

Michael Robin

Michel Poulin

Second Edition

McGraw-Hill

New York   San Francisco   Washington, D.C.   Auckland   Bogotá
Caracas   Lisbon   London   Madrid   Mexico City   Milan
Montreal   New Delhi   San Juan   Singapore
Sydney   Tokyo   Toronto

Library of Congress Cataloging-in-Publication Data

Robin, Michael, date.
   Digital television fundamentals : design and installation of video and audio systems /
Michael Robin, Michel Poulin.—2nd ed.
      p.   cm.
   Includes index.
   ISBN 0-07-135581-2
   1. Digital television. I. Poulin, Michel. II. Title.

TK6678.R63   2000
621.388—dc21   00-032904

McGraw-Hill

A Division of The McGraw·Hill Companies

1 2 3 4 5 6 7 8 9 0   PBT/PBT   0 6 5 4 3 2 1 0

ISBN 0-07-135581-2

*The sponsoring editor for this book was Stephen S. Chapman, the editing supervisor was David E. Fogarty, and the production supervisor was Pamela A. Pelton.
It was set in Century Schoolbook by Paul Scozzari of McGraw-Hill's Professional
Book Group Hightstown composition unit.*

Printed and bound by Phoenix Book Technology.

McGraw-Hill books are available at special quantity discounts to use as premiums and sales promotions, or for use in corporate training programs. For more
information, please write to the Director of Special Sales, Professional
Publishing, McGraw-Hill, Two Penn Plaza, New York, NY 10121-2298. Or contact your local bookstore.

Contents

Preface

Analog television, as it has been commercially available to the public for some 60 years, is undergoing major changes. These changes have to do with the manner in which the electrical equivalent of the image is generated, processed, recorded, and transmitted to the intended viewer.

Young professionals and recent graduates have a very sketchy and limited appreciation of the analog television technology on which all-digital television developments are based. However, mostly owing to job-related pressures, more mature professionals, deeply anchored in the analog television technology, have had only a limited opportunity to acquire an adequate understanding of the rapidly evolving digital technology.

The authors were fortunate in being able to make a smooth transition from the analog to the digital technology mostly during their tenure at the Canadian Broadcasting Corporation's Engineering Headquarters in Montreal, Canada. This book reflects the experience they have acquired in the areas of television production systems design and implementation, evaluation of equipment and technologies, active participation in the development of SMPTE standards, development of test procedures and equipment, training of engineering and technical personnel, presentations at various SMPTE conferences, and consulting to broadcasters and equipment manufacturers. This activity culminated in their active involvement in the development and implementation of the all-digital Toronto Broadcasting Center of the Canadian Broadcasting Corporation, the first major teleproduction complex using the bit-serial 270-Mbps signal distribution concept.

This book is intended for practicing engineers and technicians in the field of broadcasting and studio/video productions as well as for students of broadcast technology. It revisits the basics of analog television, video and audio, for the benefit of neophytes. It then proceeds to gradually introduce the reader to the basics of digital video and audio, reviewing, detailing, and explaining the international standards and their practical implementations.

Chapter 1 revisits the basic principles of television, irrespective of the scanning standards. It reacquaints the reader with some of the basic physical and physiological constraints, which have influenced the adoption of cost-effective analog television standards, given the technology of the times. Several aspects of the basic television standards such as scanning, resolution, video signal makeup, baseband spectrum, and transmission constraints are discussed in detail.

Chapter 2 deals with analog video fundamentals. It covers such topics as the general principles of color television (colorimetry, transducer transfer characteristics, the basic ingredients of the color television signal, and a detailed discussion of the reference color television signal: the color bars signal); the composite video concept (common characteristics of various color television systems such as compatibility with monochrome television and the principle of frequency-division multiplexing, summary of the contemporary standard definition composite analog color television systems NTSC, PAL, SECAM, performance-indicative parameters and measurement concepts, the distribution and recording of composite video signals); and the component video concept (detailed descriptions of component video signals, performance-indicative parameters and measurements concepts, the distribution and recording of component video signals).

Chapter 3 introduces the reader to the digital video world. It covers such topics as general considerations (historical background, the typical black box digital device, sampling of the signal, quantizing of the sampled values, dynamic range and headroom concept, quantizing error, and digital-to-analog conversion), the composite video digital standards (the $4f_{SC}$ NTSC and $4f_{SC}$ PAL standards detailing the general specifications, sampling structure, quantizing range and its implications, digital raster structure, performance-indicative parameters and test concepts, and a description of the bit-parallel signal distribution concept) and the component digital standards (the sampling rates, coded signals, sampling frequencies, quantizing range, sampling structure, time-domain multiplexing of data, timing reference signals, ancillary data concept, bit-parallel signal distribution concept, a review of component digital formats, and a discussion of the performance-indicative parameters and test concepts).

Chapter 4 revisits some basic principles of acoustics as they relate to the human hearing process and discusses such topics as the sound pressure level (SPL), loudness, as well as the dynamic range and spectral resolution of the ear.

Chapter 5 deals with analog audio topics such as electric signal levels, typical signal levels and level monitoring, performance-indicative parameters, dynamic range, and performance targets.

Chapter 6 introduces the reader to the digital audio world. It covers such topics as general concepts of digital audio, A/D conversion, D/A conversion, biphase mark encoding, general structure of the interface and its implementation, digital audio signal distribution, the MADI and SDIF interfacing protocol formats, audio synchronization, and digital audio recording.

Chapter 7 deals with the bit-serial signal distribution concept and the ancillary data multiplexing. It covers such topics as Shannon's theorem, channel coding, the eye diagram, the bit-serial distribution standard (interface characteristics, $4f_{SC}$, and 4:2:2 applications), performance-indicative parameters and measurement concepts (transmitter-related, medium-related, receiver-related), special test signals, digital audio multiplexing, digital videotape recording, and system considerations.

Chapter 8 covers with digital signal compression and distribution. It discusses such topics as general concepts, video data reduction techniques (with emphasis on the DCT coding process, video compression standards such as JPEG and MPEG, and video BRR performance and applications), audio data reduction techniques (with emphasis on audio compression standards and audio BRR performance and applications), and the distribution of compressed signals.

Chapter 9 deals with computers and television. It covers such topics as the computer architecture, internal communication data buses (main, local, over the top, switched, router), computer display monitors, and expansion cards (controller, V/A, and PCMCIA).

Chapter 10 treats the subject of multimedia and television. It covers such topics as the concept, technologies, hardware, and systems (PC workstations, A/V signal processing, disk storage, servers, cameras, VCRs, CD-ROM, the DVI); the interconnections (interfaces, networks); multimedia software, systems, and applications (VOD, NVOD, Photo CD, CD-I, CTI); and multimedia standardization activities.

Chapter 11 covers advanced television (ATV) concepts. It treats such topics as the need for ATV, the ATV emergence, standardization efforts, the digital television (DTV) solution (interoperability, flexibility, compression, progressive versus interlaced scanning, image and pixel aspect ratio, production and clean aperture, audio considerations, and compatibility with film). A detailed analysis of the most important DTV formats follows, including the serial distribution concept and data embedding. The Grand Alliance system and the European approach are explained.

Chapter 12 introduces the reader to new infrastructure concepts and transport technologies that are needed in intra- and interfacility applications to transfer audio-video data streams and media files. Concepts of TV program conversion between all current video formats are developed. The remainder of the chapter is devoted to technical and practical considerations concerning the implementation of DTV systems, and cost-effective transition scenarios to implement a DTV architecture.

The book can be used as a tutorial, allowing the reader to proceed from basics to more advanced topics. Alternatively, an informed reader can select the chapter of particular interest.

This book could not have been written without the experience gained by the authors in the laboratories of CBC Engineering. The authors wish to thank SONY Corporation and Tektronix for their support and encouragement and the SMPTE for permission to use its standards as a reference source.

Last, but not least, the authors wish to thank their spouses for allowing them to scatter large amounts of paper in their respective homes in preparation for the publication of this book.

Michael Robin

Michel Poulin

1

Basics of Television

Conventional television, as is currently broadcast to home viewers, was developed in the 1930s, which was a time of rapid advance in the various techniques of telecommunication, among them, the transmission of sound and pictures. Conventional television standards are the result of these early developments. They reflect the technological limitations of the times, as well as human vision characteristics, and were a compromise between cost and performance. Given the large number of television receivers throughout the world, any technological advance has to be compatible in some manner with the existing standards.

1.1 Historical Background

After experimentation with unsatisfactory mechanical image-scanning methods, the electronic scanning method was adopted in the middle 1930s. Regular "high-resolution" television transmissions began almost simultaneously in England, Germany, and France. The picture definition of the day was about 400 lines per picture, for example, 441 lines in Germany and France and 405 lines in England. The horizontal-to-vertical aspect ratio of the picture was 4:3 and is still being used today in conventional television systems.

In 1941, after years of experimenting with various 300-line and 400-line picture formats, the United States adopted the 525-line National Television System Committee (NTSC) standard. This standard is still in use today with minor backward-compatible modifications.

After World War II, England continued with its 405-line broadcasts and France with its 441-line broadcasts. In 1948, France adopted the 819-line national television standard. The rest of Europe adopted the 625-line standard. For a while, there were no fewer than three scanning standards, two color standards [phase-alternating line (PAL) and séquential couleurs à mémoire (sequential colors with memory, or SECAM)], and seven incompatible transmission standards in simultaneous operation in Europe. The situation was corrected in the early 1980s when the French 819-line transmissions and the English 405-line transmissions were phased out. Currently, Europe shares a single scanning standard (625/50), two color standards (PAL and SECAM), and only four incompatible transmission standards.

1.2 The Eye-Brain Mechanism

1.2.1 The characteristics of visible light

Light is usually identified by wavelength rather than by frequency. Visible light is confined to a relatively narrow range of wavelengths, from about 380 to 760 nm (1 nm = 10^{-7} cm). The eye perceives various wavelengths as color hues. The wavelengths corresponding to the three primary colors are

- Red: 700.0 nm
- Green: 546.1 nm
- Blue: 435.8 nm

1.2.2 The light perception

The retina, upon which the image looked at is focused, consists of two types of receptors known as rods and cones. There are between 110 million and 130 million rods and between 6 and 7 million cones.

- The rods predominate in the periphery of the retina, are more sensitive to light than the cones, and are responsible for night (scotopic) colorless vision. The rods have limited visual acuity.
- The cones predominate in the central area (fovea) of the retina, respond to higher levels of light intensity than the rods, and are responsible for day-light (photopic) color vision. At high light intensity levels the cones have a high colorless visual acuity and a diminished color visual acuity. As the light intensity decreases, the perception is shifted to the periphery of the retina where rods are more numerous.

The information received by the retina is transmitted to the brain through the optic nerve, which consists of about 800,000 individual fibers. Each fiber is fed by a dedicated ganglion cell. Almost every ganglion cell has connections to hundreds of rod cells and tens of cone cells. In the fovea region each cone has a direct connection to a dedicated ganglion cell in addition to sharing other ganglion cells with groups of cones and rods. This accounts for the high acuity of vision in the center of the visual field. This acuity diminishes as the light intensity decreases.

The information generated by the rods and cones is fed simultaneously to the brain, where the process of perception takes place. Figure 1.1 shows a simplified "block diagram" of the eye-brain mechanism outside of the fovea region.

The eye-brain mechanism results in two consequences:

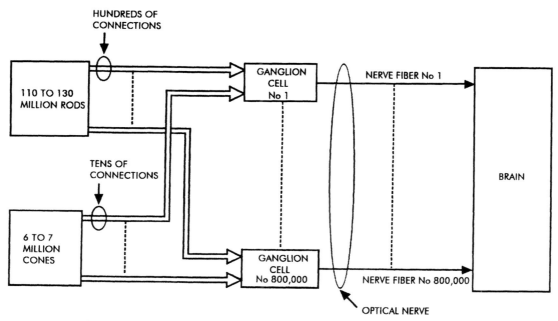

Figure 1.1 Simplified block diagram of eye-brain mechanism outside the fovea region.

- The highest visual acuity occurs in the center of the image.
- Night vision is colorless.

1.2.3 Visual acuity

Visual acuity is measured as the angle subtended by the smallest visible detail in an object. Figure 1.2 illustrates the concept of visual acuity.

Television system design takes as a reference a visual acuity of the eye of the order of 1 minute of arc. The extent to which a picture medium such as television can reproduce fine detail is expressed in terms of resolution. Television resolution is equal to the number of alternately white and black horizontal lines that can be resolved vertically over the full height of the screen. It is expressed in lines per picture height (LPH). It is determined by the rod-and-cone structure of the eye and depends upon the brightness level and contrast ratio. The 525-line and 625-line standards were developed taking into consideration the visual acuity of the eye (1 min), assumed viewing conditions in the average home (viewing distance six times the picture height), and transmission-spectrum-saving concerns. The relationship between the number of picture elements that can be resolved given a specified picture height and viewing distance is given by

$$Nv = \frac{1}{\alpha n}$$

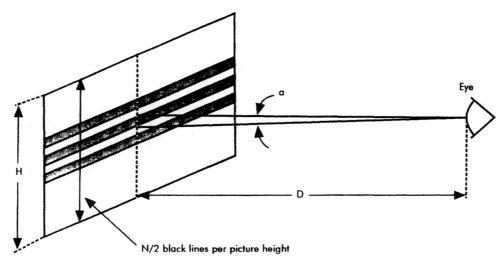

Figure 1.2 Visual acuity concept.

where N_v = Total number of elements to be resolved in the vertical direction
   α = Minimum resolvable angle of the eye (in radians)
   $n = D/H$ (viewing distance divided by picture height)

Given α = 1 min of arc, or 2.91×10^{-4} radians, and $n = 6$, we have

$$N_v = \frac{1}{(6 \times 2.91 \times 10^{-4})} \approx 572 \text{ lines}$$

This ballpark figure is at the origin of the number of lines specified for the two conventional television systems, namely, the 525-line system used mainly in North America and Japan, and the 625-line system used elsewhere in the world. The actual resolution is smaller than 525 or 625 lines for reasons explained in Sec. 1.4. High-definition-television standards, with 1125 or 1250 lines per picture, require shorter viewing distances (e.g., $n = 3$) or larger screen sizes to enable the eye to resolve all picture details.

When color images are viewed, the visual acuity depends on the color. The acuity for blue and red is about 75% of that of a white image of the same brightness. The acuity for green is about 90% of that of a white image of the same brightness.

1.2.4 Persistence of vision

Persistence of vision is the ability of the viewer to retain or in some way to remember the impression of an image after it has been withdrawn from view. When light entering the eye is shut off, the impression of light persists for about 0.1 s. Ten still pictures per second is an adequate rate to convey the illusion of motion.

Motion pictures and television use higher rates than 10 still pictures per second in order to reduce the visibility of flicker. The critical flicker frequency is the minimum rate of interruption of the projected light that will not cause the motion picture to appear to flicker. The perceptibility of flicker varies widely with viewing conditions. Among the factors affecting the flicker threshold are luminance of the flickering area, the color of the area, the solid angle subtended by the area at the eye, the absolute size of the flickering area, the luminance of the surrounding area, the luminance variation with time and position within the flickering area, and the adaptation and training of the observer. In a constant viewing situation, that is, no change in the image or surrounding area, the luminance at which flicker just becomes perceptible varies logarithmically with luminance (the Ferry-Porter law). Empirical data indicate that increasing the flicker frequency by 12.6 cycles per second raises the flicker threshold level 10 times.

Motion pictures consist essentially of a sequence of 24 photographs (frames) of a single subject that are taken every second and projected in the same sequence to create an illusion of motion. Each successive image of a moving object is slightly different from the preceding one. When projected, each frame is presented twice, through the use of a mechanical shutter, resulting in a flicker rate of 48 cycles per second.

In television the picture elements are laid down on the screen one after the other through a process of scanning, but are perceived at the same time because of the persistence of vision. Scanning consists of breaking down the picture into a series of horizontal lines, for example, 525 or 625 in conventional television. In a process called interlaced scanning, each image is analyzed and synthesized in two sets of spaced lines. Each of the two sets comprises one half of the total number of lines (262.5 or 312.5) and fits successively within the spaces of the other. Each successive set of lines is called a *field*. Two consecutive (interlaced) fields constitute a frame. The field repetition frequency is nominally 60 Hz in the 525-line standard and 50 Hz in the 625-line standard. The frame repetition frequency is 30 and 25 Hz, respectively. In television the applicable flicker frequency is the field frequency. Two adjacent lines of two consecutive fields may not be identical, resulting in interline flicker. Interline flicker is tolerable because the eye is relatively insensitive to flicker when the variation of light is confined to a small part of the field of view.

Table 1.1 gives the flicker threshold for commonly encountered flicker frequencies. The low flicker threshold typical of motion pictures explains why

TABLE 1.1 Flicker Threshold for Commonly Encountered Flicker Frequencies

Picture source	Flicker frequency, Hz	Frames per second	Flicker threshold, cd/m^2
Movies	48	24	68.5
50-Hz television	50	25	99.4
60-Hz television	60 (nominal)	30 (nominal)	616.7

they are projected in darkened rooms. The flicker threshold of the 525/60 scanning standard is considerably higher than that of the 625/50 scanning standard, resulting in more comfortable viewing in brightly lit rooms.

1.2.5 Spectral visibility

The physical quantity that primarily determines the sensation of light is its wavelength. What is physically defined as wavelength is subjectively perceived as color. Ordinary white light contains a continuum of wavelengths throughout and beyond the range of visibility. Any visible radiation of uniform wavelength is perceived by the eye as a single (monochromatic) color. Under photopic viewing conditions, the brightest part of a spectrum, consisting of equal amounts of energy at all wavelengths, corresponds to a wavelength of about 560 nm. From this maximum, visibility falls off toward both ends of the spectrum. Under scotopic viewing conditions, the maximum perceived brightness shifts down to about 500 nm, resulting in a drastically reduced visibility in the red region. Figure 1.3 shows the relationship between scotopic and photopic vision.

1.3   The Scanning Standards

The scanning standards define the manner in which a television scene is explored for its luminance and chrominance values. They specify the number of lines per frame and the number of frames per second. Technical and economic considerations have led to certain compromises in the transmission of the essential information required by the eye. The first important consideration is the fact that any electronic system is capable of transmitting only one bit of information at a time. Consequently, the picture has to be broken down into

Figure 1.3 Visibility curves of the human retina.

small elements transmitted sequentially and then reconstructed at the receiver. In the end, all the elements of the reconstructed picture have to appear simultaneously to the eye.

1.3.1 The scanning process

The conventional television standards reflect the image pickup and display technology of the 1930s. This assumes that the camera uses a pickup tube where the image is focused onto a photoconductive layer. Electrical charges, proportional to the illuminated scene at each point, are developed and stored capacitively on this layer. An electron beam is used to convert the charge image into an electrical current. This beam is focused to a circular spot and deflected continuously over the image in two consecutive fields of horizontal lines. Each consecutive field contains half of the total number of scanning lines into which the picture is scanned. Two consecutive fields (field 1 and field 2) are displaced vertically such that their scanning lines are interlaced, and together they form a frame. The image is scanned from left to right, starting at the top and tracing successive lines until the bottom of the picture is reached. The beam then returns to the top and the process is repeated. The continuous deflection of the electron beam is achieved by subjecting it to two perpendicular (vertical and horizontal) magnetic fields that result from repetitive sawtooth-shaped currents flowing through a pair of (horizontal and vertical) deflection coils. The process is called *linear interlaced scanning*. The repetition rate of the horizontal component is related to the vertical component by the factor n, resulting in the formation of n lines during a complete vertical period. The retrace times involved (both horizontal and vertical) are a result of the physical limitations of early scanning systems. The retrace times are not utilized for the transmission of a video signal but for the transmission of auxiliary information such as horizontal and vertical scanning synchronization.

In the display device, a cathode-ray tube (CRT) re-creates the original picture. A focused electron beam, deflected horizontally and vertically in synchrony with the pickup tube electron beam, is projected onto a phosphor-coated viewing screen. The CRT beam current is, ideally, proportional to the beam current in the pickup tube, and the deflection currents through the deflection coils are in synchrony with those of the pickup tube. In reality, the CRT electron beam current versus control voltage transfer characteristic is not linear. To correct for this condition, the camera video amplifier introduces an opposite nonlinearity, called *gamma correction,* resulting in a linear relationship between original picture brightness and CRT-reproduced brightness. This subject will be discussed further in Chap. 2. Figure 1.4 shows a simplified block diagram of the monochrome television system from signal source to CRT display.

1.3.2 Lines per frame

This parameter was chosen to provide a value of vertical resolution appropriate to the acuity of normal vision at a distance of about six times the screen height.

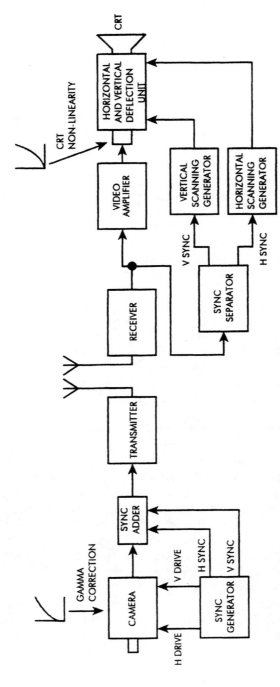

Figure 1.4 Simplified block diagram of monochrome television system from signal source to CRT display.

There is an odd number of lines per frame. The standardized values for conventional broadcast television presently are 525 and 625 lines per frame.

1.3.3 Pictures per second

The pictures per second standard is chosen to provide a sufficiently rapid succession of complete pictures (frames) to avoid flicker at levels of image brightness appropriate to viewing images in domestic surroundings. The frames are made up of two consecutive fields, each containing half of the total number of lines (262.5 or 312.5). The lines in two consecutive fields are interlaced, resulting in a frame made up of the total amount of lines (525 or 625). Historically, the values for the field repetition frequency were chosen to be equal to the power-line frequency, 60 Hz in the United States, Canada, and Mexico, and 50 Hz in other parts of the world. Under extreme conditions, a display in synchrony with the power-line frequency reduces the visibility of scanning distortions caused by stray magnetic fields and hum components, should they exist. This reduced visibility is obtained when the receiver and the transmitter operate from the same power source, which is not always the case. Consequently, the practice of synchronizing the field rate to the power line frequency has long been discontinued and, today, the vertical scanning frequency is only nominally equal to the power-line frequency, since it is obtained by counting down from a highly stable crystal-controlled high-frequency oscillator.

1.3.4 The conventional scanning standards

Two conventional television scanning standards coexist in the world today. These are the 525/50 standard and the 625/50 standard. These standards represent a cost versus performance choice based on the technology of the 1930s. Table 1.2 summarizes their characteristics.

1.4 The Resolution Concept

Historically, *resolution* was understood to mean "limiting resolution," or the point at which adjacent elements of an image cease to be distinguished. Various disciplines measure and specify resolution differently. Resolution can be specified as

- The number of units (i.e., lines or line pairs) per unit distance along the vertical and horizontal axis, such as lines per millimeter.
- The number of units (i.e., lines) for a full display, such as lines per picture height (LPH).

In television, the resolution is specified in terms of LPH. The various conventional television systems in use today were designed to achieve equal horizontal and vertical resolution, better known as *square pixels*.

TABLE 1.2   Significant Parameters of Conventional Scanning Standards

Parameter	525/60 Standard	625/50 Standard
Number of lines per frame	525	625
Number of lines per field	262.5	312.5
Number of frames per second	29.97	25
Number of fields per second (f_V), Hz	$2f_H/525 = 59.94$	$2f_H/625 = 50$
Horizontal scanning frequency (f_H), Hz	$3 \times 5 \times 5 \times 7(f_V/2) = 15{,}734.25$	$5 \times 5 \times 5 \times 5(f_V/2) = 15{,}625$
Field blanking duration (lines)	20	25
Frame blanking duration (lines)	40	50
Number of active lines per frame	485	575
Vertical resolution (N_V), LPH	$485 \times 0.7 \approx 339$	$575 \times 0.7 \approx 402$
Total line duration, μs	63.556	64
Horizontal blanking duration, μs	10.7 ± 0.1	12 ± 0.3
Active line duration, μs	52.856	52
Horizontal pixels for equal H/V resolution	$339 \times (4/3) \approx 452$	$402 \times (4/3) = 536$
Line-pair cycle duration (T), μs	$52.85/226 \approx 0.2338$	$52/268 = 0.194$
Bandwidth for equal H/V resolution, MHz	$1/T \approx 4.28$	$1/T = 5.15$
Horizontal resolution factor, lines/MHz	$339/4.28 \approx 79.2$	$402/5.15 = 78$
Horizontal resolution (N_H), LPH	333 (@4.2-MHz bandwidth)	390 (@5-MHz bandwidth)
H/V resolution ratio	0.98	0.97

1.4.1   Vertical resolution

The vertical resolution is independent of the system bandwidth and defines the capability of the system to resolve horizontal lines. It is expressed as the number of distinct horizontal lines, alternately black and white, that can be satisfactorily resolved on a television screen. Vertical resolution depends primarily on the number of scanning lines per picture and the combined effects of the camera pickup tube and the CRT scanning spot size and shape.

Ideally, the vertical resolution would be equal to the number of active lines per frame. This would happen if the scanning lines were centered on the picture details as shown in Fig. 1.5. The scanning lines cannot be assumed to occupy a fixed position relative to vertical detail at all times. Complete loss of vertical resolution will occur when the scanning spot straddles picture details as shown in Fig. 1.6. From subjective data, it has been found that raster lines in excess of the number of elements to be resolved are necessary, as shown in Fig. 1.7. This can be expressed by

$$N_V = kN_{AL}$$

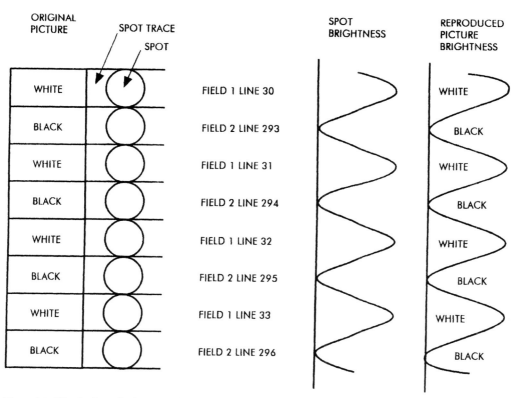

Figure 1.5 Vertical resolution equals number of active lines when the raster lines are centered on the picture details.

where N_V = Number of active vertical picture elements (pixels) to be resolved.

N_{AL} = The number of active lines (excluding lines formed while the beam is returning to the top of the picture).

k = Constant obtained from subjective measurements. This is called the *Kell factor* and is usually taken as 0.7.

In the 525/60 scanning standard there is a total of 525 lines per frame, of which 40 are blanked, leaving 485 active lines per frame. Given a Kell factor of 0.7, the effective vertical resolution of the 525/60 scanning standard is:

$$N_V = 0.7 \times 485 \approx 339 \text{ LPH or pixels}$$

In the 625/50 scanning standard there is a total of 625 lines per frame, of which 50 are blanked, leaving 575 active lines per frame. Given a Kell factor of 0.7, the effective vertical resolution is:

$$N_V = 0.7 \times 575 \approx 402 \text{ LPH}$$

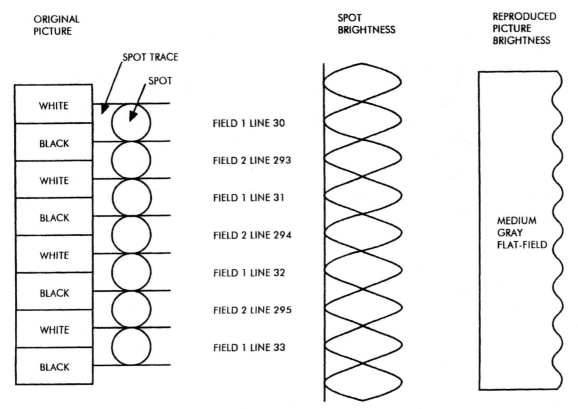

ORIGINAL
PICTURE

SPOT TRACE

SPOT

WHITE

BLACK

WHITE

BLACK

WHITE

BLACK

WHITE

BLACK

FIELD 1 LINE 30

FIELD 2 LINE 293

FIELD 1 LINE 31

FIELD 2 LINE 294

FIELD 1 LINE 32

FIELD 2 LINE 295

FIELD 1 LINE 33

SPOT
BRIGHTNESS

REPRODUCED
PICTURE
BRIGHTNESS

MEDIUM
GRAY
FLAT-FIELD

Figure 1.6 Loss of vertical resolution resulting from scanning spot straddling picture details.

1.4.2   Horizontal resolution

The horizontal resolution is directly related to the system bandwidth and defines the ability of the system to resolve vertical lines. It is expressed as the number of distinct vertical lines, alternately black and white, that can be satisfactorily resolved in three quarters of the width of a television screen. The horizontal resolution depends on the combined effects of the camera pick-up tube and CRT scanning spot dimensions as well as the high-frequency amplitude and phase response of the transmission medium. A system with a horizontal to vertical aspect ratio of 4/3, as in conventional television, needs to allow for $(4/3)N_v$ horizontal pixels to be resolved. In the 525/60 scanning standard, this results in $339 \times 4/3 \approx 452$ horizontal pixels.

Because of the finite size of the scanning spot, a beam exploring a pair of contiguous white-and-black pixels (line pair) results in a sine wave with a positive half-wave corresponding to the white pixel and a negative half-wave corresponding to the black pixel (see Fig. 1.8). A scanning beam exploring a

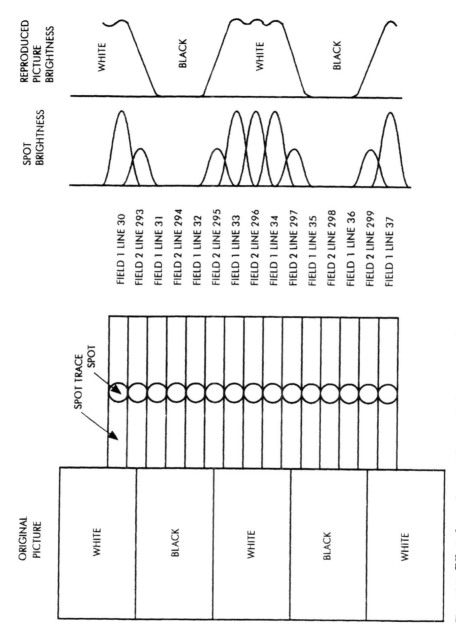

Figure 1.7 Effect of scanning spot shape and size on vertical resolution.

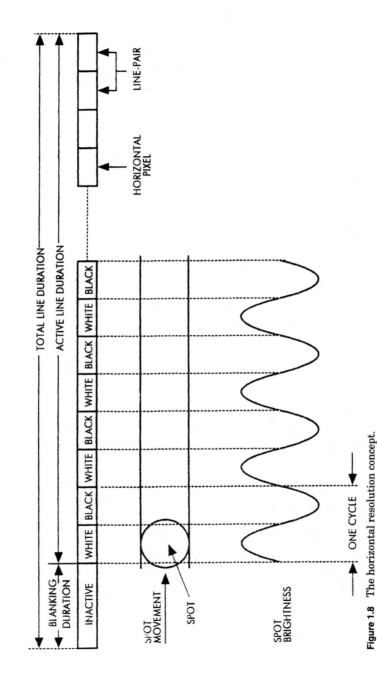

Figure 1.8 The horizontal resolution concept.

picture made up of 452 horizontal pixels results in an electrical signal with 226 complete cycles during the active horizontal scanning line.

In the 525/60 scanning standard the total horizontal scanning line duration is 63.5 μs and the horizontal blanking duration is 10.7 μs, resulting in an active line duration of 52.85 μs. The duration of a single cycle is

$$T = \frac{52.85 \ \mu s}{226} \approx 0.2338 \ \mu s$$

The fundamental frequency resulting from scanning 452 horizontal pixels is

$$F = \frac{1}{T} = \frac{1}{0.2338 \ \mu s} \approx 4.28 \ \text{MHz}$$

This is the bandwidth required for equal horizontal and vertical resolution. The horizontal resolution factor for a 4.28-MHz bandwidth is

$$\frac{339}{4.28} \ \text{MHz} = 79.2 \ \text{lines/MHz}$$

In countries using the 525/60 scanning standard (CCIR M) the maximum transmitted baseband video frequency is 4.2 MHz, resulting in a transmitted horizontal resolution of

$$N_H = 4.2 \ \text{MHz} \times 79.2 \ \text{lines/MHz} \approx 333 \ \text{lines}$$

The resulting horizontal versus vertical resolution ratio is therefore $333/339 \approx 0.982$. From an analog point of view, this represents a quasi-square pixel.

The minimum video bandwidth for equal horizontal and vertical resolution in the 625/50 scanning standard is 5.15 MHz, and the resulting horizontal resolution factor is 78 lines/MHz. Various countries have adopted different maximum transmitted baseband video frequency values, resulting in different transmitted horizontal resolutions as shown in Table 1.3.

Table 1.2 lists relevant figures of significant parameters for the 525/60 and 625/50 scanning standards.

TABLE 1.3 Horizontal Resolution Capability of Various 625/50 Transmission Standards

Standard	Bandwidth, MHz	N_H, LPH	N_H/N_V
CCIR N	4.2	327	0.81
CCIR B,G	5	390	0.97
CCIR I	5.5	429	1.067
CCIR K,L	6	468	1.16

1.5   The Composite Video Signal

The satisfactory reproduction of a picture requires the transmission of several types of information combined into a single waveform called the *composite video signal*. This signal is composed of

- Video information
- Synchronizing information

1.5.1   The video information

Video signals generated by a camera are suitably amplified to a standard level. The video signal conveys information concerning

- The blanking level
- The black reference level
- The average scene brightness level
- Picture details
- Color values (see Chap. 2)

The video signal is unipolar with one direct current (DC) level (nominally 0 V) representing black and a second level (nominally +700 mV) representing white. Any level between 0 and 700 mV represents a degree of gray.

1.5.2   The synchronizing information

The synchronizing information consists of

- Horizontal scanning synchronization
- Vertical scanning synchronization
- Chrominance decoder synchronization (see Chap. 2)

The horizontal and vertical synchronizing information is used to trigger the horizontal and vertical deflection circuits in the receiver. It consists of pulses having a specific amplitude, duration, and shape best suited to the task at hand. The synchronizing pulses are unipolar with a reference level of 0 V and a peak negative level of nominally −300 mV.

1.5.3   The makeup of the composite video signal

The video signal waveform, with a nominal peak-to-peak amplitude of 700 mV, and the synchronizing signal waveform, with a nominal peak-to-peak amplitude of 300 mV, are added together to form a composite video signal with a peak-to-peak amplitude of 1 V. The synchronizing pulses are placed in parts of the composite video signal that do not contain active picture information. These parts are blanked (forced to or below the black level) to render invisible the retrace of scanning beams on a correctly adjusted display.

1.5.4 Interface characteristics

The video equipment interconnections are made using an unbalanced coaxial cable. The source impedance, terminating impedance, and coaxial cable characteristic impedance is 75 ohms.

The peak-to-peak value of the luminance (monochrome) signal plus the synchronizing signal, measured from the peak of the synchronizing pulse (sync tip) to the peak white level, is 1 V. This standard video signal level applies to both conventional television scanning standards. Composite color signal levels are discussed in Chap. 2. Even though the peak-to-peak composite video signal levels are identical, operational practices have resulted in slightly different significant signal levels as shown in Table 1.4 and Figs. 1.9 and 1.10 as follows:

- Signal levels are expressed in millivolts in the 625/50 scanning standard. In the 525/60 standard, and specifically in North America, composite video signal levels are expressed in IRE units [Institute of Radio Engineers now known as Institute of Electrical and Electronics Engineers (IEEE)]. A standard composite video signal with a peak-to-peak amplitude of 1 V is said to have an amplitude of 140 IRE units of which 100 IRE units of luminance (monochrome signal) and 40 IRE of sync.

- The peak white level is 700 mV in the 625/50 scanning standard and 714.3 mV (100 IRE) in the 525/60 scanning standard.

- The blanking level is equal to the black level (0 V) in the 625/50 scanning standard. In the 525/60 standard, as used in North America, the black level is nominally 7.5% of the peak white level (53.5 mV or 7.5 IRE). In a monochrome signal there are no video signal components below the black (blanking) level.

- The synchronizing signal level is 300 mV in the 625/50 scanning standard and 285.7 mV (40 IRE) in the 525/60 scanning standard.

Analog distribution equipment usually has a headroom adequate to accept video signal levels of up to 2 $V_{p\text{-}p}$. Headroom limitations occur in analog videotape recorders, common-carrier links, and television transmitters. To avoid equipment overload, the camera controls are adjusted continuously by the operator to limit the signal-level excursions to the specified upper and lower limits. Specialized video oscilloscopes, called *waveform monitors,* are used for this pur-

TABLE 1.4 Monochrome Composite Video Signal Levels

Parameter	525/60 standard	625/50 standard
White level	100 IRE (714.3 mV)	700 mV
Black (setup) level	7.5 IRE (53.5 mV)	0 mV
Blanking level	0 IRE (0 mV)	0 mV
Sync level	−40 IRE (−285.7 mV)	−300 mV

Figure 1.9 Significant amplitude and timing values of composite video signal relative to picture characteristics in 525/60 standard.

Figure 1.10 Significant amplitude and timing values of composite video signal relative to picture characteristics in 625/50 standard.

pose. Digital processing equipment, using analog input-output ports, requires special care because of the limited amount of overhead of the A/D converters.

1.5.5   Blanking intervals and structure

The horizontal and vertical blanking intervals are periods during which the scanning beam retrace occurs and in which synchronization information is located.

1.5.5.1   Horizontal blanking interval. Each horizontal line outside the vertical blanking interval is divided into an active line period and a horizontal blanking interval. The horizontal blanking interval contains a negative-going horizontal synchronizing pulse followed by the color synchronization burst (see Chap. 2). The remainder of the horizontal blanking interval is kept at the blanking level to delineate the synchronizing signal properly. The details of the horizontal blanking interval related to the two conventional scanning standards are described in Table 1.5 and Figs. 1.11 and 1.12.

1.5.5.2   Vertical blanking interval. Both conventional scanning standards feature an interlaced raster. Each television frame is divided into two fields. Each field contains half of the total number of scanning lines. The fields carry every other scanning line in succession and the following field carries the lines not scanned by the previous field. Each field is divided into an active picture area and a vertical blanking interval. The vertical blanking interval contains the vertical synchronizing information surrounded by blanking periods and auxiliary equalizing pulses. This technique permits the unambiguous recovery

TABLE 1.5   Details of Line-Synchronizing Signals (See Figs. 1.11 and 1.12)

Symbol	Parameter	525/60 standard	625/50 standard
H	Nominal line period, μs	63.556	64
A	Line blanking interval, μs	10.7 (derived)	12.05 ± 0.25
B	Horizontal reference point to horizontal blanking end, μs	9.2 +0.2/−0.1	10.5 (derived)
C	Horizontal blanking start to horizontal reference point (front porch), μs	1.5 ± 0.1	1.5 ± 0.3
D	Horizontal synchronizing pulse duration, μs	4.7 ± 0.1	4.7 ± 0.1
E	Horizontal synchronizing pulse end to blanking pulse end (back porch), μs	4.5 (derived)	5.8 (derived)
F	Horizontal blanking pulse rise-time, ns	140 ± 20	300 ± 100
G	Horizontal synchronizing pulse risetime, ns	140 ± 20	200 ± 100

Figure 1.11 525/60 scanning standard details of horizontal blanking interval.

of the vertical synchronization information required to trigger the vertical scanning circuits of the receiver adequately. The vertical synchronizing signal is organized as follows:

- 525/60 scanning standard: A 9-line block divided into three 3-line-long segments. The first of the segments contains six preequalizing pulses. The second segment contains the vertical synchronizing pulse with six serrations. The third segment contains six postequalizing pulses.

- 625/50 scanning standard: A 7½-line block divided into three 2½-line-long segments. The first segment contains five preequalizing pulses. The second segment contains the vertical synchronizing pulse with five serrations. The third segment contains five postequalizing pulses.

The remainder of the vertical blanking interval not used for the vertical synchronizing block is available for special vertical interval signals. When such signals are carried on a particular line, they are confined to the active period between the horizontal blanking intervals. When such signals are not carried on a particular line, the line is maintained at blanking level. Color-synchronizing-burst considerations are covered in Chap. 2.

Figure 1.12   625/50 scanning standard details of horizontal blanking interval.

TABLE 1.6   Details of Field-Synchronizing Signals (See Figs. 1.13 and 1.14)

Parameter	525/60 standard	625/50 standard
Field period, ms	16.6833	20
Frame period, ms	33.3667	40
Vertical blanking start to front edge of first equalizing pulse, μs	1.5	Not specified
Vertical (field) blanking interval duration, lines	20 +1.5 μs	25
Preequalizing pulse sequence duration, lines	3	2.5
Preequalizing pulse width, μs	2.3 ± 0.1	2.35 ± 0.1
Vertical synchronizing pulse sequence duration, lines	3	2.5
Vertical serration pulse width, μs	4.7 ± 0.1	4.7 ± 0.2
Postequalization pulse duration, lines	3	2.5
Postequalizing pulse width, μs	2.3 ± 0.1	2.35 ± 0.1

The details of the vertical synchronizing waveforms related to the two conventional scanning standards are detailed in Table 1.6 and Figs. 1.13 and 1.14.

Details of the vertical blanking interval related to the two conventional scanning standards are shown in Figs. 1.15 and 1.16.

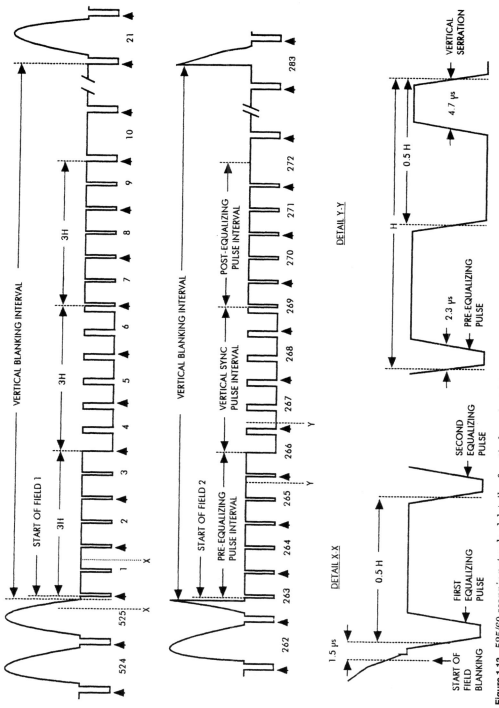

Figure 1.13 525/60 scanning standard details of vertical synchronizing waveforms.

23

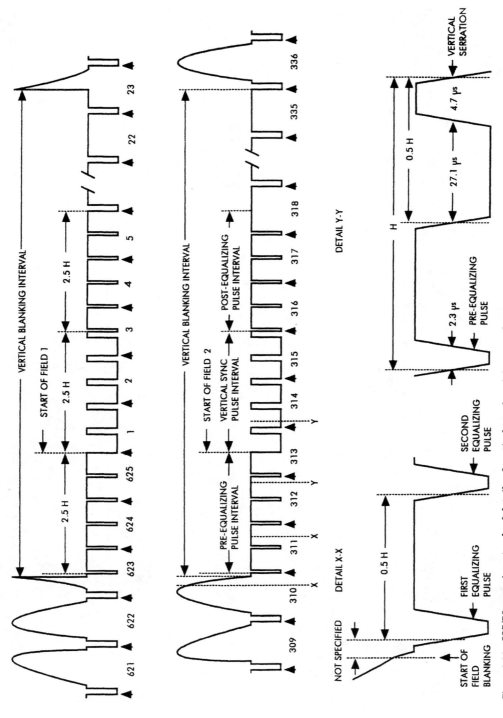

Figure 1.14 625/50 scanning standard details of vertical synchronizing waveforms.

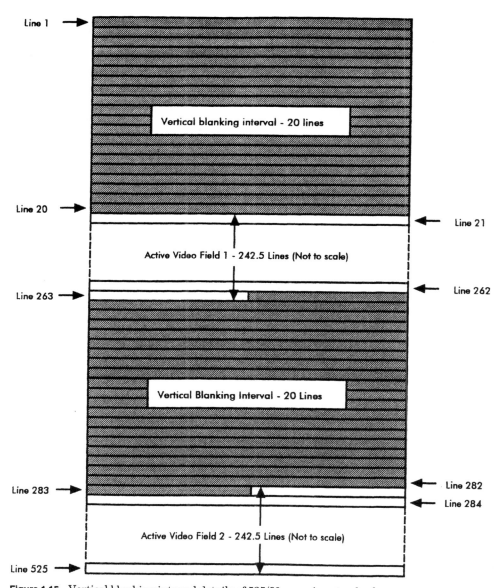

Figure 1.15 Vertical blanking interval details of 525/60 scanning standard.

1.6 The Spectrum of the Video Signal

The frequency components of a picture occupy a wide frequency band; however, the spectrum is not continuous. The spectrum components are spaced at intervals shown in Fig. 1.17, which represents the spectrum of a stationary picture for the 525/60 scanning standard. There is a strong component at the horizon-

Figure 1.16 Vertical blanking interval details of 625/50 scanning standard.

tal scanning frequency f_H and components at multiples of this frequency. There is also a strong component at the vertical scanning (field repetition) frequency f_V and components at multiples of this frequency. The 625/50 scanning standard has a similar spectrum except that $f_H = 15,625$ Hz and $f_V = 50$ Hz. If there are details that change from frame to frame, there is, in addition, a component at the frame frequency (30 or 25 Hz). The components at the horizontal scanning frequency and its multiples have sidebands at the frame and field frequencies and multiples thereof. Finally, there is a strong zero-frequency

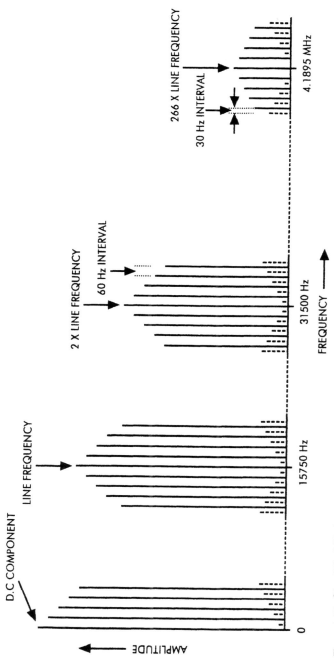

Figure 1.17 Simplified 525/60 monochrome video spectrum of a stationary scene.

Figure 1.18 Details of 525/60 scanning system spectrum as seen on a spectrum analyzer. Center frequency, 15734 Hz; horizontal resolution, 50 Hz/division. Note sideband components at 59.94 Hz and 29.97 Hz intervals.

component representing the average picture brightness. If the picture contains movement, the spectrum becomes more complex because of the addition of further components. Theoretically, the spectrum extends to infinity but is band-limited to the maximum video modulating frequency specified by the transmission standard in effect. Figure 1.18 shows a detailed spectrum analyzer display of spectral components around $f_H = 15{,}734.25$ Hz, the horizontal scanning frequency of the 525/60 NTSC color television system. Note the sideband components spaced at multiples of 59.94 Hz (the field repetition frequency) and 29.97 Hz (the frame repetition frequency).

1.7   Transmission Standards and Constraints

Transmission standards describe the characteristics of the signals radiated by television transmitters.

1.7.1   Video carrier modulation

Video transmitters are modulated in amplitude. All countries, with the exception of France, use negative modulation. In a negative-modulation system, increasing brightness in the transmitted picture produces a decrease in the modulation envelope amplitude. In a positive-modulation system, increasing the brightness in the transmitted picture produces an increase in the modulation envelope amplitude.

Figure 1.19 Significant video signal levels shown as a percentage of carrier amplitude in negative amplitude modulated systems.

The picture modulation envelope has three reference levels. These are

- Peak white level
- Blanking (black) level
- Sync level

When the black level differs from the blanking level, as in North America, a fourth reference level, the black level, is added. Figure 1.19 shows the reference levels for the negative amplitude modulation NTSC system as broadcast in North America. Other countries have slightly different standards.

1.7.2 Audio carrier modulation

All countries transmit the accompanying sound by modulating a separate carrier, situated at the high-frequency end of the television channel. All countries, with the exception of France (standard L), use frequency modulation of the audio carrier. The use of frequency modulation permits the use of the advantageous "intercarrier" audio reception method.

The peak-to-peak carrier deviation is ±25 kHz in North America (CCIR M) and ±50 kHz in most other standards (CCIR B, G, I, and K). These values are smaller than the value of ±75 kHz used in FM sound broadcasting. The relatively low energy associated with the high-frequency components of speech and music makes it appropriate to preemphasize the amplitude of the high frequencies prior to modulation at the transmitter and to deemphasize them correspondingly subsequent to demodulation at the receiver. In North America (CCIR M) the preemphasis time constant is 75 μs and in most other standards (CCIR B, G, I, and K) it is 50 μs.

1.7.3   Channel bandwidth and structure

Given a specific video baseband bandwidth, for example, 4.2 MHz in the 525/60 scanning standard, the conventional amplitude modulation of the video carrier results in upper and lower sidebands containing identical information. The resulting transmitted bandwidth is twice the baseband bandwidth. The 525/60 scanning standard would require an 8.4-MHz transmitted video bandwidth for the video information. The 625/50 scanning standard would require at least 10 MHz of transmitted bandwidth for the video information. To these figures has to be added the spectrum required by the audio carrier and its sidebands.

Spectrum conservation and frequency allocation concerns have led various countries to specify a reduced television transmission channel bandwidth. In 525/60 countries the specified transmission channel bandwidth is 6 MHz. In 625/50 countries the specified transmission channel bandwidth is 7 or 8 MHz. The method used to accommodate the required baseband video bandwidth in a reduced-bandwidth transmission channel is to transmit the upper sideband with full bandwidth and the lower sideband with reduced bandwidth. A portion of the lower sideband, corresponding to the higher video frequencies, is eliminated by a filter at the transmitter. The method is called *vestigial lower sideband transmission*. Various transmission standards have different vestigial sideband bandwidths.

Figure 1.20 shows in detail the channel structure corresponding to the CCIR M standard as used in North America. The channel bandwidth is 6 MHz. The video carrier is situated 1.25 MHz above the lower transmission channel edge. A full upper sideband of 4.2 MHz and a reduced (vestigial) lower sideband of 0.75 MHz are accommodated. In addition, a frequency-modulated audio carrier, with a center frequency 4.5 MHz above the video carrier, is also transmitted in the same channel.

Because of the vestigial lower sideband transmission, detected video signals whose frequencies are below 0.75 MHz have an amplitude twice that of signals whose frequencies lie between 0.75 and 4.2 MHz. The receiver compensates for this effect with a selectivity curve as shown in Fig. 1.21.

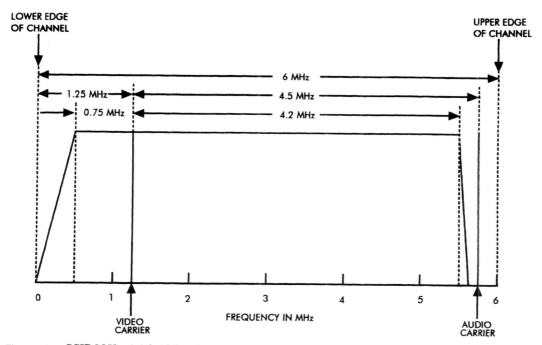

Figure 1.20   CCIR M Vestigial sideband characteristics and channel occupancy.

Figure 1.21   CCIR M Receiver selectivity characteristics to compensate for vestigial sideband effects.

Figure 1.22 shows the channel structure of several transmission standards. Note that the channel bandwidth, the vestigial lower sideband bandwidth, as well as the spacing between the video carrier and the audio carrier vary from system to system. Standards B, G, and K have a nominal vestigial sideband of 0.75 MHz whose significant spectral components are completely contained in the channel bandwidth. Systems I and L have a nominal vestigial sideband of 1.25 MHz and some significant spectral components extend below the lower channel limit. Various transmission standards for 625/50 signals have different spacings between the video and audio carriers. Standards B and G specify 5.5-MHz spacing, standard I specifies 6 MHz and standards K and L specify 6.5 MHz. Figure 1.22 also details the accommodation of the chrominance subcarrier and its sidebands in the transmitted channel.

1.7.4   Transmission constraints

The transmission of the video and audio information requires undistorted handling of the video and audio signals. Some of these concerns will be discussed in Chap. 2. One of the often misunderstood and neglected problems has to do with the modulation depth of the video transmitter. As shown in Fig. 1.19, the peak white signal level, 100 IRE in North America, results in a 12.5% modulation of the video carrier. The video carrier is completely cancelled at a video signal level of 120 IRE, resulting in a safety margin of 20 IRE. In color transmission, the peak positive signal levels could reach up to 131 IRE and create additional problems, which will be discussed in Chap. 2.

The overmodulation of the video transmitter has an obvious effect on the fidelity of the video signal. It also has a not-so-obvious and less-understood effect on the reception of the accompanying sound. This is because of the peculiar manner in which television receivers process and demodulate the audio and video carriers. Normally, since video and audio information is handled by two separate carriers, it would seem logical that two separate receivers be required to recover the original information. Television receivers, however, operate in a different manner, as shown in Fig. 1.23.

Because of the fact that the video carrier is amplitude-modulated and the audio carrier is frequency-modulated, it is possible to use the so-called intercarrier audio reception method. Here, the video and audio carriers are tuned, frequency-converted, and amplified by a single tuner. The standardized converted carriers are respectively 45.75 MHz (video) and 41.25 MHz (audio) in North America. The intermediate frequency amplifier has the adequate bandwidth and spectrum-shaping characteristic required to carry both converted carriers and their sidebands, as well as to correct for the vestigial sideband effects. Two detectors are used to recover the original information. The video detector recovers the original video information but also generates a 4.5-MHz spurious signal resulting from the interference between the audio and video carriers. This spurious signal, superimposed on the recovered video signal, is amplitude-modulated by the video signal and frequency-modulated by the audio signal. It is removed by a 4.5-MHz notch filter before further video processing.

Figure 1.22 Transmission channel occupancy of several CCIR television systems.

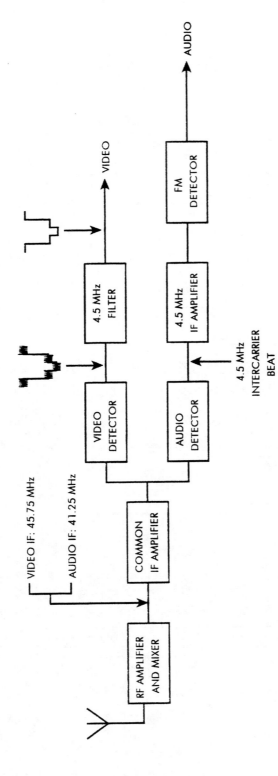

Figure 1.23 Simplified block diagram of CCIR M intercarrier receiver.

The audio detector recovers the same information as the video detector. The spurious 4.5-MHz signal is filtered by a bandpass filter to remove the video signal component and treated as a frequency-modulated carrier. It is amplified by an intermediate frequency amplifier with a center frequency of 4.5 MHz and signal amplitude variations are removed by a limiter. The resulting constant-amplitude FM signal feeds a standard FM detector to recover the original audio information. This reception method works very well with all systems employing frequency modulation of the audio carrier. Problems occur when the video transmitter is overmodulated, resulting in clipping of the video carrier. Under these extreme circumstances the derived 4.5-MHz audio carrier is periodically canceled at video horizontal and vertical scanning rates. This results in the so-called intercarrier buzz effect. This audio reception problem can be eliminated by carefully monitoring and adjusting the video signal levels feeding the transmitter to avoid transmitter overmodulation.

Analog Video Fundamentals

2.1 Color Television

Color television relies on the light properties that control the visual sensa-
tions known as brightness, hue, and saturation.

- *Brightness* is defined as the characteristic of a color that enables it to be
  placed on a scale of from dark to light.

- *Hue* is defined as the characteristic of a color that enables it to be
  described as red or yellow or blue or any other identifiable color.

- *Saturation* is defined as the extent to which a color departs from the white
  of the neutral condition. Pale colors (pastels) are low in saturation, whereas strong, or vivid, colors are high in saturation.

Virtually any color can be matched by a proper combination of three primary
colors. Green, blue, and red have been chosen as the primary colors for television. The proper combination of green, blue, and red produces white. Practical
considerations, such as the relative ease with which relatively efficient color
phosphors could be made, played an important role in the choice of television's primary colors.

In its simplest form, a color television system consists of three sensors that
receive filtered green, blue, and red images. The three sensors scan the three
respective images horizontally and vertically, in the conventional manner,
and generate green, blue, and red electrical signals. These signals are transported in some manner to the receiver for display on three CRTs with, respectively, green, blue, and red phosphors. The three CRTs reverse the process
and create a representation of the original picture by superimposing the
three images. The reproduction of pictures with a diagonal measurement of
less than about 40 in (100 cm) relies on the use of tricolor CRTs using three
dedicated electron guns and an array of triad green, blue, and red phosphor

dots or stripes on which the respective electron beams are caused to converge. In this way, three primary color images are produced on the screen of a tricolor CRT. The small dimension of the colored phosphor dots or stripes causes the observer to see a full-color image.

2.1.1 Colorimetry

The principles of colorimetry are based on Grassman's laws. These laws are

- The eye can distinguish only three kinds of differences or variations.
- In a two-component light mixture, the color mixture will change gradually if one component is steadily changed and the other is held constant.
- Sources of the same color produce identical visual effects in a mixture regardless of their spectral composition.
- The total luminance of the color is the sum of the luminances of each of the components.

In modern colorimetry, the colors are represented in a three-dimensional coordinate system (the color space) known as the *tristimulus chromaticity coordinate system*. The coordinates are designated X, Y, and Z and each possible color occupies a position in the three-dimensional system. The unit plane of this system $(X + Y + Z = 1)$ contains all the coordinates of the various colors. The total area covered by the colors is called the *Planckian locus*. The color coordinates depend on the spectral characteristics of the illuminating light, the human eye response, and the spectral reflectance (or transmittance) of the observed color. The XYZ values of a color are defined by the following equations:

$$X = K \sum_{\lambda\,=\,300}^{700} \bar{x}(\lambda)\, C(\lambda)\, L(\lambda)\, \Delta(\lambda)$$

$$Y = K \sum_{\lambda\,=\,300}^{700} \bar{y}(\lambda)\, C(\lambda)\, L(\lambda)\, \Delta(\lambda)$$

$$Z = K \sum_{\lambda\,=\,300}^{700} \bar{z}(\lambda)\, C(\lambda)\, L(\lambda)\, \Delta(\lambda)$$

where K = a normalizing factor given by

$$K = \frac{1}{\displaystyle\sum_{\lambda\,=\,300}^{700} L(\lambda)\, \bar{y}(\lambda)\, \Delta(\lambda)}$$

$L(\lambda)$ = Light spectrum characteristics
$\Delta(\lambda)$ = Wavelength increase, nm

$C(\lambda)$ = Color spectrum characteristics used
$\bar{x}, \bar{y}, \bar{z}$ = the 1931 Standard Observer characteristics
λ = Light wavelength

In practice, the three-dimensional coordinate system is replaced by its projection on the $X0Y$ plane and is designated as the Planckian locus of the x-y coordinate system. The projection equations are

$$x = \frac{X}{(X + Y + Z)} \quad \text{and} \quad y = \frac{Y}{(X + Y + Z)}$$

Figure 2.1 shows a display of the two-dimensional Planckian locus along with the triangle formed by the SMPTE/EBU phosphors. Colors are represented by two coordinates, x and y.

As indicated by Grassman's laws, all visible colors are located inside, or on the edge of, the triangle formed by the three light sources. Color television uses three light sources in a CRT designated green, blue, and red phosphors. The respective phosphor coordinates (x, y) are located on the Planckian locus.

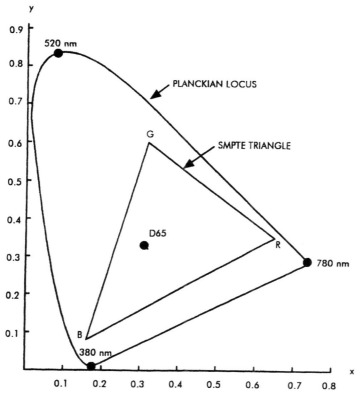

Figure 2.1 2D Planckian locus showing the location of the SMPTE phosphors and the D6500 reference white.

In the past there used to coexist two standard sets of phosphors known as SMPTE, used in North America, and EBU, used in Europe. Recently, the SMPTE and EBU phosphors have been normalized to identical values. The coordinates of the universal standard phosphors are

$$\text{Green:} \quad x = 0.310 \quad y = 0.595$$

$$\text{Blue:} \quad x = 0.155 \quad y = 0.070$$

$$\text{Red:} \quad x = 0.630 \quad y = 0.340$$

An important consideration is the reproduced system reference white. This white, referred to as D6500, has been standardized to the following x-y coordinate values:

$$x = 0.3127 \quad \text{and} \quad y = 0.3290$$

2.1.2 Transfer characteristics

The electrooptical transfer characteristic of the CRT is inherently nonlinear. In order to achieve an overall linear transfer characteristic, it is necessary to compensate for the CRT nonlinearity by introducing a compensating nonlinearity, usually referred to as *gamma correction,* elsewhere in the system. Historically, the compensation has been carried out in the camera, where the green, blue, and red signals are predistorted to match the CRT. This results in signals that can be satisfactorily viewed on relatively simple color monitors or receivers as well as improved signal-to-noise ratio (SNR) under less than ideal receiving conditions. In order to achieve uniform results, it is necessary to define the reference CRT electrooptical characteristics. Various organizations use different reference CRT electrooptical characteristics and compensation methods. The standardized North American transfer characteristics of the reference reproducer and of those of the compensating reference camera, extracted from the ANSI/SMPTE 170M-1994 standard, are presented below.

2.1.2.1 Electrooptical transfer characteristic of the reference reproducer

$$L_T = \left[\frac{(V_r + 0.099)}{1.099} \right]^\gamma \quad \text{for } 0.0812 \leq V_r \leq 1$$

$$L_T = \frac{V_r}{4.5} \quad \text{for } 0 \leq V_r \leq 0.0812$$

where V_r is the video signal level driving the reference reproducer normalized to the system reference white and L_T is the light output from the reference reproducer, normalized to the system reference white and $\gamma = 2.2$.

2.1.2.2   Optoelectronic transfer characteristic of the reference camera

$$V_C = 1.099 \times L_C^{(1/\gamma)} - 0.099 \qquad \text{for } 0.018 \leq L_C \leq 1$$

$$V_C = 4.500 \times L_C \qquad\qquad\qquad \text{for } 0 \leq L_C \leq 0.018$$

where V_C is the video signal output of the reference camera, normalized to the system reference white, and L_C is the light input to the reference camera, normalized to the system reference white and $\gamma = 2.2$.

Figure 2.2 shows the nonlinear transfer curve of a typical CRT and its correction.

2.1.3   The basic ingredients

All color television systems use the principle of additive colors with green, blue, and red as primary colors.

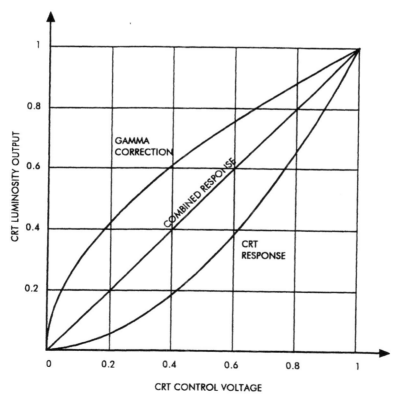

Figure 2.2   Correction of CRT nonlinear transfer curve.

2.1.3.1 Brightness information. Monochrome compatibility requires the generation and transmission of a full-bandwidth signal representing the brightness component of the televised scene. This component is called *luminance*. The mathematical expression for the luminance signal is

$$E'_Y = 0.587\, E'_G + 0.114\, E'_B + 0.229\, E'_R$$

where E'_Y = The gamma-corrected voltage corresponding to the luminance information

E'_G = The gamma-corrected voltage corresponding to the green information

E'_B = The gamma-corrected voltage corresponding to the blue information

E'_R = The gamma-corrected voltage corresponding to the red information

In a studio environment the bandwidth of the luminance signal is restricted only by the state of the art of the equipment used. Normally, the bandwidth of the luminance signal generated by a camera is at least 8 MHz, or a horizontal resolution in excess of 600 LPH. The typical analog composite videotape recorder bandwidth is 4.2 MHz in NTSC and 5 MHz in PAL. The luminance bandwidth of analog component videotape recorders is slightly worse. The transmitted luminance bandwidth is reduced by the analog transmission channel specification to 4.2 MHz in NTSC, and 5 MHz or 5.5 MHz in 625/50 PAL.

2.1.3.2 Chrominance information. The chrominance information is conveyed by two of the three primary signals minus the brightness component. These signals are known as the blue and the red color-difference signals. They are

$$E'_B - E'_Y = -0.587\, E'_G + 0.889\, E'_B - 0.299\, E'_R$$

and

$$E'_R - E'_Y = -0.587\, E'_G - 0.114\, E'_B + 0.701\, E'_R$$

The $E'_G - E'_Y$ signal can be re-created in the receiver by a suitable combination of the blue and red color-difference signals.

The color-difference signals are scaled in amplitude by suitable multiplication factors. The scaling of the color-difference signals varies with the application. The NTSC and PAL composite color television systems use identical scaling factors to avoid transmitter overloading. Component analog and digital standards use different scaling factors.

The NTSC and PAL scaled color-difference signals are:

$$E'_{B-Y} = 0.493\, (E'_B - E'_Y) \qquad \text{(Called } E'_U \text{ in PAL)}$$

and

$$E'_{R-Y} = 0.877 \, (E'_R - E'_Y) \qquad \text{(Called } E'_V \text{ in PAL)}$$

The detail of the color-difference information reflects the resolving capability of normal vision. The bandwidth of the color-difference signals used in analog composite and component applications varies from system to system but never exceeds 1.5 MHz. Receivers and monitors rarely exceed 0.5-MHz chrominance bandwidth or a horizontal resolution of about 40 lines. The main difference between the analog composite systems lies in the manner in which the color-difference signals are modulating the respective subcarriers.

2.1.4 The color bars signal

The color bars signal is widely used and often misinterpreted. This book will make many references to color bars signals and waveforms and therefore a description of the basic signal is appropriate.

There are various versions of the color bars signals in general use. They all share a common overall form. The color bars signal provides a sequence of vertical bars in the picture area showing the saturated primaries and their complements as well as black and white. The active line is thus divided in eight equal parts. The first is occupied by a luminance reference bar, that is, a white bar of a standard amplitude. The last is a black bar, that is, black level only. In between are six bars representing the three primary colors and their complements. They are, in order, yellow, cyan, green, magenta, red, and blue. The standard order of presentation has been chosen to give a descending-order sequence of luminance values.

A color bar generator has three outputs corresponding, respectively, to the green, blue, and red primary color signals, E'_G, E'_B, E'_R. These signals consist of a sequence of flat-top pulses. By a suitable overlap of the pulses in certain portions of the raster and nonoverlap in others, the three saturated primary colors as well as the three saturated complementary colors are produced. These signals may be used in their original form, matrixed into E'_Y, $E'_B - E'_Y$, and $E'_R - E'_Y$ or encoded into an analog (NTSC, PAL, SECAM) or digital (component or composite) signal.

A number of different color bar signals exists. Many of them are application-specific, that is, reflect the operational requirements of the specific organization, like color bars optimized for use with amplitude-modulated transmitters. Others are typical of the respective analog composite encoding standard and contain, in addition to the color bars, various signals serving the purpose of color monitor alignment or various performance measurements.

Two groups of saturated color bar signals will be described in this chapter. They reflect the peculiarities of the two conventional television scanning standards (525/60 and 625/50) in terms of white and black signal levels. Each group comprises a full-amplitude (100%) and a reduced-amplitude (75%) color bar signal. The various pulse levels are described in percentages of the peak white level. Figure 2.3 shows the details of the four types of color bars.

Each type of color bar is identified with four numbers with an oblique stroke between them as follows:

- The first number describes the primary color signal level during the transmission of the white bar, that is, the maximum value of E'_G, E'_B, and E'_R.

- The second number describes the primary color signal level during the transmission of the black bar, that is, the minimum value of E'_G, E'_B, and E'_R.

- The third number describes the maximum level of the primary color signal during the transmission of the colored color bars, that is, the maximum value of E'_G, E'_B, and E'_R.

- The fourth number describes the minimum level of the primary color signal during the transmission of the colored color bars, that is, the minimum value of E'_G, E'_B, and E'_R.

Figure 2.3a presents a fully saturated color bar signal with maximum signal levels of 100% and minimum signal levels of 0%. This type of color bar signal is called 100/0/100/0. Figure 2.3b shows a fully saturated color bar signal with maximum signal levels of 75% and minimum signal levels of 0%. This type of color bar signal is called 75/0/75/0. These types of color bar signals are typical of those used in 625/50 countries. They are representative of the signals used to feed a PAL encoder and would be obtained at the output of a properly adjusted PAL decoder.

Figure 2.3c presents a fully saturated color bar signal with maximum signal levels of 100% and minimum signal levels of 7.5%. This type of color bar signal is called 100/7.5/100/7.5. Figure 2.3d shows a fully saturated color bar signal with maximum levels of 75% and minimum signal levels of 7.5%. This color bar signal is called 75/7.5/75/75. These types of color bar signals are typical of those used in 525/60 countries except Japan. They are representative of the signals used to feed an NTSC encoder reflecting the original philosophy behind the 1953 NTSC standard. In those days the primary signals (E'_G, E'_B, and E'_R) had the black level set at 7.5% of the peak white level (7.5 IRE setup) and their peak level was 714.3 mV (100 IRE). Current NTSC encoders use E'_G, E'_B, and E'_G without setup and, very often, with a peak amplitude of 700 mV, leaving it to the encoder to normalize the encoded signal to NTSC specifications. This allows for standard camera circuit designs irrespective of the analog composite encoding standard. The signal at the output of an NTSC decoder will, however, be as shown in Fig. 2.3c or d.

Figure 2.4 shows the position of the eight bars on a television screen.

Figure 2.5 shows a graphic representation of the formation of the 100/0/100/0 color bars luminance component waveform E'_Y from the primary E'_G, E'_B, and

Figure 2.3   Relative amplitudes of components for four types of color bars.

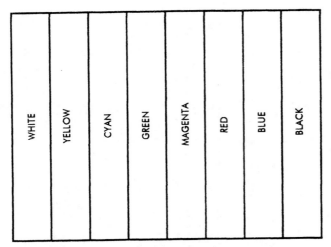

Figure 2.4 The position of the eight bars on the television screen.

E'_R signals. The signal is unipolar and has a descending-order sequence of luminance values as indicated earlier in this chapter. The amplitudes are expressed in percentages of the full-amplitude signal (that is, 700 mV). As shown, it has no synchronization (sync) added to it.

Figure 2.6 shows a graphic representation of the formation of the 100/0/100/0 color bars nonscaled blue color-difference component waveform $E'_B - E'_Y$ from the primary E'_G, E'_B, and E'_R signals. The scaled signal, E'_{B-Y}, has the same shape but a lower peak-to-peak amplitude. The signal is bipolar and has equal maximum positive and negative excursions with respect to the zero reference. The amplitudes are expressed in percentages of the full-amplitude signal (i.e., 700 mV).

Figure 2.7 shows a graphic representation of the formation of the 100/0/100/ color bars nonscaled red color-difference component waveform $E'_R - E'_Y$ from the primary E'_G, E'_B, and E'_R signals. The scaled signal, E'_{R-Y}, has the same shape but a lower peak-to-peak amplitude. The signal is bipolar and has equal maximum positive and negative excursions with respect to the zero reference. The amplitudes are expressed in percentages of the full-amplitude signal (i.e., 700 mV).

Figure 2.8 shows a graphic vector display representation of a 100/0/100/0 color bar signal. This display is obtained by feeding the E'_{B-Y} (scaled blue color-difference) signal to the horizontal input and the E'_{R-Y} (scaled red color-difference) signal to the vertical input of an oscilloscope.

A subset of the 75% color bar signals, identified, respectively, as 100/0/75/0 and 100/7.5/75/7.5, is being used by certain organizations. These signals are identical to the regular 75% color bar signals except that the luminance bar has an amplitude of 100%. They are generally being used to feed television transmitters and older analog composite videotape recorders.

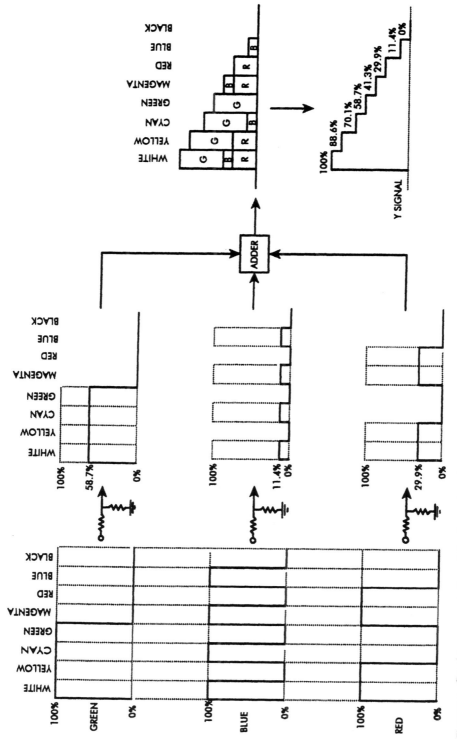

Figure 2.5 Graphic representation of the formation of 100/0/100/0 color bars Y signal from the primary green, blue, and red signals.

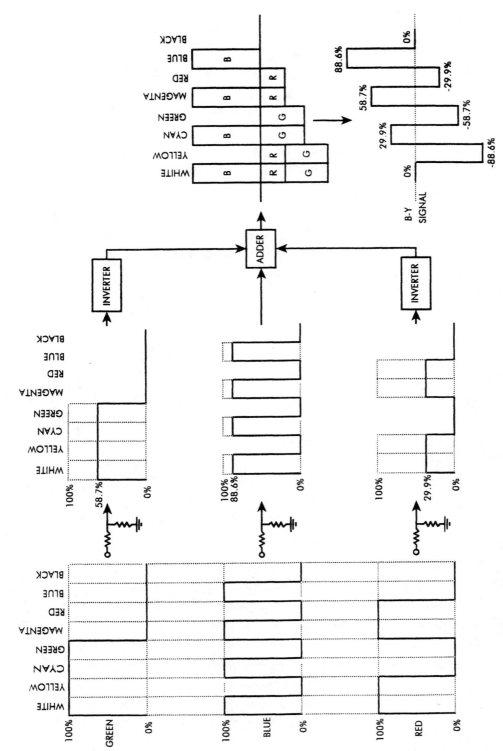

Figure 2.6 Graphic representation of the formation of 100/0/100/0 color bars blue color-difference signal from the primary green, blue, and red signals.

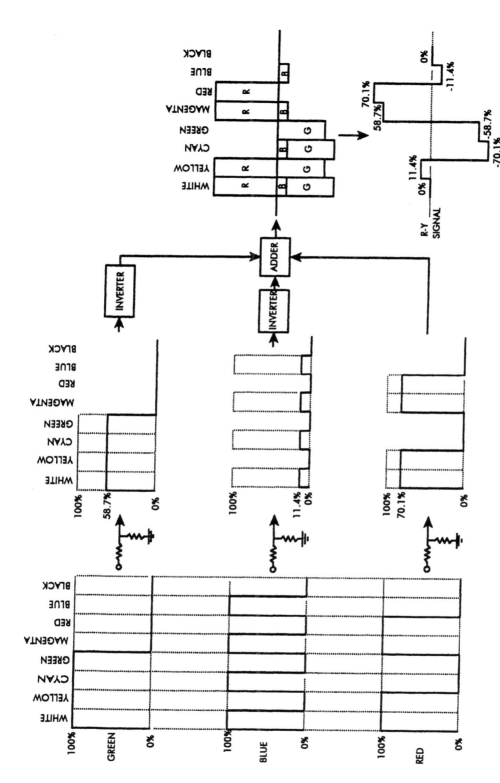

Figure 2.7 Graphic representation of the formation of 100/0/100/0 color bars red color-difference signal from the primary green, blue, and red signals.

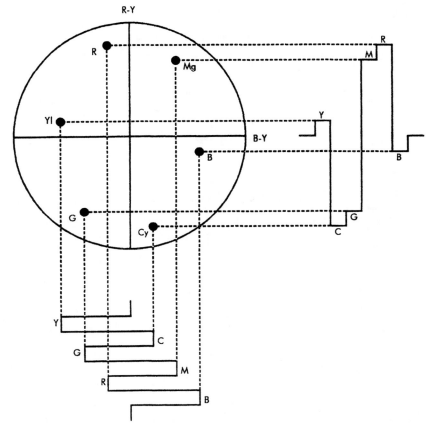

Figure 2.8 Graphic representation of the formation of a vector display of the · 100/0/100/0 color bars chrominance components.

2.2 Composite Video

Composite video describes a signal in which luminance, chrominance, and synchronization information are multiplexed in the frequency, time, and amplitude domain for single-wire distribution.

2.2.1 Common characteristics

All current broadcast color television systems, namely NTSC, PAL, and SECAM, share a number of basic features.

2.2.1.1 Compatibility requirements. The main consideration in the development of the major color television standards (NTSC, PAL, and SECAM) was compatibility with the existing monochrome television standard. There are several aspects to this concept. These are

- Monochrome compatibility: A monochrome receiver must reproduce the brightness content of a color signal correctly in black and white with no visible interference from the color information.

- Reverse compatibility: A color receiver must reproduce a monochrome signal correctly in shades of gray with no spurious color components.

- Scanning compatibility: The scanning system used for color transmissions must be identical to the one used by the existing monochrome service.

- Channel compatibility: The color system must fit into the existing monochrome TV channel and use the same spacing between the vision and sound carriers.

2.2.1.2 Frequency division multiplexing. In current broadcast usage, primary green, blue, and red (G,B,R component) signals, generated by a camera, are processed to produce an analog composite video signal (NTSC, PAL, or SECAM). All systems use as the main ingredients a wideband luminance (Y) signal and two narrowband color-difference signals (B-Y and R-Y). The chrominance signals each modulate an assigned subcarrier in a manner peculiar to the specific television system. The frequency of the subcarriers is relatively high for reduced visibility. The approximate values are 3.58 MHz for the 525/60 scanning system and 4.43 MHz for the 625/50 scanning system. The chrominance and luminance signals are frequency-division-multiplexed to obtain a single-wire composite video signal with a total bandwidth suited to the specific transmission standard.

Figure 2.9 shows the simplified block diagram of a generalized encoder. The matrix serves to derive the luminance and color-difference signals from the primary signals through a process of linear amplification, addition, and subtraction. The delay introduced in the luminance path matches the delays

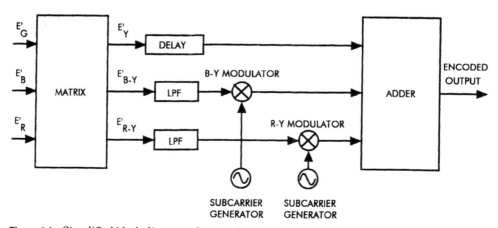

Figure 2.9 Simplified block diagram of generalized encoder.

Figure 2.10 Simplified block diagram of generalized decoder.

suffered by the reduced-bandwidth color-difference signals and helps achieve proper timing of the luminance and chrominance signals at the input of the adder. Each of the band-limited color-difference signals is fed to a modulator. The two radiofrequency (RF) signals are added to the luminance signal to obtain the composite color signal. This basic encoder diagram is deliberately simplified, and in this form applies equally well to each of the three color television systems.

Figure 2.10 shows a simplified block diagram of a generalized decoder. To decode a color television signal into green, blue, and red components, the reverse generalized process is required. First, the composite signal has to be separated by filters into chrominance and luminance components. The chrominance component is subsequently demodulated in a manner peculiar to the respective color television system to yield the color-difference signals. The luminance signal is delayed to achieve proper timing with respect to the narrow-bandwidth color-difference signals. The three signals are fed to an active matrix. By a process of amplification, addition, and subtraction, the primary color signals are recovered at the output of the matrix. This basic decoder diagram is deliberately simplified, and in this form applies equally well to each of the three color television systems.

2.2.2 The NTSC system

The NTSC (National Television System Committee) color television system is a single-channel television concept. Luminance, chrominance, and synchronization information are combined to be transmitted in a 6-MHz RF channel originally specified for monochrome transmissions. The transmission of color takes advantage of the characteristics of the spectrum of monochrome video as detailed in Chap. 1. Essentially, the chrominance information is transmitted in the spectrum "holes" of the monochrome information. The concept uses a wideband (4.2-MHz) luminance signal and two narrowband chrominance signals.

The contemporary characteristics of this system, as defined in the SMPTE 170M standard, are summarized in Table 2.1.

The encoder processes a wideband ($\geq$4.2-MHz) luminance (monochrome) signal and two narrowband color-difference signals of equal bandwidth. The color-difference signals may be B-Y and R-Y or I and Q, as in the original

TABLE 2.1 Summary of NTSC Signal Characteristics

		x	y
1. Assumed chromaticity coordinates for primary colors of receiver			
	Green	0.310	0.596
	Blue	0.155	0.070
	Red	0.630	0.340
2. Chromaticity coordinates for equal primary signals	Illuminant D_{65}: $x = 0.3127$; $y = 0.3290$		
3. Assumed receiver gamma value	2.2		
4. Luminance signal	$E'_Y = 0.587\, E'_G + 0.114\, E'_B + 0.299\, E'_R$		
5. Chrominance signals	$E'_{B-Y} = 0.877\,(E'_B - E'_Y)$ and $E'_{R-Y} = 0.493\,(E'_R - E'_Y)$ or $E'_Q = E'_{B-Y}\cos 33° + E'_{R-Y}\sin 33°$ and $E'_I = -E'_{B-Y}\sin 33° + E'_{R-Y}\cos 33°$		
6. Equation of complete color signal	$E_M = 0.925\, E'_Y + 7.5 + 0.925\, E'_{B-Y}\sin(2\pi f_{SC}t)$ $+ 0.925\, E'_{R-Y}\cos(2\pi f_{SC}t)$ or $E_M = 0.925\, E'_Y + 7.5 + 0.925\, E'_Q\sin(2\pi f_{SC}t + 33°)$ $+ 0.925\, E'_I\cos(2\pi f_{SC}t + 33°)$		
7. Type of chrominance subcarrier modulation	Suppressed-carrier amplitude modulation of two subcarriers in quadrature		
8. Chrominance subcarrier frequency, Hz	Nominal value and tolerance: $f_{SC} = 3,579,545 \pm 10$ Relationship to line frequency f_H: $f_{SC} = (455/2)f_H$		
9. Bandwidth of transmitted chrominance sidebands, kHz	$f_{SC} \pm 620$ or $f_{SC} +620/-1300$		
10. Amplitude of chrominance subcarrier	$G = \sqrt{(E'_{B-Y}{}^2 + E'_{R-Y}{}^2)}$ or $G = \sqrt{(E'_I{}^2 + E'_Q{}^2)}$		
11. Synchronization of subcarrier	Subcarrier burst on blanking backporch		

1953 specifications of the NTSC system. The bandwidth of each of the color-difference signals may be 600 kHz or 1.3 MHz. The wider bandwidth is useful in studio environments where there is no significant bandwidth limitation. Transmission and reception constraints result in color television receivers using a 600-kHz chrominance bandwidth, hence the excess chrominance bandwidth is wasted. Figure 2.11 illustrates the chrominance bandwidth versus transmitted bandwidth relationship.

Each of the scaled color-difference signals modulates a subcarrier. The two subcarriers are of identical frequency but of different phase. The phase difference between the two subcarriers is 90°, so the original signals modulating the two carriers can be recovered without crosstalk. The two subcarriers are obtained from a common crystal-controlled oscillator. The type of modulation used is suppressed-carrier amplitude modulation. The modulation system is consequently called suppressed-carrier quadrature amplitude modulation. Since the subcarrier is suppressed, only the sidebands are obtained at the output of the modulators. This results in the complete cancellation of the chrominance signal when no colors are present.

Figure 2.11 NTSC composite signal bandwidth.

The frequency of the chrominance subcarrier is an odd multiple of the half horizontal scanning frequency. This results in the interleaving of the luminance and chrominance spectra. The type of spectrum interleaving used in NTSC is called half-line offset. The frequency of the subcarrier is equal to

$$f_{\mathrm{SC}} = \frac{455}{2}\, f_H = 3{,}579{,}545 \pm 10 \text{ Hz}$$

This leads to slightly modified horizontal (15,734.25 Hz) and vertical (59.94 Hz) scanning frequencies. These frequencies are within the capture range of the receiver scanning circuits.

Figure 2.12 shows the spectrum of an NTSC signal. Note a peak of energy around the suppressed subcarrier at 3.58 MHz. Figure 2.13 shows a detailed view of the spectrum around the suppressed chrominance subcarrier. Note the chrominance sideband components spaced at f_H intervals. Low-level luminance spectral components are interleaved at $f_H/2$ intervals. The subcarrier amplitude is 20 dB lower than its significant sideband components. Normally the subcarrier should not be visible. Its low-level presence is due to the low-energy subcarrier burst transmitted as a frequency and phase reference.

A burst of 9 cycles of frequency and phase reference subcarrier is transmitted during the backporch of the horizontal blanking interval. Figure 2.14 shows details of the horizontal blanking period with the subcarrier burst. The purpose of the burst is to synchronize the receiver local crystal oscillator. This oscillator feeds a reconstituted subcarrier to the synchronous B-Y and R-Y demodulators used for the recovery of the color-difference signals. The phase of the burst is 180° with respect to the system phase reference $(E'_B - E'_Y)$.

Figure 2.15 shows a simplified block diagram of an NTSC encoder using B-Y/R-Y color-difference signals. Green, blue, and red signals are fed to a resis-

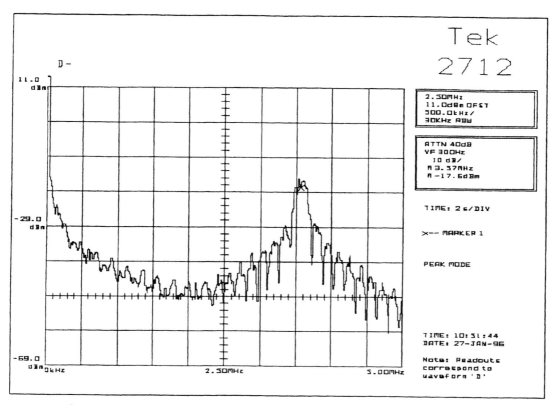

Figure 2.12 Spectrum of 100/7.5/100/7.5 NTSC color bars signal. Note a peak of energy around the suppressed 3.58-MHz subcarrier. Horizontal resolution: 500 kHz/division.

tive matrix that algebraically combines percentages of these primary color signals to form the luminance (Y) signal and the two color-difference signals. Each of the color-difference signals is band-limited before being fed to the respective balanced modulators. A 3.58-MHz subcarrier feeds the B-Y modulator and, through a 90° phase-shift network, the R-Y modulator. The Y signal is delayed to compensate for the chrominance delay introduced by the color-difference low-pass filters. The adder combines the luminance, chrominance sidebands, composite sync, and a 180° phase-shifted gated subcarrier burst into a composite color signal.

Figure 2.16 shows a phase-domain representation of the E'_{B-Y} (reference) subcarrier and the E'_{R-Y} subcarrier (+90°). A third subcarrier identifies the synchronizing burst (+180°).

Figure 2.17 shows a vector representation of the chrominance subcarrier modulation process. A given color, described by a given set of E'_{B-Y} and E'_{R-Y} signal values, is represented by two amplitude-modulated subcarriers in phase quadrature. The instantaneous values of the two modulated subcarri-

Figure 2.13 Detailed spectrum of NTSC 100/7.5/100/7.5 color bars signal around the suppressed chrominance subcarrier. Note chrominance sideband components at f_H intervals. Low level luminance spectrum components are interleaved at $f_H/2$ intervals. Horizontal resolution: 15.7 kHz/division.

ers result in a vector described by its amplitude and phase angle with respect to the B-Y subcarrier reference phase. The vector amplitude represents the color saturation and its phase angle represents the hue.

Figure 2.18 shows a 100/7.5/100/7.5 (100%) color bar-signal waveform resulting from the addition of luminance and chrominance components. Note that the peak positive signal excursion is 130.8 IRE, which is beyond the overload level of a television transmitter. Figure 2.19 shows a vectorscope display of the 100% color bar signal.

Figure 2.20 shows the waveform of a 75/7.5/75/7.5 (75%) color bar signal resulting from the addition of luminance and chrominance components.

Tables 2.2 and 2.3 list the details of the color bar signal luminance and chrominance values as well as the phase angles of the six colors.

The 75/7.5/75/7.5 color bar signal is used for transmitter tests. Studio equipment can accept either of the two color bar signals. It is important to remember that peak amplitude green, blue, and red primary signals will generate composite color signal levels equivalent to the 100% color bar signal. Since

Figure 2.14 NTSC horizontal blanking interval showing details of color burst.

Figure 2.15 Simplified block diagram of NTSC B-Y/R-Y encoder.

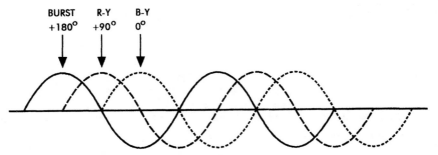

Figure 2.16 Phase domain representation of the two significant equal-frequency sub-carriers. The third subcarrier represents the synchronizing burst.

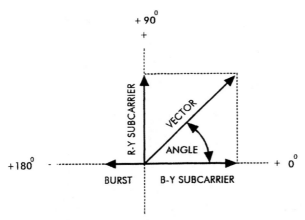

Figure 2.17 The instantaneous amplitudes of the subcarrier result in a vector whose amplitude represents saturation and phase represents hue.

there are no highly saturated yellow and cyan colors in nature, the probability of transmitter overload under normal operating conditions is very low. Problems occur, however, with synthetic signal sources, such as character generators and graphic systems, which can create primary signals resulting in excessive amplitude composite color signals and lead to transmitter overload.

Figure 2.21 shows a simplified block diagram of an NTSC B-Y/R-Y decoder. The chrominance sidebands are separated through a bandpass filter and fed to two synchronous demodulators as well as to a burst separator. The burst separator is gated by a burst key derived from the horizontal sync. Its output synchronizes a local crystal-controlled subcarrier generator through a phase-locked loop. The subcarrier generator feeds the B-Y demodulator and, through a 90° phase shift network, the R-Y demodulator. A "hue" control allows for the manual adjustment of the phase of the reconstituted subcarrier with respect to that of the color burst to obtain the correct phase relationship. The demodulated color-difference signal as well as the delayed luminance signal are fed to

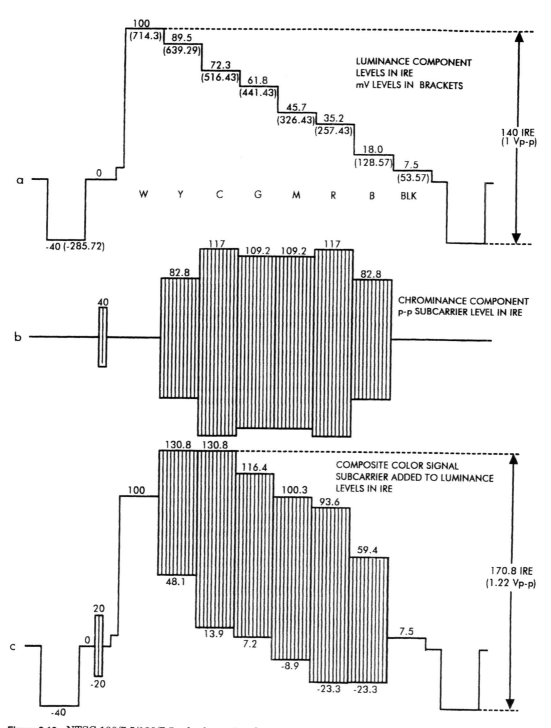

Figure 2.18 NTSC 100/7.5/100/7.5 color bars signal waveform.

Figure 2.19 Vectorscope display of NTSC 100/7.5/100/7.5 color bars signal.

a matrix that reconstitutes the three primary signals. A notch filter is used to reduce the subcarrier visibility. This simple decoder suffers from the cross-color effects resulting from high-frequency luminance spectrum components being misinterpreted as chrominance information.

Figure 2.22 shows a simplified block diagram of a comb filter decoder. This decoder takes advantage of the 180° line-to-line subcarrier phase change. The composite color signal feeds an adder directly as well as through an inverter and a one-line-duration (1 H = 63.555 µs) delay line. The adder cancels out the luminance components, which are constant from line to line, and passes the combed chrominance components. The synchronous demodulators operate in the conventional manner. The combed chrominance as well as the composite color signal are fed to a subtractor, whose output is combed luminance. This simple comb filter decoder works well with pictures having a high correlation from line to line and results in a considerable reduction of the cross-color effect at the expense of halving the vertical color resolution because of line averaging effects. The comb filter decoder is inadequate when the color information changes abruptly from line to line. More sophisticated adaptive comb filter decoders automatically switch to notch filter operation when they detect a sharp vertical color transition, and return to normal operation when this condition disappears.

One of the main faults of the NTSC system is its sensitivity to nonlinear distortions, which result in dynamic subcarrier phase changes known as *differential phase*. Figure 2.23 shows a modulated ramp waveform consisting of a ramp signal steadily rising from black to white on which is superimposed a constant-phase and constant-amplitude subcarrier signal. Figure 2.24 shows a vectorscope display of the constant-phase subcarrier signal.

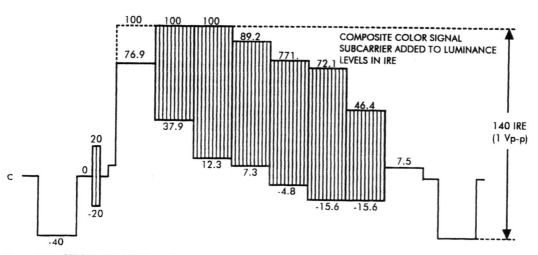

Figure 2.20   NTSC 75/7.5/75/7.5 color bars signal waveform.

TABLE 2.2   NTSC 100/7.5/100/7.5 Color Bars Signal
Waveform Characteristics*

Color	Luminance, IRE/mV	Chrominance, IRE/mV	Angle, degrees
White	100/714.3	0/0	•••
Yellow	89.5/639.29	82.8/591.44	167.1
Cyan	72.3/516.43	117/835.73	283.5
Green	61.8/441.43	109.2/780.01	240.7
Magenta	45.7/326.43	109.2/780.01	60.7
Red	35.2/251.43	117/835.73	103.5
Blue	18/128.57	82.8/591.44	347.1
Black	7.5/53.57	0/0	•••
Burst	0/0	40/285.72	180

*Luminance levels and chrominance peak-to-peak amplitudes
are expressed in IRE and millivolts. Chrominance phase angles are
expressed in degrees with respect to the B-Y reference.

TABLE 2.3   NTSC 75/7.5/75/7.5 Color Bars Signal Waveform
Characteristics*

Color	Luminance, IRE/mV	Chrominance, IRE/mV	Angle, degrees
White	100/714.3	0/0	•••
Gray	76.9/549.29	0/0	•••
Yellow	69/492.86	62.1/443.58	167.1
Cyan	56.1/400.72	87.7/626.44	283.5
Green	48.2/344.29	81.9/585.01	240.7
Magenta	36.2/258.57	81.9/585.01	60.7
Red	28.2/201.43	87.7/626.44	103.5
Blue	15.4/109.00	62.1/443.58	347.1
Black	7.5/53.57	0/0	•••
Burst	0/0	40/285.72	180

*Luminance levels and chrominance peak-to-peak amplitudes are
expressed in IRE and millivolts. Chrominance phase angles are
expressed in degrees with respect to the B-Y reference. The "white"
bar is representative of a 100/7.5/75/7.5 color bars signal waveform.

Figure 2.25 shows the same waveform after being severely distorted by a
nonlinear amplifier. Figure 2.26 shows a vectorscope display of the distorted
subcarrier. Note the relative change of the subcarrier phase (differential
phase) because of the signal distortion.

2.2.3   The PAL system

The PAL (phase-alternating line) color television system was developed to be
compatible with the 625/50 monochrome television system used in Europe
and transmitted in 7- or 8-MHz RF channels with baseband video bandwidths of 5, 5.5, or 6 MHz. The chrominance subcarrier chosen is common to
all European versions of PAL. Depending on the transmission system used,

Figure 2.21 Simplified block diagram of NTSC B-Y/R-Y decoder.

Figure 2.22 Simplified block diagram of NTSC decoder with comb filter.

Figure 2.23 Oscilloscope display of undistorted NTSC modulated ramp signal.

Figure 2.24 NTSC vectorscope display of constant-amplitude and constant-phase subcarrier component part of the undistorted NTSC modulated ramp signal. The phase of the subcarrier riding on the ramp is identical to that of the burst. The vectorscope gain is increased to expand vector amplitudes to the outer trace.

the European PAL systems are identified as B-PAL, D-PAL, G-PAL, H-PAL, and I-PAL. The main differences among these various versions of PAL are the transmitted luminance and upper chrominance sideband bandwidths. At the studio level there is only one common PAL system. Figure 1.18 showed details of various CCIR television channels and the frequency domain restric-

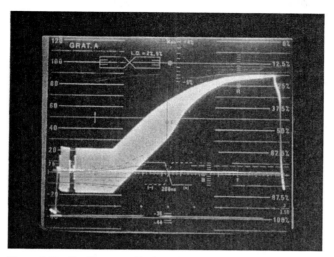

Figure 2.25 Oscilloscope display of distorted NTSC modulated ramp signal.

Figure 2.26 NTSC vectorscope display of distorted NTSC modulated ramp signal showing a subcarrier phase change relative to the burst phase. This phase change is called differential phase.

tions imposed on the transmitted PAL signal. In addition to these compatible versions of PAL, there are two incompatible PAL versions, namely M-PAL, a 525/60 version used mainly in Brazil, and N-PAL, a 625/50 narrowband version used mainly in Argentina. The characteristics of these versions of PAL are tailored to fit in a 6-MHz transmission channel. The current standard practice in these countries is to use NTSC (in Brazil) and European PAL (in

Argentina) in studio operations and transcode to the local PAL standard prior to on-air transmission.

Apart from some minor details, the PAL system is the same as the NTSC system. The main difference between the NTSC and PAL color television systems lies in the fact that the R-Y subcarrier phase is reversed from line to line of each field. The consequences of this minor modification are quite extensive. The main reason for the development of the PAL color television system was to overcome some studio and transmission equipment deficiencies. In the 1960s, these deficiencies appeared to be insurmountable to the European engineers who set out to develop a phase-tolerant system. On hindsight, it is debatable whether a new system was worth developing, since the advances in the art have eliminated most of the early problems with NTSC. The characteristics of the basic B-, D-, G-, H-, and I-PAL signals are summarized in Table 2.4.

The PAL encoder processes a wideband ($\geq$5-MHz) luminance (monochrome) signal and two narrowband color-difference signals of equal bandwidth. The chrominance signals are called E'_U and E'_V and are identical to the E'_{B-Y} and E'_{R-Y} signals of NTSC. The polarity of the E'_V is reversed every line. This is expressed in the third term of the equation of the complete color signal, E_M, in Table 2.4.

TABLE 2.4   Summary of PAL Signal's Characteristics

1. Assumed chromaticity coordinates for primary colors of receiver		x y
	Green	0.29 0.60
	Blue	0.15 0.06
	Red	0.64 0.33
2. Chromaticity coordinates for equal primary signals	Illuminant D_{65}: $x = 0.3127$; $y = 0.3290$	
3. Assumed receiver gamma value	2.8	
4. Luminance signal	$E'_Y = 0.587E'_G + 0.144E'_B + 0.299E'_R$	
5. Chrominance signals	$E'_U = 0.877 (E'_B - E'_Y)$ and $E'_V = 0.493 (E'_R - E'_Y)$	
6. Equation of complete color signal	$E_M = E'_Y + E'_U \sin (2\pi f_{SC}t) \pm E'_V \cos (2\pi f_{SC}t)$	
7. Type of chrominance subcarrier modulation	Suppressed-carrier amplitude modulation of two subcarriers in quadrature	
8. Chrominance subcarrier frequency, Hz	Nominal value and tolerance: $f_{SC} = 4,433,618.75$ Hz ± 5 (CCIR B, D, G, H) ± 1 (CCIR I) Relationship to line frequency f_H: $f_{SC} = (1135/4 + 1/625) f_H$	
9. Bandwidth of transmitted chrominance sidebands, kHz	f_{SC} +570/−1300 (CCIR B, D, G, H) f_{SC} +1066/−1300 (CCIR I)	
10. Amplitude of chrominance subcarrier	$G = \sqrt{(E'^2_U + E'^2_V)}$	
11. Synchronization of subcarrier	Subcarrier burst on blanking backporch	

The line-by-line phase alternation of the E'_V signal results in its chrominance spectral components being spaced at $f_H/2$ (7.8125-kHz) intervals from the E'_U spectral components. A half-line offset between luminance and chrominance, as used in NTSC, would result in crosstalk between the E'_V spectrum components and those of the luminance spectrum. To avoid this problem, PAL uses a combination of quarter-line offset and frame offset. The frequency of the subcarrier is

$$f_{SC} = (284 - \tfrac{1}{4})f_H + f_V \text{ Hz} = 4{,}443{,}618.75 \text{ Hz}$$

where $f_H = 15{,}625$ Hz
$\quad\ f_V = 25$ Hz

The transmitted bandwidth of the upper sideband of the chrominance signal depends on the transmission channel characteristics as shown in row 9 of Table 2.4.

Figure 2.27 shows the spectrum of a PAL signal. Note a peak of energy around the suppressed carrier at 4.43 MHz. Figure 2.28 shows a detailed view of the spectrum around the chrominance subcarrier.

A burst of 10 cycles of reference subcarrier is transmitted during the back-porch of the horizontal blanking interval. Figure 2.29 shows details of the horizontal blanking period with the subcarrier burst. The phase of the burst alternates in rhythm with the V component with reference to the U vector. Its phase alternates between $+135°$ and $-135°$. The burst is not transmitted during 9 lines of the vertical blanking interval. The burst vertical blanking timing follows a specific sequence, called *Bruch blanking,* to ensure that the first burst after the end of burst blanking has a $+135°$ phase.

The role of the burst is to synchronize the local oscillator of the receiver, which feeds the U and V synchronous demodulators as well as the switching of the V axis. The subcarrier oscillator feeds a reconstituted subcarrier to the synchronous U and V demodulators used for the recovery of the color-difference signals. The recovered subcarrier burst phase is $+180°$ with respect to the reference U vector.

Figure 2.30 shows a simplified block diagram of a PAL encoder. Gamma-corrected green, blue, and red signals are fed to a matrix that produces the luminance (Y) signal and the two color-difference signals. Each of the color-difference signals is bandwidth-limited to 1.2 MHz before feeding the respective balanced modulators. A 4.43-MHz subcarrier feeds the U modulator and, through a switchable $\pm 90°$ phase-shift network, the V modulator. A $\pm 135°$ phase-switched gated subcarrier burst is also generated. The U and burst subcarrier phase switching occurs at a frequency of $f_H/2 = 7812.5$ Hz (PAL trigger). The Y signal is delayed to compensate for the chrominance delay introduced by the color-difference low-pass filters. The adder combines the luminance, chrominance sidebands, composite sync, and gated subcarrier burst into a composite color signal.

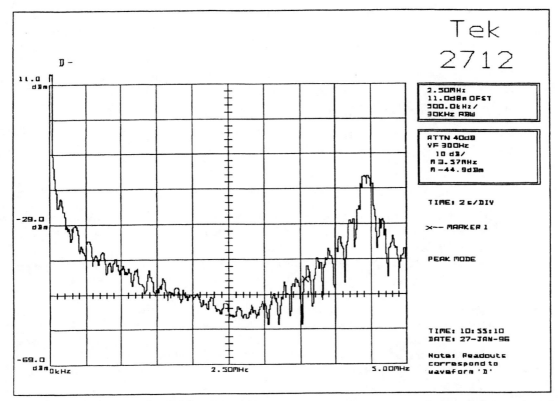

Figure 2.27 Spectrum of PAL 100/0/100/0 color bars signal. Note a peak of energy around the suppressed 4.43 MHz subcarrier. Horizontal resolution: 500 kHz/division.

Figure 2.31 shows the drawing of a 100/0/100/0 (100%) color bar signal waveform resulting from the addition of luminance and chrominance components. Figure 2.32 shows the drawing of a 75/0/75/0 (75%) color bar signal waveform resulting from the addition of luminance and chrominance components. The dotted outline of the luminance bar represents a 100/0/75/0 color bar signal.

Tables 2.5 and 2.6 list the details of the color bar signal luminance and chrominance values as well as phase angles of the six colors on alternate lines.

Figure 2.33 shows a vectorscope display of a 100% color bar signal with a +V component. Figure 2.34 shows a vectorscope display of a 100% color bar signal with a −V component. Figure 2.35 shows a vectorscope display of the superimposed +V and −V vectors.

Transmission constraints with respect to the use of 100% PAL color bars vary from country to country. The CCIR I transmission standard allows the transmission of 100% color bar signals by setting the white level at 20% of the video carrier amplitude.

Figure 2.36 shows a simplified diagram of a PAL decoder. The chrominance sidebands are separated through a bandpass filter and fed to the chromi-

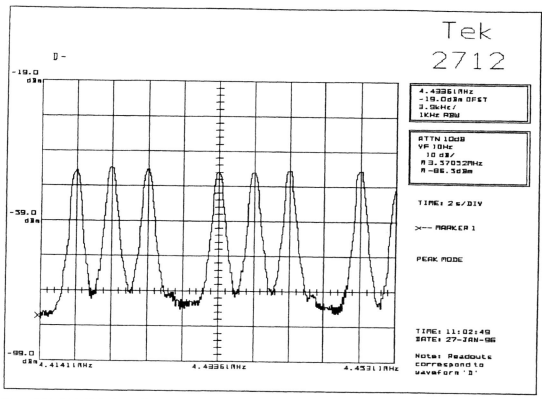

Figure 2.28 Detailed view of spectrum around the 4.43 MHz subcarrier. This picture was obtained with a synthetic signal and, therefore, the subcarrier is not suppressed in order to enhance the details of the spectrum. The spectrum is characterized by triplets consisting of the luminance (Y) in the center with the constant phase subcarrier (U) on its left (spaced at $-\frac{1}{4}f_{H}$) and the phase-alternating subcarrier (V) on its right (spaced at $+\frac{1}{4}f_{H}$). The spacing between the V component of one triplet (e.g., the third peak from the left) and the U component of the consecutive triplet (e.g., the fourth peak from the left) is $\frac{1}{2}f_{H}$. The horizontal resolution of the spectrum analyzer display is $\frac{1}{4}f_{H}$ = 3.9 kHz/division.

nance demodulators as well as to a burst separator. The burst separator is gated by a burst key derived from the horizontal sync. Its output synchronizes the local crystal-controlled subcarrier generator through a phase-locked loop (PLL). The burst phase varies from line to line and is alternately ±135° with respect to the reference U vector. The high time constant of the PLL integrates the resulting voltage variations, and the reconstituted subcarrier is +180° with respect to the U reference. The output of the bandpass filter feeds a 1-H (64-µs) delay line as well as an adder and subtractor. The signal at the output of the adder consists of U sidebands. The signal at the output of the subtractor is alternately ±V sidebands. The two signals are fed to two synchronous demodulators. The phase of the subcarrier fed to the U demodulator is constant. The phase of the subcarrier fed to the V demodulator is alternated ±90° every line (7.8125-kHz rate) with respect to that fed to the

Figure 2.29 PAL horizontal blanking interval showing details of color burst.

Figure 2.30 Simplified block diagram of PAL encoder.

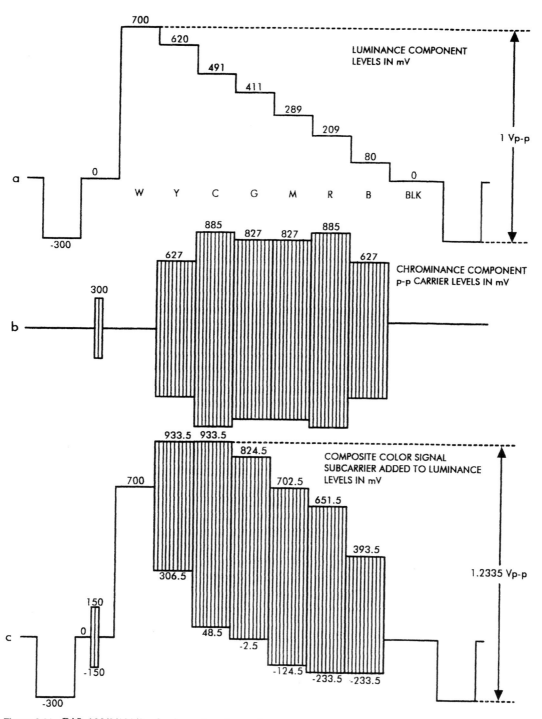

Figure 2.31   PAL 100/0/100/0 color bars signal waveform.

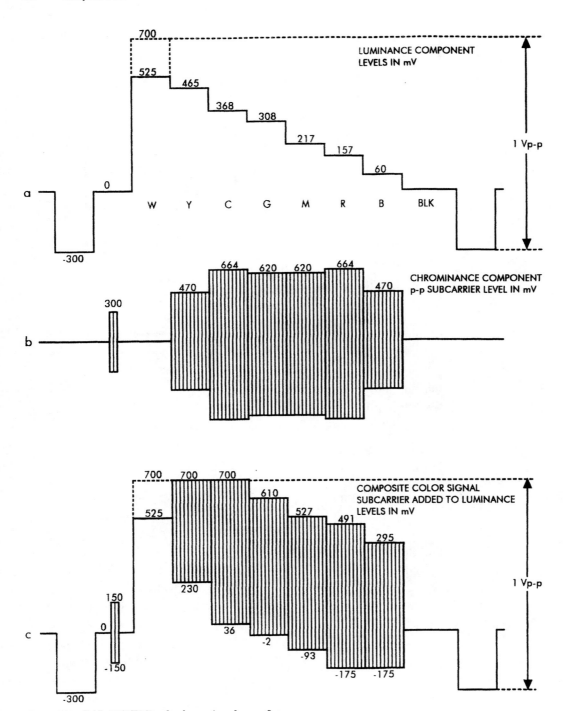

Figure 2.32 PAL 75/0/75/0 color bars signal waveform.

TABLE 2.5 PAL 100/0/100/0 Color Bars Signal Waveform Characteristics*

Color	Luminance, mV	Chrominance, mV	Angle line n, degrees	Angle line $n+1$, degrees
White	700	0	•••	•••
Yellow	620	627	167	193
Cyan	491	885	283.5	76.5
Green	411	827	240.5	119.5
Magenta	289	827	60.5	299.5
Red	209	885	103.5	256.5
Blue	80	627	347	13
Black	0	0	•••	•••
Burst	0	300	135	225

*Luminance levels and chrominance peak-to-peak amplitudes are expressed in millivolts. Chrominance phase angles on alternate scanning lines are expressed in degrees with respect to the B-Y reference.

TABLE 2.6 PAL 100/0/75/0 Color Bars Signal Waveform Characteristics*

Color	Luminance, mV	Chrominance, mV	Angle line n, degree	Angle line $n+1$, degrees
White	700	0	•••	•••
Yellow	465	470	167	193
Cyan	368	664	283.5	76.5
Green	308	620	240.5	119.5
Magenta	217	620	60.5	299.5
Red	157	664	103.5	256.5
Blue	60	470	347	13
Black	0	0	•••	•••
Burst	0	300	135	225

*Luminance levels and chrominance peak-to-peak amplitudes are expressed in millivolts. Chrominance phase angles on alternate scanning lines are expressed in degrees with respect to the B-Y reference.

U demodulator. The demodulated color-difference and the delayed luminance signals are fed to a matrix that reconstitutes the original primary signals. A notch filter is used to reduce the subcarrier visibility. The PAL chrominance vertical resolution is half that of the luminance as a consequence of the line averaging of the decoder. Comb filter variations of the PAL decoder exist but will not be described in this book.

One of the main faults of the NTSC color television system is its sensitivity to nonlinear distortions that result in differential phase. The PAL process transforms the differential phase, which results in color hue changes, into differential gain, which results in less noticeable color saturation changes. Figure 2.37 shows a modulated ramp, consisting of a ramp signal steadily rising from black to white on which is superimposed a constant-phase and constant-amplitude subcarrier signal. Figure 2.38 shows a vectorscope display of the constant-phase signal. The vectorscope gain is increased to expand the vector amplitudes to the outer graticule. Note the two ($\pm135°$) bursts on the

Figure 2.33 PAL vectorscope display of the PAL 100/0/100/0 color bars signal +V component. This display is similar to an NTSC display with the exception of the +135° burst phase.

Figure 2.34 PAL vectorscope display of the PAL 100/0/100/0 color bars signal −V component.

Figure 2.35 PAL vectorscope display of a PAL 100/0/100/0 color bars signal with superimposed $+V$ and $-V$ components from consecutive scanning lines.

Figure 2.36 Simplified block diagram of PAL decoder.

Figure 2.37 Oscilloscope display of undistorted PAL modulated ramp signal.

Figure 2.38 PAL vectorscope display of constant amplitude and constant-phase subcarrier component part of the undistorted modulated ramp signal. The vectorscope gain is increased to expand the vector amplitudes to the outer graticule. Unlike NTSC, the subcarrier riding on the ramp has a different phase than that of the associated burst. Consequently, there are two pairs of vectors displayed on the vectorscope screen: two burst vectors, on the left-hand side, and two ramp subcarrier components, on the right-hand side. The two sets of burst/ramp vectors correspond to the information carried by two consecutive scanning lines.

Figure 2.39 Oscilloscope display of distorted PAL modulated ramp.

left-hand side of the vector display and the $+V$ and $-V$ subcarrier vectors on the right-hand side of the vector display.

Figure 2.39 shows the same waveform after being distorted by a nonlinear amplifier. Figure 2.40 shows a $+V$ and $-V$ vectorscope display of the distorted subcarrier. Note that the phase change results in a similar "bow shape" of both vectors. Figure 2.41 shows the result of adding the $+V$ vector to the inverted $-V$ vector. The resulting vector has a constant phase but a variable amplitude (differential gain) depending on the phase deviation from the nondistorted phase.

2.2.4 The SECAM system

SECAM [an acronym for the French séquentiel couleurs à mémoire (sequential colors with memory)] is now mainly a transmission system and its description will be less detailed. In many organizations studio productions are carried out in analog component video or PAL and the signals are transcoded to SECAM prior to transmission. Unlike the PAL color television system, which is closely related to the NTSC system, the SECAM system uses a completely different approach to transmitting color information over a monochrome transmission channel. As the name implies, SECAM is a sequential color transmission system as opposed to the NTSC and PAL simultaneous color transmission systems. Whereas in NTSC and PAL the luminance and chrominance information is transmitted simultaneously by frequency-division multiplexing, SECAM transmits the luminance information continuously and the two-color difference signals sequentially on a line-by-line basis. As a result, the vertical color resolution is half that of the luminance. Instead of quadrature modulation of the subcarrier

Figure 2.40 PAL vectorscope display of distorted modulated ramp signal showing a phase change of the $+V$ and $-V$ subcarrier vectors relative to the associated burst. The phase change results in a similar bow-shape of both vectors.

Figure 2.41 PAL vectorscope display of the result of adding the $+V$ vector to the inverted $-V$ vector, similar to the process occurring in a PAL decoder. The resulting vector has a constant phase but a variable amplitude (differential gain) depending on the phase deviation from the nondistorted phase. Differential phase is thus transformed into differential gain.

by two color-difference signals, the SECAM system uses two color-difference signals to frequency-modulate two subcarriers with different rest frequencies. The red color-difference subcarrier has a rest frequency (f_{0R}) of 4.4 MHz and the blue color-difference subcarrier has a rest frequency (f_{0B}) of 4.25 MHz.

The significant characteristics of the SECAM signals are listed in Table 2.7 and will be summarily explained below.

The SECAM encoder processes a wideband (>5-MHz) luminance (monochrome) signal and two narrowband (1.5-MHz) color-difference signals of equal bandwidth. They are

$$D'_B = 1.5 \, (E'_B - E'_Y) \quad \text{and} \quad D'_R = -1.9 \, (E'_R - E'_Y)$$

The minus sign for D'_R indicates that negative values of $(E'_R - E'_Y)$ are required to give rise to positive frequency deviations when the subcarrier is deviated. As is usual with frequency-modulated signals, the SECAM chrominance-modulated signals are preemphasized before they are transmitted. SECAM uses two distinct types of preemphasis as follows:

TABLE 2.7 Summary of SECAM Signal Characteristics

1. Assumed chromaticity coordinates for primary colors of receiver	$\quad\quad\quad\quad x \quad\quad y$ Green $\quad$ 0.29 $\quad$ 0.60 Blue $\quad\quad$ 0.15 $\quad$ 0.06 Red $\quad\quad$ 0.64 $\quad$ 0.33
2. Chromaticity coordinates for equal primary signals	Illuminant D_{65}: $x = 0.3127$; $y = 0.3290$
3. Assumed receiver gamma value	2.8
4. Luminance signal	$E'_Y = 0.587E'_G + 0.114E'_B + 0.299E'_R$
5. Chrominance signals	$D'_B = 1.505 \, (E'_B - E'_Y)$ and $D'_R = -1.902 \, (E'_R - E'_Y)$
6. Low-frequency precorrection of color-difference signals	$D'_{B*} = A_{BF}\,(f) \, D'_B$ and $D'_{R*} = A_{BF}\,(f) \, D'_R$ where $A_{BF}(f) = [1+j(f/f_1)]/[1+j(f/3f_1)]$ $\quad\quad f$ = signal frequency, kHz f_1 = 85 kHz
7. Equation of complete color signal	$E_M = E'_Y + G \cos 2\pi \, (f_{0B} + \Delta f_{0B} f_0 D_{B*} dt)$ or $E_M = E'_Y + G \cos 2\pi \, (f_{0R} + \Delta f_{0R} f_0 D_{R*} dt)$ alternately from line to line
8. Chrominance subcarrier modulation	Frequency modulation
9. Chrominance subcarrier frequency, Hz	Nominal value: $f_{0B} = 4,250,000 \pm 2000$ $\quad\quad\quad\quad$ and $f_{0R} = 4,406,250 \pm 2000$ Relationship to line frequency f_H : $f_{0B} = 272f_H$ and $f_{0R} = 282f_H$
10. Maximum subcarrier deviation, kHz	$\Delta f_{0R} = +350/-506$ and $\Delta f_{0B} = +506/-350$
11. Amplitude of chrominance subcarrier	$G = M_0 \, [(1+j\,16F)/(1+j\,1.26F)]$ where $F = (f/f_0) - (f_0/f)$, $f_0 = 4286$ kHz and the peak-to-peak amplitude, $2M_0$, is 23% of the peak luminance amplitude
12. Subcarrier switching synchronization	By undeviated chrominance subcarrier reference on the line-blanking backporch

- Preemphasis of the color-difference video signals before they modulate the subcarrier as per row 6 in Table 2.7, resulting in improved chrominance SNR.

- Preemphasis of the modulated subcarrier as per row 11 in Table 2.7 and Fig. 2.42. This type of preemphasis is named *Bell curve* after its peculiar shape, and results in a reduced subcarrier visibility.

The undeviated subcarrier frequencies are even multiples of the horizontal scanning frequency ($f_H = 15,625$ Hz). There is, consequently, no frequency-domain interleaving of luminance and chrominance spectral components as in NTSC and PAL. The situation is further complicated by the fact the subcarriers are frequency-modulated. The two subcarriers and their sidebands occupy the baseband spectrum between 3 and 6 MHz. Figure 1.18 showed how CCIR systems K and L accommodate the subcarrier spectrum. In an attempt to reduce the visibility of the subcarrier and improve compatibility, the phases of the undeviated subcarrier frequencies are varied in a complex sequence spread over 12 consecutive fields. In order to reduce the visibility of the chrominance signal, the SECAM decoder luminance channel uses a low-pass filter with high-attenuation poles at the two undeviated subcarrier frequencies (f_{0B} and f_{0R}).

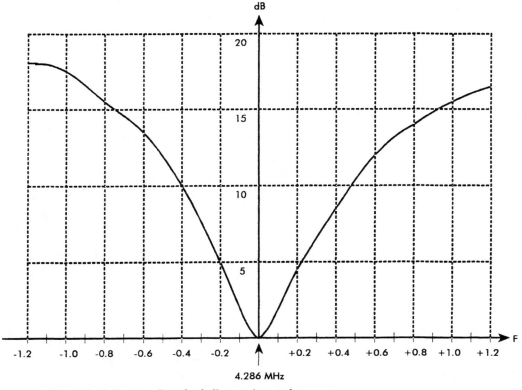

Figure 2.42   Encoder bell curve. Decoder bell curve is complementary

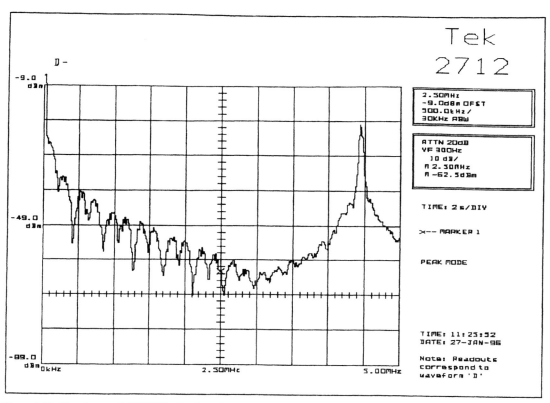

Figure 2.43 Spectrum of SECAM signal with blue modulation and an unmodulated red subcarrier at 4.406 MHz. Horizontal resolution: 500 kHz/division.

Figure 2.43 shows the spectrum of a SECAM signal with B-Y information but no R-Y information. Note the undeviated red carrier at 4.4 MHz. Figure 2.44 shows the spectrum of a SECAM signal with R-Y information but no B-Y information. Note the undeviated blue carrier at 4.25 MHz. Figure 2.45 shows the spectrum of a SECAM signal with a 3.5-MHz luminance sweep and undeviated subcarriers. The sweep is attenuated beyond 3.2 MHz, indicating that this encoder does not pass luminance above this frequency.

Since two consecutive lines carry alternate chrominance signals, it is necessary to identify them. To this end an undeviated subcarrier is transmitted during the backporch of the line-blanking interval. Unlike NTSC and PAL, where a burst of subcarrier is transmitted, in SECAM the subcarrier is transmitted continuously over the duration of the line and is blanked out only during the horizontal sync and during the vertical blanking interval. In the absence of chrominance information, its amplitude is constant over the duration of the line. Figure 2.46 shows details of the SECAM horizontal blanking interval.

Figure 2.47 shows a simplified block diagram of a SECAM encoder. Gamma-

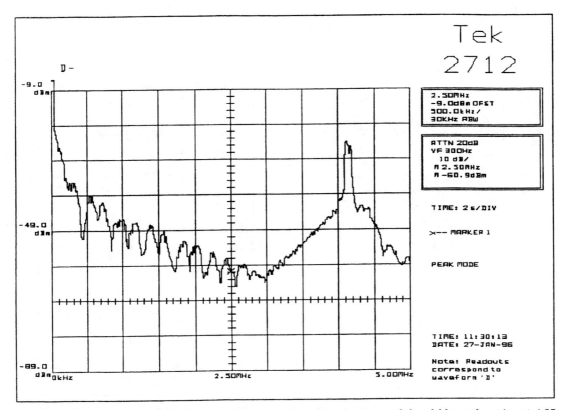

Figure 2.44 Spectrum of SECAM signal with red modulation and an unmodulated blue subcarrier at 4.25 MHz. Horizontal resolution: 500 kHz/division.

corrected green, blue, and red signals are fed to a matrix that produces the luminance (Y) signal and the two color-difference signals. Each of the color-difference signals is bandwidth-limited to 1.5 MHz, preemphasized, and feeds a frequency modulator. The outputs of the modulators are selected sequentially on a line-by-line basis by a switch controlled by the SECAM identification signal, having a frequency of $f_H/2 = 7.8125$ kHz, and fed to a bell filter. The Y signal is delayed to compensate for the chrominance delay. The adder combines the luminance, the frequency-modulated chrominance subcarrier with its sidebands, and composite sync signals into a composite color signal.

Drawings of 100/0/75/0 color bars waveforms are shown in Fig. 2.48 for a red line and in Fig. 2.49 for a blue line. The numbers indicate the maximum and minimum chrominance signal values for the various colors. Unlike NTSC and PAL, the subcarrier is present during black- and white-picture transmission. The chrominance overshoots at transitions between color bars caused by the preemphasis.

Table 2.8 lists the SECAM 100/0/75/0 color bar signal waveform character-

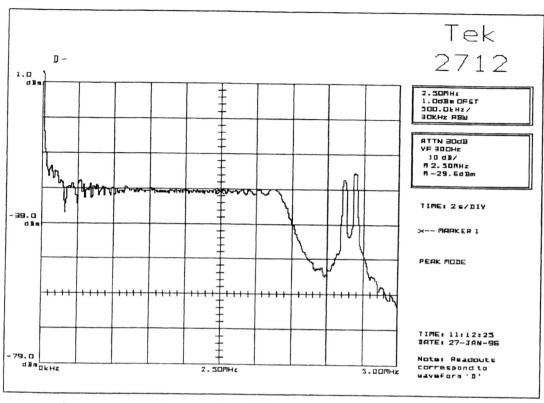

Figure 2.45 Spectrum of SECAM signal with 3.5-MHz sweep in the luminance channel and unmodulated chrominance subcarriers at 4.25 MHz (blue) and 4.406 MHz (red). Note luminance channel cutoff at 3.2 MHz in encoder. Horizontal resolution: 500 kHz/division.

istics for red and blue lines in terms of luminance amplitude as well as chrominance peak-to-peak (p-p) amplitude and associated frequency.

Transmission constraints with respect to the transmission of 100% SECAM color bars are less stringent in France because the use of positive modulation for video transmissions and amplitude modulation for sound transmissions precludes the use of the intercarrier sound reception method and the possible intercarrier "buzz" problems of NTSC and PAL. Intercarrier buzz problems do occur, however, in countries using negative modulation for video transmissions and frequency modulation for sound transmissions.

Figure 2.50 shows a simplified block diagram of a SECAM decoder. The luminance subcarrier is obtained by passing the composite signal through a low-pass filter with high-attenuation poles at 4.25 and 4.4 MHz. The chrominance subcarrier and its sidebands are obtained by passing the composite signal through a bandpass and reversed bell filter. It feeds a line-sequential 2×2 switch directly and through a 1-H (64-μs) delay line. The switch is controlled

Figure 2.46 SECAM horizontal blanking interval showing details of color subcarrier.

Figure 2.47 Simplified block diagram of SECAM encoder.

by a 7.8125-kHz pulse obtained by demodulating gated undeviated subcarrier samples present on the backporch of the horizontal blanking interval. The switch ensures that the correct line-sequential frequency-modulated subcarrier feeds the respective chrominance subcarrier demodulator. The subcarriers are amplitude-limited and feed a dedicated frequency discriminator. The

Figure 2.48 SECAM 100/0/75/0 color bars red line signal waveform.

Figure 2.49 SECAM 100/0/75/0 color bars blue line signal waveform.

demodulated color-difference signals are deemphasized and, together with the delayed luminance signals, feed a matrix that reconstitutes the original primary signals.

The SECAM color television system is relatively insensitive to nonlinear distortions inherent in analog active circuit elements. It is, therefore, ideally suited

TABLE 2.8   SECAM 100/0/75/0 Color Bars Signal Waveform Characteristics

Color	Luminance, mV	Red line		Blue line	
		Chroma, mV	Frequency, MHz	Chroma, mV	Frequency, MHz
White	700	217	4.406	166.6	4.250
Yellow	464	182	4.360	364	4.020
Cyan	368	476	4.686	168	4.327
Green	308	434	4.640	280	4.097
Magenta	217	210	4.172	210	4.402
Red	157	252	4.126	210	4.172
Blue	60	252	4.452	280	4.480
Black	0	217	4.406	166.6	4.250

Figure 2.50   Simplified block diagram of SECAM decoder.

to land-line or over-the-air transmissions if the inherent reduced luminance bandwidth is accepted. Its studio use is generally limited to switching between signal sources. This is because of the frequency-modulated subcarriers, which do not lend themselves to mixing of composite video signals. Sophisticated production techniques using SECAM signals require demodulation to analog component signals unlike NTSC and PAL, which can be processed in their analog composite format. The advent of digital composite standards, with sampling at a multiple of the stable subcarrier frequency, has all but eliminated SECAM as a production tool because of its time-varying subcarriers.

2.2.5   Performance-indicative parameters and measurements concepts

Signal distribution and processing of composite analog video signals is characterized by less than ideal performance in terms of

- Linear distortions
- Nonlinear distortions
- Noise

As a consequence, the shape of the video signals is affected, resulting in picture impairments. These impairments can be judged subjectively by observing the picture quality on a picture monitor. Various international bodies have developed subjective picture-quality grading criteria. Objective equipment performance measurements use standardized test waveforms. These waveforms are tailored to contain frequency domain components that are best suited for the measurement of the specific type of impairment.

2.2.5.1 Linear distortions. A video channel ideally resembles a perfect low-pass filter, which does not exist in practice. A practical low-pass filter, fulfilling most of the requirements of an ideal filter, would have the following characteristics:

- Flat amplitude versus frequency response all the way up to the maximum frequency of interest. Beyond the maximum frequency of interest, the frequency response tapers off smoothly.

- Constant group delay all the way up to the maximum frequency of interest. Beyond the maximum frequency of interest, the group delay increases gradually.

All systems require a specific bandwidth to transmit all significant luminance and chrominance information. The lowest frequency of interest is the frame-repetition frequency (25 or 30 Hz). The correct reproduction of slowly changing luminance values requires the transmission of a DC component of the video signal. To avoid difficulties associated with amplification and distribution of DC voltages, the low-frequency end of distribution equipment is limited to 25 or 30 Hz. The DC component is "restored," where required, through the use of clamping circuitry. The upper frequency limit of modern studio distribution equipment (e.g., distribution amplifiers, production switchers, routing switchers) is well in excess of the maximum transmitted video frequency. This is required to ensure adequate performance in situations involving a large number of system elements and passes, typical of the operational requirements of large studios. Typically, the single-pass frequency response extends to 8 MHz with a smooth rolloff beyond. Analog composite VTRs are band-limited to 5 MHz or less.

Assuming freedom from nonlinear distortions, the behavior of studio equipment in terms of linear distortions can be assessed as follows:

- In the frequency domain: By measuring the amplitude versus frequency response and group delay versus frequency characteristic.

- In the time domain: By comparing the waveform present at the output of the equipment under test with its corresponding input waveform.

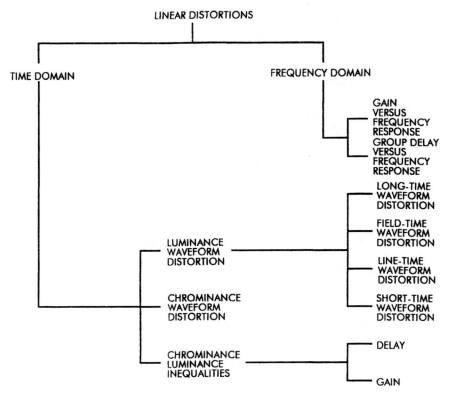

Figure 2.51   Classification of linear distortions.

Figure 2.51 shows a classification of linear distortions. Several standard test signals are commonly used to verify the performance of video equipment. Each test signal has a specific associated frequency spectrum suited to the frequency domain being investigated.

Table 2.9 summarizes significant linear distortions and gives a short description of the test concept.

Frequency response measurements are usually carried out with the multiburst signal. The multiburst signal is a "quantized" sweep signal containing a reference luminance bar followed by six bursts at discrete frequencies. It allows for a crude frequency response measurement given that the response between the frequency samples is unknown. Figures 2.52 and 2.53 show typical versions of the multiburst test waveform for the two conventional television standards.

The field-time distortion is measured with a field-time square wave. This parameter is a measure of the effect of gain and phase versus frequency effects in the frequency domain extending from 20 Hz to 2 kHz. Figures 2.54 and 2.55 show typical versions of the field-time square wave for the two conventional television standards.

TABLE 2.9   Summary of Significant Linear Distortions and Test Methods

Parameter	Definition	Test signal	Test method
Gain versus frequency response	Gain variation over band from 500 kHz to maximum frequency	Multiburst or sweep	Measure signal level at selected frequencies in dB wrt level at 500 kHz
Long-time distortion	Damped signal-level oscillations resulting from sudden picture-level changes	Any test signal allowing a controlled change in average picture level	Measure overshoot peak as % of peak luminance level
Field-time distortion	Change in shape of a field-time square wave	Field-rate square wave	Measure peak-to-peak tilt of top of bar in % wrt centerer of bar excluding first and last 250 μs
Line-time distortion	Change in shape of a line-time square wave	Line-rate square wave	Measure peak-to-peak tilt of top of bar in % wrt centerer of bar excluding first and last 1 μs
K factor	Quantified subjective impairment rating	2T sine-squared pulse	Read K rating in % on special graticule
Chrominance to luminance gain	Change in ratio of chrominance versus luminance amplitude	Modulated sine-squared pulse	Measure amplitude of chrominance component in % wrt the luminance component
Chrominance to luminance delay	Change in timing of chrominance versus luminance	Modulated sine-squared pulse	Measure delay of chrominance component in ns wrt to the luminance component

The line-time distortion is measured with a line-time square wave. This parameter is a measure of the effect of gain and phase versus frequency in the domain extending from 5 to 500 kHz. Figures 2.56 and 2.57 show typical versions of the line-time square wave, also called a bar signal, for the two conventional television standards. The risetime of the square wave is chosen such that its spectrum is confined within the boundary of the maximum transmitted video bandwidth (4.2 or 5 MHz) to avoid excessive ringing.

The short-time waveform distortion is usually expressed in terms of the K rating. The K rating is a subjective assessment of the picture quality. The test signal used is a sine-squared pulse. The significant parameter of a sine-squared pulse is the half-amplitude duration (HAD). HAD is expressed in terms of the Nyquist interval T. The relationship between T and the upper video frequency f_{max} is given by

$$T = \frac{1}{2f_{max}}$$

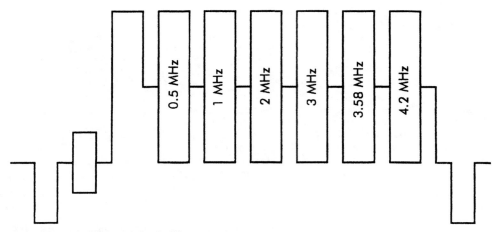

Figure 2.52   525/60 Multiburst test signal.

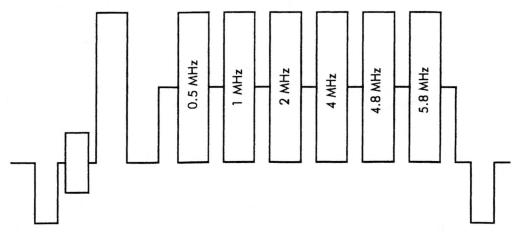

Figure 2.53   625/50 Multiburst test signal.

In the NTSC system f_{max} is nominally 4 MHz, resulting in $T = 125$ ns. In the PAL system f_{max} is nominally 5 MHz, resulting in $T = 100$ ns. The spectrum of the sine-squared pulse extends to

$$f_{max} = \frac{1}{\text{HAD}}$$

The HAD of the typical sine squared pulse is $2T$. The spectrum of the $2T$ pulse is thus limited to 4 MHz, respectively 5 MHz, with no significant components beyond either 4 MHz (NTSC) or 5 MHz (PAL). The distorted $2T$ pulse, a result of short-time distortions, is displayed inside a special K rating graticule,

Figure 2.54 Field-time square wave for 525/60 system.

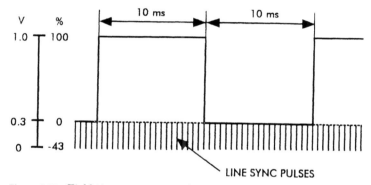

Figure 2.55 Field-time square wave for 625/50 system.

and the rating depends on the magnitude of the baseline irregularities. Figures 2.58 and 2.59 show the typical $2T$ sine-squared pulse used in the two conventional television standards.

Figure 2.60 shows the spectrum of a $2T$ pulse.

The chrominance to luminance delay and gain are measured using the modulated sine-squared pulse. This signal is formed by the linear addition of a sine-squared pulse and a chrominance subcarrier modulated by the sine-squared pulse. In the 525/60 standard the pulse HAD is $12.5T$ (1.57 μs) and in the 625/50 standard the pulse HAD is $20T$ (2 μs). Less-than-ideal equipment affects the amplitude, HAD, and the pulse baseline shape. Special nomographs are available for the calculation of the chrominance to luminance gain (in dB) or amplitude ratio (in %) and delay (in ns). Figures 2.61 and 2.62 show the modulated sine-squared pulse used in the two conventional television systems.

Figures 2.63 and 2.64 show the frequency spectrum of the modulated sine-squared pulse for the two conventional television systems.

Figure 2.56 Typical line-time square wave (bar) for 525/60 system.

Figure 2.57 Typical line-time square wave (bar) for 625/50 system.

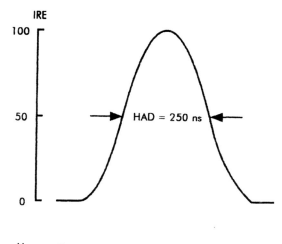

Figure 2.58 2T sine squared pulse for 525/60 system.

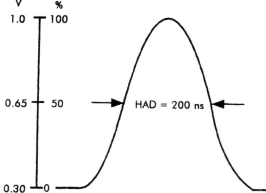

Figure 2.59 2T sine squared pulse for 625/50 system.

The 2*T* sine-squared pulse, the line-time square wave, and the 12.5*T* (or 20*T*) modulated sine-squared pulse are usually combined in a single composite video test signal.

Table 2.10 summarizes the causes and effects of linear distortions.

Contemporary studio-quality equipment is characterized by smooth and ripple-free frequency response as well as controlled group delay characteristics all the way up to the maximum frequency of interest. Consequently, many of the single-pass linear distortions are insignificant. Table 2.11 summarizes the single-pass performance of contemporary studio-quality equipment in terms of linear distortions.

2.2.5.2 Nonlinear distortions. A video channel ideally has a linear transfer curve. A linear transfer curve means that the spectral content of the signal exiting a specific piece of equipment is identical to that of the corresponding

Figure 2.60   Spectrum of sine squared 2T pulse.

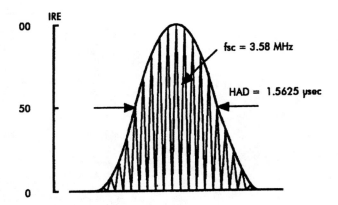

Figure 2.61   12.5T modulated sine squared pulse for NTSC.

input signal irrespective of the signal level, assuming that the signal level does not exceed the linear operating range specified for that piece of equipment. Stated differently, this means that the equipment does not generate harmonic distortions of discrete spectrum components or intermodulation distortions between pairs or groups of discrete spectrum components.

The transfer curve of real-life analog equipment is not completely linear. The nonlinear distortion introduced depends on

- The average picture level (APL)
- The instantaneous value of the luminance signal

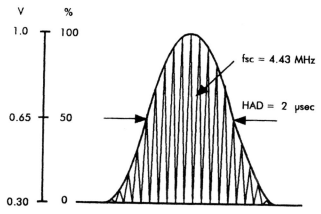

Figure 2.62 20T modulated sine squared pulse for PAL.

Figure 2.63 Spectrum of the 12.5T modulated sine squared pulse for NTSC.

Figure 2.64 Spectrum of the 20T modulated sine squared pulse for PAL.

TABLE 2.10   Causes and Effects of Linear Distortions

Cause	Test signal distortion	Picture impairment
Poor gain/frequency response above 500 kHz	Poor C/L ratio	Loss of luminance detail and color saturation
H. F. group delay	Delay of chrominance with respect to luminance	Color fringing
Limited bandwidth with sharp cutoff	Poor K rating	Echolike disturbances
Poor response in 20-Hz to 2-kHz band	Field-time waveform distortion	Vertical smearing
Poor response in 5- to 500-kHz band	Line-time waveform distortion	Horizontal smearing

TABLE 2.11   Single-Pass Performance of Studio-Type Equipment in Terms of Linear Distortions

Parameter	Wideband equipment	Analog VTR (NTSC)
Frequency response	±0.1 dB, 0.5 to 8 MHz	±0.2 dB, 0.5 to 4.2 MHz
Field-time distortion	<1%	<1%
Line-time distortion	<1%	<1%
K rating	<1%	≤2%
Chrominance/luminance gain	1%	1%
Chrominance/luminance delay	≤10 ns	≤20 ns

■ The amplitude of the chrominance signal

Depending on the specific signal processing of a particular piece of equipment and the manner in which the luminance and chrominance are treated, the nonlinearity may affect the luminance and/or the chrominance signals individually or cause interaction between them. The performance of studio-type equipment in terms of nonlinear distortions can be assessed as

■ Luminance signal amplitude distortion

■ Chrominance signal amplitude and phase distortion

This leads to a classification of the nonlinear distortions as shown in Fig. 2.65.

All systems require that the linear operating range of equipment be such as to accept composite video signals with an amplitude of up to at least 1.2 $V_{p\text{-}p}$. In addition, adequate headroom is necessary to accommodate variations in signal levels due to gain misadjustments of various elements making up a complex distribution chain. This puts very stringent requirements on the linearity of individual elements in a teleproduction facility. This is necessary to ensure adequate performance in the case of complex signal distribution pat-

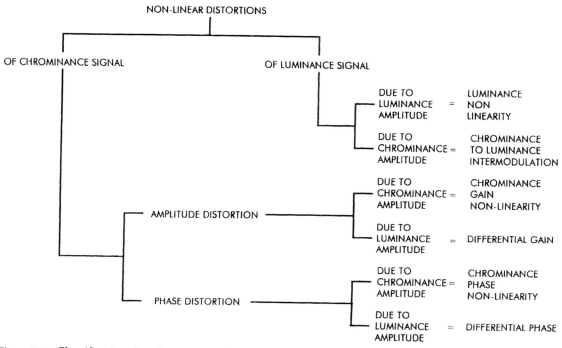

Figure 2.65 Classification of nonlinear distortions.

terns involving a large number of system elements and passes, because the nonlinear distortions are additive.

Contemporary analog teleproduction system elements such as video distribution amplifiers, production switchers, and routing switchers introduce very low nonlinear distortions, in some cases almost unmeasurable with currently available test equipment and methods. The limiting element is the analog videotape recorder (VTR), which, because of the modulation and demodulation process, introduces relatively high single-pass nonlinear distortions. The situation is further deteriorated by multiple-generation recordings typical of complex teleproductions requiring extensive editing.

Several standard video test signals are commonly used to verify the performance of video equipment in terms of nonlinear distortions. These signals simulate real-life video signals in terms of signal amplitude and spectral content.

The luminance staircase signal is used to measure luminance nonlinearities. It consists of 5 or 10 risers of equal amplitude. Figures 2.66 and 2.67 show typical staircase signals developed for the two conventional television systems.

The modulated staircase signal is used to measure differential phase and differential gain. It consists of 5 or 10 risers on which is superimposed a small constant-amplitude and phase chrominance signal. A luminance ramp is sometimes used in place of the staircase. The peak staircase amplitude and

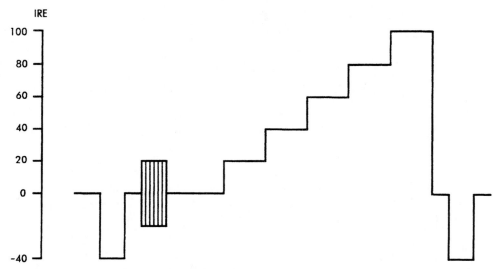

Figure 2.66 Five riser staircase for 525/60 systems.

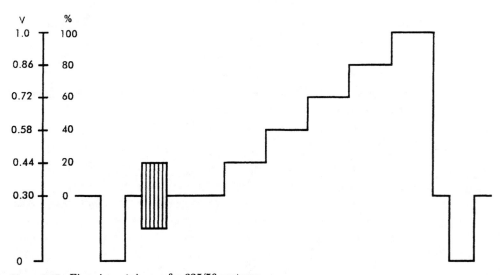

Figure 2.67 Five riser staircase for 625/50 systems.

superimposed subcarrier vary depending on the application. A reduced-amplitude signal is used for transmitter tests. Figures 2.68 and 2.69 show typical modulated staircase test signals for the two conventional television systems. System nonlinearities will affect the amplitude and phase of the subcarrier on various steps, resulting in differential gain and phase.

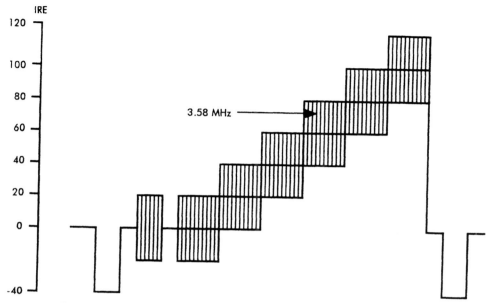

Figure 2.68 Five riser modulated staircase for NTSC system.

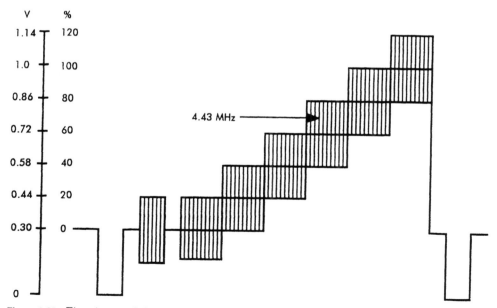

Figure 2.69 Five riser modulated staircase for PAL system.

The modulated pedestal is used to measure chrominance phase and amplitude distortions as well as chrominance to luminance intermodulation. It consists of a luminance pedestal on which are superimposed three levels of constant-phase chrominance subcarrier. Figures 2.70 and 2.71 show typical modulated pedestals for the two conventional television systems. System nonlinearities will affect the chrominance step amplitude and phase as well as modify the normally constant amplitude of the luminance pedestal.

Figure 2.70 Modulated pedestal signal for NTSC system.

Figure 2.71 Modulated pedestal signal for PAL system.

The assessment of the magnitude of the specific type of nonlinearity is performed in the following manner:

- The equipment under test is fed the test signal specified for the type of measurement to be carried out.

- The test equipment connected to the output of the equipment isolates the chrominance or the luminance component of the distorted test signal and measures the amplitude or phase deviation with respect to the reference at specific luminance and chrominance levels.

Table 2.12 summarizes and defines the nonlinear distortions and gives a short description of the measurement concepts.

The most common measurable nonlinear distortions encountered with studio-type equipment are

- Luminance nonlinearity
- Differential gain
- Differential phase

The other types of nonlinear distortions are typical of high-power television transmitters where there is a higher likelihood of interaction between the luminance and chrominance spectra.

The visible effects of nonlinear distortions as they affect NTSC composite video signals are shown in Table 2.13. The tolerance of the most visible nonlinear distortion, the differential phase, is about $10°$ for NTSC and $40°$ for PAL and SECAM. The tolerance of differential gain is about 25% for NTSC and PAL and about 55% for SECAM.

Table 2.14 lists the single-pass performance of contemporary studio-quality equipment in terms of nonlinear distortions.

2.2.5.3 Noise. Noise is defined as an unwanted parasitic signal superimposed on the wanted signal. Figure 2.72 shows a classification of noise. Depending on the mechanism of noise generation, two categories of noise can be distinguished, random noise and coherent noise.

Random noise is generated by circuit elements and is the result of the combined effects of thermal noise and Schottky noise. Being the consequence of the granular structure of matter, it cannot be eliminated, and a given circuit will have a specific signal-to-noise ratio (SNR) that cannot be improved. Figure 2.73 shows the types of random noise encountered in practice. Flat frequency response amplifying devices generate white noise, whose power density is constant with frequency. Devices that emphasize the higher frequencies with respect to the lower frequencies of the spectrum, such as input stages of cameras with photoconductive sensors, exhibit triangular noise, whose power density is proportional to the square of the frequency. The noise spectrum is modified by the characteristics of the discrete elements that constitute a system. As a consequence, white and triangular noise are seldom encountered in pure form.

TABLE 2.12 **Summary of Nonlinear Distortions and Test Methods**

Parameter	Definition	Test signal	Test method
Luminance non-linearity	Output/input amplitude proportionality change of a small unit step function as the step level is shifted from blanking level to white level. The average picture level is kept constant.	Luminance staircase	Measure % of largest step amplitude variation
Chrominance gain nonlinearity	Output/input chrominance subcarrier amplitude proportionality change as the subcarrier amplitude is varied from a minimum to a maximum specified value. Luminance and average picture levels are kept constant.	Modulated pedestal	Measure % of steps 3 and 1 chrominance amplitude change with reference to step 2
Chrominance phase nonlinearity	Chrominance subcarrier phase variation when the subcarrier amplitude is varied from a minimum to a maximum specified value. Luminance and average picture levels are kept constant.	Modulated pedestal	Measure largest phase difference (in degrees) of steps 3 or 2 with reference to step 1
Chrominance to luminance intermodulation	Variation of luminance signal amplitude resulting from the superimposition of a specified amplitude chrominance signal. The average picture level is kept constant.	Modulated pedestal	Measure % of largest luminance level change due to chrominance level
Differential gain	Amplitude change of a constant small-amplitude chrominance subcarrier superimposed on a luminance signal level that changes from blanking to white. The average picture level is kept constant.	Modulated ramp or staircase	Measure % of largest chrominance subcarrier level change with reference to burst amplitude
Differential phase	Change in phase of a constant small-amplitude chrominance subcarrier without phase modulation superimposed on a luminance signal level that changes from blanking to white. The average picture level is kept constant.	Modulated staircase or ramp	Measure largest chroma phase change in degrees with reference to burst phase

Equalizing distribution amplifiers are characterized by hypertriangular noise where the distribution is approximately triangular at low frequencies and approaches a parabola at higher frequencies. Analog VTRs are characterized by intermediate triangular noise where the power spectrum is decreased with increasing frequency as the result of emphasis and deemphasis.

The effect of random noise on the luminance component of the picture is seen

TABLE 2.13   Visible Effects of Nonlinear Distortions

Distortion	Picture impairment
Luminance nonlinearity	Distortion of picture brightness values
Chrominance/luminance intermodulation	Chrominance-level-related picture brightness changes
Chrominance gain nonlinearity	Distortion of color saturation values
Differential gain	Luminance-level-related color saturation changes
Chrominance phase nonlinearity	Chrominance-level-related color hue changes
Differential phase	Luminance-level-related color hue changes

TABLE 2.14   Single-Pass Performance of Studio-Quality Equipment in Terms of Nonlinear Distortions

Parameter	Distribution amplifier	Routing switcher	Production switcher	1-in VTR
Luminance nonlinearity	≤0.2%	≤0.5%	≤1%	≤2%
Chrominance/luminance intermodulation	Nonmeasurable	Nonmeasurable	Nonmeasurable	Nonmeasurable
Chrominance gain nonlinearity	Nonmeasurable	Nonmeasurable	Nonmeasurable	Nonmeasurable
Differential gain	≤0.2%	≤0.5%	≤1%	4%
Chrominance phase nonlinearity	Nonmeasurable	Nonmeasurable	Nonmeasurable	Nonmeasurable
Differential phase	≤0.1°	≤0.5°	≤1°	≤2°

as a pattern of black-and-white specks. The pattern is relatively coarse in the case of white noise and fine-grained in the case of triangular noise. Given the low sensitivity of the eye for fine luminance details, moderate amounts of triangular noise affecting the luminance are quite tolerable. The effect of random noise on the chrominance component of the picture is rather different. Noise components that have comparatively little impact on the luminance assume a greater importance in a color signal. Essentially, the chrominance information is transmitted as amplitude and phase modulation of a suppressed subcarrier in NTSC and PAL, hence as double-sideband components in the spectrum centered around 3.58 or 4.43 MHz. White and triangular noise components in the spectrum around the color subcarrier are demodulated in the receiver or monitor as comparatively low-frequency signal amplitude fluctuations (<500 kHz) and are seen as a coarse pattern of colored specks. The SECAM system is relatively immune to chrominance noise because of the frequency modulation of the chrominance subcarriers. Table 2.15 summarizes the types of noise encountered in a studio environment and gives a short description of the test methods.

The test signal for the measurement of noise is a flat field at the black level (i.e., a luminance level of 0 IRE). The luminance random SNR is defined as

$$SNR(dB) = 20 \log_{10} (\text{peak-to-peak video signal voltage/RMS noise voltage})$$

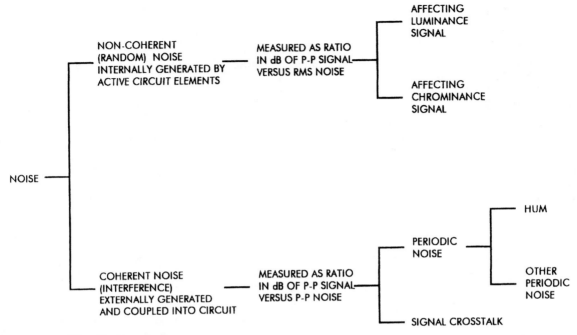

Figure 2.72 Classification of noise.

where the peak-to-peak amplitude of the video signal is 714 mV for the 525/60 system and 700 mV for the 625/50 system. The noise is measured in a frequency band extending from 10 kHz to the upper limit of the video band (4.2, 5, 5.5, or 6 MHz). The lower frequency limit ensures that hum is not measured as random noise. The upper limit ensures that irrelevant high-frequency noise is removed from the measured signal. Certain authorities use also a weighting filter approximating the eye response. Specialized equipment gives a direct reading as per the formula above. Oscilloscope measurements of luminance SNR necessitate a conversion from peak-to-peak noise voltage, as displayed on the oscilloscope screen, to root mean square (RMS) noise voltage. To measure the SNR, the "quasi" peak-to-peak amplitude of the noise is carefully estimated and a conversion factor is used to obtain the RMS noise value. Various organizations use conversion factors between 14 and 18 dB. The authors use a conversion factor of 15 dB. The SNR is consequently calculated as per the formula:

$$\text{SNR(dB)} = 20 \log_{10} (\text{p-p video signal voltage/p-p noise voltage}) + 15$$

The measurement of chrominance SNR has not been internationally standardized. Several manufacturers have developed de facto measurement methods. A full-field signal with a set chrominance level and phase is used as a test signal. The noisy signal at the output of the equipment under test is bandpass-filtered to extract the chrominance information. The chrominance information is

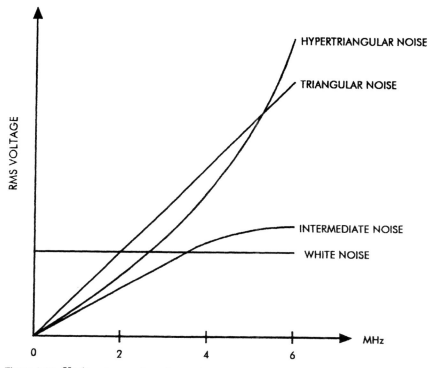

Figure 2.73 Various types of random noise spectra.

demodulated and its random variations of amplitude and phase, with respect to a standard reference, are measured as amplitude noise and phase noise.

The coherent noise is generated outside the equipment and coupled in some manner into it. This type of noise does not possess the statistical properties of random noise. It usually consists of one or more spurious signals of a periodic or quasi-periodic nature. The coherent noise can be avoided by good engineering practice, unlike the random noise. The picture impairment it causes is measurable as peak-to-peak noise. The signal to coherent noise ratio is defined as

$$SNR(dB) = 20 \log_{10}(\text{p-p video signal voltage/p-p coherent noise voltage})$$

The coherent noise may be hum, other periodic frequencies, or video signal crosstalk. The measurement of hum uses a 10-kHz low-pass filter. Higher-frequency periodic noise measurements use a 10-kHz high-pass filter.

Crosstalk is defined as the injection of an unwanted video signal into a neighboring circuit. It is measured by feeding a video signal into the crosstalking path and measuring its effect at the output of the wanted path.

Table 2.16 summarizes the effects of different types of noise on the picture.

Table 2.17 summarizes the single-pass performance of studio-type equipment in terms of noise.

TABLE 2.15   Summary of Significant Types of Noise and Test Methods

Parameter	Definition	Test signal	Test method
Continuous random noise	Ratio, expressed in dB, of the nominal amplitude of the luminance signal (714.3 or 700 mV) to the RMS amplitude of the noise in a frequency band extending from 10 kHz to the upper frequency of the video band (4.2, 5, 5.5, or 6 MHz).	Flat field	Feed test signal to input of equipment under test and measure SNR at output with specialized RMS-reading instrument. Alternately, use an oscilloscope preceded by a band-limiting filter and synchronized to display single line.
Hum	Ratio, expressed in dB, of the nominal amplitude of the luminance signal (714.3 or 700 mV) to the peak-to-peak amplitude of the noise after band-limiting to 10 kHz.	Flat field	Feed signal to input of equipment under test and measure SNR at output with specialized peak-to-peak reading instrument. Alternately, use an oscilloscope synchronized to display a single field and measure peak-to-peak hum amplitude.
Other periodic noise	Ratio, expressed in dB, of the nominal amplitude of the luminance signal (714.3 or 700 mV) to the peak-to-peak amplitude of the noise in a frequency band extending from 1 kHz to the upper frequency of the video band (4.2, 5, 5.5, or 6 MHz).	Flat field	Feed test signal to equipment under test and use an oscilloscope to isolate, identify, and measure peak-to-peak amplitude of interfering signal in the specified frequency band.
Crosstalk	Ratio, expressed in dB, of the nominal amplitude of the luminance signal (714.3 or 700 mV) to the peak-to-peak amplitude of the interfering signal.	Sweep signal and flat field	Feed sweep to unwanted channel and flat field to wanted channel. Normalize gains and measure the peak-to-peak value of the interfering signal at the wanted channel output.

TABLE 2.16   Effects of Different Types of Noise on Picture

Type of noise	Picture Impairment
Random	Colored and monochrome specks
Hum	Slow-moving horizontal black bar
Other periodic	A steady or slow-moving vertical or diagonal pattern or group of patterns affecting the luminance and/or the chrominance
Signal crosstalk	Faint steady or slow-moving pictures of interfering signal or signals

2.2.6   The distribution of video signals

The signal impairments of individual elements making up a video signal distribution system are additive. Consequently, the overall performance of a system depends on the individual performance and number of specific individual component elements assembled in a typical operational configuration.

The NTSC color television system was developed before the advent of video-

TABLE 2.17   Effect of Noise on Single-Pass Performance of Studio-Quality Equipment

Type of noise	Distribution amplifier	Routing switcher	Production switcher	1-in VTR
Random	≥70 dB	≥65 dB	≥65 dB	≥48 dB
Hum	≥60 dB	≥60 dB	≥60 dB	≥60 dB
Periodic	Nonmeasurable	Nonmeasurable	Nonmeasurable	≥60 dB
Crosstalk	≥60 dB	≥60 dB	≥60 dB	Not applicable

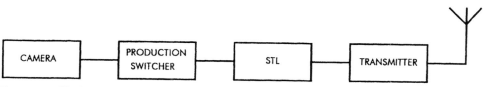

Figure 2.74 Block diagram of simple signal distribution pattern.

tape recording. The television shows were mostly live, meaning that the signal distribution followed a relatively simple path consisting, typically, of a camera, a production switcher, a studio-to-transmitter link (STL), and the television transmitter itself. The video signal impairments, large by today's standards, depended on the single-pass performance of each of the individual signal distribution elements. A typical simple signal distribution path is shown in Fig. 2.74.

The advent of the VTR created a revolution in television production methods. The shows are now prerecorded, edited, and rerecorded on tape. This results in complex signal distribution patterns involving multiple passes through various types of distribution and production equipment and, most importantly, several generations of videotape recordings. Figure 2.75 shows a signal distribution pattern of a large teleproduction center involving two videotape recording generations.

Note the use of several equalizing distribution amplifiers to compensate for high-frequency losses due to long coaxial cables, a situation typical with large studios. The signal at the output of the camera is fed to the production switcher, where it is mixed with other sources not shown in the diagram. The output of the production switcher feeds a routing switcher, which, in turn, feeds a VTR. At a later date the tape is played back on the VTR and the signal feeds a routing switcher. The routing switcher output feeds a production switcher that mixes signals from several sources. Its output is fed to the editing VTR through a routing switcher. Eventually, the final tape is played back and, through a routing switcher and an on-air switcher, feeds the STL. This typical signal distribution pattern results in

- Four routing switcher passes
- Two production switcher passes

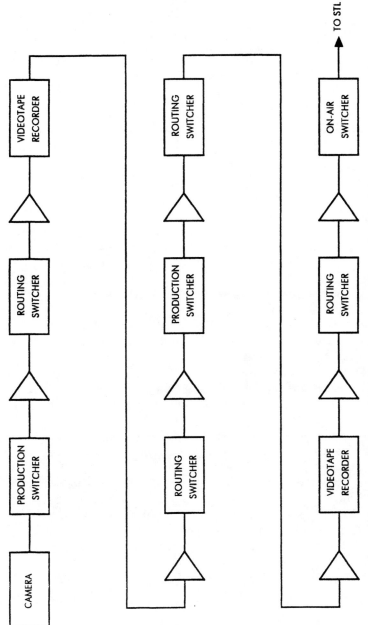

Figure 2.75 Block diagram of a complex signal distribution pattern in a large teleproduction center.

- Eight equalizing distribution amplifier passes
- Two videotape recording generations
- One on-air switcher pass

This example serves to illustrate the importance of selecting high-quality signal distribution and production equipment in assembling a large teleproduction center.

The overall performance of a distribution system can be predicted with a certain degree of accuracy by applying a formula developed by the CCIR for predicting the performance of international video signal distribution networks. The total distortion D_t of a number of individual distortions of the same type $(D_1, D_2,..., D_n)$ is given by

$$D_t = (D_1{}^h + D_2{}^h + \cdots + D_n{}^h)^{1/h}$$

where $h = 2$ for chrominance to luminance inequalities, noise, and line-time linear distortion

$h = 1.5$ for differential gain and differential phase

The authors have applied this formula with good results in predicting the worst-case video signal distribution path performance in large teleproduction centers.

2.2.7 The recording of video signals

The recording of composite color video signals has been a challenge ever since the emergence of the first VTRs in the 1950s. The major problem is the wide bandwidth of video signals. Some of the implications will be discussed in this section.

SNR considerations rule out the standard audio method of recording using a high-frequency bias signal to obtain a linear transfer curve. The main reason is the 6dB/octave response of the playback head in a tape recorder. The NTSC video signal, with a frequency range of 30 Hz to 4.2 MHz, spans over 17 octaves. This represents a playback signal amplitude range in excess of 100 dB (17 octaves × 6dB/octave), which is totally impractical. The problem of reducing the number of octaves is solved by modulating a high-frequency carrier. A video signal with a bandwidth from DC to 5 MHz modulating an 8-MHz carrier results in a recorded bandwidth of from 3 to 13 MHz, assuming a double-sideband system. This represents a frequency span slightly greater than two octaves. Frequency modulation is used because of its tolerance to amplitude variations inherent in tape recording. Figures 2.76 and 2.77 show the standardized characteristics of the frequency modulation process for NTSC and PAL.

In both standards the frequency modulator is clamped at the sync tip. The total carrier deviation is 2.94 MHz in NTSC and 2.14 MHz in PAL. The significant sideband occurs at the color subcarrier frequency. The lower subcarrier sideband frequency is the frequency corresponding to the black level less the

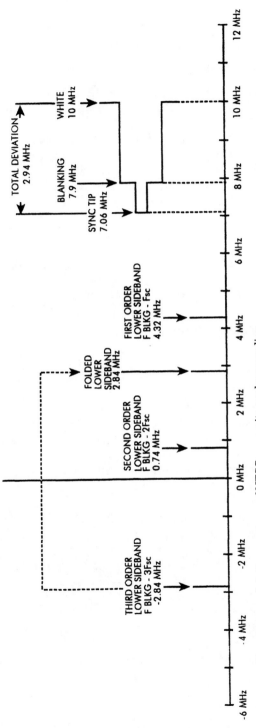

Figure 2.76 Frequency modulation spectrum of NTSC composite analog recording.

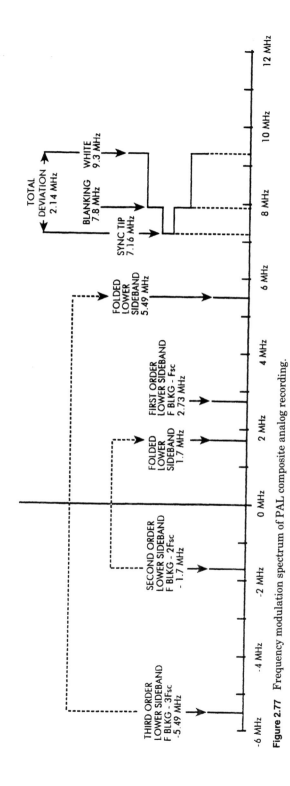

Figure 2.77 Frequency modulation spectrum of PAL composite analog recording.

111

subcarrier frequency. This frequency is 4.32 MHz in NTSC and 2.73 MHz in PAL. The energy in the second-order color subcarrier sidebands is low and the loss does not cause a noticeable distortion. The second-order sideband is, however, not lost. In NTSC it can be accommodated by the system, but in PAL it results in a frequency of −1.7 MHz. The negative sign has little meaning in terms of frequency and the sideband is "folded back" and appears as a positive frequency of 1.7 MHz. After demodulation it appears as a spurious frequency and the visual effect is known as *moiré*. The problem is worse in PAL (with a signal-to-moiré ratio of less than 40 dB) than in NTSC (with a signal-to-moiré ratio better than 40 dB). All above considerations have resulted in a limited video bandwidth leading to poor multigeneration performance.

A second major problem with videotape recordings is the time-base stability. Television signal sources, such as cameras, have very stable signals. Each signal source is synchronized to a master sync generator, ensuring that all equipment inside a teleproduction center is synchronous. In addition to being synchronized, the signal sources need to be timed. Two synchronized NTSC video signals are considered timed with respect to one another when the following conditions are simultaneously met:

- The relative delay between the beginning of corresponding fields in a four-field sequence does not exceed one line.

- The relative delay between corresponding horizontal synchronizing pulses does not exceed ±35 ns.

- The relative phase difference between the corresponding color subcarrier bursts does not exceed 1°.

- The subcarrier to horizontal sync phase (SCH) difference does not exceed 10°. This specification relates to the phase relationship between the leading edge of the horizontal sync and the zero crossing of the subcarrier burst. This requirement is especially important in videotape editing in order to avoid horizontal picture shifts at editing points.

The video signal at the output of a VTR is very unstable because of unavoidable head-to-tape speed instabilities and does not meet the requirements listed above. Professional VTRs use timebase correctors to reduce the signal instabilities, allow the signal to be synchronized to the station master synchronizing generator, and meet the signal-timing requirements listed above.

2.3 Component Video

Component video describes a system in which a color picture is represented by a number of video signals, each of which carries a component of the total video information. In a component video facility, the component video signals are processed separately and, ideally, encoding into a composite video signal occurs only once, prior to transmission. There are two choices of studio component video signal sets that can be selected. These are

- G, B, R signals: This requires the generation, distribution, and processing of three signals of equal bandwidth. These signals are readily available with all cameras.

- Y, B-Y, R-Y signals: This requires the generation, distribution, and processing of three signals obtained by matrixing the primary G, B, R signals. Usually the Y signal has a wider bandwidth than the B-Y and R-Y (color-difference) signals. These signals are readily available with some cameras.

There have been limited attempts at standardizing analog component video signals. Currently, a number of de facto "standards" coexist, making interconnection of equipment difficult.

Components, in some form, are the basic ingredients of any color television system. Since practical color cameras generally have three separate sensors, one for each primary color, a green, blue, red (GBR) system will exist at some stage inside the camera, even if it does not emerge in that form. GBR consists of three signals, each having the same bandwidth. It is used where the highest accuracy is needed, often for the production of still pictures. Examples of this are paint systems and computer-aided design (CAD) displays.

Some savings in bandwidth can be obtained by using color-difference signals. The human eye relies on luminance to convey picture detail, and much less resolution is needed in the color information. GBR signals are matrixed together to form a luminance (and monochrome compatible) signal Y, which has full bandwidth. The matrix also produces two color-difference signals, B-Y and R-Y, which do not need to have the same bandwidth as the Y signal. One half or one quarter of the Y bandwidth is usually acceptable, depending on the application.

2.3.1 The GBR signals

GBR component signals are essentially three monochrome video signals, each representing one of the primary colors. Possible sources of GBR signals include cameras, telecines, composite video decoders, character generators, and graphics systems.

2.3.1.1 Signal characteristics.
Figure 2.78 presents several sets of GBR signals encountered in practice and their characteristics. The signal amplitudes are typical of 100% color bars. All signals are shown with sync added. Some sets of signals have sync added to the green component only, whereas others carry sync on a separate (fourth) wire.

The first column shows the characteristics of the "NTSC-related" GBR signals generally available at the output of an NTSC camera. Setup is usually added in the NTSC encoder, so these signals do not have setup but have a 714-mV peak amplitude and, if sync is added, -286 mV of sync.

The second column shows the characteristics of NTSC-related GBR signals such as would be available at the output of an NTSC decoder. They are similar to the signals in column 1 except for the presence of setup.

The third column shows the characteristics of SMPTE/EBU signals. Note the

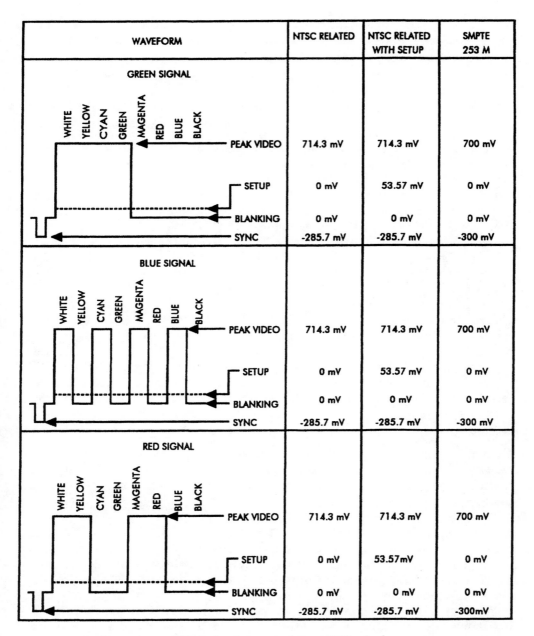

WAVEFORM	NTSC RELATED	NTSC RELATED WITH SETUP	SMPTE 253 M
GREEN SIGNAL			
PEAK VIDEO	714.3 mV	714.3 mV	700 mV
SETUP	0 mV	53.57 mV	0 mV
BLANKING	0 mV	0 mV	0 mV
SYNC	-285.7 mV	-285.7 mV	-300 mV
BLUE SIGNAL			
PEAK VIDEO	714.3 mV	714.3 mV	700 mV
SETUP	0 mV	53.57 mV	0 mV
BLANKING	0 mV	0 mV	0 mV
SYNC	-285.7 mV	-285.7 mV	-300 mV
RED SIGNAL			
PEAK VIDEO	714.3 mV	714.3 mV	700 mV
SETUP	0 mV	53.57mV	0 mV
BLANKING	0 mV	0 mV	0 mV
SYNC	-285.7 mV	-285.7 mV	-300mV

Figure 2.78 Analog component G,B,R signal characteristics—100% color bars.

absence of setup, 700-mV peak signal, and −300 mV of sync. A similar set of signals would be available at the output of a PAL decoder.

These "standards" are similar but not strictly compatible. In extreme cases, white clipping and noticeable errors in displayed light levels will occur when noncompatible equipment is interconnected. It is, therefore, essential to normalize each GBR feed to the input specifications of the equipment into which it is fed.

2.3.1.2 Performance-indicative parameters and measurements concepts.

Component GBR signals are sensitive to linear distortions, nonlinear distortions, and noise affecting the distribution system. In this respect they behave like three monochrome video signals.

Linear distortions manifest themselves as amplitude versus frequency and group delay versus frequency problems. Two additional problems typical of analog component video signals are

- Channel gain differentials: Incorrect blue and red signal amplitudes relative to the green signal amplitude. This will affect color rendition and luminance color temperature.

- Channel delay differentials: Blue and red signal delays relative to the green signal. This will affect the risetime of the matrixed luminance signal. A special test signal, the bowtie, has been developed for the measurements of channel delay differentials.

Figure 2.79 shows a classification of linear distortions as they affect component video GBR signals.

Nonlinear distortions of each of the three component signals are measured in the same manner as the luminance nonlinearity of a composite video signal. Since there is no chrominance subcarrier, all other types of nonlinearities, typical of composite video, are nonexistent.

Random noise is the most prevalent type of noise encountered with analog component GBR equipment. The noise-measurement bandwidth is 5.5 MHz. Other types of noise, like periodic noise and crosstalk, are measured in the same manner as that used for composite video signals. Crosstalk in component video equipment is a complex phenomenon. It includes crosstalk between the three component channels as well as crosstalk into and from other sets of component signals. Figure 2.80 shows a classification of noise as it affects analog component GBR signals.

Table 2.18 shows a list of typical video test signals cross-referenced to the type of test to be carried out.

There are three identical sets of test signals (G, B, and R) similar to those used with composite video systems with some slight differences as detailed below:

- The signal amplitudes are normalized to 700 mV/300 mV with no setup and there is sync on all three signals.

- The nominal bandwidth is assumed to be 5 MHz. Consequently, the bursts in the multiburst signal are at 0.5, 1, 2, 3, 4, and 5 MHz. A continuous linerate sweep in the frequency range of 0.5 to 6 MHz is also available with some test signal generators. In addition $T = 100$ ns, resulting in a $2T$ pulse HAD = 200 ns.

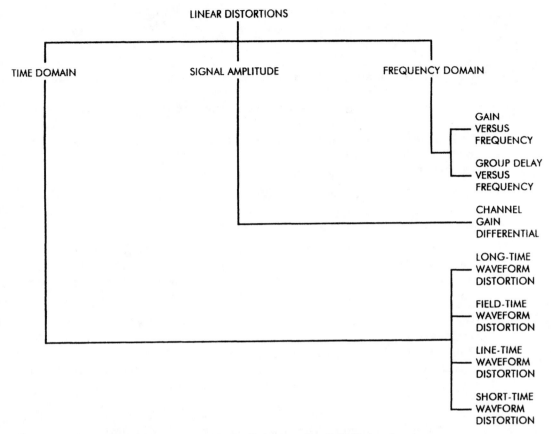

Figure 2.79 Classification of linear distortions affecting analog G, B, R signals.

The measurement of relative time delay of the red and blue signals with respect to the reference green signal is carried out with a special test signal, the bowtie, developed by Tektronix. The bowtie is made up to two sinewave packets with slightly different frequencies. The reference green channel is fed with a wave packet of 500 kHz and either of the blue and red channels is fed with a wave packet of 502 kHz. Figure 2.81 shows the waveform carrying the wave packet. The center of the 500-kHz wave packet is exactly in phase with the center of the 502-kHz wave packet.

The signal at the output of the blue (or red) channel is subtracted from the signal at the output of the reference green channel. If the two outputs are perfectly timed, the result of the signal subtraction is a bowtie-shaped signal with a cancellation in the center of the display as shown in Fig. 2.82. If the signal is delayed with respect to the reference green signal, the cancellation is shifted toward the right of the display center. It is shifted toward the left of the display center if the signal is in advance of the green reference. Suitable 20-ns markers are inserted in the display to help quantify the timing difference.

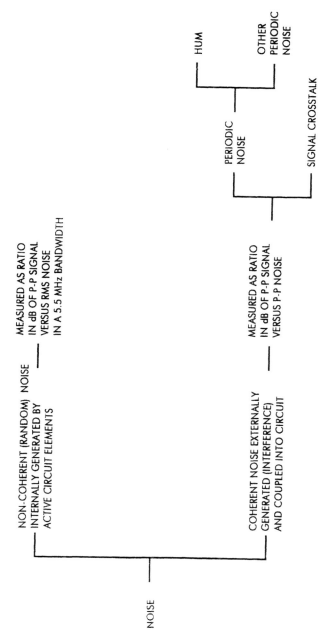

Figure 2.80 Classification of noise affecting analog G, B, R signals.

117

TABLE 2.18   G, B, R Test Signals

Test signal	Test
Color bars	Channel gain, encoding accuracy
Multiburst (bursts at 0.5, 1, 2, 3, 4, 5 MHz)	Frequency response
Field square wave	Field-time distortion
2T pulse and bar (T = 100 ns)	Line-time distortion, K rating
Bowtie	Relative time delay
Staircase	Transfer nonlinearity
Flat field	Signal-to-noise ratio

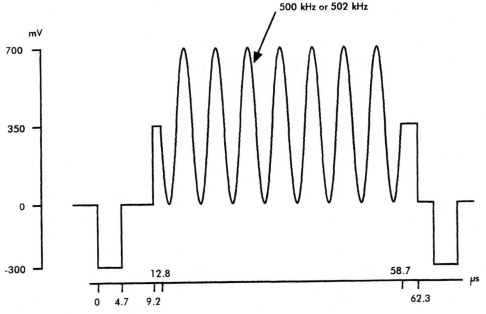

Figure 2.81 Bowtie test signal.

2.3.1.3 The distribution of signals. Discrete GBR analog component signal distribution ensures the highest picture quality. Ideally, all television production activities would be carried out using the analog GBR format and the encoding to composite video would occur only prior to transmission. The reason the wide scale use of analog GBR component signals has not gained favor is the lack of VTRs capable of recording three simultaneous wideband video signals. Consequently, analog GBR signal distribution is confined to short

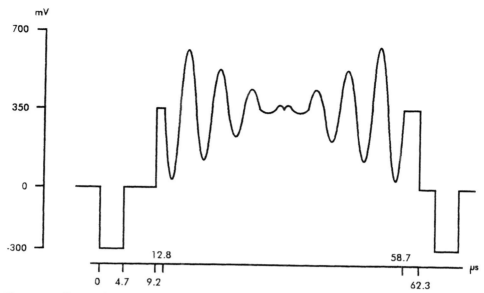

Figure 2.82 Bowtie formed by subtracting the two test signals.

distances such as between a character generator and an encoder. The main requirements are negligible differential delay between channels and identical channel gains. Differential channel delays result in loss of luminance resolution, whereas channel gain differentials give rise to incorrect matrixed luminance signals and result in corrupted colors as well as a colored background.

2.3.1.4 Signal monitoring. Video signal waveform continuity monitoring requires the use of special three-channel waveform monitors to ensure that the GBR signals do not exceed the peak level. In order to ensure that the resulting composite signal is legal, it is customary to encode the GBR signals to the local composite video standard and use a standard single-channel waveform monitor. Failure to perform adequate monitoring could result in illegal encoded signals.

2.3.2 The Y, B-Y, and R-Y signals

Y, B-Y, and R-Y signal components are linear combinations of signals representing the three primary colors: green, blue, and red. Possible sources of Y, B-Y, and R-Y signals are cameras, analog component videocassette recorders, and composite video decoders.

2.3.2.1 Signal characteristics. The luminance signal has the following mathematical expression:

$$E'_Y = 0.587E'_G + 0.114E'_B + 0.299E'_R$$

This expression is identical to that used with all composite (NTSC, PAL, and SECAM) as well as component systems.

The color-difference signals are different from the NTSC, PAL, and SECAM color-difference signals in terms of peak-to-peak signal amplitude and bandwidth. In NTSC and PAL the color-difference scaling factors were aimed at limiting the encoded signal amplitude to 130.8 IRE (+935 mV) for the cyan and yellow components of a 100% color-bar signal. This signal amplitude results in severe transmitter overload. The developers of the NTSC system relied on the fact that saturated 100% yellow and cyan colors are not commonly encountered in nature and assumed, therefore, that transmitter overmodulation would never occur in practice. The advent of character generators and other digital equipment resulted in synthetic component signals that, when encoded into NTSC or PAL, could result in composite signals that will overload a transmitter.

Figure 2.83 presents several sets of Y, B-Y, and R-Y signals encountered in practice. The signal amplitudes are typical for 100% color bars. Normally the Y (luminance) signal has sync added.

The first column shows the characteristics of the NTSC-related Y, B-Y, and R-Y signals as would be obtained at the output of an NTSC decoder. Note that the luminance signal has a 714-mV nominal peak amplitude, 53.55 mV of setup, and −286 mV nominal of sync. The color-difference signals are bipolar and have unequal peak-to-peak amplitudes: 611.52 mV for B-Y and 860.6 for R-Y. This is a result of the peculiar amplitude scaling of the NTSC color-difference signals as per the expressions below:

$$E'_{B-Y} = 0.493 \, (E'_B - E'_Y)$$

$$E'_{R-Y} = 0.877 \, (E'_R - E'_Y)$$

The second column shows the characteristic component signals conforming to the EBU N10 standard. There is no equivalent official SMPTE standard. However, EBU N10 is a de facto SMPTE standard. Note that the luminance signal has a 700-mV peak-to-peak amplitude, no setup, and −300 mV of sync. The color-difference signals are bipolar, symmetrical, and have identical 700-mV$_{p-p}$ signal amplitudes. To differentiate them from the NTSC or PAL color-difference signals, they are called P_B and P_R, respectively. Their identical peak-to-peak signal amplitudes result from the modified scaling factors as in the expressions below:

$$P_B = 0.564 \, (E'_B - E'_Y)$$

$$P_R = 0.713 \, (E'_R - E'_Y)$$

The third column shows the characteristics of the component video signals obtained at the output of a Sony Betacam SP videocassette recorder (VCR) as marketed in North America, when a 100% color bar signal is played back. The

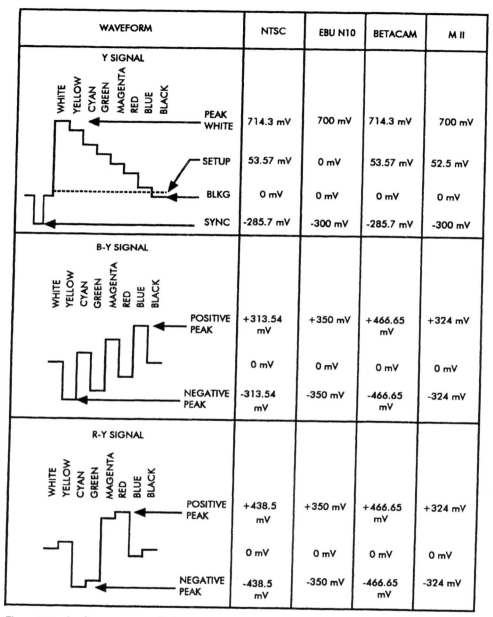

WAVEFORM	NTSC	EBU N10	BETACAM	M II
Y SIGNAL (WHITE, YELLOW, CYAN, GREEN, MAGENTA, RED, BLUE, BLACK)				
PEAK WHITE	714.3 mV	700 mV	714.3 mV	700 mV
SETUP	53.57 mV	0 mV	53.57 mV	52.5 mV
BLKG	0 mV	0 mV	0 mV	0 mV
SYNC	-285.7 mV	-300 mV	-285.7 mV	-300 mV
B-Y SIGNAL (WHITE, YELLOW, CYAN, GREEN, MAGENTA, RED, BLUE, BLACK)				
POSITIVE PEAK	+313.54 mV	+350 mV	+466.65 mV	+324 mV
	0 mV	0 mV	0 mV	0 mV
NEGATIVE PEAK	-313.54 mV	-350 mV	-466.65 mV	-324 mV
R-Y SIGNAL (WHITE, YELLOW, CYAN, GREEN, MAGENTA, RED, BLUE, BLACK)				
POSITIVE PEAK	+438.5 mV	+350 mV	+466.65 mV	+324 mV
	0 mV	0 mV	0 mV	0 mV
NEGATIVE PEAK	-438.5 mV	-350 mV	-466.65 mV	-324 mV

Figure 2.83 Analog component Y, B-Y, R-Y signal characteristics—100% color bars.

luminance signal has a peak-to-peak amplitude of 714 mV, 53.57 mV of setup, and −285.7 mV of sync. The color-difference signals are bipolar, symmetrical, and have an identical peak-to-peak amplitude of 933.3 mV. With a 75% color bar signal, the luminance has a peak-to peak amplitude of 535.72 mV and the two color-difference signals have a peak-to-peak amplitude of 700 mV. The expressions for the Betacam color-difference signals are

$$\text{Betacam (B-Y)} = 0.75\,(E'_B - E'_Y)$$

$$\text{Betacam (R-Y)} = 0.95\,(E'_R - E'_Y)$$

It is to be noted here that the Sony Betacam SP videocassette recorders marketed in 625/50 countries have signals according to EBU N10 as shown in column 2. Sony Betacam SP VCRs marketed in Japan have signals as shown in column 3, with the exception of the luminance signal, which has no setup.

The fourth column shows the characteristics of the component video signals obtained at the output of an MII VCR when the 100% color bar signal is played back. Note that the luminance signal has a peak-to-peak amplitude of 700 mV, 53.55 mV of setup, and −300 mV of sync. The color-difference signals are bipolar, symmetrical, and have an identical peak-to-peak amplitude of 648 mV. The expressions for the MII color-difference signals are

$$\text{MII (B-Y)} = 0.522\,(E'_B - E'_Y)$$

$$\text{MII (R-Y)} = 0.66\,(E'_R - E'_Y)$$

In 625/50 countries the MII component signals conform to the EBU N10 standard.

Note that the color-difference scaling factors are different in the four standards discussed above. Strictly speaking, these "standards" are incompatible. Interconnections between different types of equipment require special "transcoders" available on the market. Good housekeeping dictates the need to transcode into and from the EBU set of component signals when a variety of types of equipment have to be interconnected.

In addition to the amplitude incompatibility, the various sets of component video signals have different bandwidths. The bandwidths are related to the associated recording or transmission medium. Studio distribution equipment features identical bandwidths for the Y and color-difference signals.

2.3.2.2 Performance-indicative parameters and measurements concepts.
Component Y, B-Y, R-Y signals are sensitive to linear distortions, nonlinear distortions, and noise affecting the distribution and recording medium.

Linear distortions manifest themselves as amplitude versus frequency and group delay versus frequency problems. Two additional problems, typical of component video signals are

- Channel gain differentials: Incorrect B-Y and R-Y signal amplitude relative to the Y signal amplitude. This will affect the color rendition.
- Channel delay differentials: B-Y and R-Y signal delays relative to the Y signal, affecting the registration of the picture. A special test signal, the bowtie, has been developed for the measurement of channel delay differentials.

The luminance signal is very similar to a monochrome signal. Consequently, the measurements and the test signals are very similar to those used with monochrome and G, B, R component video signals. The nominal bandwidth is 5 MHz, resulting in $T = 100$ ns. In analog component VTRs, the luminance channel has a wider frequency response than that of the two color-difference channels. In studio distribution systems the three channels have identical bandwidths. Any bandwidth reduction occurs inside the respective equipment. The luminance bandwidth of analog component VTRs does not exceed 4 MHz. The studio distribution equipment usually has a bandwidth well in excess of 5 MHz. The test signals for the luminance channel are the multiburst (bursts at 0.5, 1, 2, 3, 4, and 5 MHz) and a pulse and bar signal with an HAD = 200 ns. A line-time sweep is also available with some test signal generators.

The analog color-difference test signals are bipolar and tailored to the reduced bandwidth of the equipment under test. The analog component VTR color-difference channel bandwidth does not exceed 1.25 MHz. The bandwidth varies from manufacturer to manufacturer. The test signals for the color-difference channel are the multiburst (with bursts at 0.5, 1, 1.5, 2, and 2.5 MHz) and the pulse and bar signal. A choice of $4T$ (400 ns HAD), $5T$ (500 ns HAD) and $8T$ (800 ns HAD) sine-squared pulse signals are usually available in test signal generators to match various types of equipment. Some test signal generators feature wideband test signals for the color-difference studio distribution equipment including a line-time sweep.

The channel delay differential is measured with the bowtie signal in a manner similar to the one used with G, B, R signals.

Figure 2.84 shows a classification of linear distortions as they affect analog component video Y, B-Y, R-Y signals.

Nonlinear distortions of the three component video signals are measured in a manner similar to the measurement of the luminance nonlinearity of a composite video signal. The test signal generator provides three simultaneous stairstep test signals. Each of these signals is intended to test the nonlinearity of the respective channel. When encoded into a composite NTSC or PAL signal, they result in illegal signals. These signals should, therefore, not be used to feed an encoder. Some component analog test signal generators feature "legal" nonlinearity test signals that result in legal encoded signals.

Random noise is the most prevalent type of noise encountered with analog component video Y, B-Y, R-Y equipment. The noise measurement bandwidth is 5.5 MHz for the luminance channel and 2.75 MHz for the color-difference channels. With equiband studio distribution equipment, the measurement bandwidth is 5.5 MHz on all three channels. Periodic noise is measured in a manner similar to that used for composite signals. Crosstalk in component video equipment is a complex phenomenon. It includes crosstalk among the three component channels, as well as crosstalk into and from other sets of component signals. Figure 2.85 shows a classification of noise as it affects analog component Y, B-Y, R-Y signals.

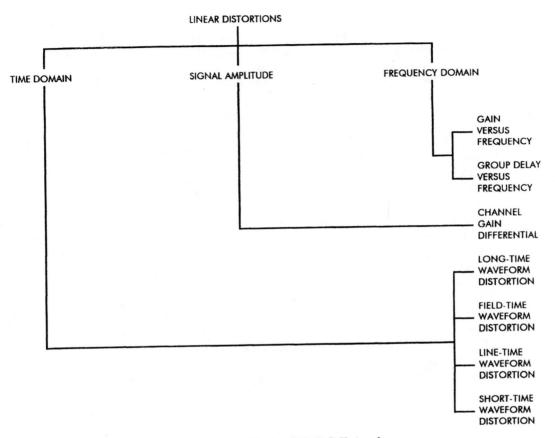

Figure 2.84 Classification of linear distortions affecting Y, B-Y, R-Y signals.

Table 2.19 shows a list of typical video test signals cross-referenced to the type of test to be carried out.

2.3.2.3 Legal and valid signals. Figures 2.78 and 2.79 describe the voltage ranges of the commonly encountered G, B, R and Y, B-Y, R-Y component analog signal formats. A set of component analog signals is considered legal if each component signal is contained within the specified voltage range of the format. A legal Y, B-Y, R-Y set of signals may result in illegal G, B, R transcoded signals as well as illegal NTSC or PAL composite signals. A set of legal component analog signals is considered valid if it results in legal signals in the format into which it is transcoded.

Television camera–originated signals are always valid because the camera generates G, B, R signals in the first place. These signals are subsequently transcoded into legal Y, B-Y, and R-Y signals and NTSC or PAL composite signals. Some composite analog video equipment, such as a paint box or a

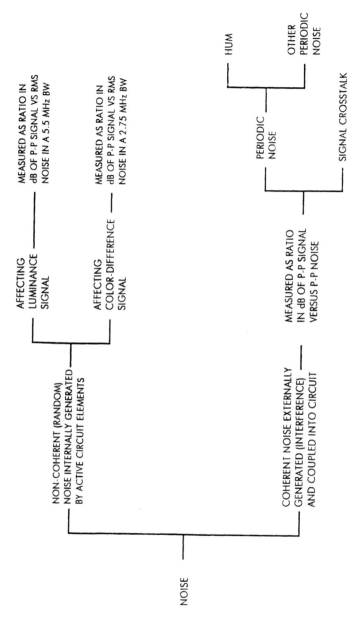

Figure 2.85 Classification of noise affecting analog Y, B-Y, R-Y signals.

TABLE 2.19   Y, B-Y, R-Y Test Signals

Test signal	Test
Color bars	Channel gain, encoding accuracy
Multiburst or sweep	Y channel: 0.5 to 5 MHz B-Y/R-Y channels: 0.5 to 2.5 MHz
Field square wave	Field-time distortion
Pulse and bar (T = 100 ns)	Line-time distortion and K rating Y channel: $2T$ pulse and bar B-Y/R-Y channels: $3T$, $4T$, $5T$, or $8T$ pulse and bar as required
Bowtie	Relative time delay
Staircase	Transfer nonlinearity
Flat field	Signal-to-noise ratio

character generator, is capable of generating nonvalid signals even though they are legal in the original format. Test signal generators are also capable of generating legal but nonvalid signals.

2.3.2.4   The distribution of signals.   It is possible but impractical and costly to design and implement a large teleproduction center where all signal distribution is in the Y, B-Y, R-Y analog component format. Consequently, the distribution of component analog Y, B-Y, R-Y signals is usually confined to small production studios or editing suites where cameras, production switchers, VTRs, and ancillary equipment are close together. This allows a tight control over the differential channel gain and delay.

2.3.2.5   Signal monitoring.   Video signal continuity monitoring requires the use of special three-channel waveform monitors to ensure that the Y, B-Y, R-Y signal levels do not exceed the legal limits. The monitoring of the Y channel is similar to the well-established monitoring practices of composite analog video signals. The monitoring of the color-difference signals will give an indication as to their peak-to-peak value but not as to their validity. In order to ensure that the resulting composite signal is legal, it is customary to encode the component analog signal into the local composite standard and use a standard single-channel waveform monitor. Some manufacturers offer three-channel component video waveform monitors with a special display allowing the user to determine if the signals are valid. Failure to perform adequate monitoring could result in illegal encoded signals.

2.3.3   The component video recording

Small-format video recording equipment, operating at low tape speed, has difficulties handling full-bandwidth composite signals. Historically, the approach has been to record the chrominance and luminance signals separately. This eliminates the need for a relatively high FM carrier signal that is required to avoid moiré effects caused by the high energy of the chrominance subcarrier and its sidebands. Three small-format recording concepts have evolved through the years are described in this section.

2.3.3.1 The color-under recording concept. An early approach was the Sony U-Matic concept using a ¾-in cassette. The standard composite NTSC signal is separated into luminance and modulated chrominance subcarrier with its sidebands. The luminance high-frequency response is limited to about 2.5 MHz and frequency-modulates a carrier with a peak-to-peak deviation of 1.6 MHz. The chrominance is down-converted (heterodyned) from 3.58 MHz to 688 kHz and is recorded directly on tape with the FM carrier acting as bias. This type of chrominance processing is called *color under* and results in an effective color-difference bandwidth of the order of 300 kHz. During playback, the process is reversed and the 3.58-MHz chrominance is recovered by a heterodyne process having as a reference a stable 3.58-MHz crystal oscillator. The picture quality matches that available with the older (1970 vintage) home television receivers. Even though the format was initially developed for home use, it did not find favor with consumers because of the limited recording time afforded by reasonably sized cassettes, as well as the size and weight of the hardware. The format became very popular with broadcasters for the purpose of news gathering. There were, however, problems with its use in a broadcasting environment. These were mainly a result of the fact that the video signal at the output of the VCR is best described as "quasi-NTSC." It features a relatively stable chrominance signal and a jittery narrow-bandwidth luminance signal. This type of signal is not suitable for broadcasting or mixing with other sources unless it is time-base-corrected. Several generations of time-base correctors were developed for use with various versions of the U-Matic recorders. Subsequent developments led to the Sony U-Matic SP hi-band concept featuring a higher-frequency FM carrier and a new tape formulation. This resulted in improved performance in terms of luminance frequency response (3-MHz bandwidth), SNR, and moiré. Consumer product developments resulted in the noncompatible ½-in cassette formats of Sony Betamax and JVC VHS. Their performance is similar to that of the Sony U-Matic. Subsequent developments led to the Betamax ED and SVHS formats, which use a higher-frequency FM carrier to achieve performance figures similar to those of the U-Matic SP format. Several 8-mm cassette color-under formats have also appeared on the market.

2.3.3.2 The chrominance frequency domain multiplex (CFDM) concept. This recording format, introduced by RCA as the M format, relies on the use of a pair of parallel tracks on tape for the separate recording of the luminance signal and the chrominance I and Q signals on a ½-in videocassette of VHS dimensions. The luminance signal is band-limited to 3.6 MHz and modulates an FM carrier with a center frequency of 5.1 MHz and a peak-to-peak deviation of 1.6 MHz. The I signal modulates an FM carrier with a center frequency of 5.5 MHz and a peak-to-peak deviation of 1 MHz. An added color sync pulse deviates the carrier downward to 4.86 MHz. The Q signal modulates an FM carrier with a center frequency of 1 MHz and a peak-to-peak deviation of 500 kHz. The I and Q modulator outputs are summed before being fed to the recording head that is writing the chrominance track. This results in frequency domain multiplexing of the two chrominance signals. To reduce I/Q crosstalk, the I FM signal is 12 dB higher in amplitude than the lower-frequency Q FM signal. As a consequence, the I signal serves as recording bias for the Q signal, which is "linearly" recorded.

2.3.3.3 The chrominance time domain multiplex (CTDM) concept.

The Sony Betacam format relies on the use of two parallel tracks for the separate recording of the luminance and the time-domain-multiplexed color-difference signals on a $\frac{1}{2}$-in videocassette of Betamax dimensions. The luminance signal is band-limited to about 3.5 MHz (4 MHz for the SP format) and modulates an FM carrier with a center frequency of 5.4 MHz (6.7 MHz for the SP format) and peak-to-peak deviation of 2 MHz. The B-Y and R-Y color-difference signals are bandwidth-limited to 1.5 MHz (−3 dB), digitally sampled, stored, and read out at twice the sampling frequency. This process doubles the bandwidth of the signal but halves its duration so that both color-difference signals can be fitted sequentially into each active line. The two time-compressed signals are multiplexed, and the resulting signal modulates an FM carrier with a center frequency of 4.2 MHz (5.3 MHz for the SP format) and a peak-to-peak deviation of 1.4 MHz. The noncompatible MII format uses a $\frac{1}{2}$-in videocassette of VHS dimensions and a signal-processing and recording concept similar to that of the Sony Betacam.

CTDM VCRs are available as compact portable-camera piggyback units or as full-fledged studio machines. The studio machines have incorporated time-base correctors and feature component analog and composite analog inputs and outputs. The component input-output signals conform to the signals described in Sec. 2.3.2.1. They can thus be used in a component analog environment, such as an editing suite, or in a composite analog environment, as a replacement of older composite analog VTRs. The CTDM videotape recording formats have proved very popular with broadcasters because of their versatility, size, performance, and cost.

Table 2.20 lists the measured luminance and color-difference frequency response of three formats.

Figure 2.86 shows the modulation spectrum characteristics of several component VTRs and helps clarify the luminance frequency response limitations of each format.

TABLE 2.20 Frequency Reponse of Some Analog Component VTRs

Format	Y Channel, dB@4 MHz	B-Y/R-Y Channels
Betacam SP	−0.8	−1 dB@1.5 MHz
Betacam	−7	−2 dB@1.5 MHz
SVHS	−10	−3 dB@300 kHz

Figure 2.86 Modulation spectrum characteristics of component videotape recorders.

Digital Video Fundamentals

3.1 General Considerations

Digital video is best defined as a means of describing the continuous analog video waveform as a stream of digital numbers. There are several advantages in using digital video equipment as follows:

- A digitized video signal is immune to analog signal impairments such as linear distortions, nonlinear distortions and noise beyond the inherent distortions generated by the analog-to-digital (A/D) and digital-to-analog (D/A) conversion process. This can be fully realized by processing and distributing the signal in digitized form.

- Digital video equipment can perform efficiently and economically tasks that are difficult or impossible to perform using analog video technology.

- Digitized video signals are amenable to the application of techniques for retention of essential information such as compression.

3.1.1 Historical background

The 1970s are characterized by a major revolution in television studio operations brought about by the use of digital video technology. Early digital technology was restricted to so-called digital black boxes. A digital black box is a device that has analog input and output ports and by using digital technology performs an essential signal-processing task. Among the early digital black boxes were timebase correctors, frame synchronizers, and standards converters.

The 1980s witnessed the emergence of digital videotape recorders (VTRs) based on CCIR recommendations. A variety of digital black boxes, such as digital video effects (DVE), graphic systems, and still stores, operating in a variety of noncorrelated and incompatible standards, also became available. Digital interconnections between various digital black boxes were thus difficult or impossible. The majority of these black boxes were interconnected with the rest of the analog or digital equipment using analog input/output

131

ports. The user was generally unaware or uninterested in the digital incompatibility of the black boxes as long as they performed their intended task and could be interconnected with the rest of the equipment. Compatible digital video equipment was assembled into a digital island, such as an editing suite, using a bit-parallel digital video interconnection.

The 1990s are characterized by an intense standardization activity led by the Society of Motion Pictures and Television Engineers (SMPTE). Among the many standards developed to date are the digital composite video standard ($4f_{SC}$) and the bit-serial digital interconnect standard. A large variety of video production, distribution, and recording equipment with standardized bit-serial input/output ports has become available, allowing for the first time the assembly of all-digital teleproduction facilities using bit-serial signal distribution and interconnection.

3.1.2 The typical black box digital device

Figure 3.1 shows a simplified block diagram of a typical black box digital device. It can represent any digital device in use in a teleproduction facility. The input is a conventional analog video signal. This signal is band-limited by a low-pass (antialiasing) filter and fed to an A/D converter where the analog signal is converted into digital form. This block is usually called a *coder*. The A/D conversion involves three steps:

- Sampling of the analog signal at a constant rate
- Quantizing the sampled values
- Coding the signal

The digitized signal is fed to a digital processor, which may be anything the designer wishes. In a timebase corrector, the processor performs a timebase correction in the digital domain. In a video recorder the processor records and plays back the video signal in digital format.

The processed digital signal is applied to a D/A converter. The output of the D/A converter is fed to a low-pass (reconstruction) filter that removes high-frequency spectral components and allows only the analog video signal to pass. This block is commonly called a *decoder* and changes the output of the device back into a conventional analog format.

The digital black box can be inserted into a conventional television operation and the operator need not know that the signal is being processed internally in a digital manner.

Figure 3.1 Block diagram of digital black box.

3.1.3 Sampling the signal

The analog signal amplitude is sampled periodically at T intervals by a stream of narrow pulses, resulting in a sampling frequency equal to

$$f_S = \frac{1}{T}$$

The sampling is equivalent to amplitude-modulating the signal (f_0) onto a carrier equal to the sampling frequency (f_S). The amplitude modulation generates lower and upper sidebands. The sampling waveform is rectangular, resulting in a spectrum with components at the sampling frequency and its harmonics as shown in Fig. 3.2.

An important signal sampling rule is known as the *Nyquist theorem,* which states that for a signal with a bandwidth of f_b, the sampling frequency must be equal to or greater than $2f_b$, that is, twice the bandwidth of the signal that is being sampled. Figure 3.3 illustrates the sampling process and shows a waveform that has been sampled twice during a complete sine wave cycle.

Figure 3.4 shows an example where the sampling rate is too low for a correct digital representation of the analog signal. The dotted lines show the "alias" signal resulting from the inadequate sampling frequency.

Figure 3.5 shows an ideal frequency spectrum where the baseband signal has a bandwidth of f_b and the sampling frequency is $2f_b$. This results in upper and lower sidebands each equal to f_b. This ideal spectrum shows no interference between the baseband and the lower sideband.

Figure 3.6 shows the sampled frequency spectrum resulting from the baseband spectrum exceeding $f_S/2$. The lower sideband of the sampling rate overlaps the baseband spectrum, resulting in aliasing.

In practical systems, an antialiasing filter is used to control the bandwidth of the baseband signal. Since low-pass filters have a sloping upper frequency response, it is customary to use a sampling frequency in excess of $2f_b$. Figure 3.7 shows the sampled spectrum of a filtered baseband.

Figure 3.2 Spectrum of sampled signal.

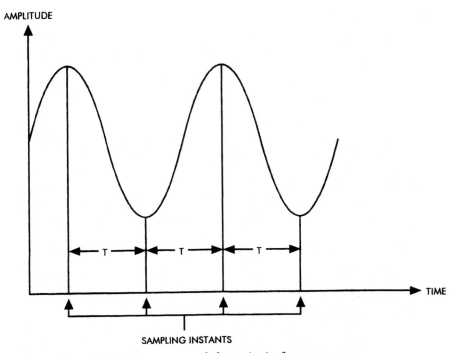

Figure 3.3 Example of a sine wave sampled at twice its frequency.

The sampling rate of video signals has evolved through the years. Analog composite video signals are sampled at a multiple of the subcarrier frequency. Early equipment used a sampling frequency of $3f_{SC}$, resulting in, nominally, 10.7 MHz for NTSC and 13.3 MHz for PAL. The current sampling rates are $4f_{SC}$, resulting in, nominally, 14.3 MHz for NTSC and 17.7 MHz for PAL. These higher sampling frequencies ease the requirements for the antialiasing and reconstruction filters and provide a better frequency response. Modern technologies have minimized the early difficulties of designing A/D converters with a high sampling frequency. The frequency modulation of the SECAM chrominance carriers makes it impractical to sample the composite color signal. SECAM signals are decoded into analog components, which are digitized separately. Analog component video signals are sampled at a multiple of the horizontal scanning frequency f_H.

3.1.4 Quantizing the sampled values

Quantizing converts all the amplitude levels of a continuously varying analog signal to one of a finite number of discrete levels Q according the expression

$$Q = 2^n$$

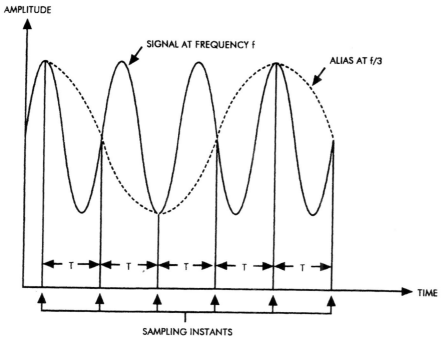

Figure 3.4   Example of a sine wave sampled at 1.33 times its frequency resulting in aliasing.

Figure 3.5   Ideal spectrum of sampled baseband.

where n is the number of bits per sample. The resulting digital signal is an approximation of the original signal, since the act of quantizing assigns discrete numerical values to each sample. Figure 3.8 shows that all sample amplitudes occurring within specific bounds are assigned a single value, that is, one of the Q levels. In most studio-quality digital video equipment, all

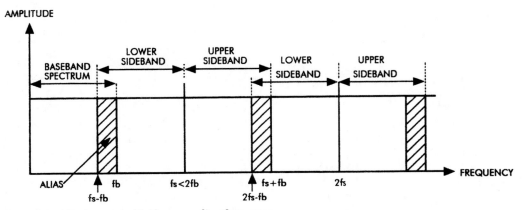

Figure 3.6 Aliasing caused by low sampling frequency.

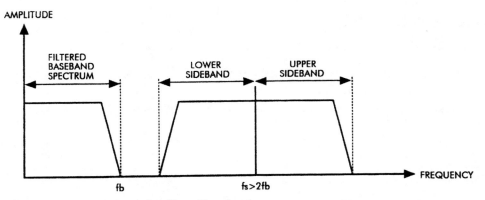

Figure 3.7 Spectrum of sampled filtered baseband.

quantizing levels are of equal amplitude and the process is called *uniform quantization*. The quantized value of an analog signal may contain an error, referred to as the *quantizing error*, not exceeding $\pm\frac{1}{2}Q$.

The number of quantizing steps, and, consequently the magnitude of the quantizing error, depends on the number of bits per sample. Early technology used 7 or 8 bits per sample, depending on the class of equipment, resulting in, respectively, 128 or 256 quantizing steps. With very few exceptions, contemporary studio-quality equipment uses 10 bits per sample, resulting in 1024 quantizing steps.

3.1.5 The dynamic range and the headroom concept

As discussed in Chap. 2, analog video signals have a well-defined peak-to-peak amplitude. The common amplitude reference is the 100% color bars signal, which assumes a peak-to-peak value of 1.22 V in NTSC and 1.2335 V, in

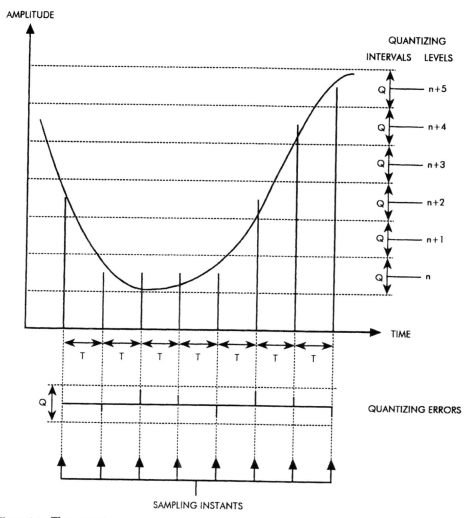

Figure 3.8 The quantizing process.

PAL. It is important that these signals be handled by the A/D converter without clipping. Similar considerations apply to the component analog video signals. Consequently, a certain amount of headroom is required to avoid A/D converter overloading and, this is specified in current digital television standards as discussed in Sec. 3.2.

3.1.6 The quantizing error

The quantizing error is a unique source of impairment in digital systems. Video signal amplitudes vary in time. The quantized values may be in error by as much as $\pm\frac{1}{2}Q$, where Q is the amplitude of the quantizing interval. For 8 bits or more per sample, the quantizing error can be interpreted as an unwanted signal (e.g., noise) added by the quantizing process to the original signal. Below 8 bits per sample, the quantizing error results in a severe distortion of the waveform and gives rise to contouring effects.

For linear quantizing and assuming that the errors are uniformly distributed, the RMS magnitude of any set of quantizing errors is equal to $Q/\sqrt{12}$. The peak-to-peak output of the D/A converter is assumed to be equal to 2^nQ. Its precise value is, in reality, $(2^n-1)Q$. The peak-to-peak signal S to RMS quantizing noise (Q_{RMS}) ratio of an ideal black box is given by

$$\frac{S}{Q_{RMS}}\text{ (dB)} = 20\log_{10}\left(\frac{2^nQ \times \sqrt{12}}{Q}\right) \approx 6.02n + 10.8$$

The above formula assumes that the analog signal occupies the complete quantizing range. Given a digital video system with 8 bits per sample, the theoretical value is

$$\frac{S}{Q_{RMS}} = 58.96\text{ dB}$$

There are several factors that affect the value of S/Q_{RMS} as follows:

- The band limiting of the spectrum to f_{max} (e.g., 4.2 or 5 MHz). Limiting the spectrum to f_{max} improves the S/Q_{RMS} ratio by a factor equal to $10\log_{10}(f_S/f_{max})$.
- The quantizing range occupied by the active video signal (i.e., 100 IRE). The S/Q_{RMS} is referenced to the peak-to-peak video amplitude (e.g., 714 or 700 mV). A 100% color bars composite analog video signal has a peak-to-peak amplitude of 1.22 V (NTSC) or 1.2335 V (PAL). Assuming that all signal values are quantized, the luminance portion of the signal, 714 mV (NTSC) or 700 mV (PAL), will occupy only a portion of the complete quantizing range. The proportion is $0.714/1.22 \approx 0.58$ in NTSC and $0.7/1.2335 \approx 0.57$ in PAL. This results in a smaller number of quantizing steps and, consequently, a reduced value for S/Q_{RMS}. If the quantizing range allows for a certain amount of headroom over the maximum peak-to-peak video amplitude, the S/Q_{RMS} ratio will be further reduced by the amount of headroom.

The modified formula for S/Q_{RMS} is

$$\frac{S}{Q_{RMS}}\text{ (dB)} = 6.02n + 10.8 + 10\log_{10}\left(\frac{f_S}{2f_{max}}\right) - 20\log_{10}\left[\frac{V_q}{(V_W - V_B)}\right]$$

where n = The number of bits per sample
f_S = The sampling frequency

f_{max} = The maximum video frequency (e.g., 4.2 or 5 MHz)
V_q = Signal voltage that occupies the whole quantizing range
V_W = White signal level, V
V_B = Blanking signal level, V
$V_W - V_B$ = 0.714 V (NTSC) or 0.7 V (PAL)

Ignoring the amount of headroom, the calculated S/Q_{RMS} ratio for an NTSC black box in a 4.2-MHz bandwidth, with a sampling frequency of 14.3 MHz and an accuracy of 8 bits per sample, is equal to

$$\frac{S}{Q_{RMS}} = 6.02n + 10.8 + 10 \log_{10}\left(\frac{14.3}{8.4}\right) - 20 \log_{10}\left(\frac{1.22}{0.714}\right)$$

$$= 56.62 \text{ dB}$$

The improved S/Q_{RMS} ratio (+2.31 dB), because of the restricted 4.2-MHz bandwidth, is offset by the effect of the reduced number of quantizing steps (−4.65 dB). If the headroom is taken into consideration, the S/Q_{RMS} ratio is further reduced by about 1 dB. These considerations will be further addressed in Sec. 3.2.

3.1.7 The D/A conversion

As shown in Fig. 3.9, the waveform at the output of the D/A converter is a train of rectangular pulses of the same amplitude as the digitized signal and with a constant width of $T = 1/f_S$. The spectrum of a rectangular pulse with a width of $T = 1/f_S$ is given by the sinc function

$$A(\text{dB}) = 20 \log_{10}\left(\frac{\sin x}{x}\right)$$

where $x = \pi f_V / f_S$
f_V = The video frequency
f_S = 1/T, the sampling frequency

Figure 3.10 shows the spectrum of a rectangular pulse with a duration of $T = 1/f_S$.

Figure 3.11 shows the spectrum of the signal at the output of the D/A converter.

The effect is a loss at the higher video frequencies and is added to the cumulative effect of the antialiasing and reconstruction filters. This high-frequency loss is usually equalized after the reconstruction filter. Figure 3.12 shows curves representing the high-frequency losses caused by the low-pass filter effect of the (sin x)/x characteristic at two sampling frequencies.

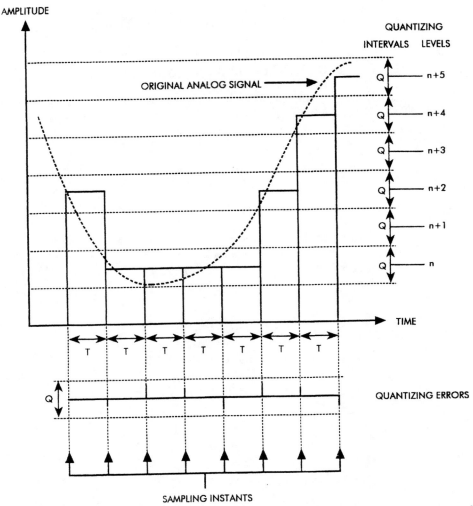

Figure 3.9 Waveform at the output of D/A converter. The outline of the original analog signal is shown in dotted lines.

3.2 The Composite Digital Standards

A long period of concept, product, and electronic component development resulted in a large number of application-specific digital black boxes operating at incompatible sample rates, number of bits per sample, and quantizing range. These products were developed to fulfill specific production needs and were designed for analog composite interconnection compatible with the all-analog composite production studios.

The trend toward a fully digital studio resulted in the need for digital video equipment industry standards. The composite digital video standards consti-

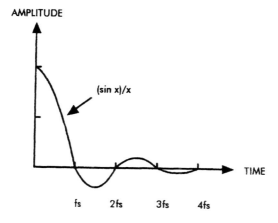

AMPLITUDE

(sin x)/x

TIME

fs   2fs   3fs   4fs

Figure 3.10 Spectrum of a rectangular pulse with a duration of $T = 1/f_s$.

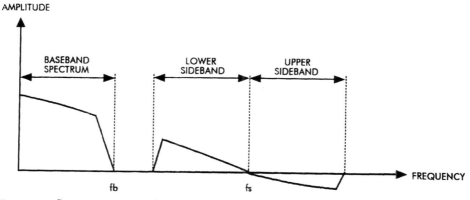

AMPLITUDE

BASEBAND SPECTRUM

LOWER SIDEBAND

UPPER SIDEBAND

FREQUENCY

fb   fs

Figure 3.11 Spectrum of signal at the D/A output.

tute a steppingstone toward the all-digital video teleproduction studio. To satisfy the needs of the industry, two sets of studio equipment composite digital standards have been developed:

- The $4f_{SC}$ NTSC standard
- The $4f_{SC}$ PAL standard

The analog composite video signal is sampled at a rate of four times the color subcarrier frequency ($4f_{SC}$). The number of bits per sample plays an important role in determining the signal quality and the economics of tape recording. The standards provide for a choice of 8 or 10 bits per sample.

In North America there was an initial interest in $4f_{SC}$ composite digital videotape recorders (VTRs). This had to do with the need to replace the obsolescent analog composite VTRs with digital VTRs featuring analog input/output ports. A number of manufacturers developed such products identified as

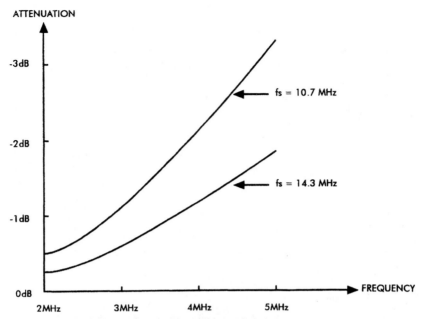

Figure 3.12 High frequency losses at two typical sampling frequencies.

D2 and D3 digital VTRs. A wide range of compatible $4f_{SC}$ video digital studio-quality equipment appeared on the market subsequently. In Europe there was limited interest in $4f_{SC}$ VTRs because they cannot handle SECAM. The development of competitively priced component digital video equipment has tilted the market toward the adoption of component digital equipment. This section will review the two $4f_{SC}$ digital video standards.

3.2.1   The $4f_{SC}$ NTSC standard

The SMPTE 244M standard defines the characteristics of the $4f_{SC}$ NTSC composite digital signals as well as the bit-parallel interconnect characteristics. This section will deal with the digital signal aspects defined by the standard and summarized in Table 3.1.

3.2.1.1 General specifications. The sampling frequency is equal to four times the subcarrier frequency or 14.3181 MHz (14.3 MHz nominal). The sampling clock is derived from the color burst of the analog signal. Figure 3.13 shows the sampling spectrum of $4f_{SC}$ NTSC. There is a significant gap between 4.2 MHz, the maximum nominal NTSC baseband frequency, and 7.16 MHz (the Nyquist frequency). The standard does not specify the characteristics of the antialiasing and reconstruction filters. The manufacturer has the choice of developing complex and costly wideband brick-wall ripple-free filters, resulting in an extended frequency response, or moderate-cost 4.2-MHz low-pass

TABLE 3.1   Summary of Coding Parameters for $4f_{SC}$ NTSC Composite Digital Signals

Input signal	NTSC
Number of samples per total line	910
Number of samples per active digital line	768
Sampling frequency	$4f_{SC} = 14.32818$ MHz
Sampling structure	Orthogonal
Sampling instant	$+33°, +123°, +213°, +303°$
Coding	Uniformly quantized
Quantizing resolution	8 or 10 bits per sample

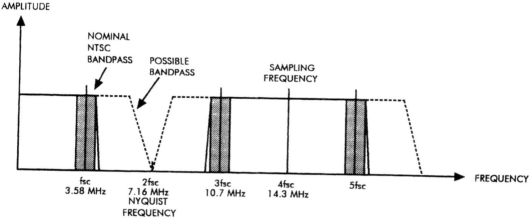

Figure 3.13   Spectrum of $4f_{SC}$ sampled NTSC signal.

filters with a gradual rolloff. As a result, various $4f_{SC}$ products have different analog bandwidths. The standard assumes that any $(\sin x)/x$ correction is carried out at the point where the digital signal is converted to analog form. It is important to note here that a digitally generated signal fed directly to a digital $4f_{SC}$ unit will have an equivalent analog bandwidth equal to $f_{SC}/2 = 7.16$ MHz. Severe overshoot and ringing of the derived analog composite signal may result unless special precautions are taken to ensure that digital blanking edges and risetimes, compatible with the analog waveforms, are included as an integral part of the digital signal.

3.2.1.2   The sampling structure.   This SMPTE standard was developed with reference to the original NTSC specifications, which used I/Q encoding instead of B-Y/R-Y encoding as is the current practice. Figure 3.14 shows that any chrominance vector can be represented by I/Q or B-Y/R-Y vectors. The original intent of the NTSC standard was to assign different bandwidths to the I signal (1.2 MHz) and to the Q signal (0.5 MHz), thus allowing for a better resolution for the orange visual information.

A typical NTSC I/Q encoder is shown in Fig. 3.15. The gamma-corrected

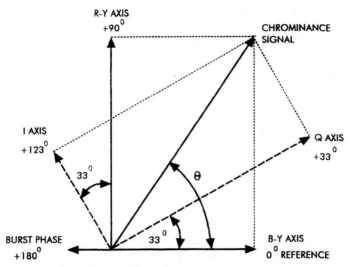

Figure 3.14 Phase diagram showing the relationship between the chrominance vector projections on the B-Y/R-Y axis system and the I/Q axis system.

Figure 3.15 Simplified block diagram of NTSC I/Q encoder.

E'_G, E'_B and E'_R primary signals are matrixed into E'_Y, E'_I, and E'_Q signals. The E'_I signal is band-limited to 1.2 MHz, and the E'_Q signal is band-limited to 0.5 MHz. The E'_Y and E'_I signals are suitably delayed to match the delayed narrow-bandwidth E'_Q signal. The two chrominance components feed dedicated suppressed carrier amplitude modulators. The subcarriers feeding the two modulators are in phase quadrature. An additional 33° subcarrier phase shift rotates the two vectors with respect to the B-Y reference as shown in Fig. 3.14. The I/Q-encoded NTSC signal can be decoded along the I/Q axes, with equal or unequal bandwidths, or the B-Y/R-Y axes with equal (equiband) wide or narrow bandwidths. Very few I/Q decoding monitors and receivers were built because of circuit complications that resulted in no visible picture improvements.

As shown in Fig. 3.16, the NTSC $4f_{SC}$ standard requires that the sampling instants coincide with peak positive and negative amplitudes of the I and Q subcarrier components. The upper part of the drawing shows that these sampling instants provide an adequate $4f_{SC}$ representation of the B-Y/R-Y information.

Given a sampling frequency f_S = 14.3181 MHz (nominally 14.32 MHz) and a horizontal scanning frequency f_H = 15,734.25 Hz, the number of samples per total line is equal to f_S/f_H = 910. The digital active line accommodates 768 samples. The remaining 142 samples make up the digital horizontal blanking interval. Figure 3.17 depicts the sample numbering for a nominal NTSC signal. The half-amplitude point of the leading (falling) edge of the analog horizontal sync signal falls between samples 784 and 785. The first of the 910 samples represents the first sample of the digital active line and is designated sample 0 for the purpose of reference. The 910 samples per line are, therefore, numbered 0 to 909. Samples 0 through 767 inclusive contain the digital active line data.

3.2.1.3 The quantizing range and its implications.

Table 3.2 lists significant composite analog NTSC signal levels of a 100/7.5/100/7.5 color bars signal and their corresponding $4f_{SC}$ NTSC hexadecimal digital values for 8- and 10-bit quantizing resolution.

Figure 3.18 shows the relationship between analog NTSC signal levels and 10-bit sample values of a 100/7.5/100/7.5 color bars signal.

The standard provides for 1024 digital levels (2^{10}), expressed in decimal numbers varying from 0 to 1023 or in hexadecimal numbers varying from 000 to 3FF. Digital levels 000, 001, 002, 003 and 3FC, 3FD, 3FE, 3FF are protected and not permitted in the digital stream. This leaves 1016 digital levels, expressed in decimal numbers varying from 4 to 1019 or in hexadecimal numbers varying from 004 to 3FB, to represent the video signal. The sync tip level is assigned the value 16 decimal or 010 hexadecimal. The highest signal level, corresponding to yellow and cyan, is assigned the value of 972 decimal or 3CC hexadecimal. The standard provides for a small amount of bottom headroom, levels 4 to 16 decimal or 004 to 010 hexadecimal, and top headroom, levels

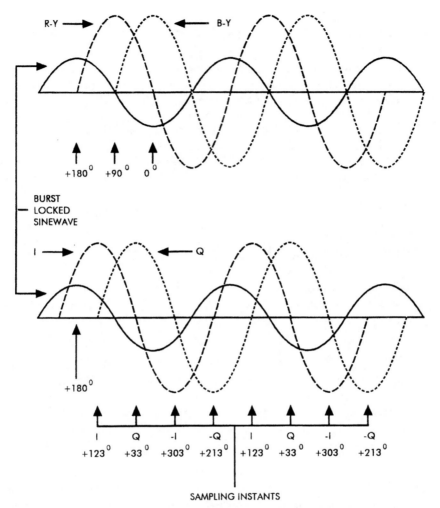

Figure 3.16 $4f_{SC}$ sampling instants of NTSC composite analog signal.

972 to 1019 decimal or 3CC to 3FB hexadecimal. The total headroom is of the order of 1 dB, and allows for misadjusted or drifting analog input signal levels. This reduces the S/Q_{RMS} ratio by the same amount. The resulting theoretical S/Q_{RMS} can be calculated using the formula developed in Sec. 3.1.5:

$$\frac{S}{Q_{RMS}}(\text{dB}) = 6.02n + 10.8 + 10 \log_{10}\left(\frac{f_S}{2f_{max}}\right) - 20 \log_{10}\left[\frac{V_q}{(V_W - V_B)}\right]$$

Given the following values for the variables in the formula:

n = 10 bits per sample

f_S = 14.32 MHz

Figure 3.17 $4f_{SC}$ NTSC sample numbering and horizontal sync relationship.

$$f_{\max} = 4.2 \text{ MHz}$$
$$V_q = 1.3042 \text{ V}$$
$$V_W - V_B = 0.7143 \text{ V}$$

The calculated value of S/Q_{RMS} is

$$\frac{S}{Q_{\mathrm{RMS}}} \approx 68.10 \text{ dB}$$

In an 8-bit system, 254 of the 256 levels (01 through FE) are used to express a quantized value. Levels 00 and FF are protected and not permitted

TABLE 3.2   Significant Composite Analog NTSC Levels of a 100/7.5/100/7.5 Color Bar Signal and Their Corresponding $4f_{SC}$ NTSC Hexadecimal Digital Values for 8- and 10-Bit Quantizing Resolution (See Fig. 3.18)

	8-bit resolution	10-bit resolution
Protected levels	FF	3FC,3FD,3FE,3FF
Highest quantized level	FE	3FB
Peak chroma level	F3	3CC
White level	C8	320
Black level	46	11A
Blanking level	3C	0F0
Sync tip level	04	016
Lowest quantized level	01	004
Protected levels	00	000,001,002,003

in the data stream. The calculated theoretical value of S/Q_{RMS} for an 8-bit system is

$$\frac{S}{Q_{RMS}} \approx 56.06 \text{ dB}$$

3.2.1.4   The digital raster structure.   The digital active field duration exceeds that of the analog active field. The digital active field period is positioned to begin before and end after analog video. In a four-field NTSC sequence, the digital vertical blanking interval extends from line 525, sample 768 to line 9, sample 767 inclusive for fields I and III and from line 263, sample 313 to line 272, sample 767 inclusive, for fields II and IV. Figure 3.19 depicts the relationship between the analog NTSC composite video signal vertical blanking and the digital $4f_{SC}$ NTSC vertical blanking.

The digital active line duration exceeds that of the analog active line. The digital active line is positioned to begin before and end after analog video. Thus, the blanking edges of the analog video are contained within the digital active line period. This effectively masks the analog overshoot and ringing generated by the limited bandwidth of the post D/A reconstruction filter and associated with the fast risetime of the digital blanking. The digital horizontal blanking interval extends from sample 768 to sample 909 inclusive on all lines outside of the vertical blanking interval. Figure 3.20 shows the location of some significant samples occurring during the horizontal blanking interval.

The standard specifies that the subcarrier phase to horizontal sync timing (SC/H) in the digital domain be equal to zero. This specification ensures unambiguous insert editing instants by VTRs operating in an editing suite. In analog NTSC, a zero SCH signal is defined as a signal in which the horizontal reference point is coincident with the zero crossing point of a burst-locked sine wave (a continuous sine wave of the same phase as the burst). The relationship between the NTSC subcarrier frequency and the horizontal scanning (line) frequency causes the direction of the zero crossing to alternate

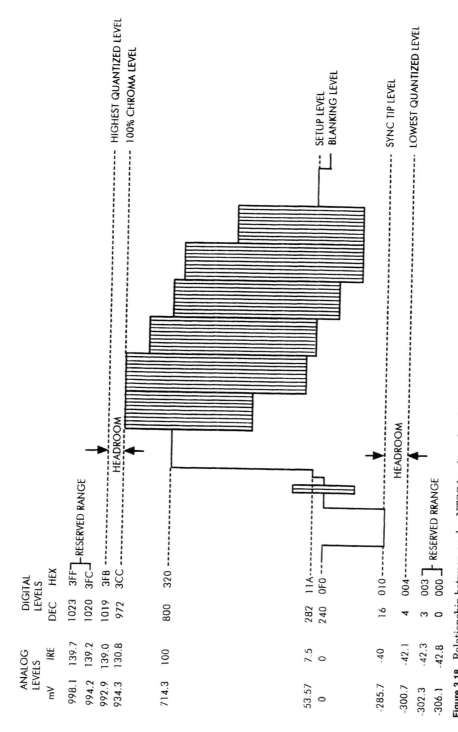

Figure 3.18 Relationship between analog NTSC levels and 10-bit digital sample values of a 100% color bar signal digital levels 000, 001, 002, 003 and 3FC, 3FD, 3FE, 3FF are protected and not permitted in the data stream.

Figure 3.19 Relationship between the analog NTSC composite video signal vertical blanking and the digital $4f_{SC}$ NTSC vertical blanking.

Figure 3.20 $4f_{SC}$ NTSC digital horizontal blanking interval showing the location of some significant samples.

on successive lines. Field I is defined as that field in which the first zero crossing on line 10 is positive-going. The first zero crossing of burst on line 10 of field III is negative-going. Color frame A describes fields I and II. Color frame B describes fields III and IV. A sequence of four consecutive fields occurs before the subcarrier zero crossing on line 10 is positive again. This very detailed specification needs to be met by all signal sources associated with a videotape editing studio. Figure 3.21 shows details of the vertical blanking interval structure and the relationship between the NTSC four-field sequence and the subcarrier zero crossing direction.

Figure 3.22 shows details of line 10, field I, color frame A and the relationship with the $4f_{SC}$ NTSC sampling instants.

3.2.2 The $4f_{SC}$ PAL Standard

The analog NTSC and PAL signals have certain similarities, but the $4f_{SC}$ digital representations are quite different. This section will deal succinctly with the digital signal characteristics as summarized in Table 3.3.

3.2.2.1 General specifications. The sampling frequency is equal to four times the subcarrier frequency or 17.734475 MHz (17.73 MHz nominal). The sampling clock is derived from the color burst of the analog signal. Figure 3.23 shows the sampling spectrum of $4f_{SC}$ PAL. There is a significant gap between 5 MHz, the maximum nominal PAL B,G baseband frequency, and 8.86 MHz

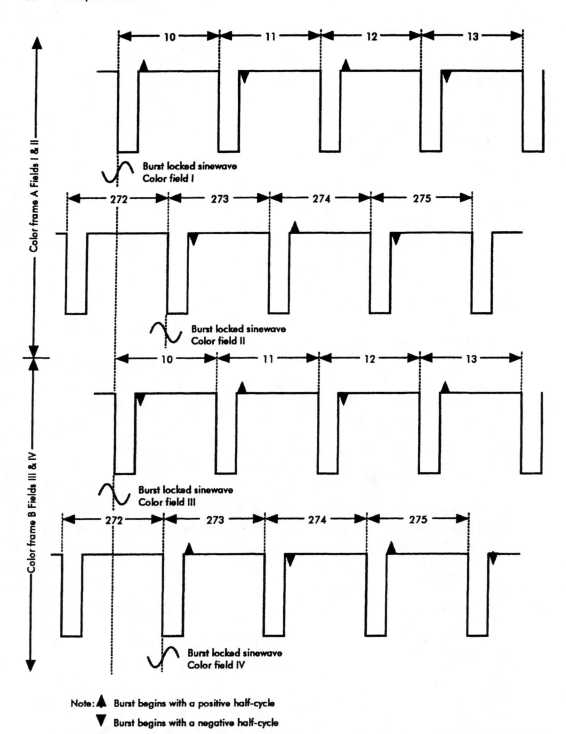

Note: ▲ Burst begins with a positive half-cycle

▼ Burst begins with a negative half-cycle

Figure 3.21 Details of the vertical blanking interval structure showing the relationship between the NTSC four field sequence and the subcarrier zero crossing direction.

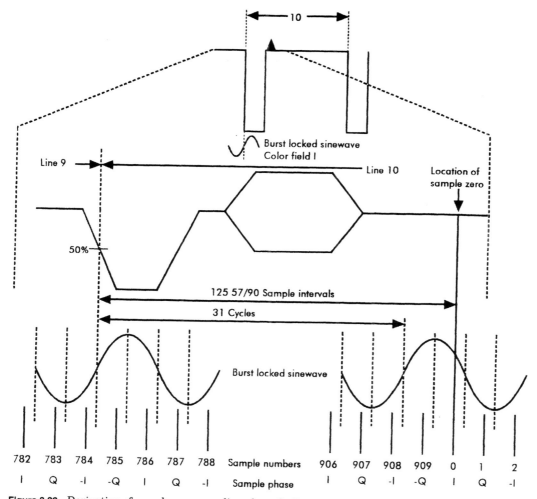

Figure 3.22 Derivation of sample zero sampling phase for line 10, field I, color frame A.

TABLE 3.3 **Summary of Coding Parameters for 4f_{sc} PAL Composite Digital Signals**

Input signal	PAL
Number of samples per total line	1135*
Number of samples per active line	948
Sampling frequency	$4f_{SC} = 17.734475$ MHz
Sampling instant	$+45°, +135°, +225°, +315°$
Coding	Uniformly quantized
Quantizing resolution	8 or 10 bits per sample

*Except lines 313 and 625, which have 1137 samples each.

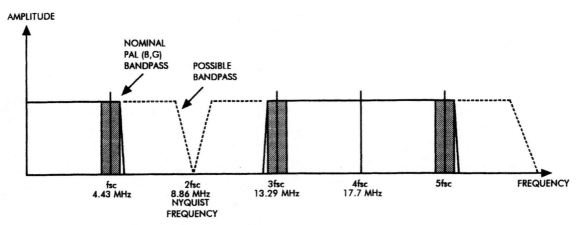

Figure 3.23 Spectrum of $4f_{SC}$ sampled PAL signal.

(the Nyquist frequency). The standard does not specify the characteristics of the antialiasing and reconstruction filters, but their design is facilitated by the high sampling frequency. Various $4f_{SC}$ products have different analog bandwidths. The $(\sin x)/x$ correction is assumed to be carried out at the point where the digital signal is converted back to analog. As with $4f_{SC}$ NTSC products, there may be problems with severe overshoot and ringing when a digitally generated signal is fed directly to the digital $4f_{SC}$ unit, as a consequence of the limited bandwidth of the reconstruction filter. To avoid such problems, digital blanking edges and risetimes compatible with the analog waveforms have to be included as part of the digital signal.

3.2.2.2 The sampling structure. Figure 3.24 shows details of the PAL vertical blanking interval structure. The color burst phase alternates between the two values of $+135°$ and $+225°$.

The PAL subcarrier contains a 25-Hz offset as per the formula

$$f_{SC} = 285.75 f_H + 25 \text{ Hz} = 4,433,618.75 \text{ Hz}$$

The number of subcarrier cycles per frame (two fields) is equal to

$$\frac{f_{SC}}{25} = 177,344.75 \text{ cycles/frame}$$

The minimum number of frames containing an integer number of subcarrier cycles is 4 ($4 \times 177,344.75 = 709,379$) or 8 fields. This is an important consideration in the design of videotape editing equipment.

The analog signal is sampled at four times the subcarrier frequency along the burst axes. The phase reference for the sample clock is the color subcarrier at $0°$ ($+U$ axis). Figure 3.25 shows the position of the sampling instants of the $4f_{SC}$ PAL standard.

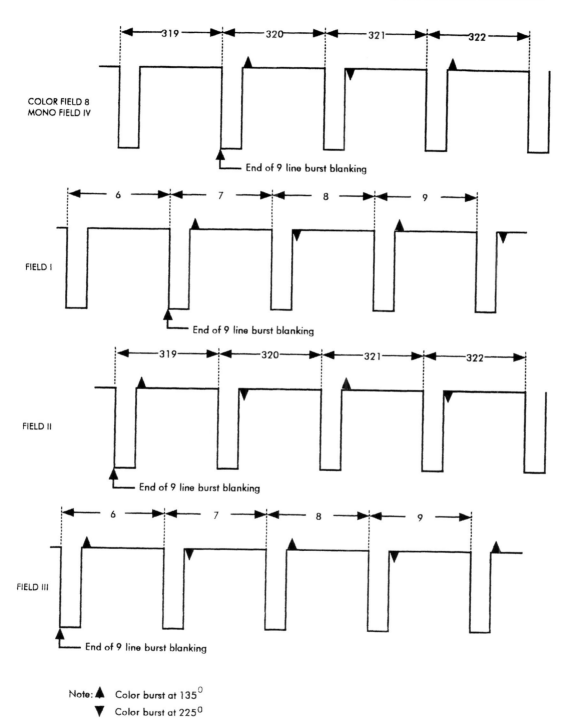

Figure 3.24 Details of the PAL vertical blanking interval structure.

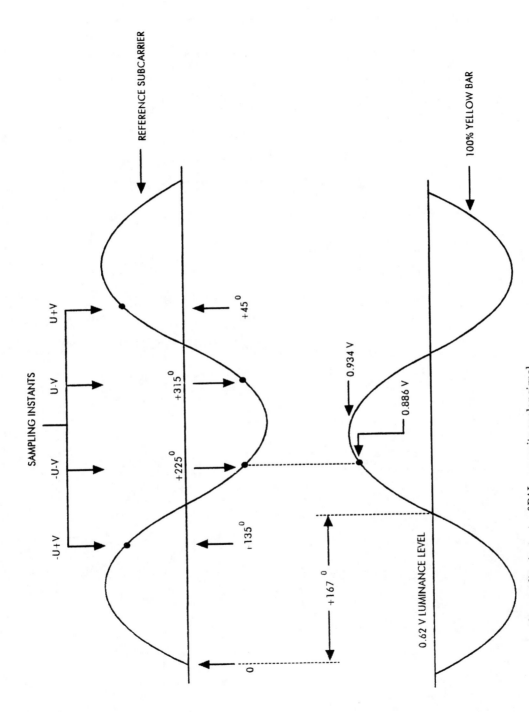

Figure 3.25 $4f_{SC}$ sampling instants of PAL composite analog signal.

The sampling frequency is given by

$$f_S = 4f_{SC} = 17,734,475.00 \text{ Hz}$$

The number of sample periods between two consecutive digitized horizontal synchronizing pulses is given by

$$\frac{f_S}{f_H} = \frac{17,734,475.00}{15,625.00} = 1135.0064$$

The total number of samples per frame is given by

$$\text{Samples/frame} = \text{Samples/line} \times \text{lines/frame} = 1135.0064 \times 625 = 709,379$$

Digital signal processing requires an integer number of samples to describe a television line. The number chosen is 1135 samples per line, resulting in 709,375 samples per frame. In order to produce the correct number of samples per television frame, two additional samples are added on lines 313 and 625. Of the 1135 samples in each active picture line (lines 313 and 615 are in the vertical blanking interval), 948 samples constitute the digital active portion of the line. The remaining 187 samples constitute the digital horizontal blanking interval. Figure 3.26 depicts the sample numbering for a typical line. The first of the 948 samples representing the active picture line in each line is designated sample 0 for the purpose of reference. The 1135 samples per line are numbered 0 through 1134. Samples 0 through 947 contain the digital active line data. Samples in the horizontal blanking interval are considered to belong to the following line. A complete analog line is described by samples 948 through 0 to 947 inclusive.

3.2.2.3 The quantizing range and its implications. Table 3.4 lists significant analog PAL signal levels of a 100/0/100/0 color bar signal and their corresponding $4f_{SC}$ PAL hexadecimal digital values for 8- and 10-bit quantizing resolution.

Figure 3.27 shows the relationship between analog PAL signal levels and 10-bit sample values of a 100/0/100/0 color bar signal.

The standard provides for 1024 digital levels (2^{10}), expressed in decimal numbers varying from 0 to 1023 or in hexadecimal numbers varying from 000 to 3FF. Digital levels 000, 001, 002, 003 and 3FC, 3FD, 3FE, 3FF are protected and not permitted in the digital stream. This leaves 1016 digital levels, expressed in decimal numbers varying from 4 to 1019 or in hexadecimal numbers varying from 004 to 3FB, to represent the video signal. The sync tip is assigned the value 4 decimal or 004 hexadecimal. Unlike in $4f_{SC}$ NTSC, there is no bottom headroom and the sync tip is the lowest quantized signal level. The highest quantized analog signal level is 908.3 mV, corresponding to a digital level 1019 decimal or 3FB hexadecimal. This is well below the 100/0/100/0 color bar signal yellow and cyan positive subcarrier excursions,

Figure 3.26 $4f_{\text{SC}}$ PAL sample numbering and horizontal sync relationship.

which equal 933.5 mV. There is, consequently, a negative headroom at the top of the signal. Figure 3.25 shows that the sampling instants are chosen such that the sampling of a signal corresponding to yellow occurs ahead of its maximum amplitude. The sampled yellow signal amplitude is 0.886 V, which is comfortably below the highest quantized level of 908.3 mV. The D/A converter is thus capable of reconstructing the original signal. This approach results in a slight improvement of the S/Q_{RMS}, of the order of 0.5 dB, without introducing any signal distortion. The resulting theoretical S/Q_{RMS} can be calculated using the formula developed in Sec. 3.1.5:

TABLE 3.4 Significant Composite Analog PAL Levels of a 100/0/100/0 Color Bar Signal and Their Corresponding $4f_{SC}$ PAL Hexadecimal Digital Values for 8- and 10-Bit Quantizing Resolution (see Fig. 3.25)

	8-bit resolution	10-bit resolution
Protected levels	FF	3FC, 3FD, 3FE, 3FF
Highest quantized level	FE	3FB
Peak chroma level	>>FE*	>>3FB*
White level	D3	34C
Blanking level	3C	100
Sync tip level	04	016
Lowest quantized level	01	004
Protected levels	00	000,001,002,003

*Digital values corresponding to yellow and cyan components of a 100/0/100/0 analog composite PAL color bar signal. These signal amplitudes are beyond the quantizing range specified by the standard. The specified sampling instants allow the re-creation of this signal in the D/A process.

$$\frac{S}{Q_{\text{RMS}}}\text{(dB)} = 6.02n + 10.8 + 10\log_{10}\left(\frac{f_S}{2f_{\max}}\right) - 20\log_{10}\left[\frac{V_q}{(V_W - V_B)}\right]$$

Given the following values for the variables in the formula:

$$n = 10 \text{ bits per sample}$$
$$f_S = 17.72 \text{ MHz}$$
$$f_{\max} = 5 \text{ MHz}$$
$$V_q = 1.2131 \text{ V}$$
$$V_W - V_B = 0.7 \text{ V}$$

The calculated S/Q_{RMS} is

$$\frac{S}{Q_{\text{RMS}}} \approx 68.71 \text{ dB}$$

3.2.2.4 The digital raster structure. The digital active field duration exceeds that of the analog field. The digital active field period is positioned to begin before and end after analog video. In a four-field PAL sequence, the digital vertical blanking extends from line 623, sample 382, to line 5, sample 947 inclusive, for fields I and III and from line 310, sample 948, to line 317, sample 947 inclusive, for fields II and IV. Figure 3.28 depicts the relationship between the analog PAL composite video signal vertical blanking and the digital $4f_{SC}$ PAL vertical blanking.

The digital active line duration exceeds that of the analog active line. It is positioned to begin before and end after analog video. The blanking edges of

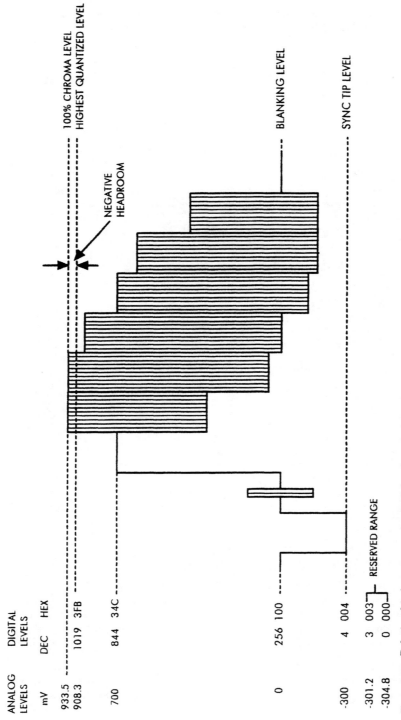

Figure 3.27 Relationship between analog PAL levels and 10-bit digital sample values of a 100% color bars signal. Digital levels 000, 001, 002, 003, and 3FC, 3FD, 3FE, 3FF are protected and not permitted in the data stream.

Figure 3.28 Relationship between the analog PAL composite video signal vertical blanking and the digital $4f_{SC}$ PAL vertical blanking.

the analog video are contained within the digital active line period. This effectively masks the analog overshoot and ringing generated by the post D/A reconstruction filter and associated with the fast risetime of the digital blanking. The digital horizontal blanking interval extends from sample 948 to sample 1134 inclusive on all lines outside the digital vertical blanking interval.

Figure 3.29 $4f_{SC}$ PAL digital horizontal blanking interval showing the location of some significant samples on a typical line. The sample positions vary from line to line.

Figure 3.29 shows the location of some significant samples occurring during the horizontal blanking interval.

3.2.3   Performance-indicative parameters and test concepts

$4f_{SC}$ digital black boxes with analog in/out ports are characterized by impairments as classified and detailed in Chap. 2. The test methods and equipment used are similar to those used with analog equipment and the reader is referred to Chap. 2 for general information. The black box is assumed to have unity in/out gain. This is verified by feeding the D/A converter with the internally generated reference color bar signal and adjusting the output gain as required. A reference analog color bar signal is then fed to the input and the input gain is adjusted, as required, for unity in/out gain.

The measurement concepts of luminance nonlinearity, differential phase, differential gain, and noise are quite different and will be discussed in some detail. Specific constant-level test signals may not be correctly represented because of quantizing errors and may thus introduce measurement errors. The measurement errors are quite substantial with 8-bit equipment and negligible with 10-bit equipment.

The measurement of some nonlinear distortions such as luminance nonlinearity, differential gain, and differential phase of analog equipment normally uses a stairstep signal with or without a superimposed subcarrier. Similar tests of $4f_{SC}$ black boxes use a ramp signal, with or without a superimposed subcarrier.

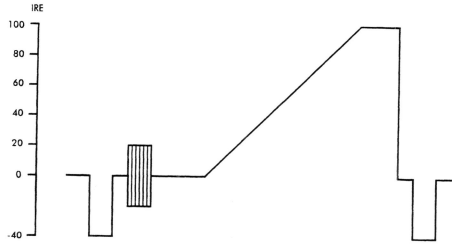

Figure 3.30 Luminance ramp for 525/60 systems.

In order to improve the accuracy, the measurement of luminance nonlinearity of $4f_{SC}$ digital black boxes uses a luminance ramp test signal. Figure 3.30 shows a luminance ramp test signal for 525/60 systems. A similar signal, not shown, is available for 625/50 systems. Unlike a stairstep signal, this type of signal activates all the quantizing levels and thus gives a better representation of the performance of the equipment under test.

The measurement method is quite different from that used with analog equipment. Figure 3.31 shows a test setup used by the authors to measure luminance nonlinearity.

The test method consists in subtracting the input signal from the output signal of the equipment under test and measuring the difference in a 1-MHz bandwidth as luminance nonlinearity. The test signal is fed to the equipment under test by "looping" through the bridging input B of the Tektronix 1480 waveform monitor. The equipment under test is assumed to have a standard 75 ohms input impedance. If this is not the case, a 75-ohm load is to be connected across the input. The equipment under test is fed a reference color burst signal used to synchronize the D/A converter. The output of the equipment under test is fed to the input A of the waveform monitor and loaded with a 75-ohm resistor. The waveform monitor input B is selected for display of a single active line of the test signal. With unity gain selected, the fine vertical gain of the waveform monitor is adjusted to display 100 IRE (700 mV in PAL) of ramp amplitude. Without readjusting the fine gain control, the waveform monitor is set to maximum gain (2% per division) and a low-pass response is selected. The A-B mode of operation is selected on the waveform monitor thus subtracting the input signal from the output signal and displaying their difference. The horizontal and subcarrier timing controls as well as

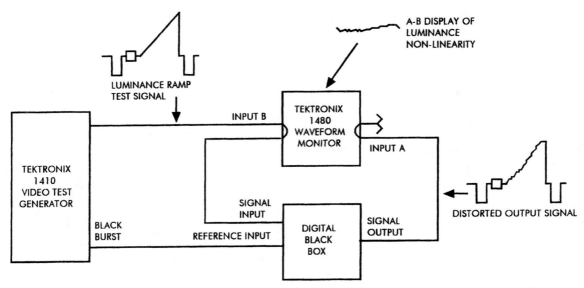

Figure 3.31 Test setup for the measurement of luminance nonlinearity of $4f_{SC}$ composite video digital equipment.

the input gain of the equipment under test are adjusted to obtain a display as nearly horizontal as possible. The luminance nonlinearity is the peak-to-peak departure from a flat horizontal display expressed as a percentage of the test signal amplitude (100 IRE in NTSC or 700 mV in PAL).

The differential gain and phase measurement of $4f_{SC}$ digital black boxes is similar to that of analog composite test equipment except for the use of a modulated ramp in place of the usual modulated staircase. Figure 3.32 shows a modulated ramp test signal for NTSC. A similar signal, not shown, is available for PAL.

The measurement of S/Q_{RMS} is similar to the measurement of luminance nonlinearity except that the measurement bandwidth is 15 kHz to 4.2 MHz (5 MHz in PAL), and an oscilloscope is used for increased measurement resolution. Figure 3.33 shows a test setup used by the authors.

The Tektronix 1480 waveform monitor is used here as a convenient A-B signal subtractor and its monitoring output is filtered by a suitable (4.2 or 5 MHz) filter and fed to a high gain (≥ 5 mV/cm sensitivity) wideband (>20 MHz) oscilloscope with delay trigger capabilities. The oscilloscope is synchronized by a vertical drive signal from the video test generator and operated in the delay trigger mode to select one active line for full screen display (time base setting, 5 μs/cm). Operating the oscilloscope in this mode effectively sets the low end of the measurement bandwidth to 15 kHz and eliminates hum and other low-frequency disturbances from the measurement. The horizontal and burst timing as well as the gain of the digital black box are fine-tuned to obtain a display as nearly horizontal as possible. The S/Q_{RMS} ratio, in dB, is given by the formula

Figure 3.32 Modulated ramp for NTSC system.

$$\frac{S}{Q}_{\text{RMS}} \text{ (dB)} = 20 \log_{10}\left[\frac{(V_W - V_B)}{\text{p-p noise}}\right] + 15$$

where S/Q_{RMS} = Signal to RMS quantizing noise ratio
V_W = White signal level, V
V_B = Blanking signal level, V
$V_W - V_B$ = 0.714 V (NTSC) or 0.7 V (PAL)
p-p noise = The "quasi" peak-to-peak noise as estimated from the oscilloscope display

Specialized automatic video testing equipment, such as the Tektronix VM700, has the capability of normalizing the luminance ramp test signal to a horizontal line and give direct readings of S/Q_{RMS} at the output of the digital black box without the need of the elaborate test setup shown in Figure 3.33.

All above considerations concerning special tests for $4f_{\text{SC}}$ composite digital equipment with analog input/output ports assume that the signal is processed in some manner without decoding it to G, B, R signals. Figure 3.34 shows the simplified block diagram of such a $4f_{\text{SC}}$ digital black box.

Some types of equipment feature analog composite NTSC (or PAL) input/output ports, but the composite signals are decoded to components and the actual internal processing is in a component digital format. Certain test signals, such as a modulated NTSC ramp, result in illegal (negative value) decoded G, B, or R signals, leading to excessive nonlinearities when encoded back into NTSC format.

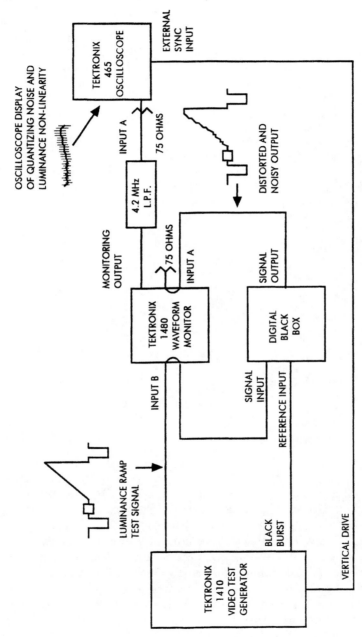

Figure 3.33 Test setup for the measurement of quantizing noise of $4f_{SC}$ composite video digital equipment.

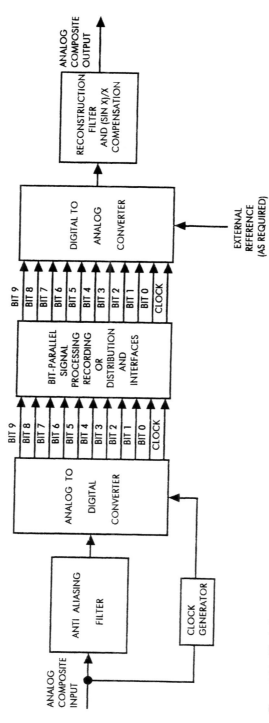

Figure 3.34 Block diagram of $4f_{SC}$ digital black box.

It has to be realized that these test signals were developed for the performance measurements of distribution networks and are not meant to be decoded to analog components. A typical example is the NTSC modulated ramp. This signal consists of a luminance ramp with a superimposed subcarrier having a phase equal to that of the burst ($+180°$ with respect to the B-Y reference). When decoded to G, B, R analog component signals, the blue signal assumes a negative value for a significant duration of the active line. Under such conditions, the measured distortions of one specific product were found to be 20° differential phase and 20% differential gain at the encoded NTSC analog component output. This would indicate excessive nonlinear distortions until it is realized that the NTSC modulated ramp with a 180° phase superimposed subcarrier is an illegal signal when decoded. This signal is sensitive to chrominance/luminance intermodulation effects (such as differential phase and gain) and is meant to be used in the performance measurements of distribution networks where no decoding to basic components is carried out. In reality the above-mentioned product operates satisfactorily when fed signals generated by a camera.

3.2.4 Bit-parallel $4f_{SC}$ digital signal distribution

The $4f_{SC}$ digital signal bit-parallel equipment interconnection is adequate for short distances and simple point-to-point signal distribution patterns. It is inadequate for large teleproduction centers with complex signal distribution patterns where the high cost of multicore cables and the large size of the multipin connectors come into play. In addition, there are serious electronic problems with multicore cables. The main problem is the difficulty of controlling the data transfer time. This is a consequence of the fact that the propagation speeds of the pulses along all the cores of the cable are not exactly the same, resulting in excessive clock to data differential delays. The problem is compounded by crosstalk between the conductors and suboptimal termination leading to marginal data integrity. The block diagram of Fig. 3.34 shows a $4f_{SC}$ bit-parallel black box. This block diagram is also representative of a studio bit-parallel digital signal distribution path.

The $4f_{SC}$ digital signal is distributed using a shielded twisted 12-pair cable of conventional design over distances not exceeding 50 m without transmission equalization or any special equalization at the receiver. The bit-serial digital signal transmission is the preferred method of interconnection of composite digital equipments, especially when the cable length exceeds 50 m.

The bits of the digital words that describe the video signal are transmitted in a parallel arrangement using 10 (8 for 8 bits/sample) conductor pairs. An eleventh (ninth for 8 bits/sample) carries a clock at $4f_{SC}$ (14.31818 MHz for NTSC and 17.73447 MHz for PAL). The interface consists of one transmitter and one receiver in a point-to-point connection. Figure 3.35 shows the diagram of a balanced interface circuit for bit-parallel digital signal distribution typical for each of the 11 (9 for 8 bits/sample) conductor pairs.

Figure 3.36 shows the pin assignment of a standard DB25 connector used for bit-parallel digital interconnection.

A, A' = DATA LINE
B, B' = RETURN LINE
Zt   = CABLE TERMINATION
A, B = TRANSMITTER INTERFACE POINTS
A', B' = LOAD INTERFACE POINTS
C   = TRANSMITTER CIRCUIT GROUND
C'   = LOAD CIRCUIT GROUND
Vg   = GROUND POTENTIAL DIFFERENCE
Zo   = CABLE CHARACTERISTIC IMPEDANCE

Figure 3.35 Balanced interface circuit for bit-parallel digital signal distribution.

3.2.4.1 Signal conventions

- The line driver and the receiver must be ECL-compatible to permit the use of standard ECL parts for either or both ends. "Standard ECL" refers to the 10.000 series of ECL logic.

- With reference to Fig. 3.35, the DATA terminal of the transmitter is positive (+) with respect to the RETURN terminal for a binary 1 (HIGH, H, or ON) state.

- The data lines are designated DATA 0 through DATA 9. DATA 9 is the most-significant bit.

3.2.4.2 Interface electrical characteristics.
The $4f_{SC}$ interface electrical characteristics are summarized in Table 3.5. Figure 3.37 shows the clock to data timing relationships at the transmitter output.

3.3 The Component Digital Standards

In an analog component system, information is conveyed by the infinite variation of G, B, R or Y, B-Y, R-Y signal amplitude inside some limits. It is a characteristic of analog component systems that the signal degradations at the output are the sum of all the degradations introduced by each stage through which the three signals have passed. This sets a limit to the number of stages through which a set of component signals can pass before they become too impaired to be worth watching. The three major impairments affecting analog

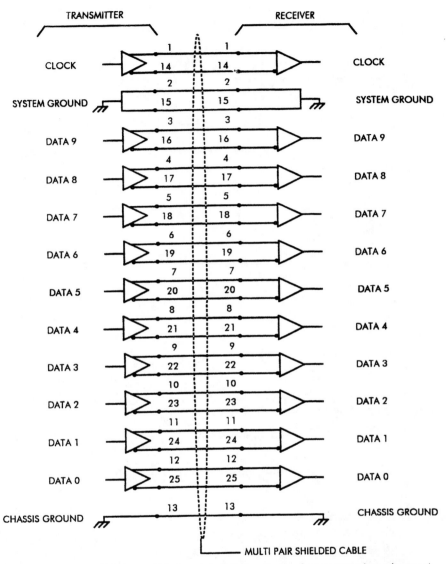

Figure 3.36 Bit-parallel transmitter to receiver connection and connector pin assignment.

signals, namely linear distortions, nonlinear distortions, and noise comple-
mented by timing instabilities, can be reduced by good engineering practices
but never completely eliminated. The use of component signals as opposed to
composite signals in the studio environment may remove some of the impairments associated with composite video signals. The penalty is an increased
cost, increased complexity, and reduced availability of production equipment.

In a digital domain such impairments can be reduced to those generated by

TABLE 3.5 $4f_{SC}$ Bit-Parallel Interface Electrical Characteristics

General	Video data are carried on 10 (8 for 8 bits) wire pairs in NRZ form. An eleventh (ninth) pair carries a synchronous $4f_{SC}$ clock
Transmitter characteristics	▪ Balanced output ▪ Source impedance: 110 ohms max. ▪ Driver output common mode voltage: -1.3 V $\pm15\%$ p-p with respect to ground terminal ▪ Signal amplitude: 0.8 to 2.0 $V_{p\text{-}p}$ across 110 ohms ▪ Rise and fall times: <5 ns and differ by no more than 2 ns, as measured between the 20% and 80% amplitude points across a 110-ohm resistor connected directly across the output terminals ▪ Clock signal jitter: <5 ns of average time of rising edges computed over at least one television field ▪ Clock signal positive (+) transition nominally occurs between the data transitions
Receiver characteristics	▪ Input impedance: 110 ±10 ohms ▪ Input sensitivity: ≥185 mV for proper operation ▪ Maximum input signal: 2 $V_{p\text{-}p}$ ▪ Common mode rejection: Common mode noise with a ±0.5-V amplitude should not affect operation ▪ Accepted clock/data differential delay: ≤5 ns

the single-pass A/D and D/A process, provided that the signal is processed, recorded, and distributed in digital form and the encoding to composite analog video is performed only once, prior to conventional VHF/UHF on-air transmission. The composite digital $4f_{SC}$ concept might be an attractive evolutionary approach, but the availability of competitively priced component digital studio-quality equipment as well as the trend towards in-plant, land-line, on-air, or satellite component digital signal distribution at full or reduced bit rates is a major incentive toward the changeover to component digital teleproduction studios.

North American and European standardization efforts resulted in CCIR Recommendation 601, Encoding Parameters of Digital Television for Studios. This recommendation established an agreement on a component digital approach that is compatible with both 525/50 and 625/50 scanning standards and is at the root of all subsequent component digital developments. The initial document recommended a resolution of 8 bits per sample, but current studio-quality equipment almost universally uses 10 bits per sample.

The recommended digital coding is based on the use of one luminance and two color-difference signals or the green, blue, and red signals. The characteristics of these two sets of component signals are described in Chap. 2.

3.3.1   The sampling rates

A family of sampling rates, based on a reference frequency of 3.375 MHz, has evolved. These are

▪ 4:1:1 sampling: The luminance signal is sampled at 13.5 MHz (4×3.375 MHz) and each of the two color-difference signals is sampled at 3.375 MHz.

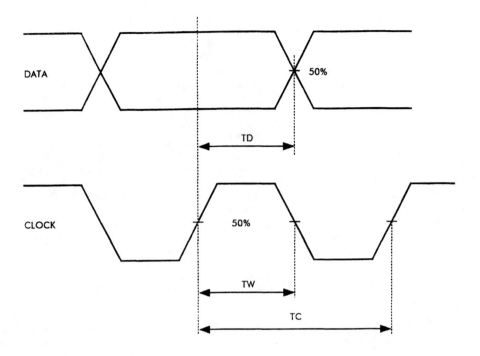

	NTSC	PAL
TW	34.92 ± 5ns	28 ± 5ns
TC	69.84 ns	56.39 ns
TD	34.92 ± 5 ns	28 ± 5 ns

Figure 3.37 $4f_{\text{SC}}$ clock and data timing relationship at transmitter output.

This method is used by some low-end equipment and is suitable for situations where the bandwidth of the color-difference signals does not need to exceed 1.5 MHz.

- 4:2:2 sampling: The luminance signal is sampled at 13.5 MHz and each of the two color-difference signals is sampled at 6.75 MHz (2×3.375 MHz).

- 4:4:4 sampling: The luminance signal as well as the two color-difference signals are sampled at 13.5 MHz. Alternately G, B, R equiband signals may be sampled. This sampling rate is typical of some upper-end studio-quality digital processing equipment.

Tables 3.6 (for scanning standard 525/60) and 3.7 (for scanning standard 625/50) summarize the 4:2:2 encoding parameters, which will be discussed in

TABLE 3.6   Summary of 4:2:2 Encoding Parameters for the 525/60 Scanning Standard

Scanning standard	525/60
Coded signals	$E'_Y = 0.587\,E'_G + 0.11\,E'_B + 0.299\,E'_R$ $E'_{CB} = 0.564(E'_B - E'_Y)$ $E'_{CR} = 0.713(E'_R - E'_Y)$
Number of samples per total line	Y: 858 C_B: 429 C_R: 429 Total: 1716
Number of samples per digital active line	Y: 720 C_B: 360 C_R: 360 Total: 1440
Sampling structure	Orthogonal Line, field, and frame repetitive C_B and C_R samples cosited with odd Y samples in each line
Sampling frequency	Y: $858 \times f_H$ = 13.5 MHz C_B and C_R: $429 \times f_H$ = 6.75 MHz
Coding	Uniformly quantized PCM
Quantizing resolution	8 or 10 bits per sample for the luminance and each color-difference signal

TABLE 3.7   Summary of 4:2:2 Encoding Parameters for the 625/60 Scanning Standard

Scanning standard	625/50
Coded signals	$E'_Y = 0.587\,E'_G + 0.11\,E'_B + 0.299\,E'_R$ $E'_{CB} = 0.564(E'_B - E'_Y)$ $E'_{CR} = 0.713(E'_R - E'_Y)$
Number of samples per total line	Y: 864 C_B: 432 C_R: 432 Total: 1728
Number of samples per digital active line	Y: 720 C_B: 360 C_R: 360 Total: 1440
Sampling structure	Orthogonal Line, field, and frame repetitive C_B and C_R samples cosited with odd Y samples in each line
Sampling frequency	Y: $864 \times f_H$ = 13.5 MHz C_B and C_R: $432 \times f_H$ = 6.75 MHz
Coding	Uniformly quantized PCM
Quantizing resolution	8 or 10 bits per sample for the luminance and each color-difference signal

some detail in this section. The 4:2:2 standard is erroneously referred to as D1, a digital VTR format.

3.3.2 The coded signals

The coded signals are a luminance signal and two color-difference signals. The expressions for the luminance signal (E'_Y) and the two color-difference signals (E'_{CB} and E'_{CR}) are as per Chap. 2 and Tables 3.6 and 3.7. The notation for the two color-difference signals is as per CCIR 601. In North American documents they may occasionally be referred to as P_B and P_R.

3.3.3 The sampling frequencies

The luminance sampling frequency of 13.5 MHz was selected to allow an integer number of sample periods in the line periods of both the 525/60 and the 625/50 scanning standards. It is obtained from the input video sync using a phase-locked-loop-controlled oscillator operating at $858 \times f_H$, in the 525/60 standard, and $864 \times f_H$, in the 625/50 standard. The specified amplitude versus frequency and group delay versus frequency response of the luminance signal antialiasing and reconstruction filters are shown in Fig. 3.38. The luminance frequency response is flat up to 5.75 MHz, which is adequate for both scanning standards.

The 13.5-MHz sampled spectrum of the filtered luminance signal is shown in Fig. 3.39. As shown, there is a very narrow and critical gap between the maximum baseband luminance frequency (5.75 MHz) and the Nyquist frequency (6.75 MHz). This explains the tight specification of the antialiasing and reconstruction filters.

The color-difference sampling frequency is obtained by dividing the luminance sampling frequency by 2. The specified amplitude versus frequency and group delay versus frequency response of the color-difference antialiasing and reconstruction filters is shown in Fig. 3.40. The color-difference signals frequency response is flat to 2.75 MHz.

The 6.75-MHz sampled spectrum of the filtered color-difference signals is shown in Fig. 3.41. As shown, there is a very narrow and critical gap between the maximum baseband color-difference frequency (2.75 MHz) and the Nyquist frequency (3.375 MHz). This explains the tight specifications of the antialiasing and reconstruction filters.

The bandwidth of the luminance and color-difference channels, as determined by the sampling frequencies and the low-pass filters, is adequate for a single pass. Multiple passes, such as when a number of component digital black boxes are connected using analog input/output ports, result in significant linear distortions related to the deteriorating amplitude versus frequency and group delay versus frequency response.

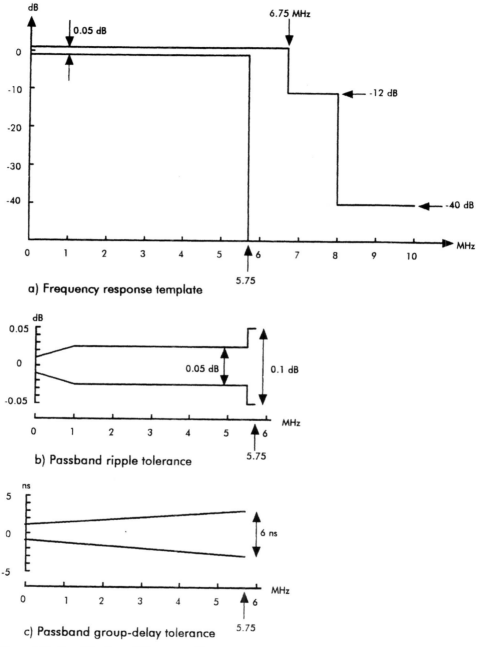

a) Frequency response template

b) Passband ripple tolerance

c) Passband group-delay tolerance

Figure 3.38 Specification of Y low-pass filter.

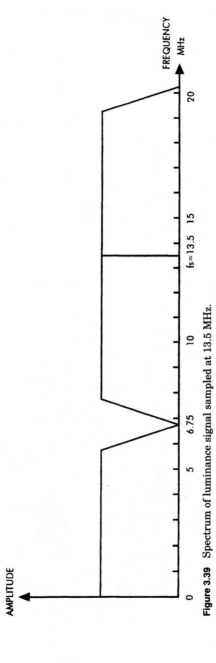

Figure 3.39 Spectrum of luminance signal sampled at 13.5 MHz.

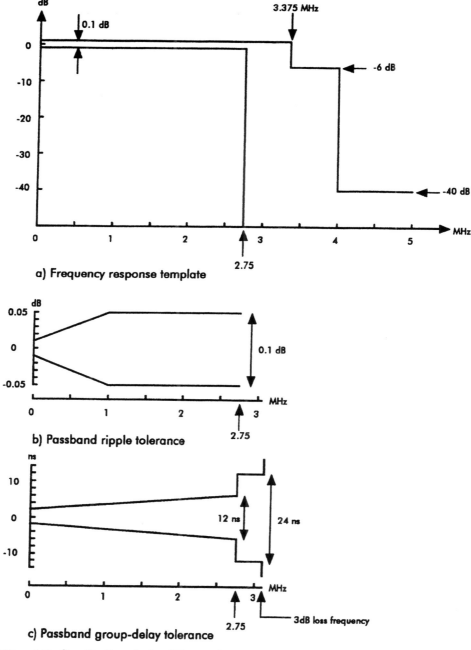

a) Frequency response template

b) Passband ripple tolerance

c) Passband group-delay tolerance

Figure 3.40 Specification of color-difference low-pass filter.

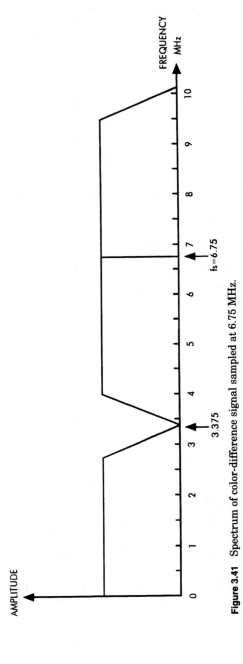

Figure 3.41 Spectrum of color-difference signal sampled at 6.75 MHz.

TABLE 3.8   Significant E'_Y Component Analog Levels of a 100/0/100/0 Color Bar Signal and Their Corresponding Y Digital Signal Hexadecimal Values for 8- and 10-Bit Quantizing Resolution (see Fig. 3.42)

	8-Bit resolution	10-Bit resolution
Protected levels	FF	3FC,3FD,3FE,3FF
Highest quantized level	FE	3FB
Peak white level	EB	3AC
Blanking level	10	040
Lowest quantized value	01	004
Protected levels	00	000,001,002,003

3.3.4   The quantizing range and the implications

Table 3.8 shows significant E'_Y component analog levels of a 100/0/100/0 color bar signal and their corresponding Y digital signal hexadecimal values for 8- and 10-bit quantizing resolution.

Figure 3.42 shows the relationship between analog signal levels of the Y component of a 100/0/100/0 color bars signal and 8-bit and 10-bit digital sample values.

In a 10-bit system, there are 1024 digital levels (2^{10}), expressed in decimal numbers varying from 0 to 1023 or in hexadecimal numbers varying from 000 to 3FF. Digital levels 000, 001, 002, 003 and 3FC, 3FD, 3FE, 3FF are reserved to indicate timing references. This leaves 1016 digital levels expressed in decimal numbers varying from 4 to 1019 or in hexadecimal number varying from 004 to 3FB to represent the luminance levels. The blanking level is assigned the value 64 decimal or 040 hexadecimal. The peak white level is assigned the value of 940 decimal or 3AC hexadecimal. The standard provides for a small amount of bottom headroom, levels 4 to 64 decimal or 004 to 040 hexadecimal, and top headroom, levels 940 to 1019 decimal or 3AC to 3FB hexadecimal. Note that the sync portion of the luminance signal is not sampled. The headroom allows for misadjusted or drifting analog signal levels. The theoretical S/Q_{RMS} of a 10-bit system can be calculated using the formula developed in Sec. 3.1.5:

$$\frac{S}{Q_{RMS}} \text{ (dB)} = 6.02n + 10.8 + 10 \log_{10}\left(\frac{f_S}{2f_{max}}\right) - 20 \log_{10}\left[\frac{V_q}{(V_W - V_B)}\right]$$

Given the following values for the variables in the formula,

$$n = 10 \text{ bits per sample}$$
$$f_S = 13.5 \text{ MHz}$$
$$f_{max} = 5.75 \text{ MHz}$$
$$V_q = 0.8174 \text{ V}$$
$$V_W - V_B = 0.7 \text{ V}$$

Figure 3.42 Relationship between analog signal levels of the Y component of a 100% color bars signal and 8 bit and 10 bit digital sample values. The digital levels 00 and FF for 8 bit and 000, 001, 002, 003 and 3FC, 3FD, 3FE, 3FF for 10 bit are not permitted in the data stream.

The calculated value for $S/Q_{\rm RMS}$ is

$$\frac{S}{Q_{\rm RMS}} \approx 70.35 \text{ dB}$$

The calculated theoretical value of $S/Q_{\rm RMS}$ for an 8-bit system is

$$\frac{S}{Q_{\rm RMS}} \approx 58.3 \text{ dB}$$

Table 3.9 shows significant E'_{CB} and E'_{CR} color-difference component analog levels of a 100/0/100/0 color bar signal and their corresponding C_B and C_R digital signal hexadecimal values for 8- and 10-bit quantizing resolution. Note that the color-difference signals are bipolar and are therefore shifted by 0.350 V to fit the A/D converter, which requires unipolar signals.

Figure 3.43 shows the relationship between the levels of the C_B component of a 100/0/100/0 color bar signal and 8-bit and 10-bit digital sample values.

Figure 3.44 shows the relationship between the levels of the C_R component of a 100/0/100/0 color bar signal and 8-bit and 10-bit digital sample values.

In a 10-bit system there are 1016 digital levels expressed in decimal values 4 to 1019 or hexadecimal levels 004 to 3FB available to represent the C_B and C_R signals. The blanking level is assigned the value 512 decimal or 200 hexadecimal. The maximum analog signal positive level is assigned the value 960 decimal or 3C0 hexadecimal. The maximum analog signal negative level is assigned the value 64 decimal or 040 hexadecimal. The standard provides for a small amount of bottom headroom, levels 4 to 64 decimal or 004 to 040 hexadecimal, and top headroom, levels 960 to 1019 decimal or 3C0 to 3FB hexadecimal, allowing for misadjusted or drifting analog signal levels. Given the following values for the variables:

$$n = 10 \text{ bits per sample}$$
$$f_S = 6.75 \text{ MHz}$$
$$f_{\rm max} = 2.75 \text{ MHz}$$

TABLE 3.9   Significant E'_{CB} and E'_{CR} Color-Difference Component Analog Levels of a 100/0/100/0 Color Bar Signal and Their Corresponding C_B and C_R Digital Signal Hexadecimal Values for 8- and 10-Bit Quantizing Resolution (See Figs. 3.43 and 3.44)

	8-Bit resolution	10-bit resolution
Protected levels	FF	3FC,3FD,3FE,3FF
Highest quantized level	FE	3FB
Maximum positive level	F0	3C0
Blanking level	80	200
Maximum negative level	10	040
Lowest quantized level	01	004
Protected levels	00	000,001,002,003

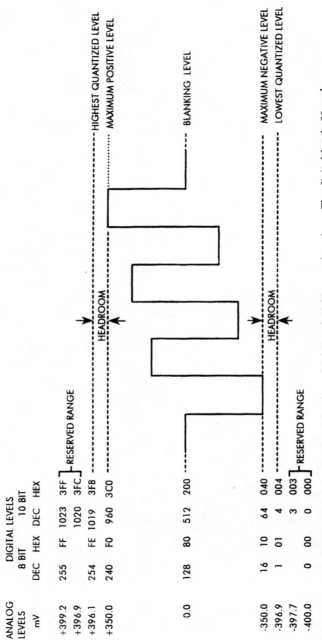

Figure 3.43 Relationship between C_B signal levels and 8 bit and 10 bit sample values. The digital levels 00 and FF for 8 bit and 000, 001, 002, 003 and 3FC, 3FD, 3FE, 3FF for 10 bit are excluded.

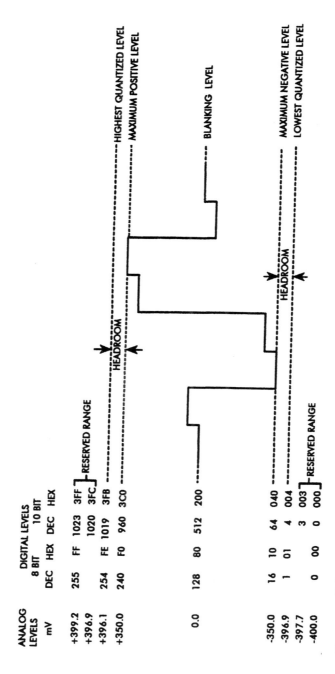

Figure 3.44 Relationship between C_R signal levels and 8-bit and 10-bit sample values. The digital levels 00 and FF for 8 bit and 000, 001, 002, 003 and 3FC, 3FD, 3FE, 3FF for 10 bit are excluded.

$$V_q = 0.7992 \text{ V}$$
$$V_W - V_B = 0.7 \text{ V}$$

The calculated theoretical value of S/Q_{RMS}

$$\frac{S}{Q_{\text{RMS}}} \approx 70.74 \text{ dB}$$

The calculated theoretical value of S/Q_{RMS} for an 8-bit system is:

$$\frac{S}{Q_{\text{RMS}}} \approx 58.7 \text{ dB}$$

3.3.5 The sampling structure

Sampling the analog video signal at a multiple of the horizontal scanning frequency results in the sampling instants being vertically aligned on a line-by-line and field-by-field basis. This is known as *orthogonal sampling*. Figure 3.45 shows details of the orthogonal sampling structure of the 4:2:2 member of the family. As a consequence of the fact that the luminance signal sampling frequency is twice that of each of the two color-difference signals, there are twice as many luminance samples as there are color-difference samples. As shown in Table 3.6 the C_B and C_R samples are cosited with odd Y samples in every line.

The number of luminance samples per total line is equal to f_S/f_H, where $f_S = 13.5$ MHz and f_H is the horizontal scanning frequency. Given the slightly different values of f_H, the number of luminance samples per total line for the two scanning standards is 858, numbered 0 through 857, (525/60 standard) and 864, numbered 0 through 863 (625/50 standard), respectively.

Given $f_S = 6.75$ MHz, the number samples per total line of each color-difference signal for the two scanning standards is 429, numbered 0 through 428, (525/60 standard) and 432, numbered 0 through 431 (625/50 standard), respectively.

There are 720 active luminance samples, numbered 0 through 719, and, respectively, 360 each C_B and C_R color-difference samples, numbered 0 through 359, in both standards.

The horizontal blanking duration is 138 clock intervals, numbered 720 through 857 (525/60 standard) and 144 clock intervals, numbered 720 through 863 (625/50 standard) in the two scanning standards.

Figures 3.46 and 3.47 show the relationship between the video samples and horizontal sync in the two standards.

Figure 3.48 shows the relationship between the 4:2:2 digital and 525/60 analog fields. The allocation of active lines in the digital fields is arranged so as to avoid the digital processing of half-lines. As a consequence, the digital field 1 has 262 lines and the digital field 2 has 263 lines. The digital blanking duration of both fields is equal to 9 lines.

Figure 3.45 Details of 525/60 scanning standard line and field repetitive 4:2:2 orthogonal sampling structure showing position of cosited $Y/C_B/C_R$ samples and isolated Y samples.

Figure 3.49 shows the relationship between the 4:2:2 digital and 625/50 analog fields. The allocation of active lines in the digital fields is arranged so as to avoid the digital processing of half-lines. As a consequence, the number of active lines is 288 in both fields. The digital blanking duration preceding the active part of field 1 is 24 lines and that preceding field 2 is 25 lines.

3.3.6 The time-division multiplexing of data

Depending on the application, the luminance and color-difference samples can be treated (distributed) separately or time-division-multiplexed. The combined (multiplexed) number of luminance and chrominance samples per total line (words per total line) is 1716, numbered 0 through 1715, (525/60 standard) and 1728, numbered 0 through 1727 (625/50 standard), respectively.

Figure 3.46 Relationship between video samples and horizontal sync in the 4:2:2 525/60 component digital standard.

The digital active line accommodates 720 Y samples, 360 C_B samples, and 360 C_R samples in both standards or a total number of 1440 words per active line, numbered 0 through 1439.

The resulting number of words in the digital blanking interval is 276, numbered 1440 through 1715 (525/60 standard) and 288, numbered 1440 through 1727 (625/50 standard), respectively. Figures 3.50 and 3.51 show the distribution of the video words in the active line interval and horizontal blanking interval for the two scanning standards.

Figure 3.52 shows a simplified block diagram of a 4:2:2 encoder with time-division-multiplexed 27 Mwords/s bit-parallel output. Y digital information at a data rate of 13.5 Mwords/s (sample duration of 74 ns) and C_B and C_R digital

Figure 3.47   Relationship between video samples and horizontal sync in the 4:2:2 625/50 component digital standard.

information at a data rate of 6.75 Mwords/s, respectively (sample duration of 148 ns) are fed to a digital data combiner that reads the digital data sequentially at a data rate of 27 Mwords/s (sample duration of 37 ns). In this drawing the sample numbering starts with 1 instead of 0 to show that C_B and C_R samples are cosited with odd (1,3,5...) Y samples.

The video data words are conveyed as a 27 Mwords/s multiplex in which the words are sent in the following order: C_B, Y, C_R, Y, C_B, etc., where the first three words (C_B, Y, and C_R) refer to cosited luminance and color-difference samples, and the following word Y refers to the following luminance-only (isolated) sample. The first video data word of each active line is C_B.

Figure 3.53 shows details of the 525/60 scanning standard horizontal blanking interval and the composition of the digital data multiplexing. Also

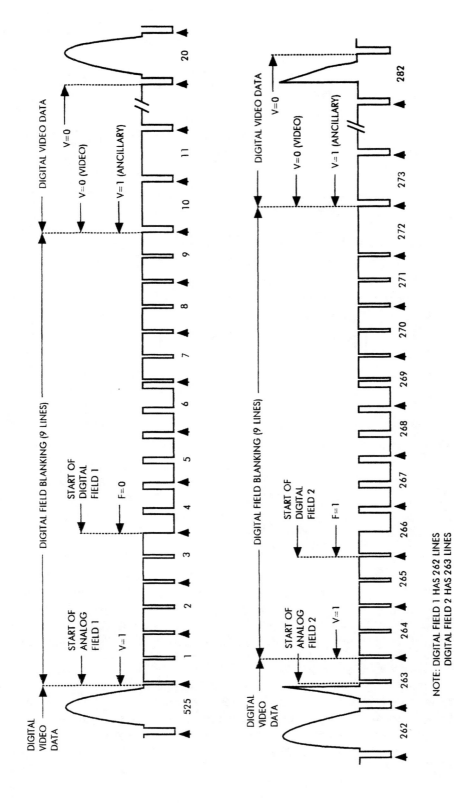

Figure 3.48 Relationship between 4:2:2 digital and 525/60 analog fields.

NOTE: DIGITAL FIELD 1 HAS 262 LINES
DIGITAL FIELD 2 HAS 263 LINES

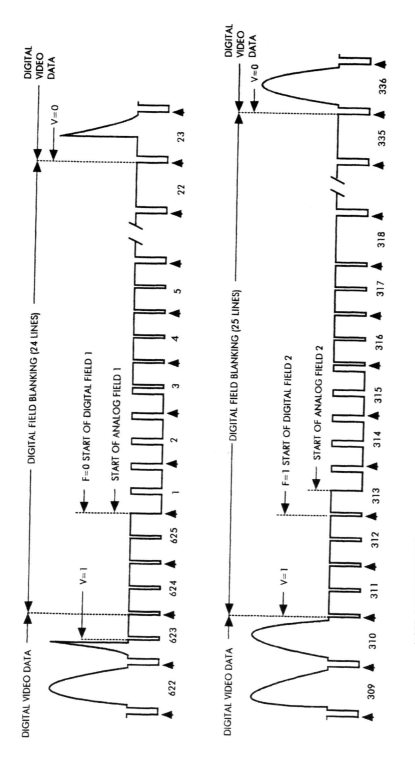

Figure 3.49 Relationship between 4:2:2 digital and 625/50 analog fields.

NOTE: BOTH DIGITAL FIELDS HAVE 288 ACTIVE LINES
DIGITAL FIELD BLANKING IS 24 LINES PRECEDING ACTIVE PART OF FIELD 1 AND 25 LINES PRECEDING THAT OF FIELD 2

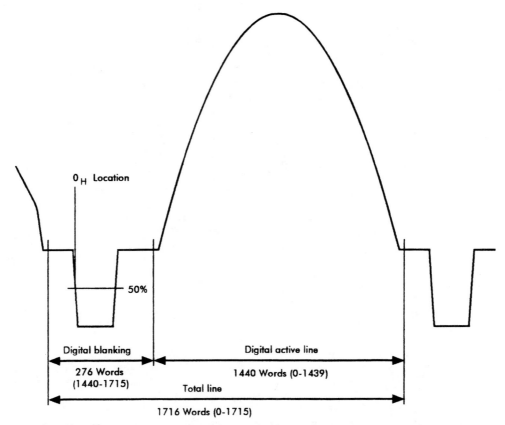

Figure 3.50 Simplified view of 525/60 scanning line showing location of video words during the active interval and the horizontal blanking interval.

shown is the position of the timing reference signal (TRS) for end of active video (EAV) and start of active video (SAV), which are discussed in Sec. 3.3.7.

Figure 3.54 shows the details of the 625/50 scanning standard horizontal blanking interval and the composition of the data multiplexing. Also shown is the position of the EAV and SAV timing reference signals.

3.3.7 Timing reference signal

The component digital standards do not provide for the sampling of the analog sync pulses. Two timing reference signals are multiplexed into the data stream on every line immediately preceding and following the digital active line data. Eight data words in the horizontal blanking interval are reserved for the transmission of timing reference signals. Their position in the data multiplex is shown in Figs. 3.53 and 3.54 for the two scanning standards. As shown in Fig. 3.53, in the 525/60 standard the words numbered 1440 through

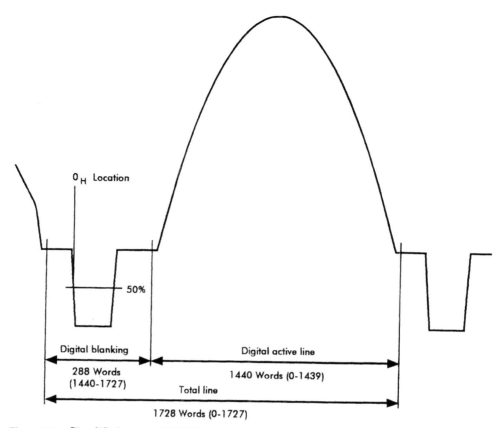

Figure 3.51 Simplified view of 625/50 scanning line showing location of video words during the active interval and the horizontal blanking interval.

1443 are reserved for the transmission of the EAV timing reference signal. Words 1712 through 1715 are reserved for the transmission of the SAV timing reference signal. As shown in Fig. 3.54, in the 625/60 standard the words numbered 1440 through 1443 are reserved for the transmission of the EAV timing reference signal. Words 1724 through 1727 are reserved for the transmission of the SAV timing reference signal. The EAV and SAV signals retain the same format during the field blanking interval.

Each timing reference signal consists of a four-word sequence. The sequence of four words can be represented, using a 10-bit hexadecimal notation, in the following manner:

$$3FF \quad 000 \quad 000 \quad XYZ$$

The first three words are a fixed preamble. The 3FF and 000 hexadecimal values are reserved for timing identification and they unambiguously identify

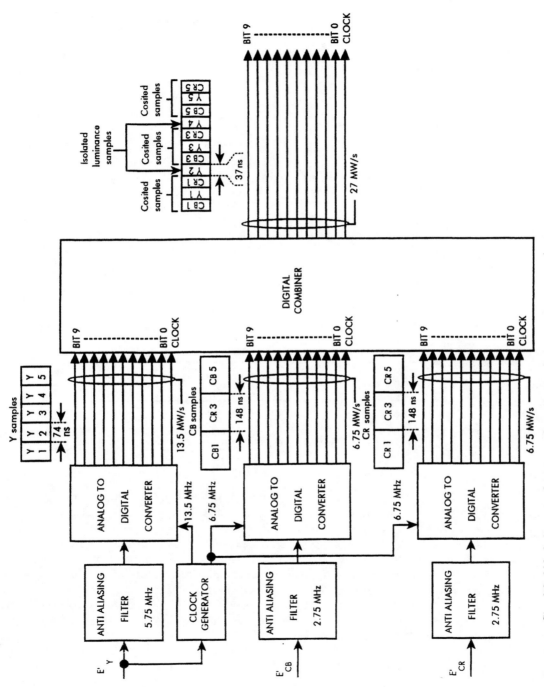

Figure 3.52 Simplified block diagram of 4:2:2 encoder with time division multiplexed 27 MW/s bit-parallel output.

Figure 3.53 Details of the 525/60 scanning standard horizontal blanking interval showing the composition of the 4:2:2 digital data multiplex and the position of the timing reference signals, EAV and SAV.

193

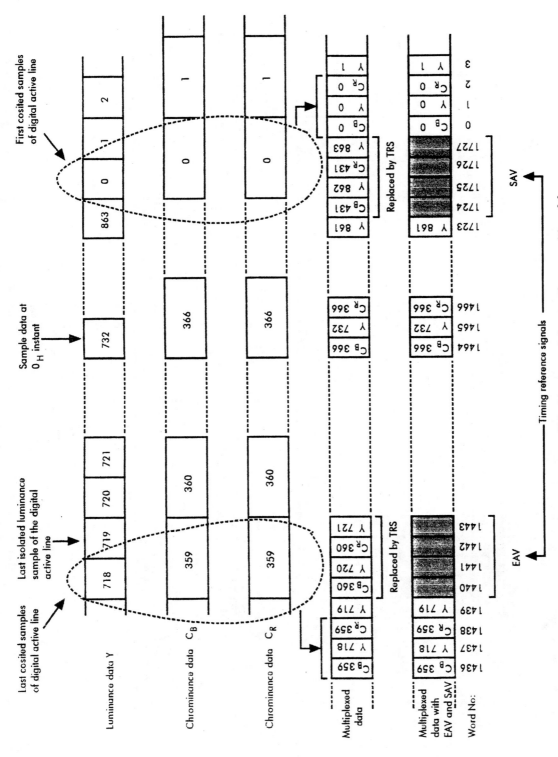

Figure 3.54 Details of the 625/50 scanning standard horizontal blanking interval showing the composition of the 4:2:2 digital data multiplex and the position of the timing reference signals, EAV and SAV.

194

the start of the SAV and EAV sync information. XYZ represents a variable word. It contains information defining:

- Field identification
- State of vertical blanking
- State of horizontal blanking

Figures 3.55 (525/60 scanning standard) and 3.56 (625/50 scanning standard) are spatial representations of the timing reference signals during a television frame.

Tables 3.10 (525/60) and 3.11 (625/50) list the binary values of the four words making up the EAV and SAV TRS sequences. The binary values of the first word of the sequence are the equivalent of the hexadecimal value 3FF. The binary values of the second and third word, respectively, are the equivalent of the hexadecimal value 000.

The fourth word has a fixed binary value of 0 for bits 0 and 1, and 1 for bit 9. Bits 6, 7, and 8 assume variable F, V, and H values conveying three types of information as follows:

- F: Field identification

$$F = 0 \qquad \text{during field 1}$$

$$F = 1 \qquad \text{during field 2}$$

- V: Vertical blanking identification

$$V = 0 \qquad \text{during active video}$$

$$V = 1 \qquad \text{during vertical blanking}$$

- H: Horizontal blanking identification

$$H = 0 \qquad \text{for SAV}$$

$$H = 1 \qquad \text{for EAV}$$

Bits 2, 3, 4, and 5 of the fourth word assume variable P0, P1, P2, and P3 values that depend on the status of the F, V, and H bits. They provide a limited error correction (single-bit errors) and detection (2-bit errors) of the F, V, and H bits. Tables 3.12 (525/60 scanning standard) and 3.13 (625/50 scanning standard) list the binary values and the related hexadecimal (XYZ) values reflecting the binary values of F, V, and H and the dependent binary values of P0, P1, P2, and P3 for specific lines and instants.

The interval starting at EAV and ending with SAV is the digital horizontal blanking period, as shown in Figs. 3.53 and 3.54. Small blocks of data, less than 268 words (525/60 scanning standard) or 280 words (625/50 scanning

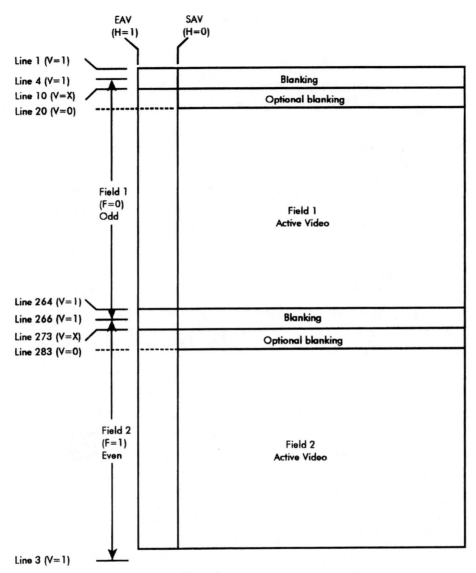

Figure 3.55 525/60 scanning standard digital timing reference signal locations.

standard) in total length, can be transmitted within the horizontal blanking period on every line.

During the vertical blanking duration, large blocks of data, up to 1440 words, including preambles, may be transmitted within the interval starting with the end of SAV and ending with the start of EAV on the lines of the vertical blanking interval. Only 8-bit words can be used in the vertical blanking interval. Certain restrictions on the lines that can be used apply in the two

Figure 3.56 625/60 scanning standard digital timing reference signal locations.

scanning standards. In the 525/60 scanning standard, ancillary data can be inserted only in the active portion of lines 1 through 19 and 264 through 282, and only if active video is not transmitted. Video data may optionally be present on lines 10 through 19 and 273 through 282, precluding the use of these lines for ancillary data transmission. In the 625/50 scanning standard, lines

TABLE 3.10   4:2:2 Timing Reference Signals for 525/60 Scanning Standard

Bit	Word 1440 and 1712	Word 1441 and 1713	Word 1442 and 1714	Word 1443 and 1715
9	1	0	0	1
8	1	0	0	F
7	1	0	0	V
6	1	0	0	H
5	1	0	0	P3
4	1	0	0	P2
3	1	0	0	P1
2	1	0	0	P0
1	1	0	0	0
0	1	0	0	0

TABLE 3.11   4:2:2 Timing Reference Signals for 625/50 Scanning Standard

Bit	Word 1440 and 1724	Word 1441 and 1725	Word 1442 and 1726	Word 1443 and 1727
9	1	0	0	1
8	1	0	0	F
7	1	0	0	V
6	1	0	0	H
5	1	0	0	P3
4	1	0	0	P2
3	1	0	0	P1
2	1	0	0	P0
1	1	0	0	0
0	1	0	0	0

20 and 333 are reserved for equipment self-checking purposes.

The words during the horizontal and vertical blanking period not used to transmit ancillary data must have the following values:

- The words corresponding to Y samples must have the hexadecimal value 040.

- The words corresponding to C_B and C_R samples must have the hexadecimal value 200.

3.3.8 Ancillary data

There are two categories of ancillary data, namely, horizontal and vertical ancillary data. Unlike the $4f_{SC}$ composite digital standard, which can carry ancillary data only in the bit-serial data stream, ancillary data may be inserted in any portion of the 4:2:2 bit-parallel digital data stream not occupied by timing reference signals or video data. The time-division multiplexing of ancillary data with the bit-serial digital data stream of the $4f_{SC}$ and 4:2:2 digital formats will be discussed in Chap. 7.

TABLE 3.12 Ten-Bit Hexadecimal (XYZ) and Binary Values of 525/60 Scanning Standard 4:2:2 TRS (SAV and EAV) Fourth Word for Specific Lines, Instants, and Related Protection Bits (P0, P1, P2, P3)

Line	Instant	Hex	9	F	V	H	P3	P2	P1	P0	1	0
20–263	Field 1 Active video SAV	200	1	0	0	0	0	0	0	0	0	0
20–263	Field 1 Active video EAV	274	1	0	0	1	1	1	0	1	0	0
4–19 264–265	Field 1 Field blanking SAV	2AC	1	0	1	0	1	0	1	1	0	0
4–19 264–265	Field 1 Field blanking EAV	2D8	1	0	1	1	0	1	1	0	0	0
283–525	Field 2 Active video SAV	31C	1	1	0	0	0	1	1	1	0	0
283–525	Field 2 Active video EAV	368	1	1	0	1	1	0	1	0	0	0
1–3 266–282	Field 2 Field blanking SAV	3B0	1	1	1	0	1	1	0	0	0	0
1–3 266–282	Field 2 Field blanking EAV	3C4	1	1	1	1	0	0	0	1	0	0

TABLE 3.13 Ten-Bit Hexadecimal (XYZ) and Binary Values of 625/50 Scanning Standard 4:2:2 TRS (SAV and EAV) Fourth Word for Specific Lines, Instants, and Related Protection Bits (P0, P1, P2, P3)

Lines	Instant	Hex	9	F	V	H	P3	P2	P1	P0	1	0
23–310	Field 1 Active video SAV	200	1	0	0	0	0	0	0	0	0	0
23–310	Field 1 Active video EAV	274	1	0	0	1	1	1	0	1	0	0
1–22 311–312	Field 1 Field blanking SAV	2AC	1	0	1	0	1	0	1	1	0	0
1–22 311–312	Field 1 Field blanking EAV	2D8	1	0	1	1	0	1	1	0	0	0
336–623	Field 2 Active video SAV	31C	1	1	0	0	0	1	1	1	0	0
336–623	Field 2 Active video EAV	368	1	1	0	1	1	0	1	0	0	0
624–625 313–335	Field 2 Field blanking SAV	3B0	1	1	1	0	1	1	0	0	0	0
624–625 313–335	Field 2 Field blanking EAV	3C4	1	1	1	1	0	0	0	1	0	0

3.3.8.1 Horizontal ancillary data (HANC). Ten-bit HANC data are permitted in all horizontal blanking intervals. Each block of data is preceded by a three-word ancillary data header

000   3FF   3FF

To accommodate equipment using 8 bits, all values in the reserved range must be considered equivalent to 000 and 3FF, respectively.

3.3.8.2 Vertical ancillary data (VANC). In the 525/60 scanning standard, VANC data are permitted only in the active portion of lines 1 to 13, 15 to 19, 264 to 276, and 278 to 282. Lines 14 and 277 are reserved for digital vertical interval time code (DVITC) and video index. VANC data are of 8-bit format. Each block of data is preceded by a three-word ancillary data header

000   3FF   3FF

To accommodate 10-bit and 8-bit equipment all values in the reserved range must be considered equivalent to 000 and 3FF, respectively.

3.3.9 Bit-parallel 4:2:2 digital signal distribution

The 4:2:2 digital signal bit-parallel equipment interconnection is adequate for short distances and simple, point-to-point signal distribution patterns. It is inadequate for large teleproduction centers with complex signal distribution patterns where the high cost of multicore cables and the large size of multipin connectors come into play. Generally, the parallel interface has been superseded by a bit-serial implementation, which is far more practical in large installations. Consequently, the bit-parallel distribution concept will be given a short treatment in this section.

The 4:2:2 bit-parallel digital signal is distributed using a shielded twisted 12-pair (balanced) cable of conventional design. The bits of the digital words that describe the video are transmitted in a parallel arrangement using 10 (8 for 8 bits/sample) conductor pairs. An eleventh (ninth for 8 bits/sample) carries a clock at 27 MHz. Figure 3.35 shows the diagram of a balanced interface circuit for bit-parallel digital signal distribution for each of the 11 (9 for 8 bits/sample) conductor pairs. Figure 3.36 shows the pin assignment of a standard DB25 connector used for bit-parallel digital interconnection.

3.3.9.1 Signal conventions

- The line driver and receiver must be ECL-compatible to permit the use of standard ECL parts for either or both ends. "Standard ECL" refers to the 10.000 series of ECL logic.

- With reference to Fig. 3.35, the DATA terminal of the transmitter is positive (+) with respect to the RETURN terminal for a binary 1 (HIGH, H, or ON) state.

- The data lines are designated DATA 0 through DATA 9. DATA 9 is the most-significant bit.

TABLE 3.14   4:2:2 Bit-Parallel Interface Electrical Characteristics

General	Video data, timing reference, and ancillary signals are time-multiplexed and are carried on 10 (8 for 8 bits) wire pairs in NRZ form. An eleventh (ninth) pair carries a synchronous 27-MHz clock
Transmitter characteristics	▪ Balanced output ▪ Source impedance: 110 ohms max ▪ Driver output common mode voltage: -1.3 V $\pm15\%$ with respect to ground terminal ▪ Signal amplitude: 0.8 to 2.0 V_{p-p} across 110 ohms ▪ Rise and fall times: <5 ns and differ by no more than 2 ns, as measured between the 20% and 80% amplitude points across a 110-ohm resistor connected directly across the output terminals ▪ Clock signal jitter: <3 ns of average time of rising edges computed over at least one television field ▪ Clock signal positive (+) transition nominally occurs between data transitions
Receiver characteristics	▪ Input impedance: 110 ±10 ohms ▪ Input sensitivity: ≥185 mV for proper operation ▪ Maximum input signal: 2 V_{p-p} ▪ Common mode rejection: Common mode noise with a ±0.5 V amplitude should not affect operation ▪ Accepted clock/data differential delay: <11 ns

3.3.9.2 Interface electrical characteristics. The 4:2:2 interface characteristics are summarized in Table 3.14. Figure 3.57 shows the clock to data timing relationships at the transmitter output.

3.3.10 Review of other component digital sampling formats

There are several other component digital formats based on the CCIR 601 Recommendation. Several of these formats will be shortly described in this section.

3.3.10.1 The 4/3 aspect ratio family. Figure 3.58 shows details of the orthogonal sampling structure of the 525/60 scanning standard 4:1:1 member of the family. As a consequence of the fact that the luminance frequency is four times that of each of the two color-difference signals, there are four times as many luminance samples (720 samples per active line) as there are color-difference samples (180 samples per active line).

Figure 3.59 shows details of the orthogonal sampling structure of the 525/60 scanning standard 4:4:4 member of the family. This format uses either equiband luminance and color-difference signals or G, B, R signals. Because of the equal sampling frequencies of the three analog component signals, there results an equal number of samples per active line.

The 4:2:2 sampling structure has been favored over many years as a minimum requirement for good chroma keying capabilities in a teleproduction environment. Currently, high-end postproduction uses the 4:4:4 format for

TW = 18.5 ns ± 3 ns

TC = 37 ns (nominal)

TD = 18.5 ns ± 3 ns

Figure 3.57   4:2:2 clock and data relationship at transmitter output.

very complex processing, whereas most of the TV production is performed with 4:2:2 equipment.

Picture processing has improved over the years and 4:1:1 sampling can now be used resulting in "acceptable-to-good" chroma keying. In news and light postproduction applications, good picture quality can be achieved with 4:1:1 equipment if all processes are performed using digital in/out interfaces to avoid cumulative A/D and D/A multiconversion picture degradations. In addition, these applications do not require extensive chroma keying and special effects.

3.3.10.2   The 16/9 aspect ratio family.   In preparation for the adoption and implementation of high-definition television (HDTV) 16/9 aspect ratio teleproduction, signal distribution, and transmission standards, there is a trend toward producing video material in the conventional scanning standards using a 16/9 aspect ratio. Two tentative approaches have resulted.

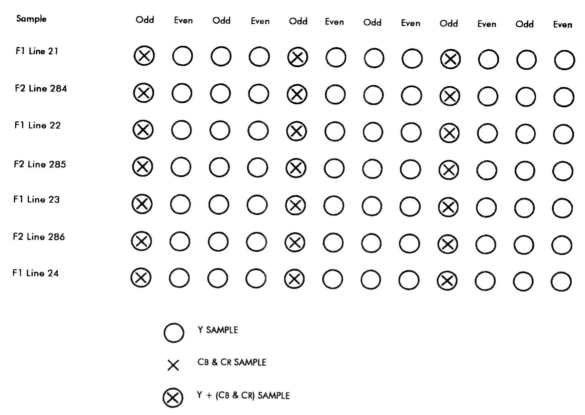

Figure 3.58 Details of 525/60 scanning standard line and field repetitive 4:1:1 orthogonal sampling structure showing position of cosited Y, C_B, C_R samples and isolated Y samples.

The 5.333…:2.665…:2.665… concept aims at obtaining a horizontal resolution equal to that of the 4:2:2 standard. This requires an increase of the 4:2:2 sampling frequencies proportional to the increase of the aspect ratio. This results in the Y signal being sampled at 18 MHz (5.333… times 3.375 MHz) resulting in 960 samples per active line. Each of the color-difference signals is sampled at 9 MHz (2.665… times 3.375 MHz) resulting in 480 samples per active line. With the exception of some routing switchers and VTRs, there is no equipment supporting this standard. The associated multiplexed bit-parallel data rate is 36 Mwords/s.

The 4:2:2, 16/9 aspect ratio approach retains the sampling rates of the 4/3 aspect ratio standard. This results in a number of samples per active line equal to that of the 4:2:2, 4/3 aspect ratio format, hence a reduced horizontal resolution. Cameras and picture monitors require redesign to support this standard. Video effects equipment designed for the standard 4/3 aspect ratio may behave erratically when handling 16/9 4:2:2 signals. Video distribution

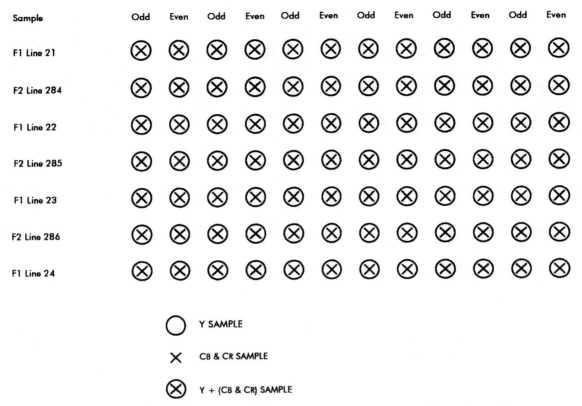

Figure 3.59 Details of 525/60 scanning standard line and field repetitive 4:4:4 orthogonal sampling structure showing position of cosited Y, C_B, C_R samples.

and recording equipment designed for the standard 4:2:2, 4/3 aspect ratio format have no difficulty handling this format. The associated multiplexed bit-parallel data rate is 27 Mwords/s.

Figure 3.60 shows a graphic comparison of the two 16/9 aspect ratio formats with the 4:2:2 4/3 aspect ratio format emphasizing the number of samples per active line.

3.3.10.3 The 4:2:0 sampling structure. Special applications, such as various MPEG bit-rate reduced digital signal distribution formats, use a special sampling structure defined as 4:2:0. The philosophy behind this sampling format is the reduction of the vertical resolution of the color-difference information (vertical subsampling) to match the reduced horizontal resolution resulting from sampling at a submultiple of the luminance sampling frequency. Figure 3.61 shows details of the 525/60 scanning standard vertically subsampled 4:2:0 structure derived from a 4:2:2 sampling structure by vertical interpolation of four line and field sequential cosited C_B/C_R samples. The interpolation

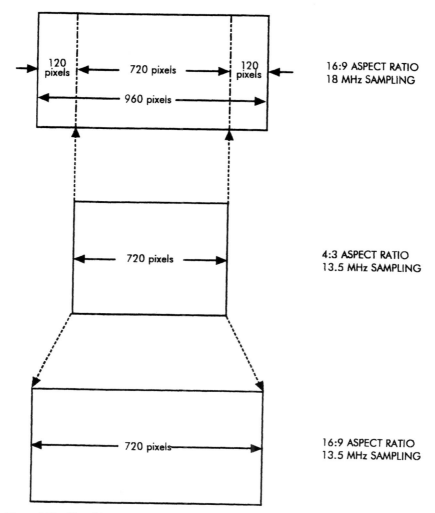

Figure 3.60 Graphic comparison between two 16/9 aspect ratio formats and the 4:2:2 4/3 format emphasizing the number of samples per active line.

creates derived color-difference samples vertically aligned with the luminance samples and located in between two adjacent interlaced scanning lines. The interpolation algorithm is given by:

$$E = \frac{1A + 3B + 3C + 1D}{8}$$

where A, B, C, D = Values of four line and field sequential luminance-cosited color-difference samples

E = Derived interpolated color-difference sample value

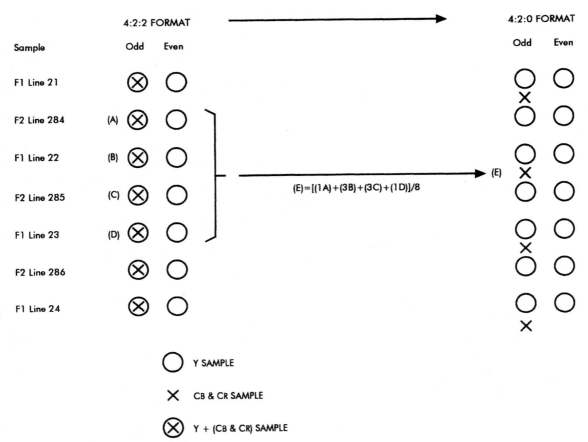

Figure 3.61 Details of 525/60 scanning standard vertically subsampled 4:2:0 structure derived from a 4:2:2 sampling structure by vertical interpolation of four line and field sequential cosited C_B and C_R samples.

3.3.11 Performance-indicative parameters and test concepts

4:2:2 digital black boxes with component analog in/out ports are characterized by impairments as classified in Chap. 2. The test methods and equipment are similar to those used for component analog equipment, and the reader is referred to Chap. 2 for general information. Figure 3.62 shows the simplified block diagram of a 4:2:2 component digital black box. Block diagrams of 4:1:1 and 4:4:4 digital black boxes are very similar to the block diagram shown in Fig. 3.62 with the exception of the band-limiting filters and sampling frequencies.

The black box is assumed to have unity in/out gain. This is verified by feeding the three D/A converters with the internally generated reference color bar signal and adjusting the gain of the three analog output channels as required. A reference analog component color bar signal is then fed to

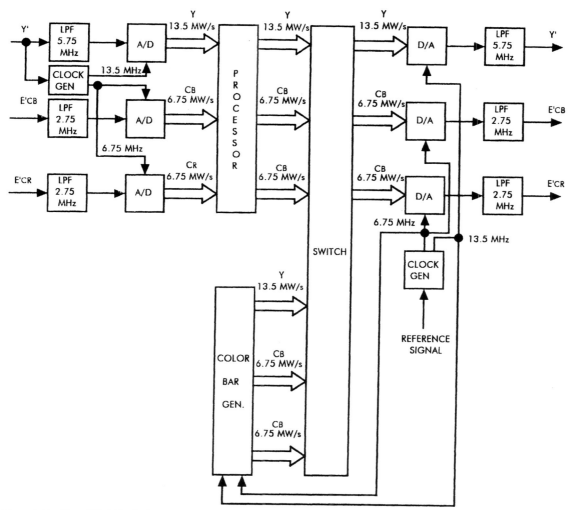

Figure 3.62 Simplified block diagram of a 4:2:2 digital black box.

the three analog inputs and the respective gains are adjusted for unity in/out gain.

The test signals for measuring linear distortions, in particular the frequency response, are tailored to the bandpass characteristics of the Y and C_B/C_R channels. The frequency response has a direct effect on the horizontal resolution of the component digital black box. There are certain misconceptions concerning the relationship between the sampling frequency, the number of samples per active line, and the horizontal resolution.

As discussed in Chap. 1, the resolution of a television picture is expressed in lines per picture height (LPH). The vertical resolution or, otherwise expressed, the number of discernible vertical picture elements (pixels), is equal to the number of active lines per frame multiplied with the Kell factor, taken as 0.7. In the 525/60 scanning standard the resulting vertical resolution is 485 × 0.7 ≈ 339 LPH (vertical pixels). In the 625/50 scanning standard this results in 575 × 0.7 ≈ 402 LPH (vertical pixels). For equal horizontal and vertical resolution, the number of pixels per picture width is equal to the vertical resolution in LPH multiplied by the picture aspect ratio (4/3). In the 525/60 scanning standard this would result in 339 × (4/3) = 452 horizontal pixels per picture width. In the 625/50 scanning standard this would result in 402 × (4/3) = 536 horizontal pixels per picture width. The horizontal resolution, and hence the number of pixels per picture width, depends on the transmission bandwidth. Transmission constraints impose a 4.2-MHz transmitted bandwidth for the 525/60 scanning standard and 5 MHz for the dominant (CCIR B,G) 625/50 scanning standard. These transmission bandwidths result in a slightly suboptimal horizontal resolution in both standards. Figure 3.63 shows the relationship between the idealized response of four typical luminance bandwidth-limiting filters, the number of samples per picture width (active line) and the horizontal resolution in LPH.

Figure 3.63 Relationship between the idealized luminance bandwidth, the resulting number of samples per active line, and the horizontal resolution.

The quoted figures are calculated using a rounded-up figure of 80 lines of horizontal resolution per megahertz of bandwidth for both scanning standards. If the antialiasing and reconstruction filters are neglected, a 4:2:2 component digital black box would be able to produce pictures with an equivalent bandwidth of 6.75 MHz (the Nyquist frequency), resulting in 720 samples (pixels) per active line or a horizontal resolution of 540 LPH. To avoid aliasing and beat phenomena, pre and post filters with a cutoff frequency of 5.75 MHz are being used. An idealized brickwall filter with a cutoff frequency of 5.75 MHz would result in 613 pixels per active line or a horizontal resolution of 460 LPH. By comparison, an idealized 5-MHz brickwall filter (CCIR B,G) would result in 533 horizontal pixels and a horizontal resolution of 400 LPH and a similar 4.2-MHz filter (NTSC) would result in 448 horizontal pixels and a horizontal resolution of 336 LPH. Practical low-pass filters result in a reduced performance in terms of pixels per active line and horizontal resolution. Table 3.15 lists the calculated values for ideal low-pass filters.

Several analog component test signals have been developed specifically for use with component digital black boxes. Of special interest is the limit ramp test signal. Figure 3.64 shows the luminance limit ramp test signal and Fig. 3.65 shows the color-difference limit ramp test signal. These signals have peak-to-peak amplitudes corresponding to 1016 digital levels expressed in hexadecimal numbers varying from 004 to 3FB. They are used to determine the quantizing range of the digital black box and verify that it is according to the specifications. They can also be used to measure the transfer nonlinearity.

A second set of component analog test signals, the shallow ramp, serves to measure the S/Q_{RMS} of the black box. Figure 3.66 shows the luminance shallow ramp test signal. It has a peak-to-peak amplitude of 70 mV and is superimposed on a pedestal with a reference level of 350 mV. The pedestal level can be varied from 0 to 700 mV and allows a detailed analysis of the quantizing range in terms of linearity and S/Q_{RMS}. The color-difference shallow ramp test signal is shown in Fig. 3.67. It has a peak-to-peak amplitude of 70 mV and is superimposed on a pedestal with a reference level of 0 mV. The pedestal level can be varied from -350 mV through 0 to $+350$ mV and allows a detailed analysis of the quantizing range in terms of linearity and S/Q_{RMS}.

The test setups for nonlinearity and S/Q_{RMS} are similar to the $4f_{SC}$ test setups shown in Figs. 3.31 and 3.33. The low-pass filter for noise measurements has a bandpass of 5.75 MHz for luminance and 2.75 MHz for color-difference signals.

TABLE 3.15   Calculated Values of Horizontal Resolution and Number of Pixels per Active Line for Ideal Low-Pass Filters

Bandwidth (MHz)	4.2	5	5.75	6.75
H resolution (LPH)	336	400	460	540
Pixels per active line	448	533	613	720

Figure 3.64 Luminance limit ramp test signal.

Figure 3.65 Color-difference limit ramp test signal.

Figure 3.66 Luminance shallow ramp test signal.

Figure 3.67 Color-difference shallow ramp test signal.

4

Elements of Acoustics

Sound is defined as an oscillation in pressure, stress, particle displacement, or particle velocity in an elastic medium. Sound is also the sensation produced through the ear by the above oscillations. Sound may be of a wanted (music or speech) and unwanted (noise) nature. Audible oscillations occur within a range of approximately 20 Hz to 20 kHz, although few people can hear frequencies at either extreme.

4.1 The Sound Pressure Level

The sound pressure is measured in dynes per square centimeter (d/cm²), microbars, or newtons per square meter (N/m²). The relationship between these measurement units is:

$$1 \text{ d/cm}^2 = 1 \text{ microbar} = 0.1 \text{ N/m}^2$$

The reference for sound pressure level (SPL) measurements is 0.0002 d/cm², which corresponds to the threshold of hearing of an average person below age 30 at a frequency of 1 kHz. The SPL of a sound is expressed in decibels above the reference sound pressure according to the formula:

$$\text{SPL(dB)} = 20 \log_{10} \left(\frac{P}{P_{REF}} \right)$$

where SPL(dB) = The number of decibels
P = The measured sound pressure, d/cm²
P_{REF} = 0.0002 d/cm²

Figure 4.1 shows some sound pressure levels encountered in practice and expressed in d/cm² as well as decibels with reference to the threshold of hearing.

In a broadcast studio environment we can identify three typical SPLs. These are:

Figure 4.1 Sound pressure levels encountered in practice.

1. 120 dB SPL: This is typical of the peak SPL of a symphonic orchestra
2. 74 dB SPL: This is the average SPL of typical voice programs and is used as a reference level by microphone manufacturers.
3. 30 dB SPL: This is typical of the ambient noise SPL.

4.2 Loudness and Loudness Level

The loudness of a sound is a subjective term. It is that attribute of an auditory sensation in terms of which sounds can be ordered on a scale extending from soft to loud. The unit of loudness is the sone. The loudness depends on the sound pressure and the frequency of the stimulus. The calculated loudness of a steady sound, in sones, is related to the loudness level, in phons, by the equation

$$n_s = 2^{(L-40)/10}$$

where n_s is the loudness in sones and L is the loudness level in phons. The sone is thus defined as the loudness of a sound having a loudness level of 40 phons. A twofold change in loudness corresponds to a change in the loudness level by 10 phons.

The loudness level of a sound, expressed in phons, is numerically equal to the median SPL, expressed in decibels relative to 0.0002 d/cm², of a 1000-Hz reference tone. The calculated loudness level of a sound, in phons, is related to the loudness, in sones, by the equation

$$L = 40 + 10 \log_2 n_s$$

where L is the loudness level in phons and n_s is the loudness in sones.

The manner in which the human ear and brain interact accounts for the fact that at frequencies other than 1000 Hz, the ear requires different SPLs for the same perceived loudness. Figure 4.2 shows the normal equal loudness contours for pure tones as per ISO standard 226. These can be viewed as

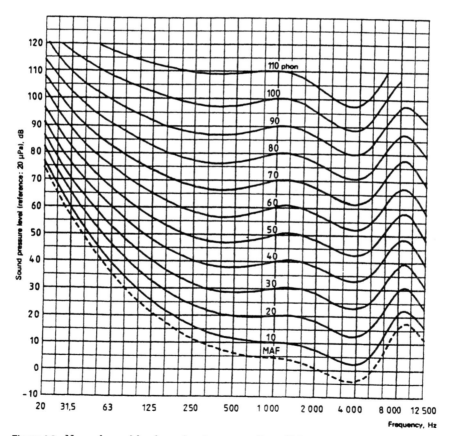

Figure 4.2 Normal equal-loudness level contours (from ISO 226: 1987).

inverted ear frequency response curves at various SPLs. A 1000-Hz tone having an SPL of 40 dB has a loudness level of 40 phons. To give the same sensation of loudness at 63 Hz, the SPL must be increased by about 20 dB. This shows that the sensitivity of the ear is considerably lower at frequencies below 1000 Hz. The equal-loudness contours have different shapes at other SPLs. As the sound level increases, the ear frequency response improves and it is reasonably flat, within ±10 dB, at an SPL of 110 dB.

4.3   The Dynamic Range of the Ear

The dynamic range of the ear is bounded at the top by the threshold of pain and at the bottom by the threshold of hearing. It is typically 120 dB and varies from individual to individual. Regular exposure to sounds of about 90 dB SPL will eventually cause hearing loss. Sounds having SPLs of 120 dB and above can cause pain and immediate and permanent loss of hearing. Sounds having SPLs below 0 dB are inaudible. Above age 30, the hearing normally deteriorates in terms of threshold of hearing and perception of the higher frequencies of the audible spectrum. The threshold of hearing depends on the level of the ambient noise, which has a masking effect. Noise masking is defined as the process by which the threshold of audibility of a wanted sound is raised by the presence of an unwanted sound, in this case noise.

4.4   The Spectral Resolution of the Ear

The human hearing mechanism is characterized by a resolving capability that can be attributed to an array of overlapping bandpass filters called *critical bands*. The existence of critical bands accounts for the sound-masking phenomenon. Sound masking is defined as the amount by which the threshold of audibility of a sound is raised by the presence of another (masking) sound. In the presence of a dominant sound of a given frequency, alternate lower-level sounds, whose frequencies are inside the same critical band, may become inaudible. This psychoacoustical characteristic is utilized in digital audio bit-rate reduction techniques.

Analog Audio Fundamentals

5.1 Electrical Signal Levels and Units of Measurement

Audio signal amplitudes are continuously variable. Given the extremely wide range of audio voltages, it is customary to express audio signal levels in decibels with reference to a specific power or voltage level. Currently, there are three commonly used units of measurement of audio levels.

5.1.1 The dBm

The dBm abbreviation is used to express the root-mean-square (RMS) power of a sine-wave signal with respect to a 1-milliwatt (mW) reference. A power of 1 mW dissipated into a 600-ohm load results in a 0.77459 V RMS voltage (rounded up to 0.775 V RMS). When dissipated into other load values, different voltages result. The power of 1 mW is defined as 0 dBm. Other audio power levels are expressed in dBm with respect to the reference 0 dBm power level according to the formula

$$N(\text{dB}) = 10 \log_{10}\left(\frac{P}{P_{\text{REF}}}\right)$$

where N (dB) = The number of decibels
P = The measured power level
P_{REF} = The reference power level of 1 mW

The formula can be extended to the measurement of voltages or currents as follows:

$$N(\text{dB}) = 20 \log_{10}\left(\frac{V}{V_{\text{REF}}}\right)$$

$$N(\text{dB}) = 20 \log_{10} \left(\frac{I}{I_{\text{REF}}} \right)$$

The above formulas assume that the voltages and currents are measured across identical impedances (e.g., 600 ohms).

5.1.2 The dBu

An alternate method of expressing audio signal levels is the dBu. The dBu concept assumes a near-zero source impedance and a near-infinite load impedance. Under these idealized open-load conditions, the source does not dissipate any measurable power into the load and the open-load source voltage is unaffected by the load. The reference signal level is 0.775 V RMS. For practical purposes, the dBu concept requires source impedances of the order of 50 ohms or less and load impedances equal to or in excess of 10 kohms.

5.1.3 The dBV

An alternate method of expressing audio signal levels is the dBV. This is an open-load voltage concept and the reference voltage is 1 V. The dBV is used by microphone manufacturers.

5.2 Typical Signal Levels and Impedances

There is a wide variety of studio-quality equipment available to the user. Normally, in terms of signal level, there are two main categories, namely, low-level devices (typically microphones) and high-level devices (everything else).

5.2.1 Microphone signal levels and impedances

Microphone sensitivity ratings, measured at 74 dB sound pressure level (SPL), are commonly expressed in open-load microvolts, or dBV. Impedances for professional-quality microphones are standardized at 150 ohms, but other values are also encountered in practice. A typical moving-coil microphone, with a source impedance of 150 ohms, generates an open-load voltage of 100 μV (-80 dBV) at 74 dB SPL. The input impedance of the microphone preamplifier bridges the microphone output, that is, has a value of 1500 ohms or higher, to avoid microphone damping and input signal-to-noise degradation due to excessive signal loss. Figure 5.1 shows a typical microphone input circuit configuration. Note that the microphone signal source is balanced with respect to ground and a balanced and shielded cable is used for connection to the preamplifier.

Since the microphone preamplifier input impedance is not infinite, the microphone will dissipate a minute power into this load. This situation is typical with moving-coil or ribbon microphones. Table 5.1 shows typical moving-coil microphone signal levels for three typical SPLs. Columns 2, 3, and 4 list the related open-load microphone signal levels in microvolts, dBV, and dBu,

Figure 5.1 Typical block diagram of microphone input circuit.

TABLE 5.1 Microphone Signal Levels

| SPL, (dB) | Open-load conditions | | | 1.5-kohm load, μV |
	μV	dBV	dBu	
120	20,000.00	−34	−31.8	18,000.00
74	100.00	−80	−77.8	90.9
34	1.00	−120	−117.8	0.909

respectively. Column 5 lists the related microphone signal level into a 1500-ohm load expressed in microvolts RMS. Evidently, for load impedances other than 1500 ohms, the values in column 5 will be different. Electrostatic microphones feature internal preamplifiers and generate higher output levels. Audio consoles are provided with variable-input attenuators to adapt to this situation. In all cases the microphone preamplifier input impedance is equal to 1500 ohms or higher.

5.2.2 Line signal levels and impedances

Audio signal levels generated by microphones are suitably amplified to line signal levels and distributed inside broadcast plants or to common carriers for land or satellite transmission.

5.2.2.1 The power-matching concept.

In North America in the 1930s, the audio distribution line impedance inside broadcast plants, as well as that of the common carriers and audio transmitters, was standardized at 600 ohms. The interconnection of various types of equipment inside broadcast plants as well as connections to common-carrier equipment followed the power-matching concept. Figure 5.2 shows a typical configuration.

This concept assumes that

- The audio signal source (i.e., common-carrier line, audio tape recorder, audio distribution amplifier, or audio mixing console) output impedance is 600 ohms, balanced and floating.

- The loading input impedance of the receiver (i.e., common-carrier line,

Figure 5.2 Typical block diagram of power matching circuit.

audio tape recorder, audio distribution amplifier, or audio mixing console) is 600 ohms, balanced and floating.

The audio signal level is expressed in dBm. A standard operating level (SOL) of +8 dBm into 600 ohms, or 1.95 V RMS, was chosen in North America. Some authorities, including sound recording studios, opted for a +4 dBm SOL inside the plant. The SOL represents the steady-state maximum level or the peak program level as measured with a standardized audio signal level meter (volume unit or VU meter). Figure 5.3 shows the relationship between the dBm level, the voltage across 600 ohms, and the power levels of audio signals. Many older plants operate in this manner.

Figure 5.3 Typical block diagram of voltage matching circuit.

POWER mW	dBm	VOLTS @600 OHMS
1000	30	24.5
800	29	21.8
600	28	19.5
500	27	17.3
400	26	15.5
300	25	13.8
	24	12.3
200	23	10.9
150	22	9.75
	21	8.69
100	20	7.75
80	19	6.90
60	18	6.15
50	17	5.48
40	16	4.89
30	15	4.36
	14	3.88
20	13	3.46
15	12	3.08
	11	2.75
10	10	2.45
8	9	2.18
6	8	1.95
5	7	1.73
4	6	1.55
3	5	1.38
	4	1.23
2	3	1.09
1.5	2	0.975
	1	0.869
1	0	0.775

Figure 5.4 Relationship between dBm and voltage at 600 ohms to power in mW.

5.2.2.2 The voltage-matching concept. Modern studios operate in a voltage-matching configuration. Figure 5.4 shows a typical setup. The signal level is expressed in dBu and the SOL in North America is +4 or +8 dBu. This considerably reduces the power requirements of the signal source, since it is required to dissipate a minute amount of power across the bridging load. An added advantage is the improved frequency and transient response of the system, since the capacitive loading of the shielded balanced audio cable has a lesser effect across a source impedance of 50 ohms than it has across a source impedance of 600 ohms. The interface with common carriers retains the power-matching philosophy. The signal-level monitoring is carried out with a peak program meter (PPM).

5.3   Signal-Level Monitoring

Audio signal levels are closely monitored in a studio environment to ensure that sudden or sustained peaks do not cause overloading and distortion in various elements of a system. There are two generic types of audio-signal-level monitoring meters that have found application in a studio environment. These are the volume unit (VU) meter and the PPM.

5.3.1   The VU meter

The VU meter was developed primarily for the control and monitoring of audio program. The specifications of the VU meter reflect the philosophy of the 1930s. Essentially the VU meter is a moving-coil RMS-type audio-signal-level measuring instrument. It is fitted with two scales:

- A VU scale marked 0 (reference deflection) at about 71% maximum scale reading, extending to +3 (maximum) and −20 (minimum).
- A percentage scale, with 100% corresponding to 0 VU reading.

The VU meter sensitivity is adjustable such that the 0 VU reference level can be made to correspond to the SOL (+4 or +8 dBm) under steady-state sinusoidal voltage conditions. Its dynamic characteristics are such that if a sinusoidal voltage of a frequency between 35 Hz and 10 kHz, and of such amplitude as to give reference pointer deflection under steady-state conditions, is suddenly applied, the pointer will take 0.3 s to reach reference deflection. This characteristic was chosen in order to approximate the assumed response of the human ear. The 0.3-s risetime characteristic of the VU meter introduces a masking effect. Essentially, the instrument is unable to give accurate audio signal level indications under complex wave, fast risetime, and input signal conditions. The instantaneous speech or music signal level may, in reality, be 10 VU and more above the readings of the VU meter. The use of the VU meter has resulted in the need for a relatively large headroom in recording and distribution system elements to avoid the clipping of sudden bursts in audio signal levels. The use of the VU meter in broadcasting and recording studios has been largely abandoned.

5.3.2   The PPM

The PPM is a peak-reading instrument capable of adequately displaying audio signal transients. Some current designs feature a 10-ms attack time (risetime) and a 2.85-s fallback time. This characteristic amounts to a "sample-and-hold" approach to audio-signal-level monitoring. It allows the user to monitor audio signal levels under steady-state and program conditions accurately and reduces the need for large amounts of headroom in amplifiers. The meter input impedance is bridging, that is, it is greater than 6000 ohms. Neither the scale nor the display are universally standardized.

5.4 Performance-Indicative Parameters and Measurement Concepts

The performance of an audio system element (e.g., audio mixing console, audio tape recorder) or a complete system (consisting of a number of individual system elements connected in a typical operational configuration) is expressed in terms of measured values of performance-indicative parameters. As shown in Fig. 5.5, audio performance-indicative parameters are grouped in three major categories, linear distortions, nonlinear distortions, and noise.

When the equipment is first installed, performance tests are carried out initially, to determine whether the equipment meets manufacturing and operational specifications, routinely, to ensure operational readiness, periodically as preventive maintenance, and after equipment repair. The performance measurements are carried out using a variable-frequency, high-quality, low-distortion audio signal generator and a precision audio analyzer. A phase meter may be required for phase versus frequency response measurements. An oscilloscope is used to display the signal to be measured. The test equipment has to be accurate, reliable, in good condition, and compatible with the equipment under test.

5.4.1 Linear distortions

Electrical audio signal waveform modifications that are independent of the signal amplitude are called *linear distortions*. It is assumed here that the amplitude of the electrical signal does not exceed the clipping level of the equipment. There are two major types of linear distortions encountered in practice: nonuniform frequency response and nonuniform phase response.

Figure 5.5 Classification of audio performance indicative parameters.

5.4.1.1 Amplitude vs. frequency response test.

Amplitude vs. frequency response is defined as the peak-to-peak variation, over a specified frequency range, of the measured amplitude of an audio signal, expressed in dB with reference to a the signal level at a specified frequency. The reference frequency is 1 kHz. Audio tape recorder reference frequency is usually 300 Hz.

The input port of the object under test is fed a signal of 1 kHz at the SOL. This would be +8 or +4 dBu for high-level inputs and, typically, −60 or −70 dBu for microphone inputs. The gain (gains) are adjusted to obtain SOL (+8 or +4 dBu) at the output. The audio analyzer is calibrated to read 0 dB at the reference frequency. The frequency of the input signal is varied in discrete steps, or continuously, and readings in dB, with reference to 0 dB, are taken at specific frequencies. Analog audio tape recorder frequency response tests are carried out at a level of 10 dB below the maximum output level (MOL). The measurement bandwidth is 20 Hz to 20 kHz. The amplitude vs. frequency response test is a routine test.

5.4.1.2 Phase vs. frequency response test.

Phase vs. frequency response is defined as the variable phase shift occurring in a system at several frequencies within a given band.

The input of the object under test is fed a signal of variable frequency. A calibrated phase meter is connected to the output of the object under test. The signal present at the input of the object under test is fed to the phase reference input of the phase meter. A plot of phase vs. frequency is carried out over the bandwidth of interest. The phase vs. frequency response test is not a routine test.

5.4.2 Nonlinear distortions

Nonlinear distortions of an electrical signal are caused by deviations from a linear relationship between the input signal and the resulting output signal of a given system. There are two types of nonlinear distortions encountered in practice: harmonic distortion and intermodulation distortion.

5.4.2.1 Harmonic distortion.

Harmonic distortion occurs when a system whose input is fed with a pure sine-wave signal of frequency f produces at its output a signal of frequency f as well as a set of signals whose frequencies ($2f$, $3f$,..., nf) are harmonically related to the input frequency.

The distortion factor of a signal is the ratio of the total RMS voltage of all harmonics to the total RMS voltage. The performance of audio amplifying devices is expressed in terms of percentage of total harmonic distortion (THD) at a specified output level. For studio-quality equipment, the output level at which THD is measured is 10 dB above SOL (+18 or +14 dBu). The percentage of THD is the distortion factor multiplied by 100. The mathematical expression is:

$$\text{THD} = \frac{\sqrt{E_2^2 + E_3^2 + \cdots + E_n^2}}{\sqrt{E_1^2 + E_2^2 + E_3^2 + \cdots + E_n^2}} \times 100$$

where THD = The percentage of total harmonic distortion
E_1 = Amplitude of fundamental voltage
E_2 = Amplitude of second harmonic voltage
E_n = Amplitude of nth harmonic voltage

To measure the THD of an amplifier, the audio analyzer removes the fundamental (first harmonic) component of the distorted signal present at the output of the object under test and all the remaining energy, including noise and harmonics, is measured. The measurement bandwidth is usually limited to 20 kHz. Because of the contribution of noise to the measurement results, the method is better described as total harmonic distortion and noise (THD + N). The tests are carried out at several frequencies, such as 50 Hz, 100 Hz, 1 kHz, 5 kHz, 7.5 kHz, and 10 kHz. THD measurements at frequencies above 10 kHz are irrelevant since the harmonics generated by the object under test are above the audio bandwidth. The THD measurement is a routine maintenance test.

The performance of analog audio tape recorders is expressed in terms of percentage of third harmonic distortion of a pure sine wave having a frequency of 333 Hz. To measure the third harmonic distortion of a tape recorder, the audio analyzer extracts the third harmonic with a bandpass filter having a center frequency of 999 Hz (three times the fundamental frequency of 333 Hz). The amplitude of the filtered 999-Hz signal is measured and the Third Harmonic Distortion (THD) is computed as:

$$\text{THD} = \left(\frac{E_3}{E_1}\right) \times 100$$

where TDH = Third harmonic distortion
E_1 = Amplitude of fundamental voltage
E_3 = Amplitude of third harmonic voltage

The measurement is carried out at SOL and MOL. Measurements at higher frequencies are irrelevant because of the effects of the recording preemphasis. The measurement of THD is a routine maintenance test.

5.4.2.2 Intermodulation distortion. Intermodulation distortion (IMD) occurs when a system whose input is fed with two signals of frequencies f_1 and f_2 generates at its output, in addition to the signals at the input frequencies, signals having frequencies equal to sums and differences of integral multiples of the input frequencies.

The SMPTE IMD test specifies the use of a test signal consisting of two separate frequencies ($f_1 = 60$ Hz and $f_2 = 7$ kHz) with a respective amplitude ratio of 4:1. The IMD causes the 7-kHz "carrier" to be modulated by the 60-Hz signal. This results in the generation sidebands above and below the 7-kHz carrier with components at 60 Hz and its harmonics. To measure the IMD, the audio analyzer amplitude demodulates the 7-kHz carrier to recover the 60-Hz modulating signal and its harmonics. The IMD is computed as

$$\text{IMD} = \left(\frac{\text{Demodulated signal}}{E_{f_2}} \right) \times 100$$

where IMD = Intermodulation distortion
E_{f_2} = Amplitude of 7-kHz component

The IMD measurement is a routine maintenance test.

5.4.3 Noise

Audio signals are affected by noise. Noise is best defined as an unwanted disturbance superimposed on a useful signal. The noise level is usually expressed in dB relative to a reference value and is commonly referred to as signal to noise ratio (SNR). In studio equipment the reference level for SNR measurements is MOL or 10 dB above SOL. In analog audio tape recorders the reference level is the tape saturation at which a third harmonic distortion of between 1 and 3% (depending on the class of equipment) occurs. The typical MOL of an audio tape recorder is 6 dB above SOL.

5.4.3.1 Random noise. The main source of random noise is thermal noise caused by the thermal agitation of the electrons. Given R, the resistive component of an impedance Z, the mean square value of the thermal noise voltage is given by

$$E_n{}^2 = 4kTBR$$

where E_n = The RMS noise voltage
k = Boltzmann's constant (1.38×10^{-23} joules/kelvin)
T = The absolute temperature in kelvin
B = The bandwidth, Hz

T is usually assigned a value such that $1.38T = 400$, corresponding to about $17°$ C. Thus the formula for thermal noise can be written as

$$E_n{}^2 = 1.6 \times 10^{-20} BR$$

The SNR at the output of a system depends on the noise generated by the resistive component of the signal source, for example, the microphone, and the noise generated by the earliest amplifier stage in the chain.

TABLE 5.2   Open-Load SNR at Dynamic Microphone Output

SPL, dB	Microphone output voltage, μV	SNR, dB
120	20,000.00	99.21
74	100.00	53.19
61	22.40	40.19
34	1.00	13.19

Assuming $B = 20$ kHz and a microphone with a resistive component $R = 150$ ohms, $E_n = 0.219$ μV. This is the theoretical thermal noise of a microphone input circuit. Table 5.2 lists the theoretical SNR at the microphone preamplifier input at four SPL levels under open-load conditions.

The microphone preamplifier contributes its own random noise, which considerably reduces the SNR of the system. The situation can be visualized as having an ideal noiseless amplifier whose input is fed by a noise generator. This fictitious noise is called the *equivalent input noise* of the amplifier. The difference between the equivalent input noise level and the calculated theoretical thermal noise level of the audio signal source is called the *noise factor* of the amplifier.

The measurement of SNR is a rather involved procedure and the accuracy of the results depends on a strict adherence to a set of rules. The following routine test procedure is suitable for the SNR measurement of an audio mixer:

- Feed a 1-kHz audio signal at the rated input level (e.g., -70 dBu) at the microphone input.

- Disable all inputs except the one in the measurement path.

- Disable all compressor and equalizers in the measurement path.

- Connect the audio analyzer at mixer output.

- Adjust input sensitivity, channel gain, and master gain to obtain SOL (+4 or +8 dBu) as measured by the audio analyzer.

- Remove the input signal source and substitute it with a low-noise 150-ohm resistor.

- Measure the noise at the output with the audio analyzer in dBu in a 20-kHz bandwidth.

- The SNR is given by the difference, in dB, between MOL, in dBu, and the measured noise, in dBu.

The measurement of SNR is a routine maintenance test.

5.4.3.2 Periodic noise.
This type of noise is generated outside of the equipment and coupled in some manner into it. Unlike random noise, periodic noise can be completely eliminated by good engineering practice. The main

type of periodic noise, commonly called *hum,* is 60 Hz and its harmonics. The power line is its source.

The measurement of signal to periodic noise ratio is similar to the measurement of signal to random noise ratio except that a spectrum analyzer or oscilloscope may be added to help identify the frequency of the periodic noise. The measurement of signal to periodic noise is a routine maintenance test.

5.4.3.3 Crosstalk. Crosstalk is defined as the injection of an unwanted signal into a circuit from a neighboring circuit via a mutual impedance. An example is the crosstalk that can occur between signal sources in an audio mixer.

The measurement of crosstalk is quite involved. It consists in feeding a signal to the unwanted (crosstalking) input and measuring its effect at the wanted path output, whose input is loaded with its characteristic source impedance. The two paths have to be adjusted for normal operating conditions. The audio analyzer is connected to the wanted path output and the input of the unwanted (crosstalking) path is fed with a signal whose frequency is varied in discrete steps or continuously in the bandwidth of interest. The signal-to-crosstalk ratio is expressed in dB with reference to MOL. A typical application is the measurement of left channel to right channel crosstalk in a stereophonic system. The measurement of crosstalk is not a routine maintenance test.

5.5 The Dynamic Range

The dynamic range is defined as the difference, in dB, between the overload level and the minimum acceptable signal level in a system.

5.5.1 The overload level and the headroom concept

The overload level, also called maximum operating level or maximum output level (MOL) is usually defined in terms of acceptable THD. Although there is no general agreement on the maximum accepted value of THD, the figure of 1% is generally quoted for audio consoles and distribution amplifiers. Analog audio equipment is adjusted such that the MOL is higher than the SOL. The difference between the MOL and the SOL, expressed in dB, is called *headroom.* The MOL of an audio mixing console or audio distribution amplifier is usually specified as 10 dB or more above the SOL. Higher values of headroom may be needed when VU meters are used for output-signal-level monitoring.

Audio mixing consoles are also specified in terms of maximum input level (MIL). The MIL of an audio mixing console is the microphone input level at which the THD due to the microphone preamplifier is 1%. The input headroom of an audio mixing console is the difference between the MIL and the rated input level (e.g., −60 dBu). Input headroom specifications for audio mixing consoles are between 20 and 35 dB.

The MOL of analog audio tape recorders is the output level at which the third harmonic distortion is between 1 and 3%, depending on the class of the equipment. The headroom is typically 6 dB.

5.5.2 The minimum acceptable signal level

The minimum acceptable level in a system is closely related with the acceptable SNR at low signal levels. This is clearly an operational decision. Ideally, the SNR at the lowest acceptable signal level should not be lower than 40 dB. Table 5.2 lists the theoretical SNR at a standard dynamic microphone output for several SPL levels under open-load conditions. This table shows that 40 dB SNR can be achieved at an SPL of 61 dB, assuming that the ambient acoustical noise in the studio is 0 dB SPL. A higher level of ambient acoustical noise will raise the minimum acceptable signal level.

5.5.3 Limits of dynamic range in a studio environment

Figure 5.6 shows how four basic elements, namely the microphone, the recording studio, the audio mixer, and the analog audio tape recorder, each contribute to a reduced dynamic range in a studio. There are a number of assumptions made as follows:

- The microphone source resistance is 150 ohms.
- The microphone sensitivity is −80 dBV @ 74 dB SPL.
- The recording studio ambient noise is 30 dB SPL.
- The peak SPL is 120 dB.
- The studio operates in a voltage-matching mode.
- The SOL is +8 dBu.
- The audio mixer MOL @ 1% THD is +18 dBu.
- The audio mixer SNR is 80 dB with respect to MOL.
- The audio mixer is lined up such that an SPL of 120 dB produces MOL at the output.
- All compressors, if any, are disabled.
- The audio tape recorder is optimized and the MOL @ 3% THD is +14 dBu.
- The audio tape recorder SNR is 60 dB with respect to MOL.
- The lowest acceptable SNR at low signal levels is 40 dB.

The assumptions made above reflect the expected single-pass performance of typical equipment available on the market. These are ideal operating conditions. In actual practice the results may be different and, possibly, worse.

Figure 5.6 Factors contributing to the dynamic range in a studio environment.

5.5.3.1 The dynamic range of a microphone.

Given SPL peaks of 120 dB, an ambient acoustical studio noise level of 0 SPL, and a standard dynamic microphone as defined above, the theoretical dynamic range at the microphone output is 59 dB. It is limited mostly by the thermal noise of its resistive component.

5.5.3.2 The dynamic range of a studio.

Figure 4.1 shows that the ambient acoustical noise in a broadcast studio is of the order of 30 dB SPL. This limits the dynamic range of the studio to 50 dB. Unlike the random noise generated by the resistive component of the microphone, the studio acoustical ambient noise has mostly low-frequency spectral noise components.

5.5.3.3 The dynamic range of an audio mixer. Top-of-the-line analog audio mixers have an SNR on the order of 80 dB with reference to the MOL. If the mixer is adjusted such that SPL peaks of 120 dB generate output-level peaks (MOL) of +18 dBu (10 dB above an SOL of +8 dBu) at 1% THD with an SNR of 80 dB, the mixer background noise level is −62 dBu as measured at its output. This limits the dynamic range of the audio mixer to 40 dB.

5.5.3.4 Analog audio tape recorders. The MOL of a typical analog audio tape recorder is usually 6 dB above the SOL at low and medium (<1000 Hz) frequencies. Inevitably, especially when the signal-level monitoring at the output of the audio mixing console is carried out using a VU meter, the audio tape recorder may be fed signals with peaks exceeding 10 dB above SOL. The situation is worsened by the use of high-frequency preemphasis in the recording amplifier to compensate for tape losses. This can lead to tape overload in the presence of high-level, high-frequency spectral components in the audio signal. The amount of high-frequency preemphasis depends on the class of equipment and the tape speed. The SNR with reference to the MOL is typically between 40 and 60 dB, depending on the class of the equipment and the quality of the tape used. The resulting dynamic range of 20 dB falls short of the performance of the audio mixer.

5.5.4 Operational approaches

In order to avoid mixer input and output overloading and audio tape recorder overloading at high SPL levels and a reduced SNR at low SPL levels, the operator is "riding the gain," that is, manually adjusting input signal levels, channel gains, and the master fader to achieve optimum operating conditions. Some audio mixing consoles with a large number of microphone inputs feature individual input channel signal compressors to ease the task of the operator.

There are very few and conflicting choices to correct the dynamic range problem of the audio tape recorder. Among them are

- Using a Dolby-type dynamic noise reducer that improves the SNR at low signal levels

- Accepting a reduced SNR at low signal levels

- Accepting a certain amount of clipping of short high level audio signal peaks

- Using premium-grade audio tape

The ultimate improvement is obtained by replacing the analog audio tape recorder with a digital audio tape recorder.

5.5.5 Transmission constraints

The ultimate compression of the dynamic range occurs at the television audio transmitter. FM transmission standards for television audio in North America specify the use of preemphasis with a time constant of 75 μs, resulting in a 14-dB boost at 10 kHz. A complementary deemphasis at the receiver results in an improvement of the SNR and a linear frequency response. The preemphasis can potentially lead to overmodulation of the transmitter in the presence of high-level, high-frequency spectral components in the audio signal. Various types of limiter/compressor combinations are used in an effort to avoid over-modulation of the transmitter and achieve an acceptable SNR.

5.6 Performance Targets

The signal impairments of individual elements making up an audio production plant are additive. Consequently, the overall performance of the complete plant or a subsystem depends on the individual performance and number of individual components assembled in a typical operational configuration. This sets a limit to the number of stages a signal can pass through before it becomes too impaired to be acceptable. The manner in which the impairments are added depends on the specific impairment. The CCIR has developed an empirical method to permit the calculation of the impairments of a chain made up of individual component elements as applied to a hypothetical 2500-km path. This method has been applied successfully by the authors in calculating the performance targets of a large teleproduction plant.

Given a typical audio signal distortion D, the total distortion D_t resulting from the contribution of n subsystems exhibiting distortions of the same kind $(D_1, D_2, ..., D_n)$ is given by

$$D_t = (D_1^{\ h} + D_2^{\ h} + \cdots D_n^{\ h})^{1/h}$$

where $h = 2$ for noise and frequency response
$h = 1.5$ for harmonic distortion

Table 5.3 lists the single-pass performance figures of typical analog audio studio equipment.

Figure 5.7 shows the block diagram of a complex audio signal distribution pattern in a large teleproduction center.

TABLE 5.3   Single-Pass Performance Figures of Typical Analog Audio Studio Equipment

	Frequency response	THD	SNR
Distribution amplifier	±0.1 dB (20 Hz–20 kHz)	0.1%*	90 dB
Routing switcher	±0.1 dB (20 Hz–20 kHz)	0.2%*	80 dB
Audio mixer	±0.1 dB (20 Hz–20 kHz)	0.3%*	80 dB
VTR audio track	±0.5 dB (30 Hz–15 kHz)	3%**	60 dB

  * Total harmonic distortion.
  **Third harmonic distortion.

Figure 5.7 Block diagram of a complex audio signal distribution pattern in a large teleproduction center.

6

Digital Audio Fundamentals

Audio has undergone major developments. The Audio Engineering Society (AES) together with the European Broadcasting Union (EBU) has developed the AES/EBU digital audio standard. This has led to the development of audio tape recorders and studio equipment where the signal processing is carried out in the digital domain, which restricts the analog-type distortions to the analog-to-digital (A/D) and digital-to-analog (D/A) tandem process, thereby improving the signal transparency.

6.1 General Concepts of Digital Audio

6.1.1 Introduction

Currently, the typical broadcast environment is predominantly based on analog technology. However, since the beginning of the 1980s, digital audio equipment has steadily been replacing analog equipment in broadcast and production systems. The major resulting advantages are the significant improvement of the audio tape recording multigeneration performance and its suitability for storage and computer-based production systems.

This digital equipment, featuring analog inputs and outputs, is designed to directly replace an analog device and operate in an analog environment. However, the trend in broadcast and production technology is toward the all-digital studio, in which all aspects of recording, processing, and transmission take place in the digital domain. To this end, a universal transmission protocol, defined in AES/EBU standard documents, has been developed to allow all equipment to transmit and receive a recognizable form of digital audio.

6.1.2 Digital audio concepts

The integration of digital equipment into an analog environment requires that analog signals be converted to digital and vice versa. The digital audio processing becomes attractive if the A/D process results in negligible signal degradations, the implementation complexity is reasonable, and the digital format is suitable for transmission and recording.

6.2 Principles of A/D Conversion

At present, the standard audio transducers, the microphone, and the loudspeaker, are analog devices. Thus, time-continuous electrical signals must be converted in a time-discrete numerical format for digital signal processing (DSP).

The A/D conversion is the main factor affecting the quality level attainable with digital audio. It also limits the dynamic range and the harmonic distortion of the recovered analog signals. Steps required to achieve a satisfactory A/D conversion and performance are examined in the following sections.

6.2.1 Ideal sampling

Sampling consists of measuring the amplitude of an analog waveform at periodic intervals. The measurement accuracy is determined by the sampling frequency, the value of which is a compromise between high measurement accuracy and complexity and cost of digitizing, storing, and transmitting a large amount of data.

The sampling process consists in multiplying the analog signal with a stream of time-repetitive pulses transmitted at the sampling frequency. To simplify the description of this process, we will consider an ideal sampling case where the sampling pulse duration is close to zero. This process will be referred to as pulse amplitude modulation (PAM) and is represented in the time domain in Fig. 6.1 and in the frequency domain in Fig. 6.2.

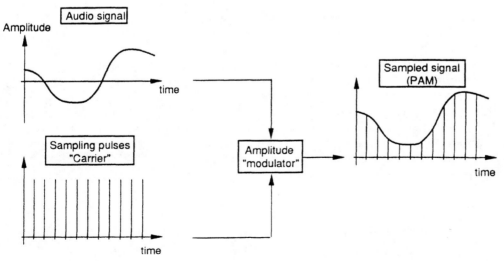

Figure 6.1 The sampling process results in a pulse amplitude modulation.

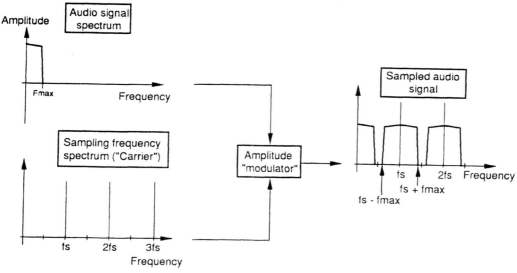

Figure 6.2 Modulated spectrum shows side bands around multiple of the sampling frequency.

Both figures exhibit similarities with the amplitude modulation process. However, in the frequency domain, multiples of the sampling (or "carrier") frequency appear with their own sidebands. This is because the carrier is not a sinusoidal signal but a series of pulses at the sampling frequency. This type of signal is represented in the frequency domain by a series of frequency components occurring at multiples of the sampling frequency.

As in the amplitude-modulation process, Fig. 6.3 shows that the analog signal is recovered ("demodulated") by using a simple low-pass filter.

If f_{max} is the maximum frequency of the analog signal spectrum and f_S the sampling frequency, the lower and upper sidebands are limited at frequencies equal to $(f_S + f_{max})$ and $(f_S - f_{max})$, respectively. From Fig. 6.2, it is easy to understand the Nyquist theorem, which stipulates that "a minimum sampling rate of two f_{max} samples per second is required to reconstruct a band-limited analog signal of bandwidth f_{max}." If the sampling frequency f_S is too low, the lower sideband overlaps the audio baseband spectrum. This phenomenon, called *aliasing,* is represented in Fig. 6.4.

6.2.2 Nyquist principle and aliasing

Violating the Nyquist theorem results in undersampling. Undersampling generates aliasing and occurs when the sampling frequency used is lower than twice the highest frequency of the analog signal or when the highest frequency in the analog signal exceeds $f_S/2$. To avoid aliasing given a sampling frequency f_S, the analog signal must be properly band-limited to $f_S/2$ before A/D conversion.

Figure 6.3 Recovery of the analog audio signal.

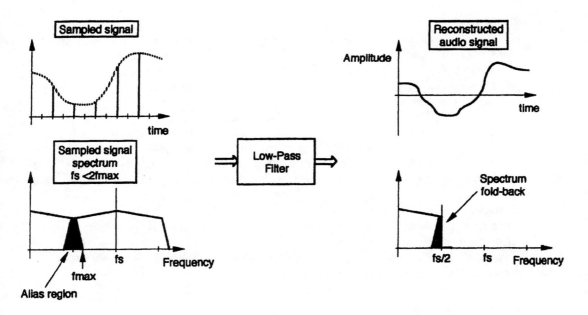

Figure 6.4 Aliasing effect in the frequency domain.

When aliasing occurs, the reconstruction of the analog signal from the PAM signal (in the D/A converter) generates a "foldback" of the frequencies higher than $f_S/2$ into the reconstructed analog baseband signal. These aliasing frequency components result in what is subjectively described as a "metallic" version of the original sound. Figure 6.4 shows the "spectrum foldback" and the distortion of the reconstructed analog signal.

6.2.3 Actual sampling

The ideal sampling described in Sec. 6.2.1 assumes that the sampling pulse has a near-zero duration. To allow time for A/D conversion, the pulse amplitude value of each sample is held until the next sample arrives, which generates a staircase representation of the analog audio signal, as shown in Fig. 6.5. This duration is equal to a sampling period $(1/f_S)$.

The resulting frequency response of the hold process is the Fourier transform of the sample pulse. It results in attenuation of the high frequencies and is referred to as the *aperture error*. The attenuation envelope is similar to a filter with a $(\sin x)/x$ frequency response. The content of the sampled signal spectrum exhibits nulls at multiples of sampling frequency, as shown in Fig. 6.6.

6.2.4 Quantization

Each sample of the original analog waveform is assigned a digital binary code value by a device called a *quantizer*. In a linear 4-bit system, for example, there are 16 possible binary values to encode the pulse amplitude of each sample. In the example shown in Fig. 6.7a, the original sine-wave audio signal is mea-

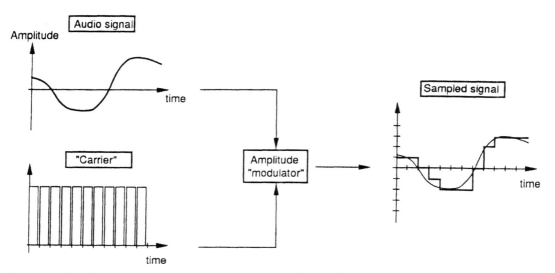

Figure 6.5 Time-domain representation of the actual sampling-and-hold process.

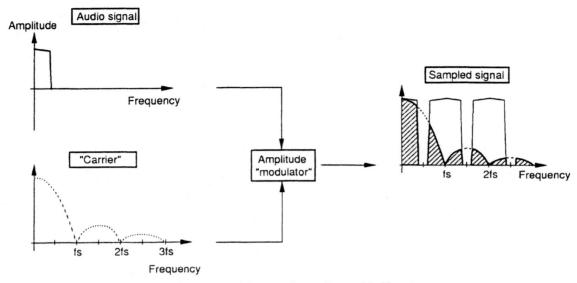

Figure 6.6 Frequency-domain representation of the actual sampling-and-hold process.

sured at each sampling time. As there are only 16 discrete binary values to describe the amplitude of each sample, the actual amplitude may fall in between two discrete binary values. In this case, the nearest value is selected and a quantization error is generated that corresponds to the difference between the original audio waveform and the staircase representation of the sampled-and-held audio signal, as illustrated in Figs. 6.7b and c.

The quantizing example of the Fig. 6.7 leads to several important conclusions:

- The represented binary value range is not symmetrical for positive and negative excursion of the original audio signal. A binary number representation known as "two's complement" allows for negative values. The most-significant bit (MSB) of the sample value represents the sign, as shown in Fig. 6.8. In a 20-bit system, the maximum amplitude is limited by the hexadecimal 7FFFF and 80000 values.

- Low-amplitude analog audio signals are quantized with very few discrete levels. This results in significant quantizing errors of low-input signals. The quantizing error amplitude can be reduced by increasing the number of discrete levels. A 5-bit system would reduce the quantization interval by half, and so on. Early 16-bit audio quantizers had the capability to provide 65,535 ($2^{16}-1$) quantizing intervals. Recent high-precision A/D converters achieve a true accuracy of 20 bits, thus reducing the 16-bit quantizing interval by a factor of 8. A different method of reducing the quantizing error amplitude is increasing the sampling frequency. This is called over-sampling and will be described in the following sections.

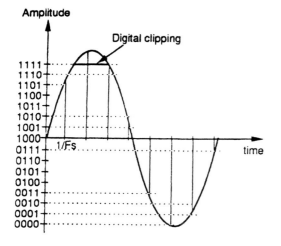

a) Sampling time and 4-bit quantizer values

b) Hold values

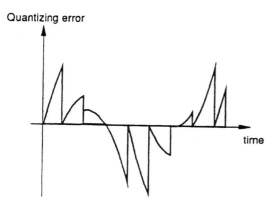

c) Quantizing errors

Figure 6.7 Sampling process and quantizing errors in a 4-bit A/D system.

- If the analog audio signal amplitude exceeds the quantizing range, digital clipping occurs.

The maximum quantizing error amplitude is equal to half a quantizing step as the quantizer assumes the nearest discrete binary value, as illustrated in Fig. 6.9. The sampling instant T_1 generates a measured value that falls exactly between two quantizing levels. At this sampling instant, the quantizing error is maximum and the quantizer may generate a binary value immediately below or above the measured value.

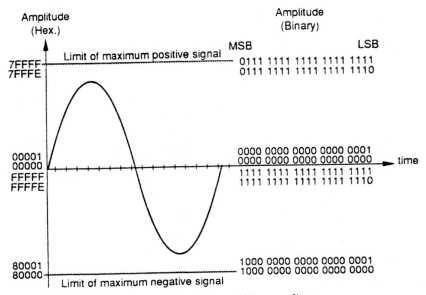

Figure 6.8   20-bit example to two's complement binary coding.

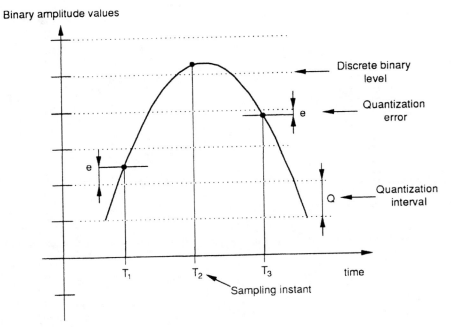

Figure 6.9   Quantizing interval and error.

The quantizing error is signal-dependent, but the slight amount of noise existing on the original signal adds random components, especially when the sampled value is halfway between two quantizing levels. The resulting error can be interpreted as noise added to the original signal and produces a roughness or granular quality in the sound. In a simplistic approach, it can be assumed that there is no predictive relationship between the error of one sample and the error of the next one. This results in a flat quantizing error spectrum with equal energy at all frequencies, which is characteristic of white noise.

The signal-to-noise ratio (SNR) is expressed as the ratio of the maximum sine-wave signal (below clipping) to the root-mean-square (RMS) value of the quantizing error. For an n-bit A/D converter, half the 2^{n-1} quantizing intervals are used for one polarity of the sine wave. The maximum RMS sine-wave value V is then

$$V = \frac{2^{n-1}Q}{\sqrt{2}}$$

where Q is the amplitude of one quantizing interval and (2^{n-1}) represents half the 2^n quantizing intervals.

An audio signal with a wide spectrum and high amplitude values generates a quantization error signal having an equal probability of assuming any value between $+Q/2$ and $-Q/2$ as shown in Fig. 6.10. There are therefore Q random values in one quantizing interval, which generate a flat error spectrum, and the probability for each value is equal to $1/Q$. The RMS value of the quantized noise is expressed as the square root of the mean of the sum of the squared errors, as follows:

- Sum of the squared errors

$$\int_{-Q/2}^{Q/2} e^2 \, de \qquad \text{(Eq. 6.1)}$$

- Mean of the sum

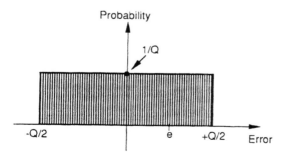

Figure 6.10 Equal probability of being any value between $-Q/2$ and $+Q/2$.

$$\frac{1}{Q} \int_{-Q/2}^{Q/2} e^2 \ de \qquad\qquad \text{(Eq. 6.2)}$$

- Square root of the mean

$$\text{Noise voltage (RMS)} = \sqrt{\left(\frac{1}{Q} \int_{-Q/2}^{Q/2} e^2 \ de\right)} = \frac{Q}{\sqrt{12}} \qquad \text{(Eq. 6.3)}$$

The ratio of the RMS values of the signal and the quantizing noise is expressed as

$$\text{SNR} = \frac{V}{N} = \frac{2^{n-1}Q}{\sqrt{2}} \ \frac{\sqrt{12}}{Q} = 2^n \sqrt{1.5} \qquad \text{(Eq. 6.4)}$$

which becomes the well-known expression in decibels

$$\text{SNR (dB)} = 6.02n + 1.76$$

As an example, a 16-bit A/D converter yields an SNR of about 98 dB. Recent 20-bit converters have an SNR of 122 dB. However, this relation is not true in the presence of sine-wave analog signals at low levels. The quantizing error becomes signal-dependent and a distortion of the incoming signal occurs. Furthermore, intermodulation may result if two or more low-input signals of different frequency are converted simultaneously.

6.2.5 Coding

Each quantized binary value is encoded in a form adapted to the type of sampled signal and the transmission and recording requirements. The most-used coding systems are pulse code modulation (PCM), pulse width modulation (PWM), adaptive delta modulation (ADM), floating point, and differential PCM (DPCM). The last two coding systems are described later in Secs. 8.2.2 and 8.7.1 because of their use in audio compression systems.

PCM is the simplest and most widely used audio coding system, but it is also the least efficient. PCM quantifies linearly all quantizing intervals by means of a fixed scale over the signal amplitude range. The A/D conversion resolution accuracy (e.g., 18-bit) determines the number of quantizing intervals (i.e., $2^{18} = 262{,}144$) available for the coding of the analog signal amplitude. All quantizing levels are assigned code words in a logical progressive order, as shown in Fig. 6.8.

6.2.6 Dither

To reduce the quantizing distortion on low-level signals, a technique called *dither* is used, which requires the addition of white noise to the signal. It is important that the RMS value of the added noise be not greater than a third of a quantizing interval.

Conversion back to analog generates a signal that has no relation to the original signal. However, the low-pass signal reconstruction filter "time averages" the signal. The glitches are filtered out and the original signal is approximately reconstructed.

Figure 6.11a shows a low-level analog sine-wave signal that has been digitized. The quantizing error is very important and will be converted back to analog. In Fig. 6.11b, noise is added. When the analog signal amplitude is close to zero, the A/D output remains zero. As this signal level increases, positive peaks of the noise added to the signal exceed the $Q/2$ limit, whereas negative peaks remain below this limit. As a result, the A/D converter toggles between two quantizing levels (see Fig. 6.11c). As the analog signal amplitude reaches about the top level of the maximum positive signal, the converter output remains high. Similar results are obtained when the analog signal decreases to the maximum negative voltage.

Figure 6.11d shows the D/A converted signal after averaging. No quantization distortion remains and the noise floor has slightly increased by 2 dB. The dither process uniformly redistributes the quantizing error into a random noise by smoothing the effect of sharp quantizing levels.

6.2.7 Dynamic range

The maximum audio level before clipping is the largest digital code the A/D converter can offer, as shown in Fig. 6.8. This level corresponds to 0 dBFS (full scale) and all digital levels are referenced to this point as negative values. The minimum level is set by the dither noise of the converter, which is about −120 dBFS for a 20-bit A/D converter. Consequently, the dynamic range is 120 dB.

The maximum signal level (MSL) is the analog RMS voltage corresponding to 0 dBFS for a particular system or device. The standard operating level (SOL) is the analog steady-state RMS voltage of a system or device that deflects a VU or PPM meter to zero. SOLs can be expressed in dBu or dBm according to the voltage-matched or 600-ohm power-matched terminating practices, which are explained in Sec. 5.2.2. The headroom is usually expressed in dB as the ratio of MSL to SOL signal levels.

A limiter can be positioned ahead of the low-pass filter to ensure that harmonics generated by clipping do not produce aliasing. The overall effect of the limiter is to prevent any harsh-sounding aliasing if an unexpected transient exceeds the dynamic range of the channel.

Manufacturers have chosen −20 dBFS as the SOL (equal to +8 dBm or 0 VU), giving 20 dB of headroom. This will ensure that the audio system's dynamic range accommodates peaks extending 20 dB above the average. However, some audio devices have a different headroom. This does not affect digital dubbing, but transfers in the analog domain are affected. Attenuator pads must be inserted between audio devices to maintain a unique operating level in broadcast operations.

Figure 6.11 Dithering of low level input signal.

Typical digital audio-processing devices have different input/output interfaces that can be divided into three categories:

- Digital audio input/output. Audio signals can be received, processed, and transmitted with 20 dB of headroom. No problems occur.

- Analog audio input/output. Audio signals received from analog equipment generally have headroom below 12 dB. After digital conversion, they are processed and back-converted to analog before transmission with the original headroom. No problems occur.

- Analog and digital audio inputs/outputs. In this type of device, only the analog output may cause problems. If the digital audio input signals are received or processed with 20 dB of headroom, the back-conversion to analog generates signals with peaks reaching +28 dBm. This will overload typical analog audio systems connected to the output of this device.

6.2.8 Standard sampling frequencies

Currently, there are three standard sampling frequencies in use:

- 32 kHz (professional transmission standard): This early digital audio sampling frequency was selected for use in land lines feeding FM stereo broadcast transmitters.

- 44.1 kHz (consumer standard): This sampling frequency was selected to allow the use of NTSC or PAL U-Matic videotape recorders (VTRs) fitted with a PCM adaptor to record and play back digital audio signals transformed into pseudo-video waveforms. Later, these VTRs were used to master compact disks (CDs) and 44.1 kHz became a de facto standard that is also used in R-DAT play-back-only applications.

- 48 kHz (broadcast audio standard): This frequency has a simple relationship to the 32-kHz frequency and facilitates standard conversion. It accepts an analog bandwidth of up to 22 kHz.

6.2.9 Preemphasis

Audio signal preemphasis is used in the digital tape-recording chain in conjunction with a deemphasis in the playback chain. This reduces the quantizing noise audibility at high frequencies.

When the preemphasis function is activated, the level of the high-frequency analog audio input signals is preemphasized during recording, while the level of any noise remains unchanged. When played back, the audio signal is automatically returned to its original level and, simultaneously, the noise is reduced to a negligible level.

This function is normally turned off, especially if the analog audio input level is abnormally high at frequencies above 3 kHz.

The preemphasis function cannot be actuated for digital input signals on digital audio recorders. However, if the digital input signal has been preemphasized by external equipment using AES/EBU format CD-type (15/50-μs) method, the analog converted signal is deemphasized when played back. If the digital input signal has been preemphasized using the CCITT method, the analog output is not deemphasized.

It is recommended for interchange of recorded tapes that no preemphasis be used. The SMPTE is presently defining a Recommended Practice on this issue. In the event that preemphasis is used, automatic sensing of this mode takes place within the digital audio recorder. Tapes that have both preemphasized audio and nonpreemphasized audio edited together exhibit some remaining frequency response artifacts that may exist for a short period of time. This condition should be avoided.

6.3   Principles of D/A Conversion

6.3.1   The D/A converter

In a D/A converter, the two's complement binary words are first decoded to find out the binary value they represent, and then they are converted into a voltage representing the amplitude of the original audio signal samples.

A simple D/A converter is made of precision resistors combined with weighted current sources that are switched according to the clocked binary code words. Figure 6.12 shows a basic D/A converter with n-bit resolution. The n-bit data control n switches. The ON and OFF positions of the digital switches correspond to the ones and zeros of the digital signal. These switch positions generate an output voltage value proportional to the weight of bits in the sample word. The D/A output signal waveform is shown in Fig. 6.13a.

6.3.2   Aperture effect

As previously explained in Sec. 6.2.3, the actual sampling process uses a wide sampling pulse to allow time for the A/D conversion. Similar considerations exist during the D/A conversion, where switching glitches appear in the converted signal. Another sample-and-hold circuit is added to resample the D/A output signal, as shown in Fig. 6.13a and b.

The implications of the use of the sample-and-hold process have already been explained and shown in Fig. 6.6. The sampled signal spectrum is amplitude-modulated by a $(\sin x)/x$ envelope signal. The envelope rolloff at high frequencies varies with the sampling pulse duration. If the pulse has a near-zero duration, the sampling is ideal, $(\sin x)/x = 1$, and there is no high-frequency rolloff. The rolloff increases with the sampling pulse duration, which is also called the *aperture time*.

The aperture effect can be expressed in terms of the aperture ratio. This is

$$V = -V_{Ref}\left(\frac{bit0}{2} + \frac{bit1}{4} + \frac{bit2}{8} + \frac{bit3}{16} + \cdots + \frac{bit\,n\text{-}1}{2^n}\right)$$

In the example above,

$$V = -V_{Ref}\left(\frac{1}{2} + \frac{0}{4} + \frac{1}{8} + \frac{1}{16} + \frac{0}{32} \cdots + \frac{0}{2^n}\right) = -V_{Ref}\ (1/2 + 1/8 + 1/16) = -V_{Ref}\ (11/16)$$

Figure 6.12 Digital-to-Analog converter.

the relation between the interval of time during which the sample pulse remains constant (t) and the sampling period (T_s).

$$\text{Aperture ratio} = \frac{t}{T_s}$$

For an aperture ratio of 100%, the attenuation at half the sampling frequency is about 4 dB, as shown on Fig. 6.14. If the aperture ratio is reduced to 12.5%, this attenuation is only 0.2 dB, but the amplitude of the sideband components has increased.

6.3.3 Low-pass filter

The output low-pass filter (also called the reconstruction filter) removes the higher harmonics (above $f_S/2$) that were added in the sampling process. This filter, also called an antiimage filter, removes the high-frequency components contained in the staircase signal at the resampling circuit output. The low-pass-filtered output signal is shown in Fig. 6.13c.

Slightly boosting frequencies immediately under $f_S/2$ compensate for the aperture effect but reduce SNR, since the high-frequency noise is boosted along with the signal.

Figure 6.15 summarizes the PCM encoding and decoding process.

a) D/A output signal

b) Resampled pulse to reduce pulse width
    (Aperture) and remove glitches

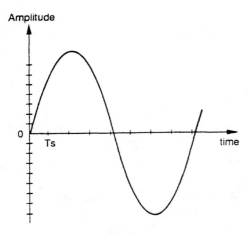

c) Low-pass filtered D/A output signal

Figure 6.13   D/A output, resampling, and low-pass filtering.

6.3.4 Oversampling

Oversampling is used to improve the A/D and D/A performance by reducing the quantizing error and aliasing component amplitude. Figure 6.16 shows two-times oversampling effects on the time-domain representation of the audio signal. With high variation of the analog audio signal amplitude, additional discrete quantizing levels are inserted and the resulting quantizing error is reduced. At low variation of the signal there is no improvement, because the same quantizing level is used for several sampling times. However, oversampling is combined with a more accurate A/D conversion resolution, such as in 18-bit systems, which divides a 16-bit quantizer interval into 4. This

a) The sampling-and-hold effect on signal spectrum components, from 0 to 3fs.

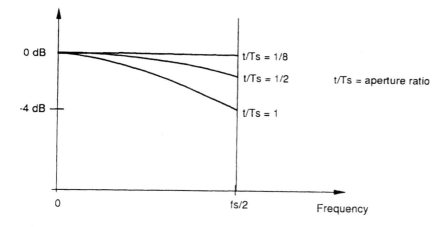

b) The sampling-and-hold effect on signal spectrum components, from 0 to fs.

Figure 6.14 Aperture ratio and high-frequency roll-off.

Figure 6.15 PCM encoding and decoding.

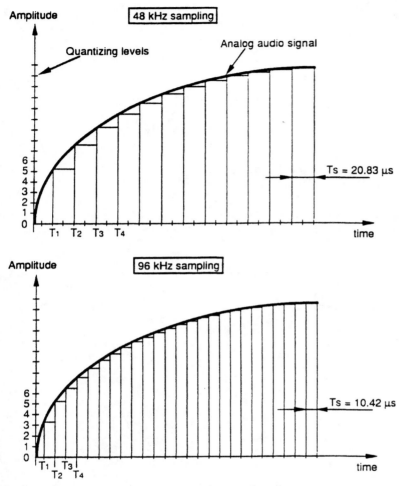

Figure 6.16 Two times oversampling in the time domain.

higher conversion resolution reduces the quantizing error by 4 and enhances the system's ability to reproduce low-level signals. Audibly, such improvements appear as increased definition, clarity, and overall realism.

A two-times oversampling system spreads the quantizing noise spectrum over twice the usual bandwidth, improving the SNR by 3 dB. The peak SNR for a sinusoidal signal becomes

$$\text{SNR (dB)} = 6.02n + 1.76 + 10 \log_{10} d$$

where n is the number of bits per quantized sample, and d is the oversampling factor. Four-times oversampling increases the SNR by 6 dB, which is equivalent to one extra bit in the quantization process.

Figure 6.17 Two-times oversampling in the frequency domain.

Figure 6.17 shows two-times oversampling effects on the frequency domain representation of the audio signal. The spectrum image of the sampled signal is shifted at twice the original sampling frequency. Consequently, both the antialiasing filter and the reconstruction filter do not need to have a sharp rolloff, which generates significant phase shifts in the analog signal. The possibility of having aliasing components in the reconstructed signal is completely eliminated.

Figure 6.18 shows a block diagram of a four-times oversampling A/D and D/A system.

Practical implementations of oversampling techniques use digital transversal filters to simplify the design of the A/D and D/A converters. Digital filters are used to remove unwanted frequency components in the sampled signal spectrum before the D/A conversion and avoid the use of sharp-cut antialiasing and reconstruction filters. Simple low-pass filters can then be used as antialiasing and reconstruction filters.

A digital filter operates by multiplying, delaying, and adding binary numbers. This process can easily be performed by computer programs using mathematical algorithms and procedures to manipulate data in memory. The digital filter performance does not rely on the electrical characteristics of

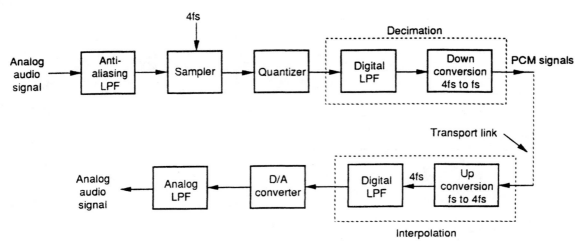

Figure 6.18   Four-times oversampling A/D and D/A conversion system.

resistors and capacitors as in a conventional analog filter, and thus it is more precise and reliable. Furthermore, it is easy and inexpensive to build.

The number of delay cells determines the digital filter order. More multipliers improve its cutoff characteristics and reduce filter ripples. A high-order digital filter is capable of sharp cutoff without causing phase distortion.

Figure 6.19 shows an example of a digital filter used in four-times oversampling conversion. Each cell (D) is a delay, such as a shift register. Each delayed sample is multiplied by coefficients, using the multipliers $X_1, X_2, ..., X_n$, and the results are summed in a binary adder.

To explain briefly how a digital filter works, first consider the digital filter shown in Fig. 6.19 with only one multiplying coefficient per delay cell. These first n coefficients $(a_1, a_2, ..., a_n)$ are derived from a perfect low-pass filter response curve [(sin x)/x]. Each 16-bit input word is shifted across all delay cells and is multiplied by each 14-bit first coefficient, which generates 30-bit length multiplier products. This digital filter is called a *transversal filter*, since input samples pass through the cascaded delays and generate a series of output pulses, spaced at a sampling period. Their amplitudes are the result of multiplying the sample amplitude by each respective value of the (sin x)/x curve taken at instants spaced at the reciprocal of the sampling frequency $(T_s = 1/f_S)$. Figure 6.19b shows the digital filter output for sample impulse responses x and $(x+1)$. The sum of all impulse responses corresponds exactly to that of an analog filter used for the reconstruction of the original waveform. Thus, a transversal filter is used to eliminate all frequency components higher than the audio baseband before D/A conversion.

In a four-times oversampling filter, additional samples are created through interpolation, in order to generate three intermediate values for every original value. These intermediate samples are multiplied by fixed impulse-response coefficients (b_1 to b_n, c_1 to c_n, d_1 to d_n) in a similar manner to that of the simple

a) n-order transversal filters used in four-times oversampling conversion

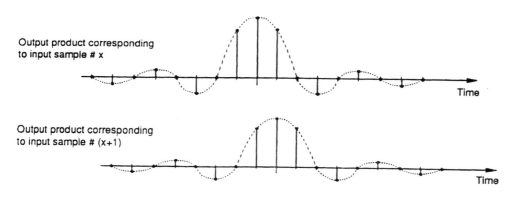

b) 17-order transversal filters output operating at nominal sampling frequency (one multiplying coefficient per delay cell).

c) Delayed and summed digital filter responses at nominal sampling frequency (one multiplying coefficient per delay cell).

Figure 6.19 Transversal filters used in four-times oversampling conversion.

Figure 6.20 Signal spectrum at the output of a four-times oversampling digital filter.

digital filter (a_1 to a_n coefficient only) explained previously. The total digital output frequency rate is then increased by 4 (192 kHz). The resulting sampled signal spectrum is shown in Fig. 6.20. Although the original has been sampled at 48 kHz, all images around 48, 96, and 144 kHz are removed by the four-times oversampling digital filter, and the quantizing noise is spread over a bandwidth equal to $4f_S/2 = 96$ kHz, that is, four times larger, reducing the noise level by 6 dB.

6.3.5 Noise shaping

Oversampling converters extend the frequency range of the sampled signals and spread the quantizing noise spectrum over the base-band range, as shown in Fig. 6.17. Noise-shaping techniques reduce the noise spectrum level within the audio base-band frequency range while increasing the noise spectrum level beyond this range. The overall amplitude of the noise is kept constant, but its spectrum has changed. Figure 6.21 shows the noise-shaping effect on a four-times oversampled signal spectrum. A significant reduction of the audible noise level can be achieved, especially if second-order or third-order filters are used for noise shaping.

Figure 6.22a shows a block diagram of a four-times oversampling A/D converter with noise-shaping feedback loop. This loop acts as a flat-response transfer function for the analog input signal coming from outside the loop, and as an integration transfer function for the quantizing error signal introduced within the loop by the A/D converter. Figure 6.22b shows the A/D converter output spectra of the input analog signal and the quantizing error signal.

6.3.6 Practical limitations of A/D and D/A conversions

Both the A/D and D/A conversion processes contribute errors to an audio signal. These errors, taken separately, are generally small, but many passes

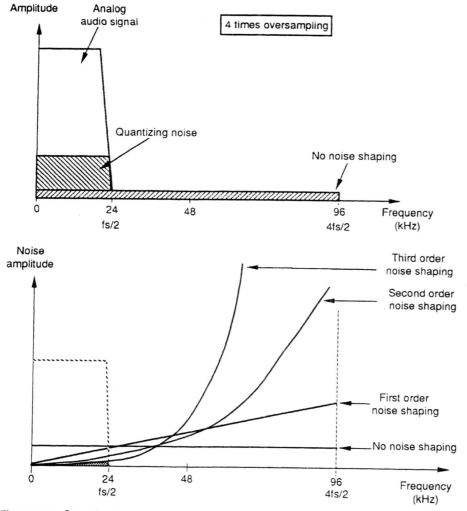

Figure 6.21 Quantization noise components in oversampled quantized signal and noise shaping performance.

through A/D and D/A converters cause an accumulation of these errors at specific signal levels or for certain groups of frequencies, resulting in an audible signal degradation. These errors can be classified into two categories:

- Low-pass filter (LPF) errors. Group delay occurs when a group of frequencies is delayed more than other groups in a signal spectrum. Group delay errors result in distortions of the reconstructed analog signal. In addition, antialiasing and reconstruction LPF ripples may accumulate and generate significant amplitude errors in the reconstructed analog signal. These

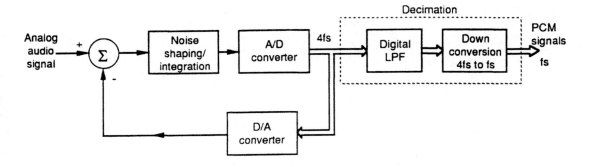

a) Block diagram of a four-times oversampling A/D with noise shaping feedback loop

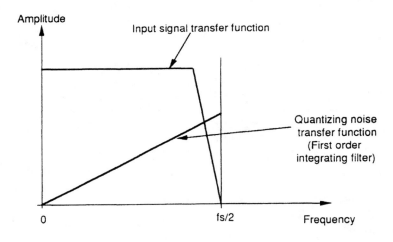

b) Spectra of analog input audio signal and quantizing noise signal at the A/D converter output

Figure 6.22   Four-times oversampling A/D with noise shaping feedback loop.

       errors are reduced with the use of high-order digital filters, as explained in Secs. 6.3.3 and 6.3.4.

- Converter errors. Erroneous quantized values are produced when the sampling clock frequency varies, resulting in inaccurate sampling-time positions. A/D and D/A converter nonlinearities, such as unequal quantizing intervals, have similar effects.

       When noise is superimposed on an analog signal, erroneous quantized values are also produced. Furthermore, noisy edges affect the signal-reshaping process by adding jitter to the digitized signal.

       The use of oversampling converters has resulted in substantial improvements of digital sound quality by avoiding the need for sharp cutoff antialiasing and reconstruction filters and providing improved conversion linearity. They also are less sensitive to low-frequency jitter, and, therefore, do not require a highly stable clock frequency. However, oversampled A/D converters can be more sensitive to high-frequency jitter than normal converters.

6.4 Description of Biphase Mark–Encoded Signal

6.4.1 Channel bandwidth

Channel coding is used in digital recording and transmission systems to match certain properties of the coded data to the characteristics of the recorder or transmission channel. In this section, some important characteristics of codes are explained through comparisons between Nonreturn-to-zero (NRZ) and Biphase-mark (BPM) encoded waveforms.

Channel coding modifies the original data to achieve the highest bit density possible within the limiting characteristics of the channel bandwidth. DC or very low frequencies will not pass through transformers.

A virtually error-free data transmission requires an information transfer rate less than the channel capacity. Potential for loss of information decreases when the transmission time or the channel bandwidth is increased. To achieve the best performance for data transmission, channel coding is necessary.

Transmission channel band-pass characteristics result in high and low frequency base-band signal distortions. It is the objective of channel coding to shape the spectrum of the digital signals and to minimize the distortions.

Many different waveshapes for encoding digital information are being used, each with its own application and limitations. In order to allow unambiguous extraction of clock and data from the read or transmitted waveform, signal transitions must occur frequently enough to facilitate synchronization of the decoder clock.

The AES together with the EBU developed a digital audio transmission standard, known as the AES/EBU standard, as well as AES3-1992, ANSI S4.40-1992, or IEC-958. The transmission medium is wire, which allows high bandwidth capability and a serial transmission of the parallel data words generated by A/D converters. The 16- to 20-bit parallel data words are serialized by sending first the least-significant bits (LSBs). Word clock data must be added to identify the start of each sample in the decoding process. The resulting data stream is BPM-encoded. Figure 6.23 shows a simplified block diagram and data sample processing of an AES/EBU encoder.

6.4.2 NRZ and BPM encoding

6.4.2.1 Digital signal encoding format. Figure 6.24 compares the timing diagrams for a data stream encoded with different schemes. The data signal shows a serial representation of parallel data words.

Figure 6.23 Simplified block diagram of an AES/EBU encoder and parallel to serial word conversion.

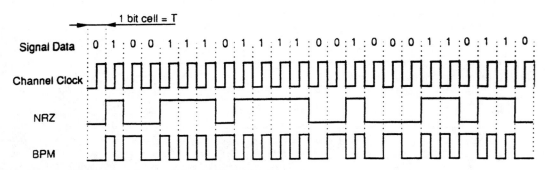

Figure 6.24 Bi-phase mark encoded signal waveform.

The rules for encoding in NRZ and BPM are defined as

- NRZ. Zeroes and ones are transmitted as positive and negative levels so that transitions only occur whenever the data bits change. NRZ coding means that during consecutive "ones," the level remains constant and does not return to zero between each data bit.

- BPM. There is always a transition at the beginning of every data bit interval and a transition in the middle of each "one" bit interval. Consequently, there will never be more than two consecutive "ones" or "zeros" in the BPM-encoded data stream. This stream signal has an important characteristic: it is polarity-free, as only the presence or absence of a transition in the middle of a data bit cell determines a "zero" or "one" datum. This code, also known as Manchester-I code, is used for recording time code on tape and coding AES/EBU format audio signals.

6.4.2.2 Signal spectrum.

For high-density digital transmission, the choice of an optimal modulation code is very important. A code that closely matches the characteristics of this transmission channel must satisfy the bandwidth and spectral component amplitude requirements.

The spectral properties of NRZ and BPM schemes are compared in Fig. 6.25. NRZ shows a large amount of energy concentrated at low frequencies. The spectrum of BPM is smaller than that of NRZ. BPM-coded data require a larger channel bandwidth. The amplitude content of BPM-coded data is minimal in the low-frequency and high-frequency range and is maximal at the bit rate value. The shape of its spectrum has zero amplitude content at very low frequencies (DC).

6.5 General Structure of the AES/EBU Interface Protocol

The AES/EBU digital audio standard is an interfacing protocol allowing digital equipment to transmit and receive digital audio signals. It specifies that

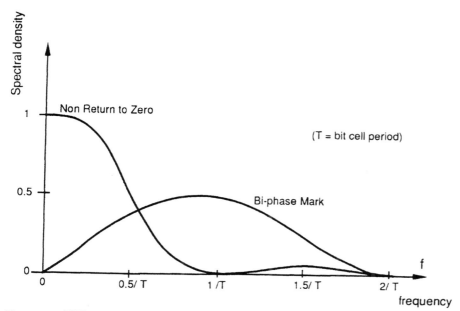

Figure 6.25 NRZ and BPM signal spectral density distributions.

the audio data must be in two's complement binary coding, as previously shown in Fig. 6.8.

6.5.1 The format structure

The AES/EBU signal format has the audio frame structure shown in Fig. 6.26. This frame is made up two subframes (subframe A and subframe B), each combining sample data from one audio source or channel, auxiliary data, preamble data, and associated data, validity (V), user (U), channel (C), and parity (P). Audio frames are grouped into blocks of 192 frames. A flag (Z) in the data stream allows the detection of the start of each block. The audio frame duration is 20.83 μs in a 48-kHz sampling system ($T_s = 1/48$ kHz). The duration of one AES/EBU audio block is

$$20.83 \ \mu s \times 192 = 4000 \ \mu s$$

The four preamble bits, also called the sync word, are used to identify the start of a new sample and a new audio block. Three different words are used:

- Sync word Z. This bit sequence indicates the start of the first frame of a new audio block. This word generates the Z flag.

- Sync word X. This bit indicates the start of all remaining frames.

- Sync word Y. This bit indicates the start of every B subframe.

These sync words are all 4 data bits in length and are not BPM-encoded as all remaining bits in subframes. The corresponding 8-bit encoded sequences are shown in Fig. 6.26. Their structures minimize the DC component on the transmission line, and facilitate clock recovery and subframe identifications as they are unique in the serial data stream. In fact, the 3-consecutive-bit sequence violates the BPM encoding rules where a maximum of only two consecutive bits is possible.

An audio frame contains two subframes of 32 bits each. Each audio sample word can be specified from 16 to 20 bits or up to 24 bits when the four auxiliary bits are assigned as sample bits. The auxiliary bit assignment is indicated in the channel status bit array, byte 2, bits 0–2.

These auxiliary bits can also carry auxiliary information such as a low-quality auxiliary audio channel for producer talk-back or studio-to-studio communication, as shown in Fig. 6.27. Every 4 ms or 192 frames, there are 4 bits × 192 = 768 bits that can be organized in 64 audio words of 12-bit resolution. These 64 words per each 4 ms provide an equivalent sampling frequency of 16 kHz. Each audio subframe can carry one separate auxiliary audio channel.

Each subframe has four associated bits as follows:

- Validity bit (V). The validity bit is set to zero if the audio sample word data are correct and suitable for D/A conversion. Otherwise, the receiving equipment is able to mute its output for defective samples. This capability has

P = Parity bit
C = Channel status bit ⎫
U = User bit   ⎬ Associated data
V = Validity bit   ⎭

Note: Sync word sequences can have positive or negative parities depending on the preceding bit state.

Figure 6.26 AES/EBU digital audio data structure.

Figure 6.27 Auxiliary data channel formatting in 12-bit sample for coordination channel.

not been implemented by all manufacturers and some equipment may not generate or verify the sample word validity.

- User bit (U). The user bit in each subframe is sent to a memory array made of 24 rows of 8 bits (or 1 byte) as shown in Fig. 6.28. This array is filled row after row with 192 user bits coming from the same audio source or type of subframe, which constitute the duration of one audio block. Possible user data formats are indicated in the channel status byte 1, bits 4–7. The recommended practice AES18-1992 specifies the format of the user data channel of the AES digital audio interface. In this format, data packets can be transmitted through the interface in blocks, which may be synchronized to television frame rates, such as in digital video tape recorders (DVTRs).

- Channel bit (C). Similar to the user bit, the channel bit in each subframe is sent to a memory array made of 24 rows of 8 bits (or 1 byte) as shown in Fig. 6.28. The content of this array, shown in Figs. 6.29 and 6.30, is very important for the identification of the content of the audio data word.

- Parity bit (P). The parity bit is always set to indicate an even parity. As a consequence, the polarity of the first BPM bit of every sync word is identi-

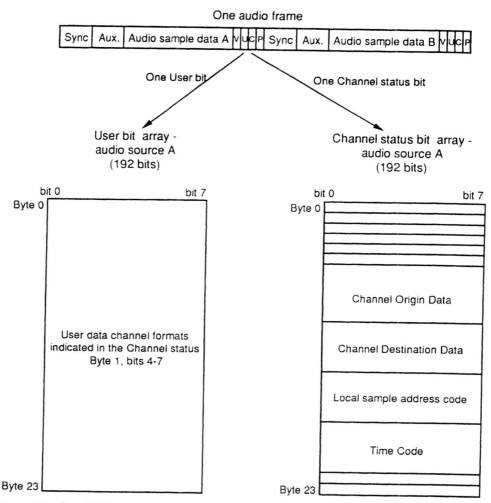

Figure 6.28 C-and U-bit data memory array formatting.

cal for each subframe, be it positive or negative. Even parity ensures that the total number of ones in the 64 BPM cells of a subframe is always even. The parity bit permits the detection of an odd number of errors resulting from transmission interface errors. Some equipment ignores this bit or may not properly handle its indication.

The channel status array shown in Figs. 6.29 and 6.30 details the bit arrangement and their significance in AES/EBU data stream communications. In byte 0, bit 0 indicates the consumer or professional channel use. If the channels are for consumer use, byte 0, bit 0 = 0 and the channel status array is limited to bytes 0 and 1, which also indicate

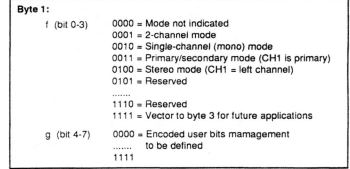

Figure 6.29 AES/EBU digital audio channel bit (C) data structure.

- Normal audio data or digital data is present
- Copy prohibiting
- Use of preemphasis in the A/D conversion process
- Indication of category codes (general, CD, PCM adaptor, DAT)

If the channels are for professional use, byte 0, bit 0 = 1, and all the channel status array content is applicable. The most used bit arrangements in this array are inserted in bytes 0 to 4 and listed below:

- a = Use of channel status channel
- b = Audio or nonaudio use

Byte 2:

h (bit 0-2)	000 = Use of auxilary sample bits not defined.	
		Sample length 20 bits
	001 = Auxilary sample bits are part of 24 bit samples	
	010 = Reserved	
	.....	
	111 = Reserved	
i (bit 3-5)	00000 = source wordlength and encoding history	
	....	
	11111 = To be defined	
r (bit 6-7)	Reserved	

Byte 3:

j (bit 0-7)   Future multichannel function description

Byte 4:

k (bit 0-1)   Digital audio reference signal
r (bit 2-7)   Reserved

Byte 22:

bit 1-3	Reserved
bit 4	Bytes 0-5; unreliable = 1
bit 5	Bytes 6-13; unreliable = 1
bit 6	Bytes 14-17; unreliable = 1
bit 7	Bytes 18-21; unreliable = 1

Figure 6.30 AES/EBU digital audio channel bit (C) data structure (*Continued*).

- c = Audio signal preemphasis mode
- d = Locking of source sample frequency
- e = Sampling frequency (32, 44.1, or 48 kHz)
- f = Channel mode (two-channel, stereo, single-channel, primary/secondary*)
- g = User bit management
- h = Use of auxiliary sample bits
- i = Source word length and source encoding history

---

*Primary/secondary mode means a single audio source in channel A and data in channel B.

- j = Future multichannel function description
- k = Digital audio reference signal
- r = Reserved

Byte 23 is a cyclic redundancy check code (CRCC) for validity verification of the channel status data array content at the receiving end.

It must be noted that some broadcast equipment such as DVTRs are not transparent to all the channel status array data content. Generally, bytes 0 and 1 are only recorded and updated as necessary in the playback mode. All remaining bytes in the channel status array are discarded by the DVTR.

The AES has recently issued an AES-2id-1996 information document, Guidelines for the Use of the AES3 Interface. The intent is to clarify the AES3 standard and avoid specifications misinterpretations and implementation problems.

6.5.2 AES/EBU data signal characteristics

At a 48-kHz sampling rate, the total data rate is $32 \times 2 \times 48000 = 3.072$ Mbps. After BPM encoding, the data stream rate is doubled at about 6.144 Mbps, which yields a Nyquist frequency of 3.072 MHz. The BPM spectrum distribution exhibits nulls at multiples of 6.144 MHz.

Sync words contain three consecutive low cells followed by three consecutive high cells and contribute a low fundamental frequency to the AES/EBU signal spectrum at $3.072/3 = 1.024$ MHz.

Each audio frame contains 64 bits, which are sent every 20.83 μs. Therefore, one frame data bit lasts 325.5 ns and one BPM bit cell has a duration of about 163 ns. Then, eye diagrams that result from the superimposition of several data stream bit cells are spaced 163 ns apart.

6.6 AES/EBU Signal Electrical Characteristics

Characteristics of the AES/EBU professional-format interface are shown in Table 6.1.

Characteristics of the AES/EBU consumer format interface are shown in Table 6.2. This consumer format is used in CDs and RDATs equipped with digital inputs and outputs. It is also known as SPDIF (Sony-Philips Digital Interface).

6.7 Digital Audio Interface Implementation

6.7.1 Digital audio input interface

Figure 6.31 shows a typical hybrid audio signal input interface of an audio tape recorder or digital processor. Analog or serial digital data input signal can be selected for processing.

TABLE 6.1   Characteristics of the AES/EBU Professional Format Interface

Format	Serial transmission of two channels of sampled and linearly encoded data

Transmitter characteristics	Balanced output
	Connector: XLR with male pins and female shell
	Pin allocations:
	    Pin 1: Cable shield, signal earth, or ground
	    Pin 2: Signal (polarity unimportant)
	    Pin 3: Signal (polarity unimportant)
	Source impedance: $110 \pm 20\%$ ohms
	Balance: < -30 dB (to 6 MHz)
	Output signal amplitude: 2 to 7 $V_{p\text{-}p}$ across 110-ohm load (balanced)
	Rise- and falltime: 5 to 30 ns
	Jitter: 20 ns

Receiver characteristics	Balanced input
	Connector: XLR with female pins and male shell
	Pin allocations:
	    Pin 1: Cable shield, signal earth, or ground
	    Pin 2: Signal (polarity unimportant)
	    Pin 3: Signal (polarity unimportant)
	Input impedance: $110 \pm 20\%$ ohms
	Common mode rejection ratio: Up to 7 $V_{p\text{-}p}$ to 20 kHz
	Maximum accepted signal level: 7 $V_{p\text{-}p}$
	Cable specifications: Shielded twisted pair, 100 to 250 m maximum
	Cable equalization: Optional

TABLE 6.2   Characteristics of the AES/EBU Consumer Format Interface

Format	Serial transmission of two channels of sampled and linearly encoded data
Transmitter characteristics	Unbalanced output Connector: RCA phono jack Source impedance: 75 ohms Output signal amplitude: 500 $mV_{p\text{-}p}$ across 75-ohm load (unbalanced)
Receiver characteristics	Unbalanced input Connector: RCA phono jack Input impedance: 75 ohms

Analog audio input channels are processed and converted to 16- to 20-bit digital data at a 48-kHz sampling rate to match the AES standard. Coding of the sample is two's complement linear pulse-code modulation (PCM).

6.7.2   AES/EBU decoder and demultiplex

A BPM decoder is used to convert the BPM-encoded serial AES/EBU digital audio input signal into a signal data stream. Then, the audio data signals from the two-channel-multiplexed data stream are separated to produce two

Figure 6.31 Typical hybrid audio input interface of a four-channel digital processor.

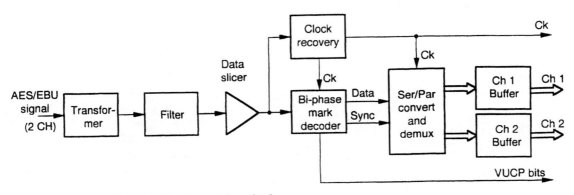

Figure 6.32 Serial audio input decoder and demultiplexer

bit-parallel audio data streams. Associated bits (V, U, C, P) are also extracted from each subframe for processing control and synchronization of subframes and frames, and for the generation of the 192-bit channel status and user blocks. Figure 6.32 shows the main decoding and demultiplexing functions for one AES/EBU serial audio data input. The data slicer, in presence of noisy signals, may introduce jitter in the receiver. The jitter can be defined as the uncertainty in the transition time of decoded bit cells. Filters should be used to remove noise before data slicing.

6.8 Digital Audio Signal Distribution

The original AES3-1992 standard defines the distribution of AES/EBU signals through a twisted-pair audio cable. Another recent distribution format has been adopted and specified in the AES-3id-1996 information document, and in an ANSI/SMPTE 276M-1995 standard document. Both define the transmission of AES3 formatted data through unbalanced coaxial cable.

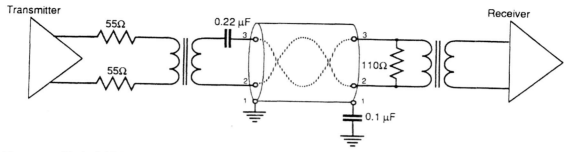

Figure 6.33 Typical AES3-1992 distribution interface.

6.8.1 110-ohm twisted-pair cable distribution

Characteristics of this distribution format have already been mentioned in Sec. 6.6. The recommended AES3-1992 distribution interface is shown in Fig. 6.33.

6.8.2 75-ohm coaxial cable distribution

This standard was developed to overcome the limitations of the twisted-pair distribution format with respect to cable length capability and the important size and cost of the XLR-type connectors, and most importantly, to allow the use of nonclamping analog video distribution amplifiers and routers for digital audio signal distribution. However, most of the audio equipment is equipped with XLR-type connectors and conversion to BNC-type connectors must be considered. Furthermore, some analog distribution amplifiers (DAs) might not have enough bandwidth, because a minimum of 12-MHz bandwidth is necessary to transmit the third harmonic of the BMP-encoded AES/EBU spectrum. Characteristics of the 75-ohm coaxial cable distribution format are listed in Table 6.3.

The recommended AES-3id distribution interface is shown in Fig. 6.34. This informative document also contains information regarding cable performance, cable equalizer characteristics, and passive adaptors to and from AES3 equipment. Figure 6.35 shows examples of 14-dB attenuators for conversion between balanced 110-ohm and unbalanced 75-ohm format systems. Figure 6.36 shows examples of 75-ohm coaxial to 110-ohm balanced matching networks.

6.8.3 Wiring practices and interconnection

Balanced cable distribution should be used in studios to avoid ground loop problems. Installed analog cables in existing studios can be used for digital audio distribution, but the maximum cable length may be limited to 100 m, depending on the cable type. Recent high-quality twisted-pair cable types allow 250 m of length.

TABLE 6.3   Characteristics of the 75-ohm AES/EBU Interfaces

Channel coding	BMP-encoded AES/EBU signal
Transmitter characteristics	Unbalanced output with BNC connector
	Source impedance: 75 ohms nominal
	Return loss: >25 dB (0.1 to 6 MHz)
	Output signal amplitude: 1 V ±10% across 75-ohm load
	DC offset: 0.0 V ±50 mV
	Rise- and falltime: 30 to 44 ns between 10 and 90% of signal amplitude points
	Data jitter: ≤20 ns$_{\text{p-p}}$
Receiver characteristics	Unbalanced input
	Input impedance: 75 ohms nominal
	Return loss: >25 dB (0.1 to 6 MHz)
	Minimum input level sensitivity: 100 mV
	Cable equalization: Optional

Figure 6.34   Typical AES3id-1996 distribution interface.

The cable shield must be interconnected as per Figs. 6.33 and 6.34 to avoid interference. Only one receiver should be connected to a digital audio device output.

6.9   Other interfacing protocol formats

In addition to the dominant AES/EBU protocol, three other interfacing formats are widely used, the MADI (Multichannel Audio Digital Interconnect), the SDIF2 (Sony Digital Interface Format), and the SPDIF (Sony Philips Digital InterFace).

6.9.1   MADI format

The MADI format is defined in the standard document AES 10-1991 and the information document AES-10id-1995. It can accommodate up to fifty-six 32-bit signals conforming to the AES3-1992 standard. MADI was originally developed as a point-to-point operation system to interconnect multitrack recorders to digital audio consoles or processors. Other applications include digital routing systems and studio-to-studio interconnections.

a) 14-dB transformer attenuator

Load impedance = 110Ω
Source impedance = 75 Ω
Return loss   >25 dB
Attenuation = 1:525 (14.5 dB)

Load impedance = 112 Ω
Source impedance = 72 Ω
Return loss   >25 dB
Attenuation = 1:5 (14 dB)

b) 14-dB resistive attenuator

Figure 6.35 Balanced to unbalanced matching networks (After SMPTE 276M standard).

Figure 6.36 Unbalanced to balanced matching networks (After SMPTE 276M standard).

MADI signals are easily converted into AES/EBU subframes, because only the first 4 bits are different. Fifty-six subframes are serialized to make up a MADI frame and then encoded with a 4/5 scheme, which takes groups of 4 bits in the serial data stream and converts them into 5-bit words by means of a lookup table, to reduce the encoded data DC content. Each encoded subframe is then $32 + 8 = 40$ bits in length. Sampling rates of 32 to 48 kHz are supported with a possible variation of $\pm12.5\%$ to allow varispeed operation of audio recorders. The transmission data rate is then fixed at 125 Mbps to provide enough bandwidth for the encoded data stream (56 channels $\times$ 40 bits $\times$ 48 kHz $\times$ 1.125 = 121 Mbps).

The first 4 bits are affected to

- Bit 1 = MADI channel 0, which indicates the MADI frame synchronization.

- Bit 2 = MADI channel active, which indicates the validity of the transmitted subframe.

- Bit 3 = Stereo right or left channel, which differentiates the left channel from the right.

- Bit 4 = MADI sync, which identifies the start of an AES3 audio block (192 AES3 frames) to ensure a proper recovery of the associated C and U bits at the receiving end.

Synchronization codes are added to every MADI frame in the encoded data stream to facilitate a proper synchronization between equipment. The transmission is intended to be asynchronous and FIFO buffers in the transmitter and the receiver are used for synchronization to an external sampling reference signal.

The transmission medium can be either a wide-bandwidth coaxial cable (up to 50 m) or a fiber optic link (over 50 m). The AES-10id-1995 document provides implementation guidelines for fiber-optical interfaces.

6.9.2 SDIF2 format

This format was developed by Sony for professional and semiprofessional mastering and recording applications. It is used for interconnection of single-channel links at both 44.1 and 48 kHz and is made of 32-bit-long audio words. The first 20 bits are reserved for the audio sample value. The following 9 bits are used to create control words and the remaining 3 bits add sync information to the 32-bit word. Control words provide audio channel information about preemphasis, normal audio or nonaudio data, copy prohibition, SDIF audio block synchronization information every 256 audio words, and user data information.

The transmission medium is a 75-ohm coaxial cable operating at transistor-transistor logic (TTL) levels and a data rate of 1.54 Mbps. It is a point-to-point operation interconnection system. Three coaxial cables are necessary to transmit left-channel data, right-channel data, and word clock signals.

6.9.3 SPDIF format

This format is a manufacturer's proprietary name for the consumer mode of the AES/EBU (AES3-1992) format protocol. Details can be found in Sec. 6.6.2. It was developed for serial transmission of digital audio data between semiprofessional and consumer devices. Format converters (data and level conversion) are necessary between AES3 professional mode and AES3 consumer mode equipment.

6.10 Audio Synchronization

Digital audio signals are made of discrete samples. Mixing, inserting, or assembling digital audio signals, from a variety of sources, requires the synchronization of samples to a reference source, in both phase and frequency. Two identical devices located in the same studio may generate a slow drift in the sample timing at the two outputs. As with video, this requires that a reference sampling rate be sent from a central generator or from one device to another. With audio sources external to the production center, this is not feasible, and consequently there will be a slow drift in timing between the internal and external sources. Consequently, solutions must be implemented to permit full digital audio production without the occurrence of pops and clicks. The synchronization complexity varies depending on whether digital audio sources or digital audio for video sources are concerned.

6.10.1 Synchronization between digital audio signals

The synchronization of different digital audio sources is achieved in two steps:

- Time alignment of sample clocks or frequency synchronization.
- Frame alignment of audio signals, that is, phase synchronization.

The AES11-1991 Recommended Practice specifies both the frequency and the phase synchronization of digital audio equipment in studio operations. Specialized generators can deliver a very stable reference for frequency synchronization such as needed in a large audio production facility. In this situation all production equipment is slaved to the master reference generator. Small studios can use the output of one device to feed and serve as a reference for all other equipment through a digital distribution amplifier.

Figure 6.37 shows the digital sample alignment to a digital audio reference signal (DARS). AES-11 specifies the digital audio samples must be in phase with a reference signal, with a tolerance of ±5% of an audio frame at a transmitter output, and with a tolerance of ±25% of an audio frame at a receiver input. The timing (or temporal) reference point is the first edge of the X or Z sync word.

When two digital audio signals are of different sampling rate or when there is no possibility to lock signals together, sample-rate converters and synchro-

Note: Timing is expressed as a percentage of the sampling period.

Figure 6.37 AES/EBU digital audio alignment to a DARS.

nizers are used. Synchronous conversion occurs when both rates are locked and then related by a ratio of integer numbers. Asynchronous conversion is much more complex and expensive because there is no temporal relationship between samples from original and converted signals. Figure 6.38 shows the synchronization to a common reference of all digital audio equipment in a production studio.

6.10.2 Synchronization between digital audio and video signals

In a television environment, the digital audio reference signal must be locked to the video reference signal to avoid a drift in the relationship between audio and video signals and allow click-free audio and video switching.

In 625/50 systems, there is an exact number of audio samples per video frame (1920 audio samples at a 48-kHz sampling rate). The phase relationship between audio and video signals can be easily maintained, as defined in EBU Recommendation R83-1996 and shown in Fig. 6.39. The AES3 audio can be locked synchronously to video by deriving a 48-kHz reference signal from the 625-line reference video signal. Figure 6.40 shows the relationship between 625-and 525-line video frequencies and the 48-kHz audio sampling rate.

Table 6.4 shows the relationship between the number of audio samples per video frame and the video frame rate for the three commonly used audio sampling frequencies.

In the 525/60 video system, the number of audio samples per video frame is not an integer number but rather a fractional number (8008/5), which can be calculated as follows:

$$\frac{33,366.67 \ \mu s}{20.8333 \ \mu s} = 1601.6 \ \text{samples}$$

Figure 6.38 Synchronization of digital audio signals in a production studio.

Note: Timing is expressed as a percentage of the sampling period.

Figure 6.39 Digital audio signal to video alignment for 625-line TV systems.

*Note: Numbers in brackets are relative to 4fsc PAL systems.

Figure 6.40   Relationship between video and audio sampling frequencies.

TABLE 6.4   Sampling Frequencies and Number of Audio Samples per Video frame

Sampling frequency, kHz	29.97-fps video	25-fps video	30-fps video
48	8008/5	1920/1	1600
44.1	147147/100	1764/1	1470
32	16016/15	1280/1	3200/3

where 33,366.67 μs is the duration of a video frame and 20.8333 μs is the duration of an audio frame. After five video frames, an integer number of audio samples (1601.6 × 5 = 8008) is obtained.

The five-phase sequence of digital audio to video alignment is shown in Fig. 6.41. The number of audio samples in each video frame is as follows:

- Frame 0   1602
- Frame 1   1601 (801 in Field 1 and 800 in Field 2)

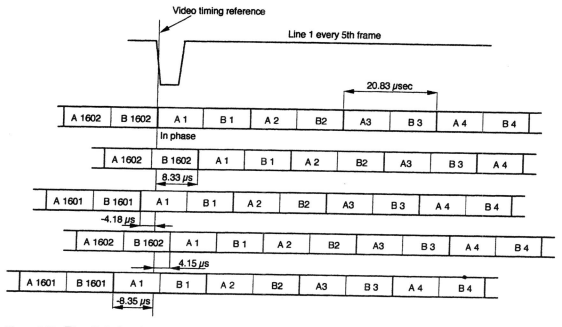

Figure 6.41 Five digital audio phases to video alignment for 525-line TV systems.

- Frame 2   1602
- Frame 3   1601
- Frame 4   1602

The relative timing of digital audio to the video reference point is calculated as follows:

- The phase of the first video frame (line 1 horizontal sync signal) is coincident with an AES Z or X sync sequence.
- The phase of the second video frame (line 1 horizontal sync signal) with respect to the nearest Z or X sync sequence is

$$20.83333 \ \mu s \times 1602 = 33375.00 \ \mu s$$

$$1 \ \text{video frame} = 1/29.97 = 33366.67 \ \mu s$$

$$\text{Phase} = 33,375.00 - 33,366.67 = 8.33 \ \mu s$$

- The phase of the third video frame (line 1 horizontal sync signal) with respect to the nearest Z or X sync sequence is

$$(20.83333 \times 1601) + 8.33 - 33,366.67 = -4.18 \ \mu s$$

Figure 6.42 Example of five-frame digital audio sequence resulting from a switch between two synchronous audio sources.

If audio signals are synchronous but not phased according to this five-frame sequence, an audio sample is injected or dropped, depending on the switching point. As an example, consider the case where there is a switch from source A to source B. The resulting switcher output signal is shown in Fig. 6.42, which exhibits three consecutive frames with 1602 samples each. One audio sample has been added in this example. This frequency modulation of the audio sampling rate increases with the number of switchings and may empty or fill the receiving buffer to the point where the audio signal is muted or produces a click. DVTRs internally resynchronize the input signal to the correct five-frame sequence before editing so that the sequence is maintained on the recorded tape and at the DVTR output in playback.

Present television equipment processes and delivers digital audio signals without any consideration to the five-frame sequence. Furthermore, the audio-video phasing may change at every power-up.

Processing equipment, such as digital audio synchronizers, can correct the repetition or omission of audio samples at the switching point or when there is a slight sampling frequency difference between the source and the receiver. The synchronizer selects quiet passages or pauses in the audio signal content and permits the removal or addition of samples only at such times.

The AES3 audio can be locked synchronously to video by deriving a 48-kHz reference signal from the 525-line reference video signal. Some generators such as the Tektronix SPG-422 provide digital audio signals that are referenced to analog and digital video signals as shown in Fig. 6.43. All equipment locked to this generator will produce the same audio to video alignment. However, a digital audio delay unit must be inserted before multiplexing audio and video signals. This is due to the delay of one or two frames that are accumulated in the video production mixer and special effects processor. The audio production mixer might also have the capability of delaying its output to obtain an exact timing to video signals.

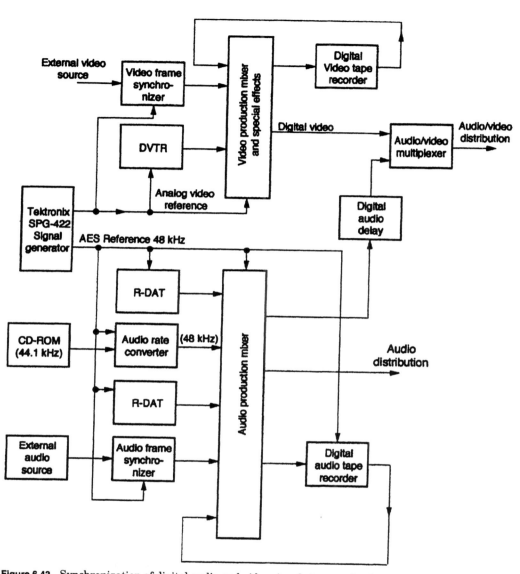

Figure 6.43 Synchronization of digital audio and video signals in a production studio.

6.11 Digital Audio Recording

The digital audio tape (DAT) format was originally developed as a consumer-type recorder, employing rotary heads, hence its well-known name R-DAT (rotary-DAT). However, DAT's possible use to make high-quality pirated digital copies has stopped its wide distribution to the consumer market. This format is now invading the professional market. The audio quality is such that this format is largely used in production and broadcast applications.

Analog input signals are digitized and channel-encoded by means of an 8–10 modulation scheme before being recorded on a 3.81-mm-wide metal tape. The playback system is a mirror image of the recording side.

The R-DAT recording format comprises four record/playback modes where the most interesting uses 48-kHz sampling, and two playback only modes at a 44.1-kHz sampling rate. Table 6.5 summarizes the general, electrical, and mechanical characteristics of the R-DAT format.

TABLE 6.5   R-DAT Characteristics

General characteristics	Number of audio channels: 2
	Recording time: 2 h (4 h in 32-kHz mode 2)
	No erase heads; record heads used also as playback heads
	Azimuth recording to minimize crosstalk between tracks
	Fast search time at 200 times normal speed
	Two optional longitudinal tracks for future use
	Cassette size: 73×54×10.5 mm
Electrical characteristics	Frequency response: Flat from 10 Hz to 22 kHz (f_s = 48 kHz)
	Dynamic range: 96 dB
	Sampling frequencies:
	32 kHz for long play
	44.1 kHz for playback only of commercial tapes
	48 kHz for high-quality record/playback
	Quantization:
	16-bit linear at 48- and 44.1-kHz sampling rate
	12-bit nonlinear at 32-kHz sampling rate (mode 2)
	Distortion: <0.005%
Mechanical characteristics	30-mm-diameter rotary head drum with 90° wrapping angle
	Drum revolution: 2000 r/min (1000 r/min @ 32-kHz mode 2)
	Tape speed: 8.15 mm/s (4.075 mm/s @ 32-kHz mode 2)
	Track pitch: 13.591 μm
	Tape width: 3.81 mm
	Tape thickness: 13 μm

Bit-Serial Signal Distribution and Data Multiplexing

Advances in technology have made it cost-effective to use bit-serial digital distribution of conventional television signals. All of the digital video data bits, synchronization information, and ancillary data, such as several channels of AES/EBU audio, can be distributed inside a teleproduction plant through a single coaxial cable. In many cases, existing cables within the facility can be used for bit-serial digital signal distribution. Figure 7.1 shows the block diagram of bit-serial digital video signal distribution.

The source encoder is the conventional analog-to-digital (A/D) converter. The transmitter channel encoder transforms the bit-parallel digital signal output of the A/D converter into a bit-serial digital signal suitable for transmission by the chosen medium (e.g., coaxial cable). The signal is corrupted by thermal noise contributed by the receiver input stage. The receiver channel decoder deserializes the received bit-serial signal and recovers the bit-parallel digital video signal. Excessive thermal noise may affect the receiver channel decoder, resulting in bits in error or missing altogether. The signal decoder is the conventional digital-to-analog (D/A) converter recovering the original analog video signal.

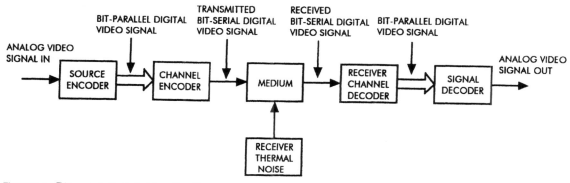

Figure 7.1 Bit-serial digital video distribution model.

The bit-serial data rate is given by

Bit-serial rate (Mbps) = Parallel bit rate (Mwords/s) × Number of
bits per word

The 4:2:2 component digital bit-serial rate is equal to

Bit-serial rate = 27 Mwords/s × 10 bits/word = 270 Mbps

The $4f_{SC}$ composite digital bit-serial rate is, nominally 143 Mbps for NTSC and 177 Mbps for PAL systems.

Figure 7.2 shows the spectrum of several conventional television bit-serial digital signals. The spectrum is typical of the nonreturn-to-zero (NRZ) coding and has nulls at the sampling frequency and its multiples. It is obvious that the distribution of bit-serial digital video signals requires very wide bandwidths that can be accommodated in a studio environment given adequate hardware technology. The distribution of bit-serial digital video signals over land lines, on-air transmitters, or by satellite requires a reduction of the bit rate to match existing standard communication channel capacities.

7.1 Shannon's Theorem

According to Shannon, a noisy communication channel has a specific capacity measured in bits per second (bps). The channel capacity is given by the following formula:

$$C\text{(bps)} = B \log_2 \left[1 + \left(\frac{S}{N} \right) \right]$$

where B = The channel bandwidth, Hz
$\quad\;\, S$ = The received signal power, W
$\quad\;\, N$ = The accompanying noise power, W

In a studio environment, the noise contribution is substantially the thermal noise of the first amplifying element of the receiver, since other types of noise are negligible. The signal-to-noise ratio (SNR) at the input of the receiver depends on the attenuation of the interconnecting coaxial cable and the equivalent input noise of the receiver. This puts a limit to the length and associated signal attenuation of the coaxial cable connecting the transmitter and the receiver. At a given room temperature T, the thermal noise power is proportional to the bandwidth, as in the formula

$$N = kTB$$

where k is Boltzmann's constant and T is the noise temperature in kelvin.

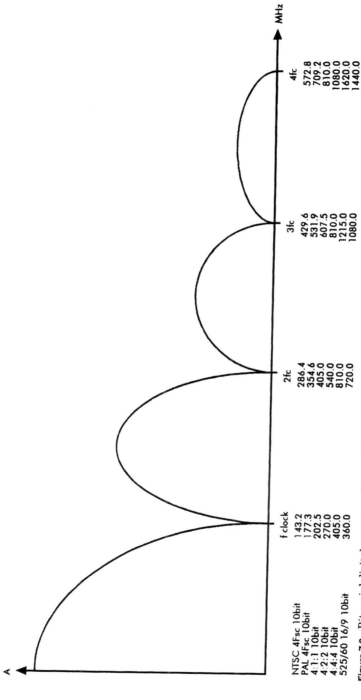

	f clock	2fc	3fc	4fc
NTSC 4Fsc 10bit	143.2	286.4	429.6	572.8
PAL 4Fsc 10bit	177.3	354.6	531.9	709.2
4:1:1 10bit	202.5	405.0	607.5	810.0
4:2:2 10bit	270.0	540.0	810.0	1080.0
4:4:4 10bit	405.0	810.0	1215.0	1620.0
525/60 16/9 10bit	360.0	720.0	1080.0	1440.0

Figure 7.2 Bit-serial digital spectrum of several formats.

The channel capacity is given by

$$C(\text{bps}) = B \log_2 \left[1 + \left(\frac{S}{kTB} \right) \right]$$

Shannon's theorem states that it is theoretically possible to transmit information with a low probability of error through a channel having a specific capacity, provided that the transmission rate is lower than the channel capacity. Shannon does not specify the means of obtaining an error-free transmission over a specific transmission channel, but simply states that there are means of achieving it.

7.2 Channel Coding

The channel coding describes the manner in which the 1's and 0's of the data stream are represented on the transmission path. There are many channel-coding standards and they all aim at optimizing some aspect of the bit-serial digital signal, such as the spectral distribution, the DC content, and the clock recovery.

The simplest and most commonly used channel code is NRZ. NRZ is characterized by logic 1 having a well-defined DC level and logic 0 having a well-defined lower DC level. The bit-serial digital signal is self-clocking. The receiver contains a clock regenerator that re-creates the clock through a phase-locked-loop (PLL) and a voltage-controlled oscillator (VCO). The PLL derives its reference from the 0-to-1-to-0 digital signal transitions. The NRZ code may result in long strings of 1's and 0's. These long "monotonous" data strings have no transitions, resulting in long periods of time during which the PLL reference is not refreshed. The accurate sampling of the bit-serial digital signal in the receiver during these periods depends on the stability of the clock VCO. In addition, the NRZ code has a zero-frequency (DC) component, which varies with the nature of the data stream, and a significant low-frequency content, making it inappropriate for AC-coupled receivers. For these reasons the NRZ code is not used in its basic form in bit-serial digital video transmissions.

Bit-serial digital video transmissions use a derivative of the NRZ code, the NRZ Inverted (NRZI) code. Figure 7.3 shows an example of an NRZ-coded digital signal and the derived NRZI-coded signal. NRZI codes logic 0's as a DC level (0 or 1) and logic 1's as a transition. When the NRZ-coded digital signal is a long string of 1's, the derived NRZI-coded signal is a square wave at one-half the clock frequency. As shown, for a given binary sequence, an NRZI-coded signal has more transitions per unit of time than an NRZ-coded signal, resulting in improved clock regenerator PLL operation. Provided that the system limits the maximum number of 0's in the data stream, the receiver clock regeneration works quite well. The standards meet this requirement by reserving the all-zero word for sync purposes only. The NRZI, although superior to the NRZ coding, still has a DC component and a significant low-frequency content.

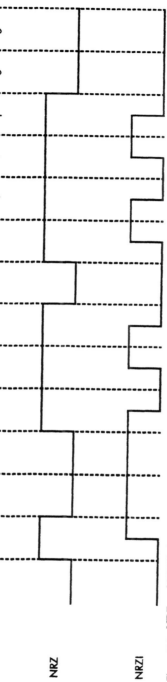

Figure 7.3 NRZ and NRZI channel coding characteristics.

A further improvement in the receiver clock recovery is obtained through scrambling. The scrambler randomizes long sequences of 0's and 1's as well as repetitive data patterns that could result in clock regeneration difficulties. It helps eliminate the DC content and provides sufficient signal transitions for reliable clock recovery. Figure 7.4 shows the block diagram of a scrambler followed by an NRZ-to-NRZI encoder.

The scrambler produces a pseudo-random binary sequence (PRBS), which in turn is combined with transmitted data in order to randomize it. It consists of a nine-stage shift register (nine sections of clocked master-slave D flip-flop marked D in the diagram) with associated feedback. The feedback signals are combined by exclusive OR (XOR) adders (marked $\oplus$ in the diagram) with the following input versus output truth table:

$$0 \text{ (at input A)} + 0 \text{ (at input B)} = 0 \text{ (at output)}$$

$$0 \text{ (at input A)} + 1 \text{ (at input B)} = 1 \text{ (at output)}$$

$$1 \text{ (at input A)} + 0 \text{ (at input B)} = 1 \text{ (at output)}$$

$$1 \text{ (at input A)} + 1 \text{ (at input B)} = 0 \text{ (at output)}$$

The scrambling function is classified using a shorthand method of describing the feedback connection, known as the *characteristic polynomial*. For the nine-stage register illustrated in Fig. 7.4 the polynomial is

$$G1(X) = X^9 + X^4 + 1$$

The scrambler can produce long runs of 1's. These are converted to transitions by an NRZ-to-NRZI converter consisting of a single-stage master-slave D flip-flop with an XOR gate. The polynomial of the NRZI converter is

$$G2(X) = X + 1$$

The original data are recovered by an NRZI-to-NRZ converter followed by a descrambler. Figure 7.5 shows the block diagram of the descrambler. The logic arrangement is identical to the one used in the scrambler except that "feedforward" is used instead of feedback. The same random sequence is added to the signal before transmission and subtracted at the decoder, resulting in the recovered data being identical to the original data.

7.3 The Eye Diagram

The eye pattern (or eye diagram) is used in specifying and verifying the characteristics of a bit-serial digital signal. The name results from the appearance on a storage oscilloscope of sections of digital symbol patterns superimposed on one another. For an infinite-bandwidth system, the transitions from 0-to-1-to-0 are instantaneous and, consequently, the "eye" is square. A practical system

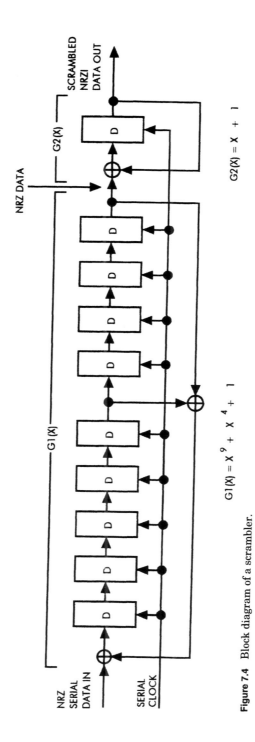

Figure 7.4 Block diagram of a scrambler.

$G1(X) = X^9 + X^4 + 1$

$G2(X) = X + 1$

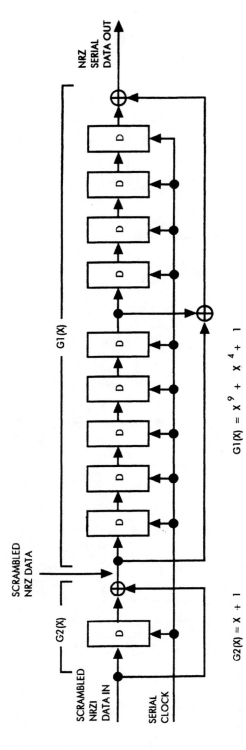

Figure 7.5 Block diagram of a descrambler.

$$G2(X) = X + 1$$

$$G1(X) = X^9 + X^4 + 1$$

has a finite bandpass, resulting in transitions with a slower risetime and the familiar eye shape. Bit-serial digital signals are specified in terms of eye amplitude, risetime and decay time, overshoot and undershoot, and jitter. Figure 7.6 shows the formation of an eye pattern from superimposed binary patterns.

7.4   Bit-Serial Distribution Standard

SMPTE Standard 259M describes the bit-serial interface for 525/60 and 625/50 digital equipment operating with either 4:2:2 component signals or $4f_{SC}$ composite signals. It has applications in a television studio using coaxial cable lengths not exceeding the amount specified by the equipment manufacturer, typically accepting a signal loss of 30 dB at either the clock frequency or half the clock frequency. The Belden 8281B coaxial cable has a loss of 30 dB at 135 MHz.

7.4.1   Interface characteristics

The interface characteristics are summarized in Table 7.1. Of particular importance are the tolerances on the parameters of the waveform present at the output of a bit-serial digital signal source (the transmitter). This signal is derived from a bit-parallel digital signal source that meets good studio practices. Figure 7.7 shows a drawing of an eye diagram waveform and identifies the measurement parameters listed in Table 7.1.

There is a great deal of confusion concerning the jitter measurements, their effect on the signal, and the measurements methods. Jitter can best be defined as a pulse-position modulation. Ideally, the jittery pulse position is compared to a stable and jitter-free primary clock, resulting in a measurement of "absolute" or total jitter. The measured value contains all pulse-position modulation (jitter) frequencies. An alternate method of jitter measurement uses a measurement reference clock derived from the jittery signal itself, resulting in a measurement of "relative" jitter. Depending on the clock extraction method, the recovered clock will contain some of the jitter spectral components of the signal to be measured. SMPTE Recommended Practice RP184 specifies two types of relative jitter in terms of the measurement bandwidth.

- Timing jitter: Jitter measured in a bandpass extending from 10 Hz to one-tenth of the clock rate.

- Alignment jitter: Jitter measured in a bandpass extending from 1 kHz to one-tenth of the clock rate.

The SMPTE jitter specifications are listed in Table 7.2.

Figure 7.8 shows a comparison of three types of jitter measurement bandpass characteristics.

SMPTE Recommended Practice RP 192 describes the recommended jitter measurement methods and the reader is referred to this document.

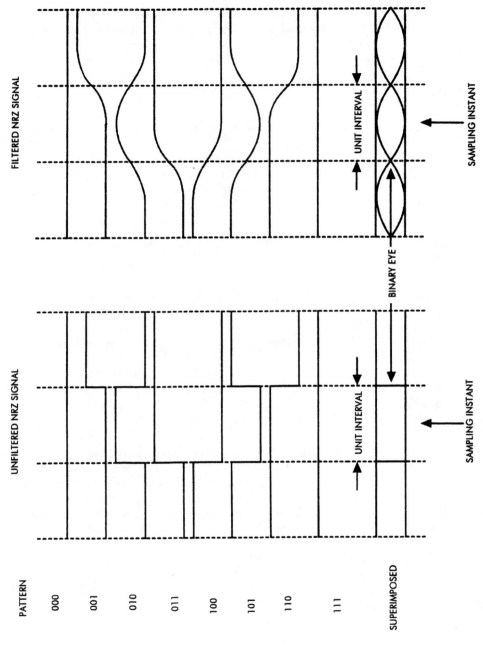

PATTERN

000

001

010

011

100

101

110

111

SUPERIMPOSED

UNFILTERED NRZ SIGNAL

FILTERED NRZ SIGNAL

UNIT INTERVAL

UNIT INTERVAL

SAMPLING INSTANT

SAMPLING INSTANT

BINARY EYE

Figure 7.6 Formation of the eye-pattern from superimposed binary patterns.

TABLE 7.1   Characteristics of Bit-Serial Interfaces

Channel coding	Transmitter characteristics (see Fig. 7.7)	Receiver characteristics
Scrambled NRZI	Unbalanced output	Unbalanced input
Input signal polarity: Positive logic	Source impedance: 75 ohms nominal	Input impedance: 75 ohms nominal
Generator polynomials: $G_1(X) = X^9 + X^4 + 1$ $G_2(X) = X + 1$	Return loss: ≥15 dB (5 MHz to clock frequency of signal)	Return loss: ≥15 dB (5 MHz to clock frequency of signal)
Data word length: 10 bits	Output signal amplitude: 800 mV$_\text{p-p}$ ±10%	Optional cable loss equalization: 30 dB at clock frequency of signal
Transmission order: LSB of any data word transmitted first	DC offset: 0.0 V ±0.5 V with reference to midamplitude of signal	
	Rise time and fall time: 0.4 to 1.5 ns between 20% and 80% of signal amplitude points. Differences not to exceed 0.5 ns	
	Overshoot of rising and falling signal edges: <10% of signal amplitude	
	Jitter: See Table 7.2	

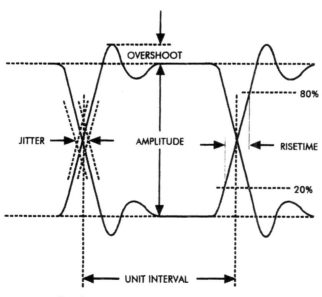

Figure 7.7 Eye diagram measurement dimensions.

TABLE 7.2   Bit-Serial Digital Signal Jitter Specifications (See Fig. 7.8)

Parameter	Magnitude	Notation
Timing jitter lower band edge	10 Hz	f1
Alignment jitter lower band edge	1 kHz	f3
Upper band edge	>1/10 clock rate	f4
Timing jitter*	$0.2\ \mathrm{UI_{p\text{-}p}}$‡	A1
Alignment jitter	$0.2\ \mathrm{UI_{p\text{-}p}}$	A2
Test signal	100% color bars	100/0/100/0
Serial clock divider†	≠10	n

*Some bit-parallel signals clock may contain jitter up to 6 ns, which could cause excessive bit-serial jitter.

†This refers to the reference signal for triggering the jitter-measuring oscilloscope. Using a times 10 division is acceptable, but it may mask some jitter components.

‡UI stands for unit interval and has the following values:

$4f_{\mathrm{SC}}$ NTSC: UI = 6.98 ns   Allowed jitter ≈ 1.4 ns
$4f_{\mathrm{SC}}$ PAL: UI = 5.63 ns   Allowed jitter ≈ 1.13 ns
        4:2:2:   UI = 3.7 ns   Allowed jitter ≈ 0.74 ns

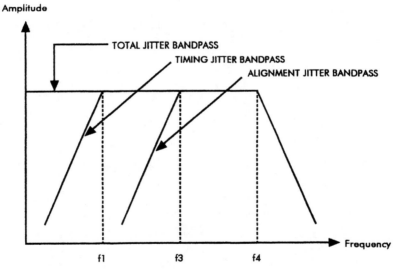

Figure 7.8   Jitter measurement bandpass.

7.4.2 $4f_{sc}$ bit-serial distribution

The bit-serial digital $4f_{\mathrm{SC}}$ data stream uses as its source the bit-parallel $4f_{\mathrm{SC}}$ data stream described in Sec. 3.2.4. The bit rate is nominally 143 Mbps for NTSC and 177.3 Mbps for PAL. Some older bit-parallel interface equipment may carry only 8 bits of video data. It is necessary for the serializer to identify this condition and add the necessary data to convert 8-bit data to a 10-bit representation.

7.4.2.1 Signal processing. Signal processing of the bit-parallel input signal is required in order to provide timing and synchronization information in the bit-serial digital domain. This information is designated TRS-ID, timing reference signal and line number identification. The TRS-ID information is present following the sync leading edge of the horizontal synchronizing pulse. It consists of two consecutive data blocks.

- The TRS data block: This block consists of four words at word number addresses 790, 791, 792, 793 in NTSC and 967, 968, 969, 970 in PAL with corresponding hexadecimal values of 3FF, 000, 000, 000. Table 7.3 lists the TRS reference signal's binary and hexadecimal values for NTSC and PAL systems.

- The ID data block: This data block consists of one word at word number address 793 in NTSC and 971 in PAL. This information helps identify the video line number and the current field (in a four-field NTSC sequence and an eight-field PAL sequence). Table 7.4 lists the values of bits 0 through 2

TABLE 7.3 $4f_{sc}$ *Timing Reference Signal's Binary and Hexadecimal Values for NTSC and PAL*

Bit	Word 790 NTSC, 967 PAL	Word 791 NTSC, 968 PAL	Word 792 NTSC, 969 PAL	Word 793 NTSC, 970 PAL
9	1	0	0	0
8	1	0	0	0
7	1	0	0	0
6	1	0	0	0
5	1	0	0	0
4	1	0	0	0
3	1	0	0	0
2	1	0	0	0
1	1	0	0	0
0	1	0	0	0
Hex	3FF	000	000	000

TABLE 7.4 $4f_{sc}$ **Line Number ID Values for Bits 0 Through 2 of NTSC Word 794 and PAL Word 971**

Field	NTSC line	PAL line	Bit 2	Bit 1	Bit 0
1	1–263	1–313	0	0	0
2	264–525	314–625	0	0	1
3	1–263	1–313	0	1	0
4	264–525	314–625	0	1	1
5	N.A.*	1–313	1	0	0
6	N.A.	314–625	1	0	1
7	N.A.	1–313	1	1	0
8	N.A.	314–625	1	1	1

*N.A. = not applicable.

TABLE 7.5 Relationship Between the Binary and Decimal Values of Bits 3 Through 7 of NTSC Word 794 and PAL Word 971 and the Corresponding NTSC and PAL Line

Dec	Bit 7	Bit 6	Bit 5	Bit 4	Bit 3	NTSC line	PAL line
0	0	0	0	0	0	Not used	Not used
1	0	0	0	0	1	1 and 264	1 and 314
2	0	0	0	1	0	2 and 265	2 and 315
3	0	0	0	1	1	3 and 267	3 and 316
⋮	⋮	⋮	⋮	⋮	⋮	⋮	⋮
29	1	1	1	0	1	29 and 292	29 and 342
30	1	1	1	1	0	30 and 293	30 and 343
31	1	1	1	1	1	≥31 and ≥294	≥31 and ≥344

cross-referenced to the NTSC and PAL lines and fields. Bits 3 (least-significant bit, LSB) through 7 (most-significant bit, MSB) identify the line. These 5 bits can represent 32 states ($2^5 = 32$). These states are used to identify the first 30 lines in each field. Table 7.5 shows the relationship between the binary and decimal values of bits 3 through 7 and the corresponding NTSC and PAL line. The binary 11111 (decimal 31) value identifies lines 31 and subsequent in each field. Bit 8 is used as a parity bit. Its value is determined by the values of bits 0 through 7. If an even number of these have the value 1, bit 8 is set at 0. Otherwise, it is set at 1. Bit 9 is the complement of bit 8.

Figures 7.9 to 7.11 (NTSC) and 7.12 to 7.14 (PAL) show the location of the TRS-ID information. The PAL TRS to H-sync edge position is reset once per field. This is due to the noninteger number of samples per line and results in the addition of two additional samples on lines 313 and 625, which have 1137 samples.

7.4.2.2 The serializer. Figure 7.15 shows the block diagram of a $4f_{SC}$ serializer. The serializer performs several functions implemented in dedicated integrated circuits (ICs). These are

Figure 7.9 $4f_{SC}$ NTSC horizontal sync period details showing location of TRS-ID and optional ancillary data.

Figure 7.10 $4f_{SC}$ NTSC vertical sync pulse details showing location of TRS-ID and optional ancillary data.

Figure 7.11 $4f_{SC}$ NTSC equalizing pulse details showing location of TRS-ID and optional ancillary data.

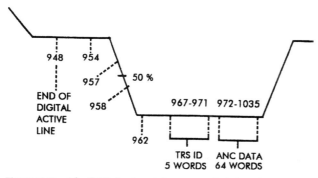

Figure 7.12 $4f_{SC}$ PAL horizontal sync period details showing location of TRS-ID and optional ancillary data.

- The TRS and ID coprocessing: This is performed by a dedicated, application-specific IC (ASIC) that generates and inserts TRS and ID data as described in Sec. 7.4.2.1.

- Parallel-to-serial conversion: This is performed by a 10-bit shift register, which is clocked at 10 times the input rate (143 MHz for NTSC and 177 MHz for PAL). If there are 8 bits in the input words, bits 0 and 1 are forced to 0 for transmission.

Figure 7.13 $4f_{SC}$ PAL vertical sync pulse details showing location of TRS-ID and optional ancillary data.

Figure 7.14 $4f_{SC}$ PAL equalizing pulse details showing location of TRS-ID and optional ancillary data.

- Scrambling: This is a process of randomization described in Sec. 7.2. This process produces NRZ signals.

- Conversion from NRZ to NRZI: This process was described in Sec. 7.2.

- Serial clock generation: The serial clock is generated using a VCO operating at the bit-serial clock frequency. Its frequency is derived from the parallel clock frequency and is controlled by a PLL circuit. The derived VCO frequency control voltage is low-pass-filtered by an unspecified filter that determines the capture range and hold range of the VCO frequency and removes high frequencies from the control voltage. This allows the serial clock to follow low-frequency jitter or drift (wander) of the parallel clock as well as correct for a temperature-related drift of its own. The VCO free-run frequency is controlled by an external potentiometer and can be preset to a selected bit rate. The potentiometer is usually coupled with a temperature-sensitive device (diode) to reduce temperature-related frequency drifts. Some manufacturers feature an automatic frequency selector that switches the VCO free-run frequency to 10 times that of the input parallel clock. The serial clock feeds the shift register, the scrambler, and the NRZ-to-NRZI converter.

- Cable driving: Following the NRZI converter there are active line drivers for each output, unlike baseband video where multiple outputs can be split from a single active driver.

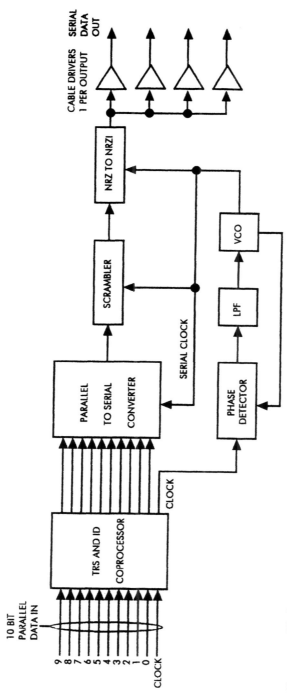

Figure 7.15 Block diagram of a $4f_{SC}$ composite digital serializer.

Figure 7.16 Spectrum of 143 Mbps $4f_{SC}$ composite digital bit-serial signal.

Some manufacturers combine the functions of serialization, scrambling, NRZ-to-NRZI conversion and clock generation in a single ASIC (e.g., Sony SBX1601). Other manufacturers use different function-packaging concepts. The cable driver may be a multiple-function IC (e.g., Sony CX1389AQ) or discrete transistors.

Figure 7.16 shows the spectrum of the 143 Mbps $4f_{SC}$ NTSC bit-serial signal.

7.4.2.3 The deserializer. Figure 7.17 shows the block diagram of a $4f_{SC}$ deserializer. The deserializer performs several functions implemented in dedicated ICs. These are

- Cable-loss equalization: An automatic cable-loss equalizer corrects for high-frequency (>8 MHz) and low-frequency (<8 MHz) losses introduced by the coaxial cable. The equalization capability is a manufacturer's choice. Some circuits are capable of automatically equalizing high-frequency losses on the order of 30 dB at 135 MHz to within ±2 dB for cable lengths

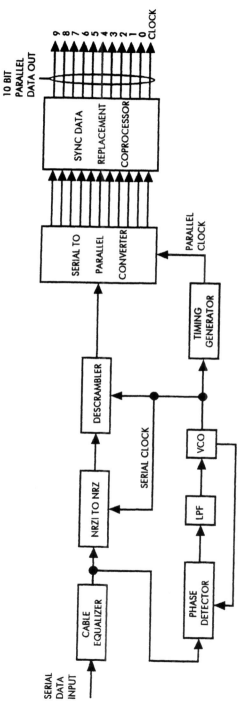

Figure 7.17 Block diagram of a $4f_{SC}$ composite digital deserializer.

between 0 and 300 m. Some deserializers have no equalization capabilities and can operate only with short coaxial cables.

- NRZI-to-NRZ conversion: This process was described in Sec. 7.2.

- Descrambling: This process was described in Sec. 7.2.

- Clock recovery: Clock recovery relies on the fact that the scrambled NRZI data stream contains a large number of transitions. The current state-of-the-art technology relies on PLL concepts for locking the receiver data extraction circuitry to the incoming data. PLLs have a specific bandwidth determined by the low-pass filter at the output of the phase detector. This bandwidth should, ideally, be very narrow to achieve a high level of noise immunity. Narrow-bandwidth PLLs have a correspondingly narrow pull-in (capture) range, requiring a highly stable crystal-controlled VCO to stop it from drifting beyond the PLL capture range. Noise immunity and capture range are conflicting requirements in the design of PLL circuitry. The current dominant technology relies on a PLL bandwidth on the order of 2 MHz. This means that the VCO will follow incoming signal jitter frequency-domain components up to a limit of 2 MHz. It means also that the VCO free-run frequency may drift up to 2 MHz from the wanted frequency and the PLL will correct for this hypothetical frequency drift. Undoubtedly, alternate technologies, characterized by different PLL bandwidths, will emerge. The VCO free-run frequency is controlled by an external potentiometer and can be set to match a selected bit rate. The potentiometer is usually coupled with a temperature-sensitive device (diode) to reduce temperature-related frequency drifts. Some manufacturers feature an automatic frequency selector that switches the VCO free-run frequency to match the bit rate of the input signal. The recovered serial clock feeds the NRZ-to-NRZI converter and the descrambler. It also feeds the timing generator, which regenerates the parallel clock required by the serial-to-parallel converter.

- Sync data replacing: A separate dedicated sync data replacement coprocessor ASIC removes the TRS-ID information and restores the bit-parallel signal to its original condition.

- Regeneration of a reclocked bit-serial signal: This function, available with some designs, permits the regeneration of a high-quality noise-free bit-serial signal from a corrupted input signal. The low-frequency jitter, inside the PLL bandwidth, will be carried through, but the high-frequency jitter will be eliminated.

Some manufacturers provide the functions of equalization, NRZI-to-NRZ conversion, descrambling, serial-to-parallel conversion, clock recovery, and regeneration of a reclocked input signal in a single ASIC (e.g., Sony SBX1602). Other manufacturers have different function-packaging concepts. The coprocessor is a separate IC.

TABLE 7.6 $4f_{SC}$ NTSC Ancillary Data Space

Horizontal ancillary data space (HANC)	55 Words/active line $\times$ 485 active lines/frame = 26,675 words/frame 26,675 Words/frame $\times$ 29.97 frames/s = 799,449.75 words/s 799,449.75 Words/s $\times$ 10 bits/word = 7.9944975 Mbps

Vertical ancillary data space (VANC)	Broad pulses:
	   376 Words/broad pulse $\times$ 6 broad pulses/field = 2256 words/field
	   2256 words/field $\times$ 59.94 fields/s = 135,224.6 words/s
	   135,224.6 words/s $\times$ 10 bits/word = 1.352246 Mbps
	
	Equalizing pulses:
	   21 Words/eq. pulse $\times$ 6 eq. pulses/field = 126 words/field
	   126 Words/field $\times$ 59.94 fields/s = 7552.44 words/s
	   7552.4 Words/s $\times$ 10 bits/word = 0.0755244 Mbps
	
	Sync tip of blanked lines:
	   55 Words/line $\times$ 20 blanked lines/field = 1100 words/field
	   1100 Words/field $\times$ 59.94 fields/s = 65,934 words/s
	   65,934 Words/s $\times$ 10 bits/word = 0.65934 Mbps

Total ancillary data space	7.9944975 Mbps (HANC) + 2.08711 Mbps (VANC) $\approx$ 10.0816075 Mbps Data formatting and exclusions may reduce this value by 10% to 20%
Total bit rate	910 Words/total line $\times$ 525 lines/frame $\times$ 29.97 frames/s $\times$ 10 bits/word = 143.1816 Mbps
Essential bit rate	143.1816 Mbps $-$ 10.0816975 Mbps $\approx$ 133 Mbps

7.4.2.4 Ancillary data capabilities. As shown in Figs. 7.9 to 7.11 for NTSC and 7.12 to 7.14 for PAL systems, there is provision for inserting ancillary data at specific instants during the vertical and horizontal blanking period.

Table 7.6 shows the ancillary data space capability of the $4f_{SC}$ NTSC bit-serial digital format. The total ancillary data space including the horizontal ancillary data (HANC) and the vertical ancillary data (VANC) is of the order of 10 Mbps. The bit rate of an AES/EBU digital audio dual-channel signal is 3.072 Mbps. It follows that the format can accommodate two AES/EBU digital audio data streams, or four audio channels, amounting to a total 6.144 Mbps. This leaves about 3.9 Mbps of ancillary data space available for the transmission of other types of information on a single coaxial cable along with the digital video signal. Alternately, the excess 10 Mbps can be eliminated, because it is nonessential to the video information, resulting in a reduced "essential" $4f_{SC}$ NTSC bit rate of 133 Mbps.

Table 7.7 shows the ancillary data space capability of the $4f_{SC}$ PAL bit-serial digital format. The total ancillary data space, including the HANC and VANC, is on the order of 11.2 Mbps. This allows for a transmission of two AES/EBU data streams amounting to 6.144 Mbps, as well as other types of information not exceeding 5 Mbps. Alternately, the excess 11.2 Mbps can be eliminated because it is nonessential to the video information, resulting in a reduced "essential" $4f_{SC}$ bit rate of 166.1 Mbps.

TABLE 7.7   **$4f_{SC}$ PAL Ancillary Data Space**

Horizontal ancillary data space (HANC)	64 Words/active line × 575 active lines/frame = 36,800 words/frame 36,800 words/frame × 25 frames/s = 920,000 words/s 920,000 Words/s × 10 bits/word = 9.2 Mbps

Vertical ancillary data space (VANC)	Broad pulses:
	  466 Words/broad pulse × 5 broad pulses /field = 2330 words/field
	  2330 Words/field × 50 fields/s = 116,500 words/s
	  116,500 Words/s × 10 bits/word = 1.165 Mbps
	
	Equalizing pulses:
	  23 Words/eq. pulse × 5 Eq. pulses/field = 115 word s/field
	  115 Words/field × 50 fields/s = 5750 words/s
	  5750 Words/s × 10 bits/word = 0.0575 Mbps
	
	Sync tip of blanked lines:
	  64 Words/line × 25 blanked lines/field = 1600 words/field
	  1600 Words/field × 50 fields/s = 80,000 words/s
	  80,000 Words/s × 10 bits/word = 0.8 Mbps

Total ancillary data space	9.2 Mbps (HANC) + 2.0225 Mbps (VANC) = 11.2225 Mbps Data formatting and exclusions may reduce this value by 10% to 20%
Total bit rate	1135 Words/total line × 625 lines/frame × 25 frames/s × 10 bits/word = 177.34375 Mbps
Essential bit rate	177.34375 Mbps − 11.2225 Mbps ≈ 166.1 Mbps

7.4.3   4:2:2 bit-serial distribution

The bit-serial 4:2:2 data stream uses as its source the bit-parallel 4:2:2 data stream described in Sec. 3.3.9. The bit rate is 270 Mbps. Some older equipment may carry only 8 bits of video data. It is necessary for the serializer to identify this situation and add the necessary data to convert 8-bit data to a 10-bit representation.

7.4.3.1   The serializer.   Figure 7.18 shows the block diagram of a 4:2:2 serializer. The serializer performs several functions implemented in dedicated ICs. These are

- Parallel-to-serial conversion: This is performed by a 10-bit shift register that is clocked at 10 times the input rate (270 MHz). If there are 8 bits in the input words, bits 0 and 1 are forced to 0 for transmission.

- Scrambling: This is a process of randomization described in Sec. 7.2.

- Conversion from NRZ to NRZI: This process was described in Sec. 7.2.

- Serial clock generation: The serial clock generation circuitry is identical to that used for the $4f_{SC}$ with the exception of clock frequency, which is 270 MHz.

- Cable driving: The circuitry used is identical to that used for the $4f_{SC}$

Some manufacturers combine the functions of serialization, scrambling, NRZ-to-NRZI conversion, and clock generation into a single ASIC (e.g., Sony

Figure 7.18 Block diagram of a 4:2:2 component digital serializer.

SBX1601). Other manufacturers use different function-packaging concepts. The cable driver may be a multiple-function IC (e.g., Sony CX1389AQ) or discrete transistors.

Figure 7.19 shows the spectrum of the 4:2:2 component digital bit-serial spectrum.

7.4.3.2 The deserializer. Figure 7.20 shows the block diagram of a 4:2:2 270-Mbps deserializer.

The deserializer performs several functions implemented in dedicated ICs. These are:

- Cable-loss equalization: The circuitry is identical to that used for $4f_{SC}$.

- NRZI-to-NRZ conversion: This process was described in Sec. 7.2.

- Descrambling: This process was described in Sec. 7.2.

- Clock recovery: A 270-MHz clock is derived from the equalized incoming signal through a 2-MHz-bandwidth PLL controlling the frequency of a VCO in a manner similar to that of a $4f_{SC}$ circuit. All considerations concerning the PLL bandwidth and the VCO frequency stability apply here as well. The VCO stability problems are compounded by the fact that the VCO operates at a higher frequency (270 MHz), making frequency drift control more difficult. The regenerated 270-MHz clock feeds the NRZI-to-NRZ converter and the descrambler. It also feeds a timing generator that regenerates the 27-MHz clock required by the serial-to-parallel converter.

A - RS-1-1

Tek
2712

500MHz
0. 2dBm
100. OMHz/
5MHz RBW

ATTN 30dB
VF WIDE
    10 dB/

TIME: 50 ms/DIV

PEAK MODE

TIME: 11: 49: 38
DATE: 07-NOV-91

Note: Readouts
correspond to
waveform 'A'

Figure 7.19 Spectrum of 270 Mbps 4:2:2 component digital bit-serial signal.

7.4.3.3 Ancillary data capabilities. As shown in Figs. 7.21 (for 4:2:2 525/60) and 7.22 (for 4:2:2 625/50), there is provision of inserting ancillary data (HANC) at specific instants during the horizontal blanking interval. In addition, 1440 eight-bit words of vertical ancillary data (VANC) may be inserted in 38 vertical-interval lines (lines 1–19 and 264–282) in the 525/60 standard and 48 vertical-interval lines in the 625/50 standard lines.

Table 7.8 shows the ancillary data space capability of the 4:2:2 525/60 standard. The total ancillary data space, including HANC and VANC, is of the order of 55 Mbps, considering the known allowed data locations. Data formatting and various exclusions may reduce the ancillary data space by up to 20%. The essential video bit rate is on the order of 214.7 Mbps.

Table 7.9 shows the ancillary data space capability of the 4:2:2 625/50 standard. The total ancillary data space, including HANC and VANC, is of the order of 57.5 Mbps, considering the known allowed data locations. Data formatting and various exclusions may reduce the ancillary data space by up to 20%. The essential video bit rate is on the order of 212.4 Mbps.

SMPTE Standard 272M defines means of embedding up to 16 audio channels or 8 AES/EBU digital audio dual-channel signals for a total of less than 25 Mbps. This leaves a considerable amount of available ancillary data (of the order of 30 Mbps) for other applications.

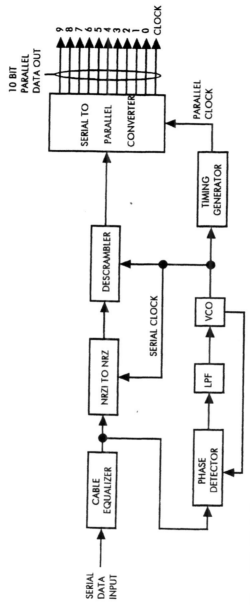

Figure 7.20 Block diagram of a 4:2:2 component digital deserializer.

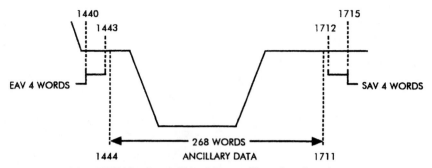

Figure 7.21   4:2:2 525/60 horizontal blanking ancillary data location.

Figure 7.22   4:2:2 625/50 horizontal blanking ancillary data location.

TABLE 7.8   4:2:2 525/60 Ancillary Data Space

Horizontal ancillary data space (HANC)	268 Words/line × 525 lines/frame = 140,700 words/frame 140,700 Words/frame × 29.97 frames/s = 4.216779 Mwords/s 4.216779 Mwords/s × 10 bits/word = 42.16779 Mbps
Vertical ancillary data space (VANC)	1440 Words/line × 38 vertical-interval lines = 54,720 words/frame 54,720 Words/frame × 29.97 frames/s = 1.6399584 Mwords/s 1.6399584 Words/s × 8 bits/word = 13.1196672 Mbps
Total ancillary data space	42.16779 Mbps (HANC) + 13.1196672 Mbps (VANC) ≈ 55.3 Mbps Data formatting and exclusions may reduce this value by 10% to 20%
Total bit rate	1716 Words/total line × 525 lines/frame × 29.97 frames/s × 10 bits/word ≈ 270 Mbps
Essential bit rate	270 Mbps − 55.3 mbps = 214.7 Mbps

7.5   Performance-Indicative Parameters and Measurement Concepts

In a studio environment the bit-serial digital signal uses coaxial cable as a transmission medium. Coaxial cable affects the signal in several ways that lead to distortion of the serialized video waveform. Bit-serial digital signal distribution technology ensures unimpaired picture quality under conditions of noise and waveform distortion that would be totally unacceptable in an analog video signal. The penalty is that there is no warning of impending dis-

TABLE 7.9   4:2:2 625/50 Ancillary Data Space

Horizontal ancillary data space (HANC)	280 Words/line × 625 lines/frame = 175,000 words/frame 175,000 Words/frame × 25 frames/s = 4.375 Mwords/s 4.375 Mwords/s × 10 bits/word = 43.75 Mbps
Vertical ancillary data space (VANC)	1440 Words/line × 48 vertical-interval lines = 69,120 words/frame 69,120 Words/frame×25 frames/s = 1.728 Mwords/s 1.728 Mwords/s × 8 bits/word = 13.824 Mbps
Total ancillary data space	43.75 Mbps (HANC) + 13.824 Mbps (VANC) = 57.574 Mbps Data formatting and exclusions may reduce this value by 10% to 20%
Total bit rate	1728 Words/total line × 625 lines/frame × 25 frames/s × 10 bits/word = 270 Mbps
Essential bit rate	270 Mbps − 57.574 Mbps = 212.426 Mbps

aster as far as picture quality is concerned. Unlike analog video technology, where cumulative waveform distortions are readily recognizable in the progressively deteriorating picture quality, the degradation of the digitally processed and distributed picture is rather sudden.

Figure 7.23 shows a classification of bit-serial signal performance-indicative parameters related to the transmitter, the distribution medium, and the receiver of bit-serial signals. The listed parameters reflect the experience gained by the authors in installing a large digital teleproduction center (CBC Broadcast Center in Toronto, Canada).

There are three areas of performance-related engineering concern as follows:

- Evaluation of equipment and technology: This activity aims to determine the performance of the equipment, verify its adherence to international and in-house standards, evaluate its robustness to pathological signals, and generate an evaluation report.

- Postinstallation acceptance tests: This activity aims to confirm that the equipment operates satisfactorily. A reduced number of parameters are tested, since the generic product has previously been tested and approved for use. These tests are also used to generate an acceptance report.

- Maintenance tests: These tests are performed to ensure that the equipment operates as installed and to verify that the equipment was restored to the original performance level after repairs.

Table 7.10 shows the application of selected performance-indicative parameters to various engineering activities.

7.5.1   Measuring transmitter-related parameters

7.5.1.1   Output signal characteristics.

The standardization of the measurements of the transmitter output signal characteristics requires accuracy, speed, and reproducibility. It is advantageous to use a digitizing oscilloscope

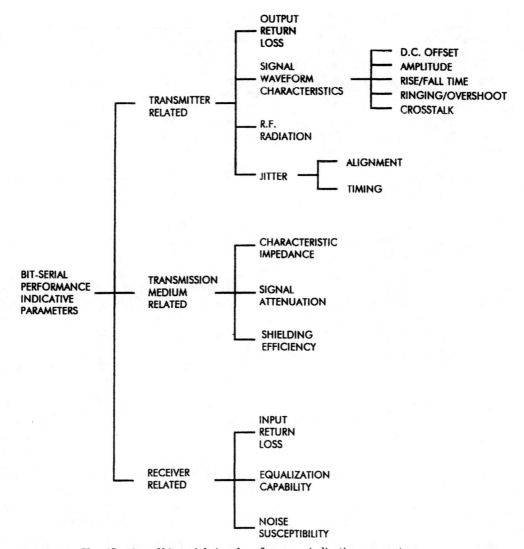

Figure 7.23 Classification of bit-serial signal performance indicative parameters.

that can be programmed to measure a set of parameters and display the results on the screen. The tests using a digitizing oscilloscope are faster and more accurate than tests using analog oscilloscopes and, in addition, a screen printout is available. A typical screen printout of a good 270-Mbps bit-serial signal waveform is shown in Fig. 7.24. The display shows the bit-serial waveform and the value of the six measured parameters. The parameters measured are amplitude, rise-time and fall time, overshoot and undershoot, and

TABLE 7.10 Performance-Indicative Parameters and Application

Parameter	Unit	Evaluation	Acceptance	Maintenance
Return loss	dB	Yes	No	No
Signal amplitude	p-p mV	Yes	Yes	Yes
Rise time/fall time	ns	Yes	Yes	Yes
Overshoot/undershoot	%	Yes	Yes	Yes
Jitter	p-p ps	Yes	Yes	Yes
Bit error rate	Error/min	Yes	Yes	Yes
Equalization capability	m@BER	Yes	No	Yes
Pathological check	Error/min	Yes	Yes	Yes
Crosstalk	p-p Jitter	Yes	Yes	No
Noise susceptibility	dB@BER	Yes	No	No
EMR	dB μV/m	Yes	Optional	No
EMR susceptibility	mV@BER	Yes	No	No

absolute jitter. The limits of acceptable performance are as per the SMPTE 259M standard.

Figure 7.25 shows a waveform with excessive overshoot and low amplitude. This type of waveform is likely to create problems with short cable lengths. With long cable lengths, the problem is less important because of the attenuation of the high frequencies.

Figure 7.26 shows the waveform shape of a 270-Mbps bit-serial signal at the end of a 30-m coaxial cable. Note a reduction of amplitude and an increase in the risetime and fall time.

Figure 7.27 shows the same waveform at the end of a 100-m coaxial cable. This signal can be reclocked and regenerated without difficulties by most equipment.

Figure 7.28 shows the same waveform at the end of a 300-m coaxial cable. Some types of equipment are capable of reclocking and regenerating this signal. The acceptable cable length for the 270-Mbps bit-serial digital signals varies between 200 and 300 m. In order to accommodate most of the equipment on the market, it is safe not to exceed a cable length of 200 m.

Figure 7.29 shows the test setup used by the authors for the measurement of six selected parameters using a digitizing oscilloscope. The output of the Tektronix TSG-422 digital component generator is fed to the input of the device under test (DUT). The oscilloscope is triggered by a 27-MHz clock regenerated from the 270-Mbps bit-serial signal of the Tektronix TSG 4:2:2 generator. This allows for an unambiguous synchronization of the oscilloscope. The accuracy of the jitter measurement depends on the oscilloscope time-base stability and the stability of the recovered clock. These are constant parameters and, consequently, their effect on the measured jitter value is unchanging. The oscilloscope is programmed for compensation of the 75- to 50-ohm matching loss, the measurement of risetime and fall time at 20% and 80% of the signal amplitude, and the measurement of peak-to-peak jitter at the eye crossing. The eye crossing is centered on the screen and the data

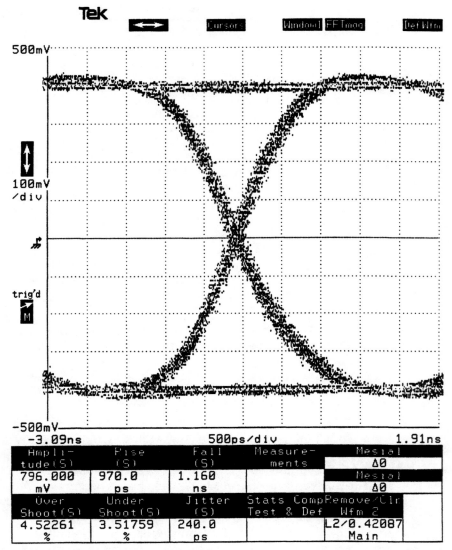

Figure 7.24 TEKTRONIX 11403 digitizing oscilloscope display printout of a good 270 Mbps bit-serial waveform.

accumulation time is 1 min. When the signal has large amounts of jitter, an alternate oscilloscope mode, the "horizontal histogram," is selected, and a display as shown later in Fig. 7.31 is obtained.

It is to be mentioned here that the SMPTE Recommended Practice RP 192 proposes various clock regenerators with selectable PLL bandwidth characteristics. Although this recommendation is implemented in some waveform moni-

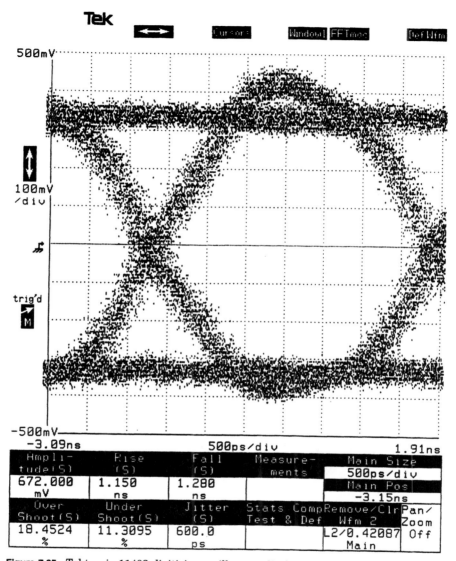

Figure 7.25 Tektronix 11403 digitizing oscilloscope display printout of a 270 Mbps bit-serial waveform with excessive overshoot and low amplitude.

tors (e.g., Tektronix WFM601i), no such clock regenerators are available for use with wideband digitizing oscilloscopes like the Tektronix 11403A. The use of an instrument like the WFM 601i is, however, useful for maintenance work.

Figure 7.30 shows a printout of the Tektronix 11403A oscilloscope operating in histogram mode and triggered by the signal itself. The waveform has a jitter of 3 ns at a frequency of 1 kHz. The oscilloscope timebase follows the

Figure 7.26 Tektronix 11403 digitizing oscilloscope display printout of a 270 Mbps bit-serial waveform at the end of a 30 meter cable.

signal jitter and measures alignment ("relative") jitter. The 1-kHz jitter is masked and the measured jitter is 280 ps, which appears to be well within the specifications.

Figure 7.31 shows a printout of the same signal but with the oscilloscope triggered by a stable reference clock. Note that the eye is almost closed and the measured timing ("absolute") jitter is 3 ns. This is clearly well above the acceptable limit.

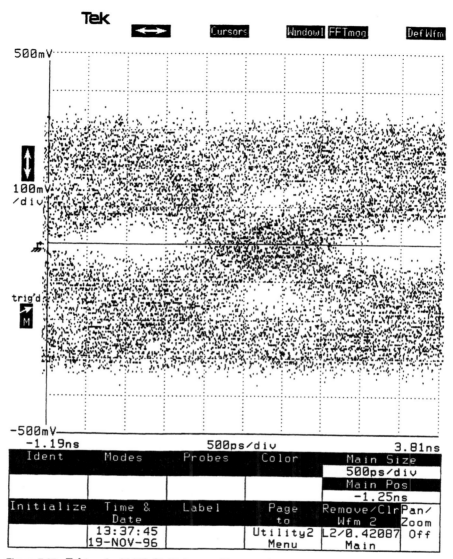

Figure 7.27 Tektronix 11403 digitizing oscilloscope display printout of a 270 Mbps bit-serial waveform at the end of a 100 meter cable.

The 1-kHz jitter in this example is well within the capture range and the hold range of the receiver clock regenerators and will not create problems as long as the signal remains in the digital domain. An analog composite NTSC or PAL signal derived directly from this signal, using the recovered clock as a reference, will exhibit a large amount of unacceptable subcarrier jitter. In order to avoid such problems, the encoder should be equipped with a frame-store synchronizer to eliminate analog jitter problems.

Figure 7.28 Tektronix 11403 digitizing oscilloscope display printout of a 270 Mbps bit-serial waveform at the end of a 300 meter cable.

7.5.1.2 Output return loss. Return loss is measured with a network analyzer, such as the Hewlett-Packard 8753A, in the range of 300 kHz to 300 MHz. Figure 7.32 shows a typical screen printout of a return loss that meets the specifications (>15 dB at 270 MHz). The measured return loss at 270 MHz is in excess of 25 dB.

7.5.1.3 Electromagnetic radiation (EMR). Various countries have their own specifications concerning acceptable levels of EMR energy. Bit-serial digital

Figure 7.29   Test setup for measuring six selected parameters using a Tektronix 11403 digitizing oscilloscope.

video equipment, especially large-capacity routing switchers, may generate high levels of EMR. Equipment design and safe installation practices help reduce the EMR to acceptable levels. Analog video technology uses a floating shield coaxial cable to avoid ground loops leading to hum problems. The coaxial cable shield and the BNC connector outer shell are not electrically in contact with the equipment frame but are isolated from it. Using this concept in digital equipment results in radiofrequency (RF) radiation problems. There are several rules to be followed in order to avoid these problems.

- The BNC connector outer shell should be in contact with the equipment frame.

- The equipment frames should have a good electrical contact with each other and the equipment rack.

- All equipment racks should be electrically connected to each other and to a reliable electrical ground.

These requirements conflict with accepted analog practices. Signal interconnections between digital areas and analog areas have to be carefully studied and floating, differential input video amplifiers have to be used at the analog receiving to avoid ground loops and analog hum problems.

The EMR level of equipment and installations is measured using a calibrated antenna and a spectrum analyzer such as the Tektronix 2712.

7.5.2   Measuring transmission-related parameters

7.5.2.1   Measurement of the characteristic impedance of a cable.   The characteristic impedance of a cable is measured using a network analyzer. A length of

Figure 7.30 Tektronix 11403 digitizing oscilloscope horizontal histogram mode display printout of a 270 Mbps bit-serial waveform with 1000 Hz jitter. The oscilloscope is triggered by the signal to be measured and follows the signal jitter. The measured alignment jitter is 280 ps.

Figure 7.31 Tektronix 11403 digitizing oscilloscope horizontal histogram mode display printout of a 270 Mbps bit-serial waveform with 1000 Hz jitter. The waveform is the same as the one displayed in Fig. 7.30 except that the oscilloscope is triggered by a stable reference 27 MHz clock. This results in a timing jitter measurement. The measured jitter is 3 ns.

Figure 7.32 Hewlett Packard 8253 network analyzer printout of return loss display meeting specifications.

cable, say 100 m, is terminated with a high-precision 75-ohm resistive load and the return loss is measured in a bandwidth of 300 kHz to 270 MHz. A screen printout is produced for the record.

7.5.2.2 Measurement of the signal attenuation of a cable. The frequency-dependent signal attenuation of a coaxial cable is measured using a spectrum analyzer. Figure 7.34 shows a Hewlett Packard 8253 network analyzer screen display printout of the attenuation versus frequency of a 300-m-long typical coaxial cable. Note a loss about 31 dB at 135 MHz and in excess of 45 dB at 270 MHz.

7.5.2.3 Measurement of the shield efficiency. The shield efficiency of a coaxial cable is measured using a spectrum analyzer. A specific length of cable is fed with a 270-Mbps bit-serial digital signal and is terminated with a precision 75-ohm load. The shield efficiency is measured as EMR using a calibrated antenna located 1 m from the stretched cable and a spectrum analyzer such as the Tektronix 2712.

CH1 A/R log MAG 5 dB/ REF .000 dB 1: -9.7104 dB

270.509 520 MHz

Figure 7.33 Hewlett Packard 8253 network analyzer printout of return loss display not meeting specifications.

7.5.3 Measuring receiver-related parameters

7.5.3.1 Input return loss. Figure 7.33 shows a screen printout of a return loss that does not meet specifications. The measured return loss at 270 MHz is about 9.7 dB. This low return loss results in a waveform distortion that can affect the receiver clock recovery when short cable lengths are used. With long cable lengths the effect of mismatch is minimized because of the high attenuation of the high frequencies.

7.5.3.2 Equalization capability. The equalization capability of the receiver is checked by feeding the receiver with a cable of increasing length and watching the output of the receiver on a color monitor for evidence of missing pixels. Errors will occur if the SNR at the input of the receiver is too low. The coaxial cable in itself is not a noise source. Rather, the thermal noise of the receiver input stage is the determining factor. It comes into play when the cable attenuates the signal to a low level such that the SNR is 20 dB or

Figure 7.34 Hewlett Packard 8253 network analyzer display printout of the attenuation versus frequency of a 300 meter typical coaxial cable.

worse. Depending on the equalizer design, the receiver is capable of accepting cable lengths between 200 and 300 m at 270 Mbps.

The performance of digital equipment is expressed in terms of the bit error rate (BER). The BER is expressed as

$$\text{BER} = \frac{\text{number of errors per second}}{\text{Bit rate}}$$

If a system operating at a bit rate of 270 Mbps has one error every television frame, the BER for 525/60 systems is equal to

$$\text{BER} = \frac{30}{270 \times 10^6} = 1.1 \times 10^{-7}$$

For 625/50 systems the BER is equal to

$$\text{BER} = \frac{25}{270 \times 10^6} = 0.93 \times 10^{-7}$$

The visual detection of errors is highly subjective. In general, if the error is

TABLE 7.11   Relationship Between Error Frequency and BER at Three Common Bit Rates

One error per	BER@143 Mbps	BER@177 Mbps	BER@270 Mbps
Frame	2×10^{-7}	2×10^{-7}	1×10^{-7}
Second	7×10^{-9}	5×10^{-9}	4×10^{-9}
Minute	1×10^{-10}	9×10^{-11}	6×10^{-11}
Hour	2×10^{-12}	2×10^{-12}	1×10^{-12}
Day	8×10^{-14}	7×10^{-14}	4×10^{-14}
Week	1×10^{-14}	9×10^{-15}	6×10^{-15}
Month	3×10^{-15}	2×10^{-15}	1×10^{-15}
Year	2×10^{-16}	2×10^{-16}	1×10^{-16}
Decade	2×10^{-17}	2×10^{-17}	1×10^{-17}
Century	2×10^{-18}	2×10^{-18}	1×10^{-18}

noticeable, the performance is unacceptable. Table 7.11 shows the calculated BER for one error over various time intervals at three common bit rates.

BER measurements can be carried out directly using equipment specifically designed for the purpose. In a properly operating system there will be no BER to measure. This is because the bit-serial distribution system operates in an environment that is free of errors. The direct measurement of BER has several disadvantages:

- The measurement has to be carried out with the equipment out of service, because typical test equipment uses a set of defined pseudo-random sequences at various bit rates.

- The sequences used differ from the bit-serial stream and some equipment will not process the test set bit patterns.

- The BER measurements are meaningless when the system is noise-free but subject to burst errors.

7.5.4   Special test signals

7.5.4.1   The EDH test signal.   Tektronix has developed an error-detection system for digital television systems called *error detection and handling* (EDH). The EDH concept has been issued as an SMPTE Recommended Practice RP 165. It is based on making cyclic redundancy code (CRC) calculations for each field of video at the serializer. Separate CRCs for the full-field and active picture, along with status flags, are sent with the other serial data through the transmission system. The CRCs are recalculated at the deserializer and, if not identical to the transmitted values, an error is indicated. Typical error-detection data are presented as errored seconds over a period of time and time since the last errored second. Figure 7.35 shows a block diagram of the EDH concept.

The EDH method can be used as an in-service test to pinpoint automatically and electronically any system failures. Depending on the sophistication of the

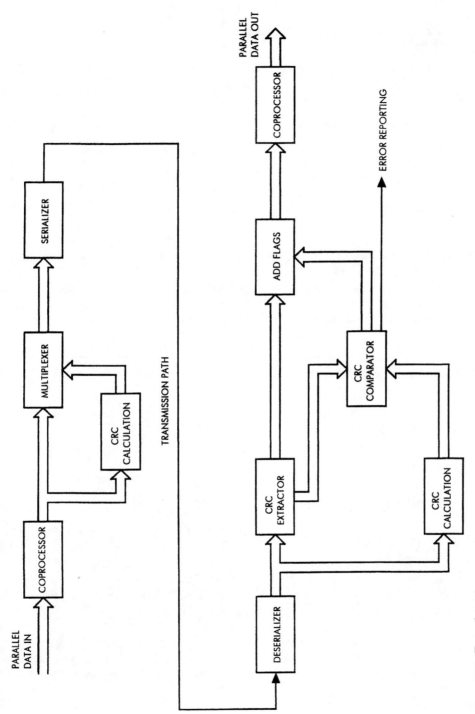

Figure 7.35 Error detection concept.

EDH monitoring and measuring equipment, errors can be indicated as a red warning light or as complex reporting of errors and their sources using a variety of error flags. The error flags are used to indicate the status for the previous field. Each flag is set or cleared on a field-by-field basis. They are defined as follows:

- EDH—Error detected here: A serial transmission data error was detected.

- EDA—Error detected already: A serial transmission error has been detected somewhere upstream.

- IDH—Internal error detected here: A hardware error unrelated to the serial transmission has been detected within a device.

- IDA—Internal error detected already: An IDH flag was received and there was a hardware device failure somewhere upstream.

- UES—Unknown error status: A serial signal was received from equipment not supporting the error-detection mechanism.

Digital data packets containing the calculated checkwords and error information flags are located in the ancillary data area of the vertical interval in a manner to complement the recommended practice of the source switching. The locations are specified in Table 7.12 for 525/60 systems and Table 7.13 for 625/50 systems.

TABLE 7.12   Location of 525/59.94 System Checkwords

Data item	Composite*	Component
Error-checking data locations		
Line 9, fields I & III, Line 272, fields II & IV		
Ancillary data header, word 1—component		1689 (000)
Ancillary data header, word 2—component		1690 (3FF)
Ancillary data header, word 3—component		1691 (3FF)
Auxiliary data flag—composite	795 (3FC)[†]	
Data ID	796 (1F4)	1692 (1F4)
Block number	797 (200)	1693 (200)
Data count	798 (110)	1694 (110)
Active picture data word 0	799	1695
Active picture data word 1	800	1696
Active picture data word 2	801	1697
Full-field data word 0	802	1698
Full-field data word 1	803	1699
Full-field data word 2	804	1700
Ancillary data error flags	805	1701
Active picture error flags	806	1702
Full-field error flags	807	1703
Reserved words (7 total)	808–814 (200)	1704–1710 (200)
Checksum for this ancillary data word	815	1711

*Values in the tables are word numbers.
[†]Values in parentheses are hexadecimal sample values.

TABLE 7.13   Location of 625/50 System Checkwords

Data Item	Composite*	Component
Error checking data locations		
Line (5‡), fields I & III, Line (318), fields II & IV		
Ancillary data header, word 1—component		Y 850 (000)
Ancillary data header, word 2—component		Cr 425 (3FF)
Ancillary data header, word 3—component		Y 851 (3FF)
Auxiliary data flag—composite	972 (3FC)†	
Data ID	973 (1F4)	Cb 426 (1F4)
Block number	974 (200)	Y 852 (200)
Data count	975 (110)	Cr 426 (110)
Active picture data word 0	976	Y 853
Active picture data word 1	977	Cb 427
Active picture data word 2	978	Y 854
Full-field data word 0	979	Cr 427
Full-field data word 1	980	Y 855
Full-field data word 2	981	Cb 428
Ancillary data error flags	982	Y 856
Active picture error flags	983	Cr 428
Full-field error flags	984	Y 857
Reserved words (7 total)	985–991 (200)	Cb429–Cr430 (200)
Checksum for this ancillary data word	992	Y 861

*Values in the tables are word numbers.
†Values in parentheses are hexadecimal sample values.
‡Line numbers in parentheses are tentative.

TABLE 7.14   Checkword Included Samples

Data item	Composite	Component
525/59.94 systems		
First full-field sample, lines 12 and 275	795	1444
First active line sample, lines 21 and 284	0	0
Last active picture sample, lines 262 and 525	767	1439
Last full-field sample, lines 8 and 271	767	1439
625/50 systems		
First full-field sample, lines (8) and (321)*	972	Cb 361
First active picture sample, lines 24 and 336	0	Cb 0
Last active picture sample, lines 310 and 622	947	Y 719

*Full-field 625/50 line numbers are tentative.

Table 7.14 shows the starting and ending samples for active picture and full-field checkword calculations. The calculation of the active picture checkword includes only the samples in the active picture area of each line. The active picture is defined as consisting only of the lines which in a composite system are full lines, thus excluding half-lines. The full-field checkwords include all samples in all lines, excluding the line containing the error data packet and the following two lines. The line immediately following the error data packet is used

TABLE 7.15 Definition of Ancillary Data Words

Data item	b9 (MSB)	b8	b7	b6	b5	b4	b3	b2	b1	b0 (LSB)
Ancillary data header, word 1—components	0	0	0	0	0	0	0	0	0	0
Ancillary data header, word 2—components	1	1	1	1	1	1	1	1	1	1
Ancillary data header, word 3—components	1	1	1	1	1	1	1	1	1	1
Auxiliary data flag —composite	1	1	1	1	1	1	1	1	0	0
Data ID	0	1	1	1	1	1	0	1	0	0
Block number	1	0	0	0	0	0	0	0	0	0
Data count	0	1	0	0	0	1	0	0	0	0
Active picture data word 0 crc<5:0>	not P	P	c5	c4	c3	c2	c1	c0	0	0
Active picture data word 1 crc<11:6>	not P	P	c11	c10	c9	c8	c7	c6	0	0
Active picture data word 2 crc<15:12>	not P	P	V	0	c15	c14	c13	c12	0	0
Full-field data word 0 crc<5:0>	not P	P	c5	c4	c3	c2	c1	c0	0	0
Full-field data word 1 crc<11:6>	not P	P	c11	c10	c9	c8	c7	c6	0	0
Full-field data word 2 crc<15:12>	not P	P	V	0	c15	c14	c13	c12	0	0
Auxiliary data error flags	not P	P	0	ues	ida	idh	eda	edh	0	0
Active picture error flags	not P	P	0	ues	ida	idh	eda	edh	0	0
Full-field error flags	not P	P	0	ues	ida	idh	eda	edh	0	0
Reserved words (7 total)	1	0	0	0	0	0	0	0	0	0
Checksum	not S8	S8	S7	S6	S5	S4	S3	S2	S1	S0

for vertical-interval switching and the following line is excluded to allow for the restoration of word framing and TRS propagation after a switch.

The checkword value consists of 16 bits of data calculated using the CRC-CCITT polynomial generation method according to the equation

$$\text{Checkword (16-bit)} = X^{16} + X^{12} + X^6 + 1$$

The definition of each ancillary data word is shown in Table 7.15. The two least-significant bits (LSBs) are assigned the value 0 to provide compatibility with 8-bit equipment. Bit b7 (V) in picture or field data word 3 is a 1 if a valid CRC has been calculated. A P in b8 provides even parity for b7 through b0 (the total number of 1's in b8 through b0 is an even number). The most-significant bit (MSB), b9, is the inverse of b8. A checksum is the last word in the error data packet as required by the ancillary data formatting for the serial digital interface. The checksum word consists of 9 bits and is the sum of the LSBs (b8 through b0) including data ID through reserved words. See Sec. 7.6 for information on ancillary data formatting.

Large bit-serial digital installations will benefit from the EDH error-detection method. When the digital signal remains in the bit-serial domain, such as in routing switchers or distribution amplifiers, it is not economical to provide the CRC calculation method. Instead, simple and inexpensive error-detection chips can be fitted to all inputs and outputs of the routing switcher, allowing the verification of the integrity of all inputs and outputs. The errors can be reported locally or to a central diagnostic computer, simplifying the identification of the faulty link or signal source. The CRC calculation method benefits the in-service monitoring or out-of-service verification of individual complex bit-serial signal distribution links consisting of a large number of interconnected elements using equipment such as a Tektronix 601i waveform monitor or equivalent test equipment.

7.5.4.2   The pathological test signal (SDI check field).   The performance of the bit-serial digital equipment depends on the correct data recovery at the receiver. There are two major drawbacks that require special attention in the design and implementation of the receiver.

- One of the drawbacks is the existence of a DC component as well as very low frequency spectral components. This makes the design of a practical equalizer very difficult.

- The second drawback is the possible occurrence of a long run length of logic 0 or 1 in the data stream. This requires a very stable VCO to avoid frequency drifts in the absence of 0-to-1-to-0 transitions in the bit-serial signal.

The fundamental idea behind the pathological test signal is to stress the transmission channel and assess the effects. Two special test signals have been developed by Sony for stressing the bit-serial receiver.

Figure 7.36 shows the signal used to test the automatic cable equalizer. The top portion of the drawing shows the signal at the output of the scrambler, consisting of two 1's followed by a run of eighteen 0's. The bottom of the drawing shows how this signal is modified by the inverter to obtain a scrambled NRZI signal consisting of a 1 followed by a run of nineteen 0's. This signal has a large DC content.

Figure 7.37 shows the signal used for testing the PLL lock-in. The top portion of the drawing shows the signal at the output of the scrambler, consisting of a 1 followed by a run of 19 0's. The bottom of the drawing shows how this signal is modified by the inverter to obtain a scrambled NRZI signal consist-

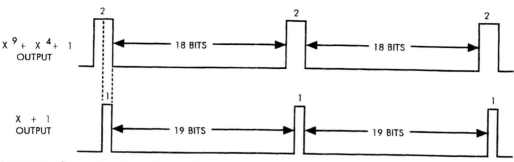

Figure 7.36 Generation of a pathological signal suitable for testing automatic cable equalization.

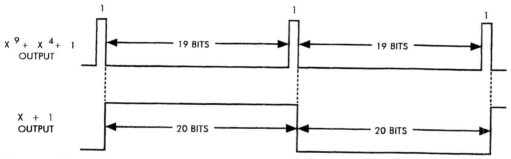

Figure 7.37 Generation of a pathological signal suitable for testing PLL lock-in.

ing of an identical sequence of 1's and 0's repeated every 20 bits. This signal provides a minimum number of crossings for clock extraction.

SMPTE RP 178 describes a recommended pathological test signal called serial digital interface (SDI) check field. The SDI check field consists of one-half field of each of the two signals described above and is shown in Fig. 7.38. The top half of the picture consists of a word combination of 300 hexadecimal for the chrominance component (C) and 198 hexadecimal for the luminance component (Y) and results in a purple shade. It is used for the cable equalizer test. The bottom half of the picture consists of a word combination of 200 hexadecimal for the chrominance component (C) and 110 hexadecimal for the luminance component (Y) and results in a shade of gray. It is used for the PLL test.

The test signal is fed to the input of the equipment under test and the output is monitored on a color monitor. Bit errors affecting the top half of the picture are a result of a malfunction of the equalizer, usually because of a coaxial cable length in excess of the equalizer correction capability. Shortening the cable will eliminate the problem. Bit errors affecting the bottom half of the picture are caused by a malfunction of the receiver clock regenerator and could indicate a condition where the free-run frequency of the PLL-controlled VCO in the receiver has drifted from the specified frequency. Excessive heat or poor

Figure 7.38   SDI Check Field.

design may be the cause of this condition. It is worthwhile verifying the performance of the bit-serial input of the color monitor with the SDI check field signal before carrying out the test to ensure that it is operating satisfactorily.

7.6   Digital Audio Multiplexing

As mentioned in Sec. 7.4, there is provision for inserting ancillary data in the digital composite and component video signals. Ancillary data include digital audio, time code, EDH, and future user and control data.

Ancillary data are formatted into data packets and multiplexed with the serial data stream. The ancillary data packets are inserted in a space defined by the SMPTE 125M and 269M documents. Figures 7.9 to 7.11 ($4f_{SC}$ NTSC) and 7.12 to 7.14 ($4f_{SC}$ PAL) show the ancillary data space for the composite digital bit-serial interface. Figures 7.21 (525/60) and 7.22 (625/50) show the ancillary data space for the component digital bit-serial interface. Figure 7.39 shows the ancillary data packet structure for the 4:2:2 component digital interface.

Each packet can carry a maximum of 262 10-bit parallel words that consist of

- A three-word ancillary data flag (ADF). Their values are 000, 3FF, 3FF, respectively, and mark the beginning of the ancillary data packet.

- An optional data identification (DID) word. This word identifies the user data content of each packet. When the content is audio data, several different DIDs are used to define the four possible groups of audio channels.

Figure 7.39 Ancillary packet structure for component digital interface.

- An optional data block number (DBN) word. This word allows a receiver to verify the transmission integrity by counting the number of ancillary data packets having a common DID. In case of data stream switching, this counter can send a flag to the audio processing system to remove transients, such as "pops" or "clicks," using an appropriate muting circuit.

- A data count (DC) word. This word indicates the number of user data words in each packet.

- A variable number of user data words (UDWs). A maximum of 255 words is permitted.

- A checksum (CS) word. This word is used by the receiver to determine the validity of the packet.

Multiple, contiguous, ancillary data packets may be inserted in any ancillary data space. They must follow immediately after the EAV, for the HANC, or SAV, for the VANC, to indicate the presence of ancillary data and the start of a packet. If there is no ADF in the first three words of an ancillary data space, it is assumed that no ancillary data packets are present.

The SMPTE Standard 272M proposes two basic modes of operation:

7.6.1 Minimum implementation

The minimum AES implementation, standardized as level A, is characterized by

- 20-bit-resolution audio words
- A 48-kHz sampling frequency
- Audio data synchronous with video data
- Only one group of four audio channels
- A receiver buffer size of 48 audio samples

The audio data packets are formatted from AES/EBU data stream information. Figure 7.40 shows the process of formatting an AES/EBU data stream into audio data packets. This example shows how 20 bits of audio data as well as related V, U, and C bits, a total of 23 bits, are extracted from subframe 1 (channel 1) of frame 0 of an AES/EBU serial data stream and mapped into three 10-bit ancillary words X, $X + 1$, and $X + 2$. The 4 preamble bits, the 4 auxiliary bits, as well as the parity bit (P) are discarded.

Table 7.16 shows the audio data structure represented by the three 10-bit data words. The channel number is indicated by 2 bits and a parity is calculated on the 26 bits, excluding all b9 address bits. These three words are inserted immediately after the ancillary data header, as shown in Fig. 7.40. A second audio channel information (channel 2) from the same AES/EBU serial data stream is inserted in a similar fashion. Frame 0 from a second AES/EBU serial data stream is subsequently inserted to complete one audio group (four channels). The notations used reflect various ways of identifying channels and frames. The AES/EBU standard specifies a sequence of 192 frames (numbered frame 0 through frame 191) forming a block. Each frame consists of two subframes (subframe 1 and subframe 2) or channels (channel 1 and channel 2). When formatted into an ancillary audio stream, two Frames (four subframes or four channels) form a group. Each group consists of two sample pairs originating from two AES/EBU serial data streams. Each sample pair can be identified in three ways.

- AES1 (CH1/CH2) and AES2 (CH1/CH2)
- CH1/CH2 and CH3/CH4
- CH00/CH01 and CH10/CH11

7.6.2 Full AES implementation

The full AES implementation, combined with several operational capabilities identified as levels B through J, is characterized by

- 24-bit-resolution audio words
- Sampling frequencies of 32, 44.1, or 48 kHz

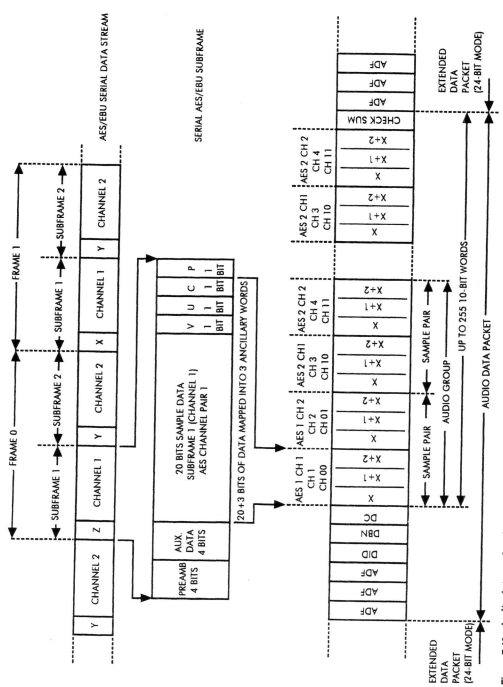

Figure 7.40 Audio data packet formatting from an AES/EBU serial audio data stream (20-bit audio mode).

TABLE 7.16 Formatted Audio Data Structure

Bit address	Word X	Word $X+1$	Word $X+2$
b9	not b8	not b8	not b8
b8	audio 5	audio 14	P
b7	audio 4	audio 13	C
b6	audio 3	audio 12	U
b5	audio 2	audio 11	V
b4	audio 1	audio 10	audio 19 (MSB)
b3	audio 0 (LSB)	audio 9	audio 18
b2	channel 1	audio 8	audio 17
b1	channel 0	audio 7	audio 16
b0	Z	audio 6	audio 15

- Audio data synchronous or asynchronous with video data
- Up to four groups of four audio channels
- A receiver buffer size of 64 audio samples
- Indication of time delay between any audio channel and the video data signal

Two additional packets are defined for carrying the extra information. For the 24-bit resolution mode, the four auxiliary bits of the two AES 1 subframes are grouped into an 8-bit word, called AUX, related to the AES 1 signal. All AUX words from other AES signals are grouped into an extended data packet as shown in Fig. 7.41. This packet has an identical header structure and a specific DID number. It is inserted in the ancillary data space immediately following the audio data packet it is related to.

An audio control packet is also defined for the transmission of information such as audio frame number, sampling frequency, active audio channels and relative audio-to-video delay of each audio channel. This packet is optional for the minimum AES implementation but is mandatory for the full AES implementation. This packet, shown in Fig. 7.42, is transmitted once per field as the first packet of the ancillary data space following line 11.

Audio data are inserted in most of the free ancillary space situated in the horizontal blanking area. Bit-serial data stream switching occurs in line 10 and audio data insertion in the following horizontal ancillary data space is

Figure 7.41 Extended data structure (24-bit audio mode).

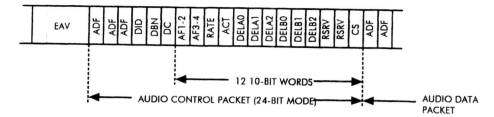

RATE: FOR EACH CHANNEL PAIR
ACT: ACTIVE CHANNEL
DELA: DELAY OF CHANNEL PAIR A WRT. VIDEO

Figure 7.42 Audio control packet structure.

not recommended. The receiver demultiplexer must contain a 64-sample buffer to allow the flawless reconstruction of the digital audio signal.

Some equipment that was manufactured and installed prior to the completion of the standardization inserts audio data in any ancillary data space. The standard recommends that receiving equipment process data inserted in any horizontal ancillary data space. Failure to do so results in data muting in each field for the duration of one horizontal blanking interval, resulting in a severe degradation of the audio signal.

DVTRs, frame synchronizers, and coder-decoders (codecs) do not record or pass the complete digital video signal transparently. Horizontal and vertical blanking areas are removed before processing in order to reduce the bit rate and are added at the device output. This results in loss of ancillary data information. In some equipment, ancillary data are properly extracted at the input, stored transparently, and reinserted at the output.

7.6.3 The audio multiplexer

Figure 7.43 shows a simplified block diagram of an audio multiplexer. Each of the two input digital audio data streams, AES/EBU 1 and AES/EBU 2, are sent to a dedicated deserializer. The deserializer decodes the biphase mark (BPM)–encoded signal and feeds it to a buffer. The buffer consists of a first-in-first-out (FIFO) circuit. The outputs of the two FIFO circuits are sent to a formatter. The 32-bit AES/EBU signal is stripped of the preamble (4 bits of X, Y, or Z), the auxiliary data (4 bits), and the 1-bit subframe parity (P). The remaining 23 bits, 20 bits of audio and 1 bit each of validity (V), user bit (U), and audio channel status (C), are formatted into three 10-bit ancillary words X, $X + 1$, and $X + 2$. The multiplexer adds the formatted audio to the delayed video. The video delay matches the delay of the digital audio processing. The audio is multiplexed into the available ancillary data space. The multiplexed audio and video are reserialized by the serializer. Note that the multiplexing of audio does not increase the bit rate, since the audio is inserted in unused sample locations. The role of the FIFO buffers is to help keep the audio data rate constant. In case of an audio buffer overflow, specific AES/EBU samples

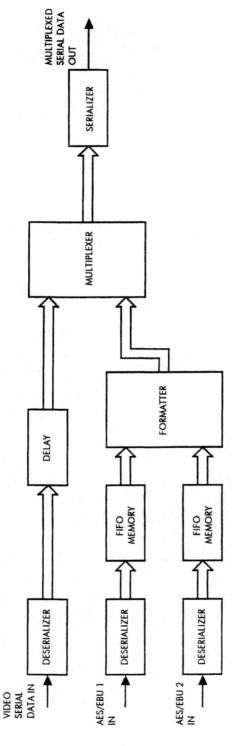

Figure 7.43 Simplified block diagram of audio multiplexer.

are deleted, as required, until the audio buffer overflow is eliminated. In case of an audio buffer underflow, specific AES/EBU samples are duplicated until the audio buffer returns to normal.

7.6.4 The audio demultiplexer

Figure 7.44 shows a simplified block diagram of a demultiplexer. The multiplexed bit-serial input signal is deserialized and fed to a video delay as well as an audio extractor. The video delay matches the delay of the audio processing. The audio extractor feeds two FIFOs, one for each AES/EBU stream. The data are clocked into the FIFO during the ancillary data intervals. The formatter adds missing information such as the preamble (4 bits), the auxiliary data (4 bits), and the subframe channel parity (1 bit) and re-creates the original 32-bit AES/EBU subframe sequence. The transmitter encodes the data into a BPM channel code as per AES/EBU specifications. The delayed video is reserialized in the usual manner. In case of an audio buffer overflow, specific audio frames are deleted, as required, until the audio buffer overflow is eliminated. In case of an audio buffer underflow, specific AES/EBU frames are duplicated until the audio buffer returns to normal.

7.7 Digital Videotape Recording

Digital videotape recording is an application of the bit-serial digital video concept. The DVTR receives a video and audio signal in one of the analog or digital formats. It reformats these signals into a bit-serial digital data stream specific to the respective recording standard. The recorded bit rate depends on such factors as

- Analog video format: Composite (NTSC or PAL) or component
- Number of bits per sample
- Bit-rate reduction, if any
- Bit-rate headroom reserved for error correction

At the time of the writing of this book, May 1997, several noncompatible recording standards coexist. They are grouped into two main categories, $4f_{SC}$ composite digital and 4:2:2 component digital.

7.7.1 $4f_{SC}$ composite DVTRs

There are two noncompatible $4f_{SC}$ videotape recording formats known as the D2 and the D3. Their mechanical characteristics are listed in Table 7.17 and their electrical characteristics are listed in Table 7.18.

The D2 VTRs are currently manufactured by AMPEX and Sony. They use a ¾-in videocassette. Recordings made on D2 VTRs manufactured by these two companies are playable on either make.

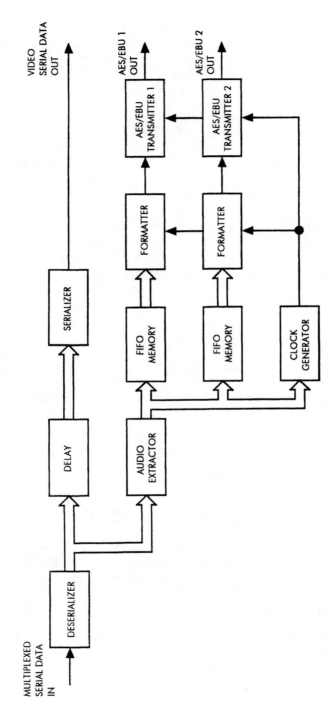

Figure 7.44 Simplified block diagram of audio demultiplexer.

TABLE 7.17   Mechanical Characteristics of $4f_{sc}$ NTSC Composite DVTRs

Parameter	Sony/AMPEX D2	Panasonic D3
Track width, μm	39.1	20
Guard band, μm	0	0
Total wrap angle, deg	188	197
Effective wrap angle, deg	178.2	178.1
Track angle, deg	6.1296	4.9192
Tracks per field	6	6
Track length, mm	150.78	117.7
Azimuth, deg	±15	±20
Tape tension, N	0.7	0.3
Tape speed, cm/s	13.17	8.34
Writing speed, m/s	27.39	21.5
Drum rotation, rps	90/1.001	90/1.001
Drum diameter, mm	96.4	76.0
Number of heads	8–10	10
Number of record heads	4	4
Number of RF channels	2	2
Tape thickness, μm	13	11/14
Tape width, in (mm)	¾ (19.01)	½ (12.65)
Coercivity, Oe	1500	1500
Maximum play time, min	208	245
Picture in shuttle	60/100 × play	±100 × play
Variable speed range	−1 to +3	−1 to +3

TABLE 7.18   Electrical Characteristics of $4f_{sc}$ NTSC composite DVTRs

Parameter	Sony/AMPEX D2	Panasonic D3
	Video	
Sampling rate, MHz	14.32	14.32
Sample resolution, bits	8	8
Bandwidth, MHz	5.5	5.5
SNR, dB	54	54
Recorded lines per frame	510	510
Tape data rate, Mbps	127	125
Video bit rate, Mbps	94	94
Min. recorded wavelength, μm	0.854	0.77
	Audio	
Sampling rate, kHz	48	48
Sample resolution, bits	16–20*	16–20*
Bandwidth	20 Hz–20 kHz	20 Hz–20 kHz
Dynamic range, dB	90	90
AES channels	2	2
RF channel coding	Miller square	8–14 code

*The 20-bit recording mode discards V, U, C, and P bits.

The D3 format is manufactured exclusively by Panasonic. It uses a ½-in videocassette and is, therefore, mechanically incompatible with D2-format VTRs.

Apart from mechanical incompatibility, the signal processing of the two formats is quite similar, resulting in similar electrical performance figures as shown in Table 7.18. Both formats use a video sample resolution of 8 bits, resulting in a 54-dB SNR. The tape data rate in both formats is 127 Mbps, of which the actual video bit rate is 94 Mbps. This bit-rate reduction is obtained by recording only "essential" video information. The sync tips are excluded and the number of recorded lines is reduced to 510. Two AES/EBU digital audio pairs (four audio channels) with a user-selectable sample resolution between 16 and 20 bits can be recorded. When using 20 bits per sample, the V, U, C, and P bits are discarded. At intermediate sample resolutions a user may choose which of the four bits is to be recorded.

The input/output analog and digital (when provided) ports are standard. Consequently, if required, a D2 and a D3 VTR could be interconnected without difficulty. The D2 and the D3 formats have been developed specifically as replacements of obsolescent analog VTRs in an analog video environment. Although their single-pass performance is superior to the machines they replace, the multigeneration performance suffers from impairments similar to, albeit smaller than, those of the analog recorders. Composite digital production equipment such as production switchers, video effects generators, and character generators are available for bit-parallel or bit-serial digital interconnection with D2 and D3 VTRs. It is, however, predictable that component digital video equipment, available at a competitive price, will eventually replace the D2 and D3 VTRs and the peripheral equipment.

7.7.2 Component DVTRs

Four noncompatible 4:2:2 component digital video recording formats presently coexist. The mechanical characteristics are listed in Table 7.19 and the electrical characteristics are listed in Table 7.20.

- The Sony D1 VTR is based on the original CCIR 601 Recommendation. It uses a ¾-in videocassette and its performance is representative of a 4:2:2 sampling structure and a sample resolution of 8 bits. The recorded tape data rate is 225 Mbps. The recorded "essential" video bit rate is 172 Mbps, resulting in 500 recorded lines. It features bit-parallel digital and component analog (CCIR and Betacam) video input/output ports. When interfaced with 10-bit 4:2:2 digital equipment, it discards the two LSBs of the data stream.

- The Sony Digital Betacam VTR uses a ½-in videocassette. The video sample resolution is 10 bits. The total recorded tape data rate is 127.76 Mbps. A 2.3:1 DCT video compression is used. The recorded video bit rate is 95 Mbps and a total of 512 lines is recorded, of which 507 lines carry video information. It features bit-serial digital, analog composite, and analog component

TABLE 7.19   Mechanical Characteristics of 4:2:2 Component DVTRs

Parameter	Sony D1	Sony Digital Betacam	Panasonic D5	AMPEX DCT 700d
Track width, μm	32–40	20	20	39.1
Guard band, μm	5–10	1.7	0	0
Total wrap angle, deg	258	196	197	188
Eff. wrap angle, deg	238	173.8	176.9	178.2
Track angle, deg	5.4005	4.62652	4.9384	6.12894
Tracks per field	10	6	12	6
Track length, mm	170	123.335	116.397	148.06
Azimuth, deg	0	±15° 15′	±20	±15
Tape tension, N	0.7	0.3	0.3	0.7
Tape speed, cm/s	28.66	9.67	16.7228	13.17
Writing speed, m/s	35.5	22.9	21.5	27.39
Drum rotation, rps	150/1.001	90/1.001	90/1.001	90/1.001
Drum diameter, mm	75	81.4	76.0	96.4
No. of heads	16	14+4	18	10
No. of record heads	4	4	8	4
No. of RF channels	4	2	4	2
Tape thickness, μm	16	13.5	11/14	13
Tape width, inch (mm)	¾ (19.01)	½ (12.65)	½ (12.65)	¾ (19.01)
Coercivity, Oe	850	1500	1500	1500
Max. play time, min	76	124	123	208
Picture in shuttle	±40×play	±50 × play	±50 × play	±20 × play
Variable speed range	−0.25 to 0.25	−1 to +3	−1 to +2	−1 to +3

(selectable CCIR or Betacam) video input/output ports. The bit-serial input accepts multiplexed (embedded) audio that is stripped and reformatted internally as required. The audio information can be multiplexed with the bit-serial video at the output. Analog Betacam videocassettes can be played back.

■ The Panasonic D5 VTR uses a ½-in videocassette. The video sample resolution is 10 bits. The total recorded data rate is about 300 Mbps. The recorded "essential" video bit rate is 220 Mbps, resulting in 510 recorded lines. In addition, it accepts 16/9 picture format signals with the 5.333:2.665:2.665 sampling structure (360 Mbps/10 bits), which it records with an 8-bit resolution. It is the only DVTR on the market capable of recording this format. It features bit-parallel and bit-serial component digital and component analog input ports and bit-parallel component and composite digital, bit-serial component and composite digital, as well as component analog output ports. D3-format videocassettes can be played back. It accepts multiplexed audio at the input and can multiplex video and audio at the output.

■ The AMPEX DCT 700d VTR uses a ¾-in videocassette. The video sample resolution is 8 bits. When interfaced with 10-bit 4:2:2 digital equipment, it performs a rounding truncation of the two LSBs. The total recorded tape data rate is 124.7 Mbps. A 2:1 discrete cosine transform (DCT) video compression is used. The recorded video bit rate is 88 Mbps, and a total of 512 lines is

TABLE 7.20   Electrical Characteristics of 4:2:2 Component DVTRs

Parameter	Sony D1	Sony Digital Betacam	Panasonic D5	AMPEX DCT 700d
Video				
Y sampling rate, MHz	13.5	13.5	13.5	13.5
B-Y/R-Y sampling rate, MHz	6.75	6.75	6.75	6.75
Sample resolution, bits	8	10	10	8
Y bandwidth	5.75	5.75	5.75	5.75
B-Y/R-Y bandwidth	2.75	2.75	2.75	2.75
Y SNR, dB	56	61	62	55*
Recorded lines per frame	500	507/512[†]	510	504/512[†]
Compression ratio	1:1	2.3:1	1:1	2:1
Tape data rate, Mbps	225	127.76	~300	124.7
Video bit rate, Mbps	172.0	95[‡]	220	88[‡]
Min. recorded wavelength, μm	0.9	0.692	0.63	0.854
Audio				
Sampling rate, kHz	48	48	48	48
Sample resolution, bits[§]	16–20	16–20	16–20	16–20
Bandwidth	20 Hz–20 kHz	20 Hz–20 kHz	20 Hz–20 kHz	20 Hz–20 kHz
Dynamic range, dB	90	90	90	90
AES channels	2	2	2	2
RF channel coding	Random NRZ	Partial resp.	8–14 code	Miller square

*NTSC in/out measurement.
[†]The first figure indicates the number of lines with video information.
[‡]The active picture area is compressed by a ratio of about 2:1.
[§]The 20-bit recording mode discards V, U, C, and P bits.

recorded, of which 504 lines carry video information. There are input ports for bit-parallel and bit-serial (optional) component digital signals and component analog signals (optional) and output ports for bit-parallel and bit-serial component digital signals as well as analog output signals. It accepts multiplexed audio at the input and can multiplex video and audio at the output.

In all formats, the bit-rate headroom, that is, the difference between the tape data rate and recorded video bit rate, is used for error-correction purposes. All formats can handle two AES/EBU digital data streams (four audio channels), which are recorded with a user-selectable resolution of 16 to 20 bits per sample. When the 20-bit sample resolution is chosen, the V, C, U, and P bits are discarded. With lower audio sample resolutions, the user may choose which of the four bits is to be carried.

7.8   System Considerations

The design of a digital teleproduction facility, whether a small studio, an editing booth, or a large multistudio plant, requires careful planning. There are several consecutive steps to be taken to ensure success. These are

- Understand the operational requirements.

- Compile a list of generic equipment required for the project.

- Contact hardware manufacturers and obtain equipment documentation with the aim of selecting the equipment that best satisfies the operational requirements at a competitive cost. Choose equipment with EDH-generating and fault-reporting capabilities.

- Obtain a sample for operational and technical evaluation. It is essential to verify that the chosen equipment satisfies the operational needs and, most importantly, that it meets the relevant bit-serial interface SMPTE standards described in this chapter. This requires an understanding of the relevant standards, access to specialized test equipment, and ability to carry out the evaluation tests. Failure to carry out a technical evaluation of the equipment prior to purchasing it can lead to costly mistakes. A rule of thumb is to select equipment that features automatic equalization and reclocking at the input even though it might, at first glance, not appear to be necessary. It may well be that at a later date the equipment may need to be fed with a longer cable and its input stage will not be able to accommodate it.

The installation of the equipment requires very few but essential precautions as follows:

- Install the equipment in well-ventilated or air-cooled racks to avoid heat buildup that could lead to equipment failure. Remember that digital equipment is usually quite compact and generates a great amount of heat.

- Use the shortest cable length for interconnecting equipment.

- If the cable length required exceeds the equalization capability of the input circuit, make provisions for a reclocking distribution amplifier situated halfway between the transmitter and the receiver.

- The inputs of production switchers are usually provided with automatic phasers capable of accommodating input feeds with timing differences up to one full line (e.g., 64 μs). Consequently, there is normally no need to time and phase the input signals.

The maintenance of digital systems with bit-serial signal distribution requires a new maintenance philosophy. Normally, if the equipment meets the standards and is well installed, it will only fail when components age. Some types of equipment may not be amenable to in-house repairs by the maintenance technicians because of the circuit complexity and/or the need for some sophisticated test equipment. In such cases the technician needs to be able to identify the defective card, replace it, and send the defective one to the manufacturer for repairs. Because of the high cost of qualified personnel, the manufacturer may dispose of the defective card and replace it with a new one. This type of maintenance activity requires a maintenance technician with the following qualifications:

- A thorough knowledge of basic analog and digital technology and standards.

- An understanding of the system and the function of each component element in the chain.

- An understanding of the operation of the system or equipment.

The operation of a digital studio also requires a new approach. Assuming that the original composite or component analog signal meets the relevant quality requirements, it will not be degraded once converted to a digital signal. Because of the ruggedness of a well-planned, well-installed, and well-maintained bit-serial digital distribution system, there is no need for extensive signal monitoring as carried out in an analog environment. This leads to the concept of source accountability, which means that signal-generating equipment [e.g., CCU (Camera Control Unit), character generator, VTR, and production switcher] operators will make sure that

- The analog information carried by the bit-serial "carrier" meets the requirements. Among the factors to be verified is the "legality" of the derived composite (NTSC or PAL) analog signal to avoid on-air transmitter overmodulation. This requires the use of a waveform monitor with a bit-serial digital input an internal decoder to analog component signals as well as a special diamond display.

- The transmitter output bit-serial (launch) signal meets the requirements. The waveform monitor needs to be connected through a very short (<10 ft) cable to the transmitter output. It is imperative to monitor the actual output since multiple outputs have dedicated drivers and monitoring one of them will not guarantee that the others are active. Some waveform monitors feature a high-return-loss passive-loop-through bit-serial input that facilitates this type of monitoring.

The analog component and bit-serial digital signal operational monitoring described above can be easily carried out with waveform monitors such as the Tektronix WFM601i.

Beyond the signal origination point there is no need for monitoring since the analog message will remain unaffected and the bit-serial waveform will be regenerated at the destination. The only operational position where analog signal monitoring will need to be carried out is in the master control room area, where outgoing composite NTSC- or PAL-encoded signals are fed to common carriers or over-the-air transmitters and incoming signals are converted to bit-serial digital signals for in-plant distribution.

The simple EDH fault monitoring of the inputs and outputs of routing switchers combined with a computerized central reporting system as well as of the inputs of the production switchers should be fully implemented and used to ensure that the bit-serial signals are not corrupted.

Digital Signal Compression and Distribution

Digital video and audio signal processing results in superior picture and sound quality and favors the introduction of new digital techniques throughout the signal chain. The penalty is the need to handle a large amount of data files in recording, computer, and transmission applications. All-digital teleproduction studios and complete broadcast plants have been built in the past five years at a very high cost. The program transmission is, however, still predominantly analog, because the high data rate of video signals (270 Mbps) cannot be delivered economically to each user.

The compression concept allows users to choose, from a variety of sampling parameters and compression ratios, the combination that best suits their specific purpose. This novel approach to signal processing promises to displace completely all the old concepts of line rates; field rates; NTSC, PAL, SECAM color encoding; as well as high-definition television (HDTV), at least as far as wideband HDTV is concerned.

Different bit rates, resulting in a large family of cost and performance levels, are recorded, processed, and transmitted digitally by copper or fiber-optic landlines, VHF/UHF transmitters, or DBS satellites. Among the spinoffs of this technology is the appearance on the market of PC-type computers used for picture generation and manipulation. In a television studio environment, these new compact, efficient, and versatile devices are revolutionizing production techniques.

This chapter will review the most-used compression techniques applied to video and audio signals. Many other techniques exist and others are under development, but they do not generate significant cost-effective applications at this time.

8.1 General Concepts of Video Bit-Rate Reduction (BRR)

The conventional NTSC, PAL, and SECAM systems use a video information compression by reducing the chrominance information bandwidth to 1.2 MHz or less, reflecting the reduced eye sensitivity to high-frequency components in the chrominance signal. Analog video component formats are based on the same principles. The CCIR-601 4:2:2 component video format standard specifies a luminance and chrominance signal bandwidth of 5.75 and 2.75 MHz, respectively (@ ±0.1 dB). After digitization, a total bit rate of 270 Mbps is generated, as described in Chap. 3. This very high data rate is not suitable for the development of low-cost image-processing applications and must be reduced.

Image information compression systems rely on human psychovisual characteristics and their limitations to remove unnecessary data components in video signals.

8.1.1 Video signal redundancies and entropy

- Statistical data redundancy. Nearly all images contain large amounts of identical data information values. With non-data-reduction systems, these identical data are repeated as necessary to reconstruct uniform areas in a picture and thus constitute redundant information in a data stream. Data redundancy exists within wide areas of the same picture (spatial redundancy) and also between a sequence of pictures (temporal redundancy). Compression systems exploit the fact that identical data need not be repeated and transmitted. For example, a blue sky in a program material exhibits a high amount of identical information that can be encoded once and repeated as necessary at the decoding end. The process of identifying identical pixel values within a frame or in a sequence of frames is called *data decorrelation*. A good example of a data decorrelation scheme is the discrete cosine transform (DCT), which concentrates most of the energy of a block of pixels in the lowest number of coefficients.

- Psychovisual redundancy. Digitized sample values in a picture are not equally perceived by the human visual system (HVS). If the HVS cannot see an error, this error does not affect the perceived quality of reproduced pictures. Consequently, some sample values can be altered or even removed without affecting the viewer tolerance to picture degradation. Experiments have shown a picture content tolerance to this information removal.

- Entropy. The entropy is a measure of the average information content of a picture that has been sampled as binary values. The occurrence of a less probable event (a sample binary value) provides more information than the occurrence of a more probable event. From that observation, it can be said that entropy defines the difficulty of encoding a given picture.

  Entropy also represents the minimum average amount of information per sample binary value that needs to be preserved in order to remove the uncertainty in the reconstructed picture. If a video compression scheme reduces the bit rate to below the entropy value of the picture, some picture information will be lost.

8.1.2 HVS characteristics

Generated and transmitted video signals are destined to be decoded by the human observer system, usually called the HVS. The human eye, in conjunction with the brain's visual cortex, is a very precise imaging system. It can operate over a wide range of light intensities, detect color differences, and perceive picture contrast as a function of spatial frequency and light intensity.

Picture width and viewer distance influence the visibility of pattern cycles. Consequently, the image frequency content is dependent on viewer position. Some assumptions have to be made about viewing conditions that involve the resolution of the display device (as explained in Sec. 1.4) and the viewing distance. This display resolution is generally expressed as a number of lines per picture height. To combine these different parameters affecting the perceived image frequency content, the image spatial frequency is defined as the number of cycles that subtends an arc of $1°$ as seen by the viewer. One cycle is made of two vertical bars or horizontal lines. The viewing conditions are shown in Fig. 8.1.

For conventional TV displays, a viewing distance of about $6H$ (H = picture height) ensures full visual resolution of image details. For an identical-size display height, this distance must be reduced to $3H$ to exploit the benefits of the higher-definition image and maintain the same level of spatial-frequency eye discrimination, also called visual acuity, expressed in cycles per degree, as

$$\text{Contrast} = \frac{I \text{ max.} - I \text{ min.}}{I \text{ max.} + I \text{ min.}}$$

$\alpha = 2 \arctan (w/2d) = 57.3 \ w/d$

Visual acuity = Number of cycles/α, in cycles per degree

Resolution = TV lines per picture width

Figure 8.1 Resolution and viewing angle.

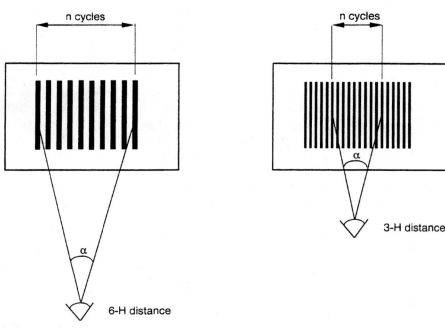

Figure 8.2 Conventional TV and high definition display viewing distance to maintain the same level of visual acuity.

shown in Fig. 8.2. Alternately, the HDTV display dimension must be increased until the eye perceives the finest details in the image at a $6H$ distance.

Visual acuity defines the eye's ability to perceive and separate detail. It varies with

- The luminance of the background. Visual acuity increases with the brightness level to a limit of 100 foot-lamberts (340 cd/m²).

- The contrast of the picture luminance and chrominance signals. The image contrast is defined as the ratio of the difference between maximum and minimum light intensities to the sum of these intensities. Picture details are visible only if there is a significant difference between them and the background. The sensitivity of the eye to luminance detail is higher than that of chrominance detail. Figure 8.3 shows eye contrast sensitivity to spatial frequencies expressed in cycles per degree. In addition, as shown in Fig. 8.4a, the eye contrast sensitivity varies with the temporal frequency of the picture. The temporal frequency above which the HVS does not perceive flicker is known as the *critical flicker frequency* (CFF). All TV standards have a temporal (field) frequency of 50 Hz or higher to reduce the

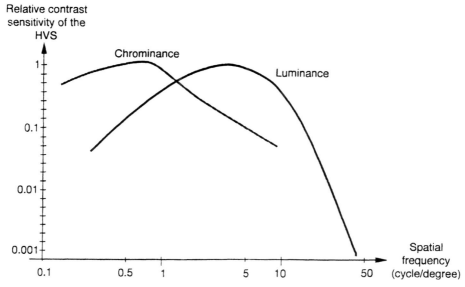

Figure 8.3 Relative variations of color and luminance contrast sensitivities of the HVS with image's spatial frequency.

perception of flicker. Figure 8.4b shows that the CFF is proportional to the picture brightness and may cause visible flicker in 50-Hz scanning systems. The CFF is proportional to the screen size and decreases with the viewing distance. This effect has determined an increase in the refresh rate of large-screen computer monitors to 70 Hz or higher.

- The luminance of the TV display receiver environment. Visual acuity falls off when the surrounding brightness is higher or significantly lower than that of the displayed image.

Fundamental HVS properties can be summarized into two categories related to the image redundancy that exists in the spatial and temporal domains.

8.1.2.1 HVS characteristics related to the spatial redundancy. Spatial redundancy is a result of the correlation between adjacent pixels values. The parameters affecting the spatial redundancy are

- Spatial frequency sensitivity. High frequencies are less visible.

- Texture masking. Errors in textured regions are difficult to see. Conversely, the HVS is very sensitive to distortions in uniform areas.

- Edge masking. Errors near the edges are harder to see.

- Luminance masking. Visual thresholds increase with background luminance. This effect is called light adaptation.

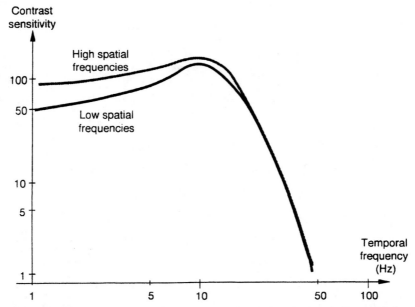

Figure 8.4a Variation of the HVS contrast sensitivity to picture temporal frequency.

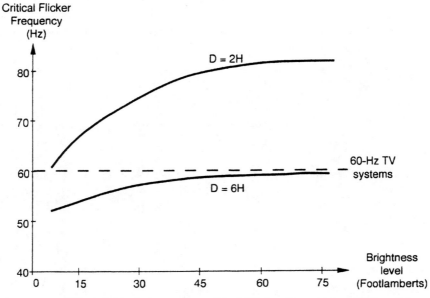

Figure 8.4b Relationship between the critical Flicker Frequency and the picture brightness level and the viewing distance.

- Contrast masking. Errors (and noise) in light regions are harder to see. This effect refers to the reduction in the visibility of one image detail by the presence of another. This masking is maximum when details are very similar in nature. Errors in image reconstruction that are below the contrast threshold are not visible.

- Frequency noise content. The HVS sensitivity to noise is reduced in the high spatial frequencies and greatest at low frequencies.

8.1.2.2 Summary of the HVS characteristics related to temporal redundancy.
Temporal redundancy is caused by the correlation between different frames in a sequence of images. The parameters affecting the temporal redundancy are

- Temporal frequency sensitivity. Below 50 Hz, flicker effects become noticeable.

- Luminance masking. High luminance levels increase the flicker effect.

- Spatial frequency content. Low spatial frequencies reduce the eye's sensitivity to image flickering.

8.2 Video Data Reduction Techniques

Data compression systems are a combination of various tools (processing techniques) used to reduce the bit rate of digital signals to a value that does not compromise the picture quality level chosen for an intended application.

Many "lossy" and "lossless" data reduction techniques have been developed over the years. Only a few are suitable for video applications. Figure 8.5 summarizes the data reduction techniques that are combined to generate Joint Photographic Expert Group (JPEG) and Moving Pictures Expert Group (MPEG) signals. Other techniques that are under development or are difficult to implement are not described in this book, such as Karhunen-Loève transform (KLT), Walsh-Hadamard transform (WHT), vector quantization, wavelet, and fractals.

Taken separately, none of these data reduction techniques can provide significant data reduction. However, using the right combination of several techniques, very efficient data reduction systems can be designed. Several systems have been standardized, such as the JPEG, the MPEG-1, and the MPEG-2 data compression systems.

8.2.1 Lossless data rate reduction

Lossless compression schemes allow the recovery of the original data information after decompression. It is a reversible coding process. Limited compression ratios (less than 2:1) are achievable with usual detailed TV pictures. The amount of data reduction is picture-content-dependent, leading to variable-bit-rate (VBR) applications such as still picture storage and transmis-

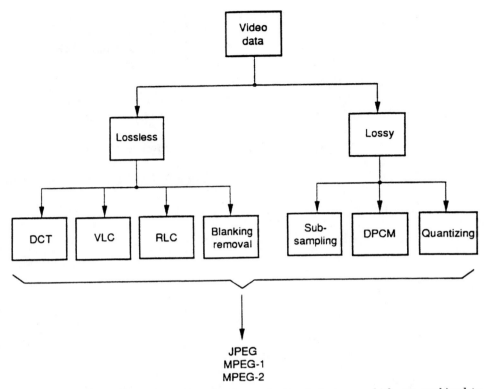

Figure 8.5 Summary of lossless and lossy data reduction techniques which are combined to generate JPEG and MPEG data signals.

sion. The best performance is achieved when a data decorrelation scheme precedes these lossless techniques.

Lossless compression techniques are

- Variable-length coding (VLC). Also called Huffman coding and entropy coding, this technique is based on the probability of identical amplitude values in a picture and assigns a short code to the values with the highest probability of occurrence and long codes to the others. At the decompression end, identical code assignments are used to reconstruct the original data values. Huffman encoding and decoding is easy to implement by means of look-up tables in hardware.

- Run-length coding (RLC). This technique relies on the repetition of the same data sample value to generate special codes that indicate the start and end of a repeated value. Only nonzero sample values are encoded, along with the number (run) of zero sample values along the scan line. Generally, strings of zeros are generated by decorrelation processes such as DCT and DPCM schemes.

- Blanking area data removal reduces the original bit stream to the active picture area content. The horizontal and vertical blanking areas of the video raster are not recorded and transmitted. They are replaced by shorter synchronization data specific to the applications.

- The forward DCT associated with an inverse DCT process is totally transparent if the coefficient word length is of 13 to 14 bits for an input video signal digitized by 8-bit-long samples. When an 11-bit-long (or less) process is used, the DCT compression becomes lossy.

8.2.2 Lossy data rate reduction

Lossy compression is realized by combining two or more processing techniques to take advantage of encoded representation of picture signals. Lossy compression uses much higher compression ratios (from 2:1 to 100:1) and results in data loss and picture degradation after decompression because of data rounding and discarding within a frame or between frames. These degradations can be minimized by exploiting the psychovisual redundancy of sample values, which is based on the HVS characteristics. The compression ratio can be made picture-content-dependent, leading to constant-bit-rate (CBR) applications such as picture storage and transmission. Data rate achievements in lossy systems depend on the requirements for acceptable picture quality.

Lossy compression techniques are

- Sample subsampling. This is a very effective data reduction method, but loss of picture resolution and aliasing components degrade the original picture content. For these reasons, sample subsampling is not applied to luminance signals. Chrominance signal subsampling methods, as exemplified by the 4:2:0 and 4:1:1 signal structure formats, are presently used in recording applications, while the 4:2:0 format is used in MPEG program production and transmission applications.

- Differential pulse coding modulation (DPCM). This is a predictive encoding scheme that transmits the sample-to-sample difference rather than the full sample value. This difference is added to the current decoded sample value, at the decoding end, to generate a reconstructed sample value. Figure 8.6 shows a simplified block diagram of the DPCM encoder and decoder.

  The DPCM process realizes an entropy reduction of the original signal as shown in Fig. 8.7. Most difference values are centered around zero because of the high probability for uniform areas in pictures. With highly detailed pictures, large difference values are possible and can be coarsely quantized (nonlinear quantization), because their occurrence is low and the HVS sensitivity in the presence of high-contrast details is reduced. Data rate reduction is realized by applying a coarse quantization of the difference signal and VLC techniques to the resulting signal. To avoid transmission error propagation, a full sample value is transmitted periodically to provide an

a) DPCM encoder

b) DPCM decoder

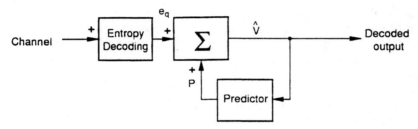

Figure 8.6 DPCM encoder/decoder.

updated reference. Adding adaptive prediction and quantization results in improvements as well as complexity to the DPCM encoding.

- Quantization and VLC of DCT coefficients. The combination of these three processes, described in detail in the following section, permits the representation of a block of pixel bytes by a small number of bits and produces the most cost-effective data reduction technique available at the present time.

8.3 DCT Coding Process and Implementation

As discussed previously, video data rate reduction techniques employ the DCT coding combined with a quantizing process and VLC to achieve a high level of compression while maintaining picture quality at a predefined level. The optimum combination is described in the following sections.

8.3.1 DCT coding process

The DCT scheme processes values of blocks of pixel data into blocks of frequency-domain coefficients. Figure 8.8 shows a summary of a one-dimensional (along one axis) DCT coding of eight consecutive luminance pixels taken from a 4:2:2 sampling digitized picture. Figure 8.8a shows the luminance

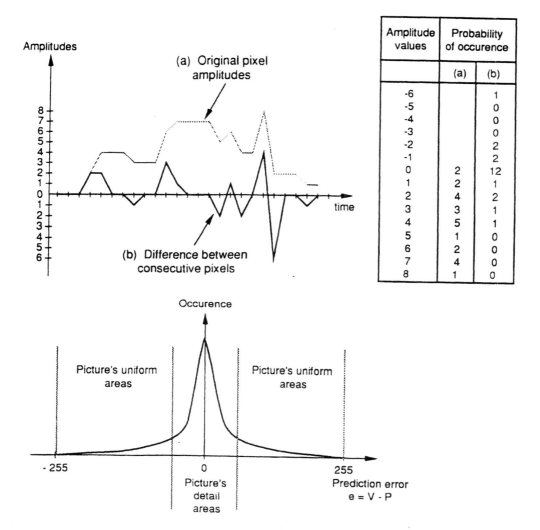

Amplitude values	Probability of occurence	
	(a)	(b)
-6		1
-5		0
-4		0
-3		0
-2		2
-1		2
0	2	12
1	2	1
2	4	2
3	3	1
4	5	1
5	1	0
6	2	0
7	4	0
8	1	0

(a) Original pixel amplitudes

(b) Difference between consecutive pixels

(c) Prediction error histogram based on a long term average of TV pictures.

Figure 8.7 DPCM encoding and entropy reduction.

amplitudes of the eight consecutive pixels. Their corresponding average DC level and the luminance pixel variance are shown in Figs. 8.8b and 8.8c. The spectrum shown in Fig. 8.8d represents the change of amplitude (or variance) along the eight pixels and ranges from 0 Hz to one-half the sampling frequency ($F_s/2 = 6.75$ MHz). The DCT coding separates this spectrum into eight frequency bands resulting in eight coefficient values that indicate the energy the waveform's spectrum has in each of these frequency bands. Figures 8.8e and f

8 consecutive luminance pixels taken from one line

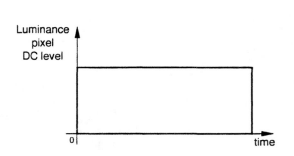

a) Luminance amplitude for the 8 consecutive pixels

b) Average DC level of the 1x8 block

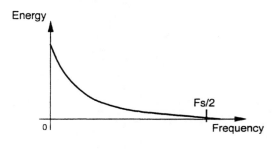

c) Variance from average level

d) 1x8 block luminance pixel spectrum

e) Luminance spectrum band separation

f) 1x8 DCT coefficient values

Figure 8.8 Summary of 1-Dimensional DCT coding (*Courtesy of Sony Corporation*).

show the luminance spectrum band separation and the respective coefficient values of each band. The first coefficient, at the very left, represents the average DC level of the waveform. From left to right, the other coefficients indicate the higher spatial frequency components of the original waveform and are called the AC coefficients. When spatial redundancy is high in a picture, many of these AC coefficients are near or equal to zero.

To achieve a higher decorrelation of the picture content, two-dimensional (along two axes) DCT coding is applied to blocks of 8×8 luminance pixel values; 8×8 DCT coefficient blocks are obtained where the top left corner number of each DCT block is the DC coefficient representing the average DC level of the corresponding 8×8 luminance pixel block. Figure 8.9 shows an example of two-dimensional DCT block coding of an 8×8 block of pixel values taken from a real picture.

In the JPEG standard, the forward DCT (FDCT) process is defined as

$$F(u,v) = \frac{C(u)C(v)}{4} \sum_{j=0}^{7} \sum_{k=0}^{7} f(j,k) \cos\left(\frac{(2j+1)u\pi}{16}\right) \cos\left(\frac{(2k+1)v\pi}{16}\right)$$

(Eq. 8.1)

where $f(j,k)$ = the original samples in the 8×8 luminance pixel block.
$F(u,v)$ = coefficients of the 8×8 DCT block.
u = the normalized horizontal frequency ($0<u<7$).
v = the normalized vertical frequency ($0<v<7$).

$$C(u), C(v) = 1/\sqrt{2} \qquad \text{if } u,v = 0$$

$$C(u), C(v) = 1 \qquad \text{if } u, v = 1, 2,..., 7$$

For the first coefficient, the normalized frequencies u and v are equal to 0. Then, this coefficient is called the DC coefficient.

$$F(0,0) = \frac{1}{8} \sum_{j=0}^{7} \sum_{k=0}^{7} f(j,k)$$

(Eq. 8.2)

This equation realizes the summation of all pixel values in the 8×8 block and divides it by 8. The result of this calculation is equivalent to eight times the mean pixel values in the block.

When the DCT process is applied to component digital signals such as Y, C_R, and C_B, the chrominance C_B and C_R signals have a maximum amplitude of ±128 binary values in an 8-bit sampling system, while the luminance Y signal has a maximum range of 0 to 255 binary values. To simplify the design of a DCT encoder, the Y signal is level-downshifted by subtracting 128 to each pixel value in the block to make the maximum range of this signal identical to that of the C_B and C_R signals. At the DCT decoding stage, the same value (128) is added to reconstructed Y pixel values.

Figure 8.9 Summary of 2-Dimensional DCT block coding (*Courtesy of Sony Corporation*).

Consequently, an 8-bit-resolution A/D converter, used to generate blocks of Y pixel values, produces binary values that are shifted between -128 and 127. The DC coefficient value of the corresponding DCT block has a range of -1024 to 1016 (eight times the binary values).

For AC coefficients, where u and v take values from 1 to 7, $C(u)$ and $C(v) = 1$ and their maximum values are in a range of ± 1020, according to the equation of the forward DCT process.

The resulting 8×8 block of DCT coefficient values exhibits a high DC value (DC = 591 in the example 1), representative of the average brightness of the 8×8 luminance block and very small values of high-frequency components in the vertical and horizontal direction, as shown in Fig. 8.9. However, horizontal AC coefficients are higher than those in the vertical direction. This is easily confirmed by the higher horizontal pixel-to-pixel value variations compared to those of the vertical direction.

Figure 8.10 shows a first example of the DCT process applied to a luminance block of pixels. The pixel values shown have already been downshifted by -128 to facilitate the encoding of the three component video signals.

As a general rule, a high pixel-to-pixel value variation in one direction of the pixel block, be it horizontal, vertical, or diagonal, will generate high coefficient values in the respective directions of the DCT coefficient block. Figure 8.11 shows a second example of the DCT process applied to a pixel block made up of alternate values of black (value = 0) and white (value = 255) pixels. After reduction by -128, the pixel values alternate between $+127$ and -128. This block exhibits the highest possible pixel-to-pixel amplitude variations, and the resulting DCT coefficients confirm this observation. Although many frequency coefficients are equal to zero, the amplitude of the high-frequency coefficients is very important and does not lead to a significant reduction of the number of AC coefficients after the quantization process. In this limited example, one can decide to remove all nonzero AC coefficients and keep only the DC coefficient value. After the inverse DCT process, a uniform gray-level picture is obtained, which is exactly what the HVS sees when the original checkerboard picture is reduced to its normal dimension and seen at a normal distance.

It must be noted that a horizontal and/or vertical low-pass filter can easily be realized by removing a certain number of horizontal and/or vertical high-frequency DCT coefficients before the inverse DCT process is applied.

Figures 8.12 and 8.13 show three-dimensional representations of original luminance pixel blocks and corresponding DCT coefficient blocks described previously in examples 1 and 2 (Figs. 8.10 and 8.11).

It is important to note that the DCT process described above assumes that the range of the Y pixel binary value is 0 to 255 after A/D conversion of the luminance signal variation from 0 to 700 mV. However, component signals complying to the CCIR-601 standard have binary values between 16 and 235, which correspond to the same variation of 0 to 700 mV. This allows for a possible headroom above the 700-mV and below the 0-mV levels, such as used in chroma-keying applications with a signal called blacker-than-black. Some

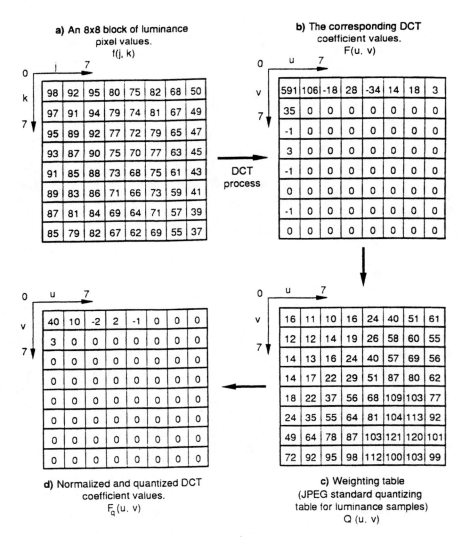

a) An 8x8 block of luminance pixel values.
f(j, k)

b) The corresponding DCT coefficient values.
F(u, v)

DCT process

d) Normalized and quantized DCT coefficient values.
F_q(u, v)

c) Weighting table (JPEG standard quantizing table for luminance samples)
Q (u, v)

Figure 8.10 8×8 DCT block formatting, example #1.

compression system implementations ignore this difference and generate incompatibility errors in the program production chain.

The DCT process does not reduce the data rate and is totally reversible. The inverse DCT process (IDCT) reconstructs the exact original pixel values if the DCT coefficients are kept unchanged, although a calculation accuracy of the order of 13 to 14 bits is necessary to avoid round-off errors. It is the combination with quantization and efficient coding techniques, such as the VLC, that makes a data rate reduction possible. DCT is employed for decorrelating the original data, compacting a large fraction of the signal energy into a relatively few low-frequency transform components, and generating zero

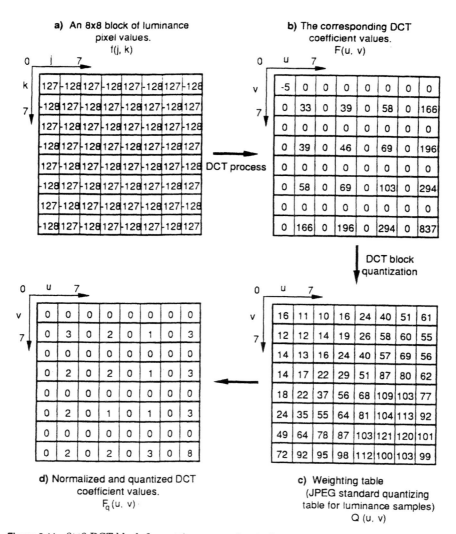

a) An 8x8 block of luminance pixel values.
f(j, k)

b) The corresponding DCT coefficient values.
F(u. v)

DCT process

DCT block quantization

d) Normalized and quantized DCT coefficient values.
F_q(u. v)

c) Weighting table (JPEG standard quantizing table for luminance samples)
Q (u. v)

Figure 8.11 8×8 DCT block formatting, example #2: checkerboard picture.

and very low values that need not be stored or transmitted. The choice of an 8×8 block size is the result of a compromise between an efficient energy compaction, requiring a large screen area, and a reduced number of real-time DCT calculations, requiring a small area. Figure 8.14 shows a simplified block diagram of a video data reduction encoder that uses a DCT scheme.

Before compression, original pictures are digitized by means of sampling structures chosen to achieve the resolution required for the specific application. Figure 8.15 shows examples of sampling structures that lead to different macroblock structures. Luminance and chrominance signal data are separated in 8×8 blocks of Y, C_R, and C_B values. Then, a macroblock is constituted

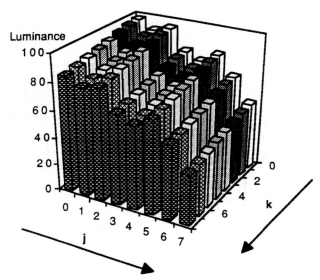

a) 3-D display of original luminance values (example #1)

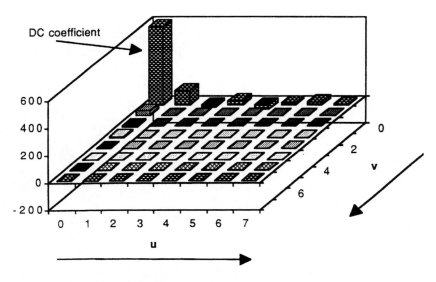

b) 3-D display of DCT coefficients (example #1)

Figure 8.12 3-D display of original luminance values and DCT coefficients (example #1).

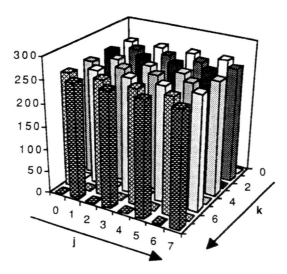

a) 3-D display of original luminance values (example #2)

b) 3-D display of DCT coefficients (example #2)

Figure 8.13 3-D display of original luminance values and DCT coefficients (example #2).

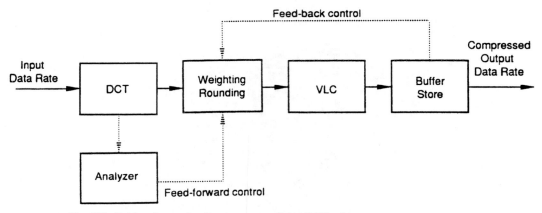

Figure 8.14 Simplified video data reduction process using a DCT scheme.

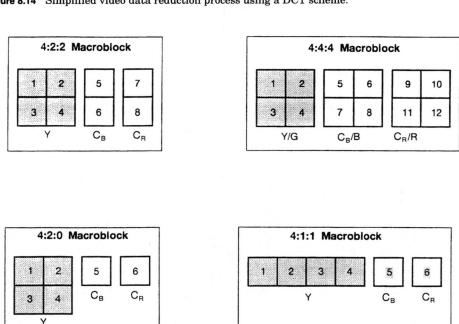

Note: All blocks are made of 8 pixels x 8 lines. The block numbering corresponds to the order in which all blocks within a macroblock are sent to the DCT process.

Figure 8.15 Macroblock structures of various sampling formats.

with four blocks of 8×8 luminance (*Y*) pixel values, forming a block of 16 pixels by 16 lines, and their associated chrominance blocks (C_R and C_B). A 4:2:0 sampling structure is made of four *Y* blocks and one C_R and one C_B block. The ordering of the blocks within a macroblock determines the sequence in which they are sent to the DCT coder.

8.3.2 DCT block quantization process

The DCT block quantization process is the most sensitive area in designing a compression system because it directly affects the reconstructed picture quality performance. The visual sensitivity of the HVS to coding artifacts mainly depends on the frequency components and the local activity in the spatial domain. Therefore, the adaptive quantization is carried out by using perceptual quantization weights that are determined by three major factors, that is, the frequency weights, the perceptual activity parameter, and the buffer status parameter, as shown in Fig. 8.16.

The basic function of the quantizer is to divide each DCT coefficient by a number bigger than one to generate numbers near to or equal to zero that can

Figure 8.16 Basic picture data reduction process made with a DCT coding scheme.

be rounded or ignored to be efficiently coded later on in a subsequent process. The idea is that low-energy coefficients, representing small pixel-to-pixel variations, can be discarded without affecting the perceived resolution of the reconstructed picture because of the psychovisual characteristics of the HVS.

The quantization process is a lossy operation and does introduce artifacts. An effective design minimizes the psychovisual effects loss by optimizing a large number of factors as a result of subjective evaluations of a variety of different pictures, representative of a wide range of TV production materials. The designer chooses between a number of factors depending on the application, performance, low coding delay, complexity, and cost targets. Figure 8.16 shows an example of a quantizer design that uses four dividing factors to better control the quantization process.

The first factor takes into account the HVS characteristics that are summarized in a quantization table, as shown in Fig. 8.17 (JPEG standard tables for Y and C_R/C_B signal DCT blocks). DC and low-frequency components are the most sensitive parameters of the original pixel block. Effective compression systems quantize the DC coefficient with a precision of 12 bits in order to avoid visible quantization errors between pixel blocks that would result in blocking artifacts of the reconstructed picture ("mosaic" effects). Conversely, high-frequency coefficients can be coarsely quantized down to 2-bit accuracy as a large high-frequency quantization error is tolerated by the HVS. Consequently, the dividing factor in a quantizing table is small for DC and low-frequency coefficients, and gradually increases for higher-frequency coefficients. Figures 8.10 and 8.11 show the resulting DCT coefficient values after quantization and rounding to the nearest integer value, which are shown in the forward quantization table $F_q(u,v)$ and calculated by the following formula:

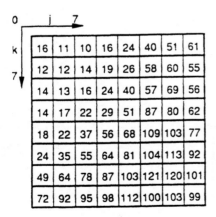

Weighting table
(JPEG standard quantizing
table for luminance samples)

Weighting table
(JPEG standard quantizing
table for chrominance samples)

Figure 8.17 JPEG standard quantization tables for luminance and chrominance signals.

$$F_q(u,v) = \text{Round}\left[\frac{F(u,v)}{Q(u,v)}\right] \approx \text{Nearest integer}\left[\frac{F(u,v)+[Q(u,v)/2]}{Q(u,v)}\right] \quad \text{(Eq. 8.3)}$$

In Fig. 8.17, a pixel block identification (block ID) allows luminance and chrominance values to be quantized differently, because chrominance information is less critical for the HVS. Quantizing noise is less visible in the chrominance components than in the luminance components, and coarser quantization can be applied.

The quantizing table feeds the quantizer with a weighting value that depends on the coefficient position in the DCT block. A highly detailed picture generates a large number of high-frequency DCT coefficients, which can result in buffer overflow if each dividing factor in the quantizing table is too low. A feedback buffer state information controls a scaling factor to optimize the buffer fullness. This scaling factor is applied to AC coefficients only, to avoid the mosaic effect generated by block-to-block average DC value errors. It is constant on a macroblock basis that contains two chrominance blocks and a variable number of luminance blocks depending on the sampling structure of the original picture. Because a macroblock comprises up to four luminance pixel blocks, picture details may put high energy in only one block among the four, resulting in coarse-quantization and high-quantization errors affecting the other blocks, and a distorted reconstruction of a ramp detail ("staircase effect" degradations). To minimize this annoying effect, the block energy content can be evaluated by looking at the largest significant AC coefficient in all blocks and divided into a certain number of classes that control the quantizer. In summary, the quantizing strategy is the key to obtaining good picture quality, and depending on manufacturer's approaches, a specific picture quality performance is achieved.

It is worth mentioning that in designing a bit-rate-reduced digital VTR, data shuffling is necessary to allow for error correction and concealment of loss of information in the recording and playing back processes. This requirement is fulfilled by shuffling DCT blocks within a field or a frame before the quantization process. Real TV pictures generally consist of a combination of large, uniform areas and small, highly detailed areas. This characteristic allows DCT blocks with high-data correlation to be shuffled among blocks with low-data correlation. The resulting quantized bit rate coming from two or more consecutive DCT blocks is made constant. Coarse quantization of the high-data-correlation blocks is unnecessary and the picture quantizing noise is reduced.

8.3.3 Zigzag scanning

The quantized DCT block is submitted to a zigzag scanning pattern to facilitate the subsequent encoding and transmission along a one-dimensional channel. Figure 8.18 shows the two-dimensional array converted into a serial

Figure 8.18 Zigzag scanning and Huffman encoding.

string of increasing spatial frequency coefficients. The zigzag pattern is chosen to first read the significant coefficients and group together the zero coefficients in as much as possible. The distribution of the nonzero coefficient depends on the original picture content. Strong pixel-to-pixel variations in the vertical direction of original pictures lead to high values of the vertical frequency components and can be better ordered by another possible scanning pattern as shown in Fig. 8.19. The type of the chosen pattern must be indicated in the encoded bit stream in order to control the decoder.

8.3.4 Run-length and level coding

Run-length and level coding (RLC) are used to encode efficiently the previously scanned quantized DCT coefficients. Each nonzero coefficient detected after the DC value is coded with a two-parameter code word, the number of zeros (run) preceding a particular nonzero coefficient and its level after quan-

Figure 8.19 Other possible DCT coefficient block scanning process (MPEG-2 Standard).

TABLE 8.1   AC Coefficient Ranges and Categories

Category	Coefficient range	
NA	0	
1	$-1,$	1
2	$-3, -2,$	2, 3
3	$-7, -6, -5, -4,$	4, 5, 6, 7
4	$-15,........,-8,$	8,.......,15
5	$-31,........,-16,$	16,........,31
6	$-63,.........,-32,$	32,.........,63
7	$-127,..........,-64,$	64,...........,127
8	$-255,............,-128,$	128,...........,255
9	$-511,............,-256,$	256,...........,511
10	$-1023,............,-512,$	512,...........,1023
11	$-2047,............,-1024,$	1024,...........,2047
⋮	⋮	

tization. One code word is generated for each (run, level) pair, as shown in Fig. 8.18.

8.3.5 Variable-length coding

The RLC code words are further encoded by allocating short code words to frequently occurring levels and long code words to infrequently occurring ones. This process is called variable-length coding (VLC). Table 8.1 shows how the AC coefficient values are grouped into categories and Table 8.2 represents an example of a Huffman code table relative to these categories. A short VLC code word signals the end of block (EOB), which indicates that all following coefficients in the block are equal to zero.

In the DCT coefficient block example shown in Fig. 8.10, the DC coefficient (40) is DPCM-encoded by using the DC value (25) of the previous DCT block. The DPCM encoding further extends the maximum Y range of -1024 to 1016 to a range of -2048 to 2032, because of the DC difference value calculation between two successive blocks. The DC coefficient difference (25) is VLC-

TABLE 8.2   Example of an AC Coefficient Huffman Code Table (JPEG Default Table)

Zero run	Category	Code length	Code word
0	1	2	00
0	2	2	01
0	3	3	100
0	4	4	1011
0	5	5	11010
0	6	6	111000
0	7	7	1111000
⋮	⋮	⋮	⋮
1	1	4	1100
1	2	6	111001
1	3	7	1111001
1	4	9	111110110
⋮	⋮	⋮	⋮
2	1	5	11011
2	2	8	11111000
⋮	⋮	⋮	⋮
3	1	6	111010
3	2	9	111110111
⋮	⋮	⋮	⋮
4	1	6	111011
5	1	7	1111010
6	1	7	1111011
7	1	8	11111001
8	1	8	11111010
9	1	9	111111000
10	1	9	111111001
11	1	9	111111010
⋮	⋮	⋮	⋮
End of Block (EOB)		4	1010

encoded by means of a lookup table (Table 8.3), which outputs a Huffman binary number based on this DC coefficient difference amplitude (category). This Huffman binary number is then completed with the binary representation of the DC coefficient difference amplitude value. Figure 8.20 summarizes the DC- and AC-coefficient VLC encoding.

AC coefficients included in RLC code words are Huffman-encoded by means of lookup tables (Tables 8.1 and 8.2). A category is first output and combined with the run value (number of zeros preceding the AC coefficient) to generate a Huffman binary number that is completed with the binary representation of the AC-coefficient amplitude value. In both DC difference and AC coefficient cases, the amplitude binary encoding uses the shortest possible binary representation of amplitude values. Negative amplitude values generate the one's-complement binary number of the positive value.

At the VLC output, all code word data coming from the same DCT block are

TABLE 8.3   Variable Length Codes for DC Coefficients

Differential DC values	Categories	Code word for luminance	Code word for chrominance
−255 to −128	8	1111 110	1111 1110
−127 to −64	7	1111 10	1111 110
−63 to −32	6	1111 0	1111 10
−31 to −16	5	1110	1111 0
−15 to −8	4	110	1110
−7 to −4	3	101	110
−3 to −2	2	01	10
−1	1	00	01
0	0	100	00
1	1	00	01
2 to 3	2	01	10
4 to 7	3	101	110
8 to 15	4	110	1110
16 to 31	5	1110	1111 0
32 to 63	6	1111 0	1111 10
64 to 127	7	1111 10	1111 110
128 to 255	8	1111 110	1111 1110

concatenated to produce the output bit stream. In the example shown in Fig. 8.18, the data corresponding to the original DCT coefficient block ($8 \times 8 \times 8$ bits = 512) is reduced to 48 bits after VLC encoding. This yields a lossy data reduction ratio of 512/48 = 10.6, made possible by the use of the quantizing table. This performance is often indicated in bits per pixel, since 48 bits are used to represent 64 pixels, which gives 48/64 = 0.75 bit/pixel. Variable-length encoding in itself is a lossless coding technique that provides an additional reduction of the bit rate, which has already been decorrelated, rounded, and reduced by the DCT quantization process.

8.3.6   Buffer memory

VLC code words are produced at variable speed, depending on the picture complexity, and written into a buffer memory (Fig. 8.16). Bits are read from this memory at a fixed bit rate that can be made user-selectable by the coder designer. A buffer occupancy control mechanism ensures that no buffer underflow (empty buffer) or overflow (full buffer) occurs by varying the scaling factor applied to the weighting table. If the buffer becomes full, the quantization is made coarser, to produce fewer bits, by increasing the scaling factor of the quantizer.

A selectable memory read clock is added to cater to different values of constant buffer output bit rate, typical of storage and transport devices for picture-compressed systems. If the buffer is emptied more slowly by a slower read clock, the buffer fullness must be reduced by increasing the scaling factor of the quantizer.

a) DC coefficient VLC encoding

(*) Category and DC Huffman binary numbers are shown in Table 8-3.

b) AC coefficient VLC encoding

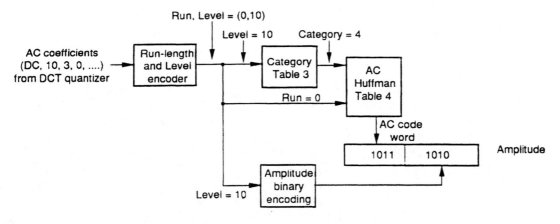

(*) Category/run and AC Huffman binary numbers are shown in Tables 8-1 and 8-2.

Figure 8.20 DC and AC coefficient variable length/Huffman encoding (JPEG Baseline standard).

Advantages of the DCT compression scheme are the use of original orthogonal signals and separable transform for each digital component signal (Y, C_R, and C_B), a near-optimum energy-compacting property, and the availability of fast algorithms and hardware.

8.3.7   DCT decoder

A reverse compression process takes place at the receiving end or in playback storing devices. Figure 8.21 shows the block diagram of a generic DCT decoder.

Huffman and quantizing tables identical to those of the DCT encoder are used to retrieve the DCT coefficient values of an 8×8 pixel block. A reverse quantization process $R_q(u,v)$ is performed as

Figure 8.21 Basic DCT decoder.

$$R_q(u,v) = F_q(u,v)Q(u,v) \qquad \text{(Eq. 8.4)}$$

An inverse DCT (IDCT) process $f*(j,k)$ is then applied to these 64 coefficients to reconstruct the original block pixel values and expressed as

$$f*(j,k) = \frac{1}{4} \sum_{u=0}^{7} \sum_{v=0}^{7} C(u)C(v)F(u,v) \cos\left(\frac{(2j+1)u\pi}{16}\right) \cos\left(\frac{(2k+1)v\pi}{16}\right)$$

$$\text{(Eq. 8.5)}$$

It is interesting to note the similarities in the calculation process of the FDCT and the IDCT in Eqs. (8.1) and (8.5). The same processing unit can be used in both the compression and the decompression hardware systems. The IDCT output is shown in Figs. 8.22 and 8.23. Differences between the block original and reconstructed values represent errors introduced in the pixel values of the original picture as a result of compression. These errors are expressed as

$$e(j,k) = f(j,k) - f*(j,k) \qquad \text{(Eq. 8.6)}$$

To evaluate the reconstructed picture quality, two error metrics are used, the root-mean-squared error (RMSE) and the peak signal-to-noise ratio, which are given by

$$\text{RMSE} = \sqrt{\frac{1}{64} \sum_{j=0}^{7} \sum_{k=0}^{7} e^2(j,k)} \qquad \text{(Eq. 8.7)}$$

$$\text{PSNR} = 20\log_{10}\left(\frac{255}{\text{RMSE}}\right) \qquad \text{(Eq. 8.8)}$$

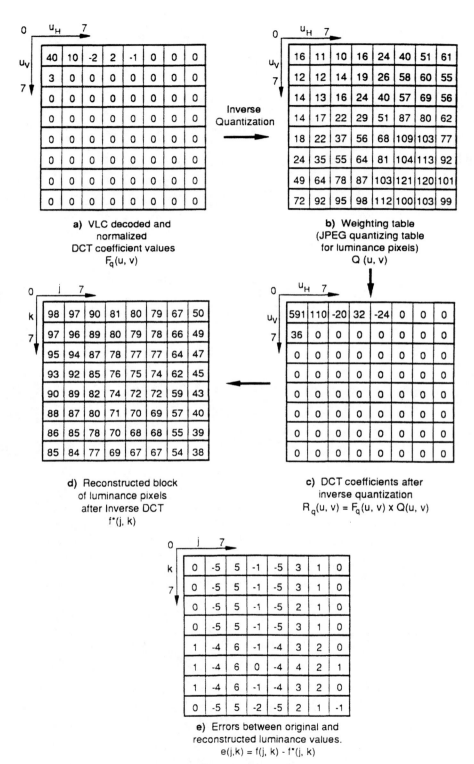

a) VLC decoded and
normalized
DCT coefficient values
$F_q(u, v)$

b) Weighting table
(JPEG quantizing table
for luminance pixels)
$Q(u, v)$

Inverse
Quantization

d) Reconstructed block
of luminance pixels
after Inverse DCT
$f^*(j, k)$

c) DCT coefficients after
inverse quantization
$R_q(u, v) = F_q(u, v) \times Q(u, v)$

e) Errors between original and
reconstructed luminance values.
$e(j, k) = f(j, k) - f^*(j, k)$

Figure 8.22 Inverse 8×8 DCT block process and resulting data errors. (Example #1).

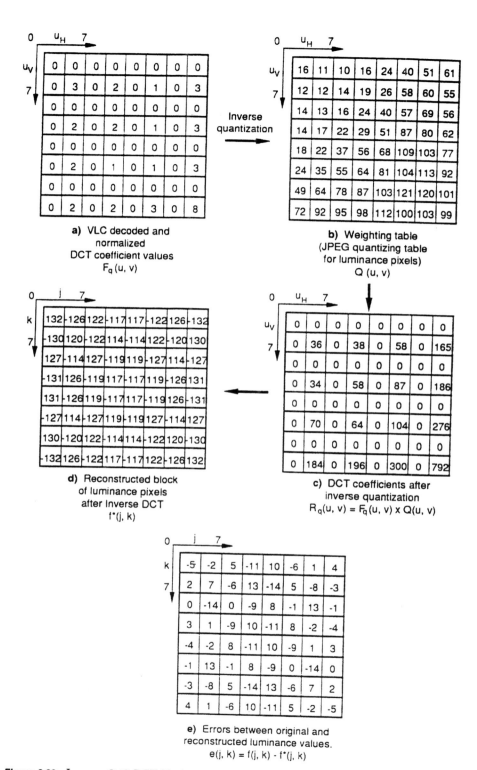

a) VLC decoded and
normalized
DCT coefficient values
$F_q(u, v)$

b) Weighting table
(JPEG quantizing table
for luminance pixels)
$Q(u, v)$

d) Reconstructed block
of luminance pixels
after Inverse DCT
$f^*(j, k)$

c) DCT coefficients after
inverse quantization
$R_q(u, v) = F_q(u, v) \times Q(u, v)$

e) Errors between original and
reconstructed luminance values.
$e(j, k) = f(j, k) - f^*(j, k)$

Figure 8.23 Inverse 8×8 DCT block process and resulting data errors. (Example #2: checkerboard picture).

The PSNR, expressed in dB, is calculated for an 8-bit (0–255) sampling resolution picture. These two error metrics can be calculated for the two previous examples:

- Example 1: RMSE = 3.26   PSNR = 37.9 dB
- Example 2: RMSE = 7.47   PSNR = 30.66 dB

The difference in PSNR confirms that the picture used in example 2, a checkerboard pattern, is more difficult to compress because of the presence of high-amplitude, high-frequency DCT components that are quantized coarsely. However, the PSNR does not give an absolute measure of the quality of a compression algorithm. All picture artifacts are weighted equally but are not perceived as such by the HVS, depending on their position and the activity in the picture. The PSNR is useful for optimizing a compression system design and evaluating the multigeneration performance by means of the same test picture.

It must be noted that specific hardware implementations may exhibit different decoded values after the IDCT process is applied. This is caused by the introduction of round-off errors that are specific to the implementation design. The IDCT calculation process accuracy is specified in the IEEE standard "Specifications for the Implementation of 8×8 Inverse Discrete Cosine Transform," IEEE Std 1180-1990.

8.3.8 Temporal data reduction techniques

Successive frames in a TV program material, or areas of them, are time invariable. The data redundancy between frames can thus be exploited to encode and transmit only pixel value differences. In the case of a DCT encoder, the temporal data reduction is obtained by adding an IDCT process, which reconstructs the difference values of an 8×8 pixel block, and a frame delay that makes possible the summation of the difference values with the pixel values belonging to the previous frame. Figure 8.24 shows a simplified block diagram of a DCT encoder exploiting temporal and spatial data redundancy to reduce the original data stream. The combination of the frame transformation into DCT coefficients with their DPCM encoding is sometimes referred as hybrid DCT/DPCM encoding. The predicted pixel block is subtracted from the original pixel block to generate a prediction-error block that is further DCT transformed, quantized, and VLC-encoded. However, the prediction is not satisfactory when there is motion between the two frames.

8.3.9 Motion compensation prediction technique

The performance of the interframe prediction scheme is good for stationary pictures, or areas of them, but is reduced for moving pictures or areas. An improved prediction scheme detects the displacement of picture details between two successive frames and issues a motion vector to indicate the new

a) Hybrid DCT/DPCM encoder

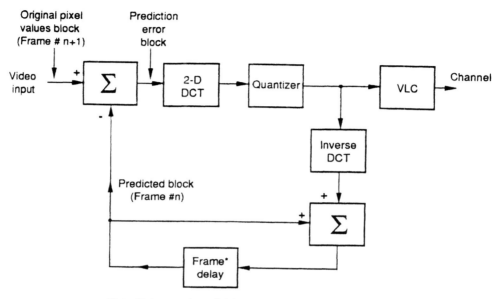

Note: Delay can be a field or a frame delay.

b) Hybrid DCT/DPCM decoder

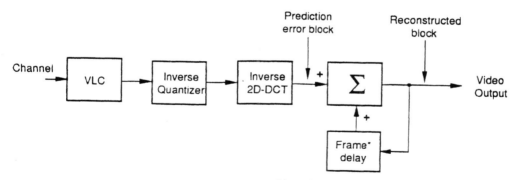

Note: Delay can be a field or a frame delay.

Figure 8.24 Hybrid DCT/DPCM encoder/decoder with temporal and spatial data decorrelation.

position of these details in the current frame, as shown in Fig. 8.25. Consequently, the motion vector indicates the coordinates of all blocks already compressed in the previous frame that will be replicated in the current frame at a new coordinate position. This scheme is known as motion-compensated interframe prediction. Motion estimation is performed using a macroblock approach

Figure 8.25 Motion vectors between two consecutive frames.

on luminance signals only. Figure 8.15 shows the constituting elements of a macroblock. A displacement vector is estimated for each macroblock. The motion vectors and the DCT-encoded difference of the current macroblock and the reference macroblock, pointed by these motion vectors, are transmitted together and provide a more efficient solution than directly encoding the current macroblock with the DCT.

Different techniques are used to determine this displacement vector. One method is called *block matching*. The selected pixel data block, called reference block, in the current frame is moved around its position within a search area in the previous frame. The reference DCT block values are then compared to 8×8 pixel block values in the search area to find the best block match. When the difference is minimum, this corresponds to the best match. A motion vector that takes into account the distance between the reference and the best match coordinate position is issued. The motion vector data information is transmitted to the decoder along with the DCT difference coefficient block.

Matched blocks in the previous frame blocks are used as a prediction by a DPCM encoder that subtracts the prediction from the input to form a low-information-content prediction-error frame, as shown in Fig. 8.26. The difference between the current block and its prediction from the previous block forms the motion-compensated prediction block.

It must be noted that there is a trade-off between the superior performance provided by the motion compensation process and the associated complexity and cost. Consequently, a 16×16 pixel block size is chosen as a unit for the motion compensation process, which corresponds to a macroblock size in all sampling structures.

The search area is defined around a macroblock in the current frame to

Figure 8.26 Motion-compensated interframe prediction DCT encoding (MPEG-2 encoder).

cover any motion between two successive frames. Then, the coordinates of this defined search area are transposed in the previous frame to find the previous position of this macroblock. The size of this area is limited by the processing complexity of the block-matching process and can be selected on a frame-by-frame basis. Over a search area of 16×16 pixels, there are 16 block positions horizontally to combine with 16 positions vertically. To reduce the motion vector calculation process, the picture resolution is downsampled in both horizontal and vertical directions to reduce the number of bits to process and estimate the major part of the displacement. For example, a 720×480 full-resolution picture is processed to a 360×240 low-resolution picture. This downsampled picture can also be used to cover a wider search area. A coarse motion estimation is first calculated using the low-resolution picture. Then, a fine motion estimator calculates the accurate displacement of objects between two successive frames. It must be noted that actual implementations of this hierarchical search method can involve three or more stages with successive picture resolution downsampling.

However, small moving objects can be ignored at the coarse motion estimation stage and result in erroneous motion vectors. For this reason, sufficiently small blocks are used. In case of a large displacement, errors in the matching process may arise between blocks containing similar gray-level patterns that are not motion-related. Furthermore, several types and levels of motion may occur within images and require conflicting specifications of the search area. The hierarchical block matching method is used to resolve this issue. One advantage of the low-resolution picture process is the reduction of noise.

Motion compensation prediction based on the past frame is basically unidirectional and is called *forward prediction*. Such predicted frames are designated as P-frames. Compared to intraframes (I-frames), P-frames allow a higher data compression. However, uncovered areas, as shown in Fig. 8.27, are not predictable from the previous picture. A given pixel block in the uncovered area of the current frame has no match in the previous frame. In this case, the pixel block (or macroblock) is coded as in an I-frame.

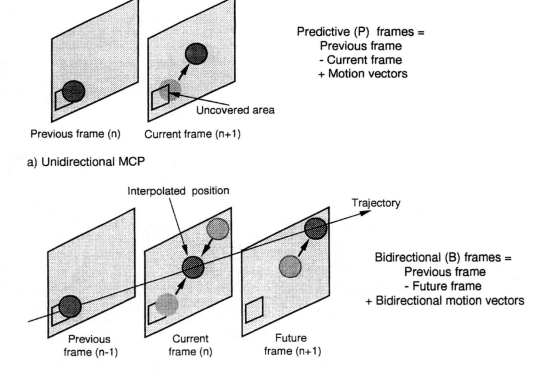

Figure 8.27 Unidirectional and Bidirectional Motion Compensation Prediction (MCP).

Bidirectional temporal prediction, also called motion compensation interpolation, uses information in a previous and in a future reference frame to predict the current interpolated frame, called the B-frame, as shown in Fig. 8.27. It is assumed that the motion trajectory is linear, that is, with no acceleration. B-frames can also be compressed through a prediction derived from a previous reference frame (forward prediction) or a future reference frame (backward prediction) in the sequence. The prediction error frame resulting from the difference between the predicted frame and the current frame is further compressed by DCT coding. The advantages and disadvantages of bidirectional prediction are

- High data compression can be achieved.

- Noise effects are decreased by averaging between the previous and the future frames.

- Uncovered areas can be properly predicted from the future reference frame.

- Edge interpolation can provide half or one-fourth pixel accuracy to reduce the edge artifacts caused by its inadequate location.

- Backward prediction from a future frame is possible if the frames are reordered and transmitted so that the future frame is received before the current B-frame. Figure 8.28 shows the generated and transmitted frame streams. This reordering process at the encoding and decoding stage introduces significant delays depending on the number of B-frames between two reference frames.

More than one bidirectional frame can be predicted between two reference frames (previous and future). There are trade-offs associated with the number of B-frames as follows:

- When this number increases, the correlation of B-frames with reference frames decreases and so increases the number of bits necessary to encode B-frames. It also decreases the correlation between reference frames as they become far apart.

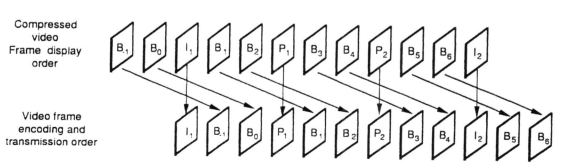

Figure 8.28   Video data frame ordering.

- The video material content requires fewer B-frames in abrupt scene changes and rapid multidirectional motion direction than in "quiet" pictures.

- The frame accuracy in editing systems, such as nonlinear editing servers and DVTRs, necessitates the decoding of B-frames with adjacent reference frames before making an edit.

Interlace video materials can be compressed field-by-field or frame-by-frame. Frame prediction coding is preferable for still or slow-motion pictures, as more spatial redundancy exists in a complete frame than between two separate fields. Field prediction coding is more efficient in scenes with high motion, as interfield motion vectors contain large high-frequency components. Figure 8.29 shows the field and frame predictions based on source scanning formats and the use of these predictions in present compression standards.

Compression schemes define the use of a combination of three frame types:

a) Field and frame predictions in interlaced and progressive scanning formats

b) Use of predictions in present compression standards

Figure 8.29 Field and frame predictions based on source scanning formats.

- I-frames: These are composed of intrablocks only, that is, without reference to other pictures, and can serve as random access points to the sequence. I-frames reduce spatial redundancy only and achieve a moderate compression.

- P-frames: These contain forward motion compensation and intra macroblocks, and provide more compression than I-frames. They can serve as reference for B-frames and future P-frames. However, coding errors can propagate between P-frames.

- B-frames: These contain forward, backward, and bidirectional motion compensation plus intra macroblocks. They provide the most compression and do not propagate errors because they are not used as a reference. However, two frames (P and/or I) within a frame sequence must be decoded first in order to reconstruct a B-frame and a processing delay occurs, which increases with the number of B-frames in the sequence. In addition, two frame stores are necessary to decode a B-frame.

MPEG algorithms allow the encoder to choose the right combination of frame types, in a repeating sequence, to meet the application's need for a given reduced data rate, random accessibility, and picture quality. This frame type sequence is called a group of picture (GOP), and some typical examples are shown in Fig. 8.30. GOPs are generally specified with two parameters:

- m, which defines the number of frames from the first I-frame to the last B- or P-frame before the next I-frame in the data stream sequence.

- n, which defines the number of B-frames between two P-frames.

8.3.10 Complementary processing techniques

Preprocessing enhancements are often applied to digital signals before being compressed.

- Prefiltering techniques: In simple words, the DCT process is comparable to an A/D conversion where a prefilter removes high-frequency components to avoid aliasing components in the converted signal. Spatial prefiltering can be used to remove high-frequency components in a video frame prior to compression. It realizes a compromise between a loss of picture resolution caused by the filtering, and the visibility of aliasing artifacts arising from the high-frequency components in this picture. This spatial filtering improves the apparent compression performance on difficult pictures originating from HDTV to SDTV (standard definition TV) conversion, computer-generated pictures, and new high-quality cameras. This prefilter can be global or locally adaptive.

- Noise-reduction techniques: Random noise is hard to encode because it reduces the pixel-to-pixel correlation in DCT blocks. The resulting encoded bit rate increases dramatically with the amount of noise in an original signal. Several techniques are being developed to reduce the invisible noise in original pictures.

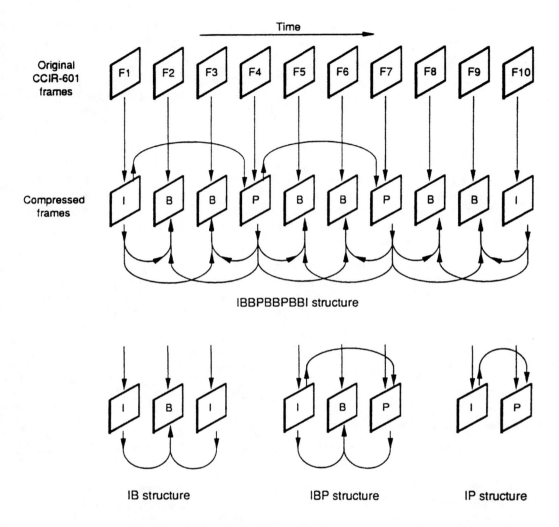

Figure 8.30 Typical Group of Pictures (GOP) in MPEG systems.

8.4 Video Compression Standards

Video compression standards are defined to achieve two main aims:

- Exploit the signal redundancy and the HVS tolerance to reduce the original picture data rate.

- Represent the compressed picture data stream in a format that can be manipulated in production environments and is resistant to transmission or storage media imperfections.

To achieve these requirements, a video data structure hierarchy is used and detailed in compression standards.

8.4.1 Video data structure hierarchy

The JPEG standard includes a data structure hierarchy with the aim of facilitating the interchange of encoded pictures. Coding parameters, such as JPEG operation mode, picture size and rate, pixel precision, compression scheme, quantization accuracy, and tables, are added to the compressed data stream. The JPEG video data structure contains six hierarchical layers that are made differently depending on the JPEG operation mode.

- Data unit (DU). This consists of a block of 8×8 component samples in the lossy operation mode.

- Minimum coded unit (MCU). This is the smallest group of interleaved DUs. It consists of two Y blocks and one C_B and one C_R block in CCIR-601 signal DCT-mode compression.

- Entropy-coded segment (ECS). This segment is made up of several MCUs. Its reduced size facilitates the recovery from limited corruption of the entropy-coded data.

- Scan. This defines the complete scanning of a picture from top to bottom.

- Frame. The frame can be made up of one or several scans, as used in the progressive DCT-based mode.

- Picture layer. The picture is at the top level of the compressed data hierarchy and includes the frame layer and codes for the start and the end of a picture.

The MPEG-1 and MPEG-2 video data structures are made up of six hierarchical layers, which are shown in Figs. 8.31 and 8.32. These are

- Block. Blocks of 8×8 pixels of luminance and chrominance signals are defined to be used for DCT compression, as explained in previous sections of this chapter.

- Macroblock. A macroblock is a group of DCT blocks which correspond to the information content of a window of 16×16 pixels in the original picture. This window size leads to different macroblock contents depending on the sampling structure used. Figure 8.15 shows four different cases. The macroblock header contains information about its type (Y or C_B or C_R) and the corresponding motion compensation vectors.

- Slice. A slice is formed of one or more contiguous macroblocks. Maximum slice dimension can be the whole picture, whereas its minimum size can be reduced to a single macroblock. The slice header contains information about its position within the picture, and its quantizer scaling factor. This

Figure 8.31 Architecture of the MPEG video data stream.

size is determined by the level of error protection needed for the intended application, as the decoder can drop erroneous slices, and the data reduction scheme performance, as the scaling factor can be more frequently adjusted with the use of small size slices. The reference DC coefficient, used in DPCM coding, is updated at the start of each slice.

■ Picture. The picture layer tells the decoder about the type of frame coding, such as I-, P-, and B-frames. The header is used to indicate the frame transmission ordering to allow the decoder to display frames in the right order. Additional information is also included in the picture header such as synchronization, the resolution, and range of the motion vectors.

■ Group of pictures. As seen previously in Sec. 8.3.9 and shown in Fig. 8.30, a group of pictures (GOP) can be made of various combinations of I-, P-, and

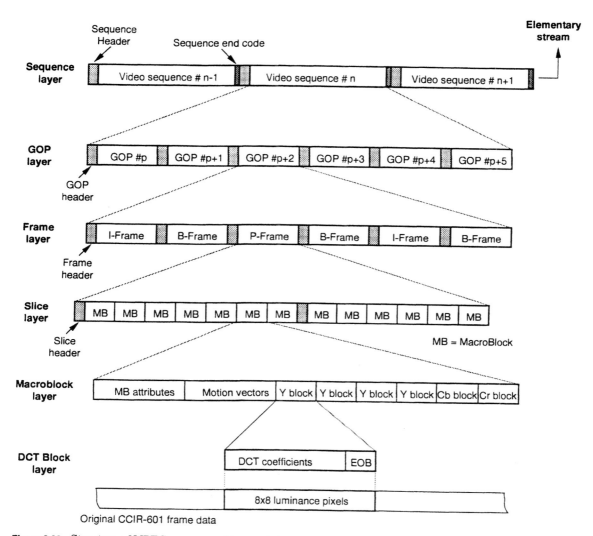

Figure 8.32 Structure of MPEG compressed-image data.

B-frames. GOP structures are described with two parameters, m and n. Each GOP starts with an I-frame and provides starting points for editing and searching. The header contains a 25-bit time and control code for VTRs and timing information.

- Video sequence. The sequence layer includes a header, one or more GOPs, and a sequence end code. The most important information contained in the header is the horizontal and vertical size of each picture, the pixel aspect ratio, the bit rate for the picture in the sequence, the picture rate, and minimum decoder buffer size requirements. The video sequence and header information constitute the encoded bit stream, also called the video elementary stream.

8.4.2 JPEG and Motion-JPEG schemes

The JPEG standard has been developed for still pictures. It operates over a wide range of compression ratios but is optimized for a ratio of about 15:1. Four modes of operations are defined.

- Sequential DCT-based (baseline system). Each image is encoded once. A subset of this mode is the baseline system.

- Progressive DCT-based. The image is encoded and decoded in multiple coarse-to-fine scans.

- Sequential lossless. The image is DPCM- and entropy-encoded, resulting in a lossless compression process, but this implies a low compression ratio.

- Hierarchical progressive (pyramidal). The image is encoded in a sequence of increasingly higher-resolution frames. The first frame is a low-resolution version of the original. Subsequent frames are coded as higher-resolution differential frames.

The main JPEG standard is characterized by

- Choice of RGB or Y, C_B, C_R color space.

- Choice of 4:4:4, 4:2:2, or 4:2:0 sampling structure.

- Picture size up to 65,536 pixels by 65,536 lines.

- Input sample precision of 8 bits in the baseline system, and 8 to 12 bits accuracy in the extended DCT-based system.

- DCT and quantization process precision of 9 bits.

- Use of a linear DPCM quantizer for the DC coefficient.

- Adaptive quantization process at the macroblock layer (16×16 pixels).

- Maximum-precision DC coefficient value is 11 bits.

- Different quantization tables for $Y, C_B,$ and C_R.

- RLC use of a modified Huffman scheme.

- Downloadable Huffman scheme.

- Variable scale factor on a block-by-block basis.

- No motion compensation, because only intraframe and intrafield compression is allowed.

- Choice of progressive or interlaced scanning.

- Channel error management. Errors in transmission channel may cause severe degradations because of the use of the Huffman coding. The JPEG standard defines several synchronization codes to limit the effects of these errors.

At relatively high compression ratios, the JPEG compression scheme can create block pixel artifacts in sharp vertical edges and may result in a subjectively detectable "softening" of the video resolution.

JPEG provides good editing capabilities with field or frame resolution. Encoding and decoding processing delays (latency) are fairly symmetric and fixed. Many chip sets are available on the marketplace, facilitating the design of low-cost encoder/decoders.

Motion-JPEG is a nonstandard extension of JPEG. Because JPEG compression is a field- or frame-based compression scheme, the development of fast processing chips has made possible the encoding of successive frames at rates of 24, 25, and 30 frames per second (fps). A buffer is added to the JPEG encoder to allow CBR operation by means of a variable quantization factor.

8.4.3 MPEG-1 video scheme

The MPEG-1 standard, also known as the ISO/IEC 11172 standard, has been primarily developed for storing moving pictures and the associated audio at about 1.5 Mbps. Other envisioned areas of application include video games, electronic publishing, and education. The quality of these pictures, based on a format known as the common source intermediate format (CSIF) at 30 fps, was found acceptable for the intended applications and rated as similar or superior to that of the VHS recorder for a bit rate of about 1.2 Mbps. However, the MPEG-1 bit-stream syntax has the capability to handle picture sizes up to 4095×4095 pixels and for bit rates as high as 100 Mbps.

8.4.3.1 The Common Source Intermediate Format (CSIF).

In December 1990, the CCITT Specialist Group on Coding for Visual Telephony established the H.621 Recommendation on low-bit-rate coding, leading to the design of worldwide-compatible codecs for videoconferencing and videophony. Given the two worldwide TV scanning formats (525/60 and 625/50), a common source intermediate format (CSIF), also called source input format (SIF), has been defined as the input format to the compression encoder, with separate specifications to suit the two formats.

Sampling frequencies are derived from those specified in the ITU-R BT-601 (CCIR-601) standard. Consequently, the number of pixels per active line are common for 525- and 625-line sources. The conversion process of ITU-R BT 601 formats into their corresponding CSIFs is easily achieved by using a horizontal decimation filter applied to Y odd fields only, and a vertical and a horizontal decimation filter applied to C_R and C_B odd fields only, as shown in Fig. 8.33. The decoding receiver must predict the even field from the decoded and horizontally interpolated odd field.

Figure 8.34 shows the decimation filter calculation process applied to all pixel positions in the CSIF picture array. The pixel value at position n is cal-

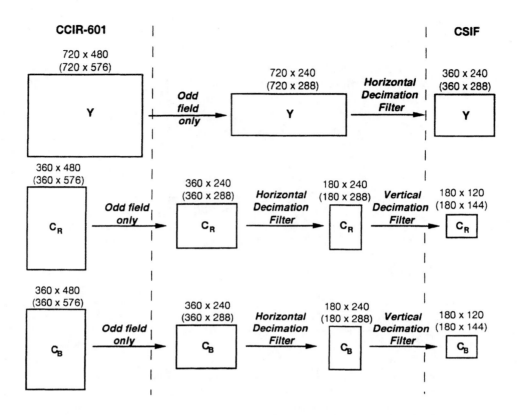

Note: Values in brackets are related to the 625 lines, 50 Hz, scanning format.

Figure 8.33 Conversion processes from CCIR-601 into CSIF and pixel array sizes.

culated from the multiplication of the pixel values at positions $(n-3)$ to $(n+3)$ by the corresponding filter coefficients shown at these positions in Fig. 8.34. The sum of these products is divided by 256, and the result is the new pixel value at position n. The next calculation is done for the pixel position $(n+2)$. The same filtering process is applied vertically to realize a vertical decimation of the C_B and C_R signals.

The number of active pixels per line (360) is reduced to a multiple of 16 (352), which is necessary to organize the active pixels of the picture in 4:2:0 macroblocks of 16×16 luminance pixels (see Sec. 8.4.1). The reduced active picture (352×240) is called the significant pixel area for CSIF. The CSIF is combined with a 4:2:0 sampling structure to further reduce the chrominance data rate, because chrominance data are less critical to the HVS. This combination achieves the necessary bit-rate reduction of video signals for applications such as videophony and videoconferencing. Table 8.4 shows the characteristics of CSIF-based image formats for 525/60 and 625/50 TV operations.

Prior to MPEG-1 encoding, a 4:2:2-to-CSIF converter is necessary to reduce

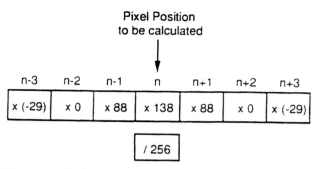

Figure 8.34 Decimation filter coefficients adjacent pixels.

TABLE 8.4   Characteristics of CSIF-Based Image Formats

	CCIR-601 525	CSIF 525 4:2:0	CCIR-601 625	CSIF 625 4:2:0
Number of pixels per active line				
Luminance (Y)	720	352	720	352
Chrominance (C_B, C_R)	360	176	360	176
Sampling frequency MHz				
Luminance (Y)	13.5	6.75	13.5	6.75
Chrominance (C_B, C_R)	6.75	3.38	6.75	3.38
Number of active lines				
Luminance Y	480	240	576	288
Chrominance C_B, C_R	480	120	576	144
Frame rate, Hz	30	30	25	25
Image aspect ratio	4:3	4:3	4:3	4:3

the original data rate, which is equal to 166 Mbps for an 8-bit system, to 31.5 Mbps. A reverse converter is also necessary at the decoding receiver.

8.4.3.2 The MPEG-1 standard. The MPEG-1 standard can be used in a wide range of applications, but its main target is the CSIF applications. To reduce decoder complexity and cost, a set of constrained parameters has been defined for the bit stream, as shown in Table 8.5. The maximum of 396 macroblocks per picture results from the array of 22 (352 pixels/16) horizontal macroblocks by 22 (240 lines/16) vertically.

The MPEG-1 is a generic algorithm that defines only the syntax for the representation of encoded bit stream and the decoding implementation. It is based on a "toolkit" approach, because not all of its features need to be used for a particular application. The bit-stream syntax is constructed in six layers, as described in Sec. 8.4.1. Two modes of operations are defined.

- Intraframe compression mode. DCT, quantization, and VLC coding for the generation of I-pictures.

**TABLE 8.5   Bit-Stream Constraint Parameters
for the MPEG-1 Encoding**

Parameters	Maximum value
Picture width	768 pixels
Picture height	576 lines
Picture rate	30 picture/s
Number of macroblocks	396
Range-of-motion vectors	±64 pixels
Input buffer size	327,680 bits
Bit rate	1.8 Mbps

- Interframe compression mode. Motion estimation, motion-compensated prediction for the generation of P- and B-pictures. MPEG-1 does not specify the motion estimation algorithm.

Main MPEG-1 standard characteristics are

- One color space is possible (Y, C_B, C_R).

- One sampling structure (4:2:0).

- A maximum picture size of 720 pixels by 576 lines using constrained parameters and 4095×4095 using full parameters.

- Input sample precision of 8 bits.

- DCT and quantization process precision of 9 bits.

- Use of a linear DPCM quantizer for the DC coefficient.

- Adaptive quantization process at the macroblock layer (16×16 pixels).

- Maximum precision DC coefficient value is 8 bits.

- Default quantization tables for intra- and intercoded frames are used if optional JPEG-style quantizing matrices are not transmitted in the video sequence header. Therefore, the matrices can be changed at the sequence level only. Quantization tables are identical for Y, C_B, and C_R data signals.

- RLC uses a modified Huffman scheme.

- VLC tables are not downloadable.

- Motion compensation allowing intraframe and intrafield compression.

- Progressive scanning capability only.

- Use of P- and B-frames.

- Motion estimation resolution of half a pixel.

- Asymmetric algorithm, as the encoder is more complex than the decoder.

- A maximum bit rate of 1.85 Mbps using the constrained parameters and up to 100 Mbps using the full parameters.

MPEG-1 allows a random access of video frames, fast-forward/fast-reverse searches through compressed bit streams, reverse playback of video, and editability of the compressed bit stream.

MPEG-1 is a subset of the MPEG-2 standard, so any MPEG-2 decoder is able to decode MPEG-1 data streams.

8.4.4 MPEG-2 video scheme

The MPEG-2 standard was developed as an extension of the MPEG-1 standard. The intent was to have a wide range of applications at higher bit rates while being downward compatible to MPEG-1 and improving some of its limitations, such as picture size and resolution, maximum bit rate, error resilience, and bit-stream nonscalability. Bit-stream scalability expresses the ability of decoding only a part of the bit stream and obtaining a decoded image at an acceptable quality level.

The standard specifications only define the bit-stream syntax and the decoding process. Consequently, the encoding performance varies with the data stream structure (GOPs), the motion compensation system accuracy, and the quantization process.

The main features of MPEG-2 are identical to those of the MPEG-1 standard. However, several improvements have been included.

- 4:4:4, 4:2:2, and 4:2:0 sampling structures
- The DC coefficient values can be expressed with extra precision.
- The quantizing matrices can be changed at the frame level.
- Different quantizing tables for the luminance and chrominance components.
- Both interlaced and progressive scanning capabilities, allowing motion compensation estimation built on a field-based predictive scheme.
- Both field and frame predictions are possible.
- Error detection and indication capabilities reduce the compressed signal sensitivity to transmission errors when high compression ratios are involved. Frequent start codes allow resynchronization of VLC decoders. Concealment motion vectors can be optionally added to I- and P-frames to help in reducing the slice loss visibility on decoded pictures. The drawback is the extra overhead caused by concealment vectors.

The MPEG-2 video data structure is made up of six layers, as described in Sec. 8.4.1. Constant bit rate (CBR) can be easily achieved, because each picture header can contain its own quantizing tables. Therefore, these tables can be adjusted to the characteristics of each picture.

Many of the MPEG-2 features and tools have been included in the standard specification to respond to many application requirements. The integration of these features into a single bit-stream syntax would be very complex. Consequently, hierarchical subsets, called profiles, have been defined. Each profile

TABLE 8.6   Maximum Constraint Parameters for MPEG-2 Levels and Profiles

Level	Profile	Simple	Main	SNR	Spatial	High	4:2:2
	Frame types:	I & P	I, P, & B	I, P, & B	I, P, & B	I, P, & B	I, P, & B
	Chroma sampling:	4:2:0	4:2:0	4:2:0	4:2:0	4:2:0 or 4:2:2	4:2:0 or 4:2:2
High	Samples/line		1920			1920	
	Line/frame		1152			1152	
	Frames/s		60			60	
	Max bit rate, Mbps		80			100	
High-1440	Samples/line		1440		1440	1440	
	Line/frame		1152		1152	1152	
	Frames/s		60		60	60	
	Max bit rate, Mbps		60		60	80	
Main	Samples/line	720	720	720		720	720
	Line/frame	576	576	576		576	608
	Frames/s	30	30	30		30	30
	Max bit rate, Mbps	15	15	15		20	50
Low	Samples/line		352				
	Line/frame		288				
	Frames/s		30				
	Max bit rate, Mbps		4				

is associated with levels describing a set of parameter constraints, such as the picture size, the frame structure (I, B, P), the maximum data rate, the maximum frame rate, and the sampling structure. Table 8.6 shows the profiles and levels as defined in the MPEG-2 standard. Maximum constrained values allow for both the 625- and 525-line scanning formats. For example, a 525-line main profile (MP)@main level (ML) MPEG-2 bit stream has a maximum bit rate of 15 Mbps for a picture size of 720×480 pixels, at 60 fps.

The low profile has the lowest number of tools and uses low bit rates and no B-frames. It is equivalent to the MPEG-1 specifications and is suitable for low-delay applications because frame reordering is not necessary.

The main profile is targeted for a large range of applications. The main level is very important because it corresponds to the resolution of conventional television. It uses B-frames and gives better picture quality for the same bit rate as the low profile, but the encoding/decoding delay increases.

The SNR profile has the tool of the main profile but adds an enhancement signal that improves the SNR of the picture.

The spatial profile is similar to the SNR profile but adds a picture resolution enhancement layer to the base layer. It provides backward compatibility to MPEG-1 encoders.

The high profile has the characteristics of the spatial profile with the addition of a 4:2:2 sampling structure capability layer. It is intended for HDTV applications, allowing HDTV receivers to decode both layers to display a HDTV picture. A standard-definition receiver would use the base layer to display a standard picture.

A recent 4:2:2 profile has been included to increase the maximum bit rate. It is similar to the main profile but allows a higher bit rate, a 4:2:2 sampling structure, and an increased vertical picture size of 576 lines for the 625/50 scanning standard and 512 lines for the 525/60 scanning standard.

In this hierarchical structure, a decoder designed for any given level and profile is able to decode all bit streams at lower profiles and levels.

8.5   Video BRR Scheme Performance and Applications

8.5.1   Video BRR scheme characteristics

The characteristics of BRR schemes are a compromise among several factors.

- Compression ratio capability (data rate). This ratio can vary from 2:1 to 150:1 and is chosen to meet the picture quality requirements for the intended application.

- Picture quality. The picture generation is the first step in the picture-processing chain and may require very high quality for high-end postproduction applications or a reduced quality for news applications, because very few picture-processing steps are made before broadcasting. Picture delivery is the last step in the picture-processing chain, and this BRR application can provide the level of picture quality for which the user is willing to pay.

- Multiple-generation capability. Many video production–related operations can only be carried out in the uncompressed domain, since suitable compression technology is not available. Original materials have to go through multiple generations of compression and decompression during the production, postproduction, and transmission processes.

- Symmetry/asymmetry. A compression scheme is symmetrical when the amount of processing on both encoding and decoding sides is identical. The MPEG-2 scheme is asymmetrical, because much less processing power is required in the decoder.

- Encoding and decoding delays. These delays vary depending on the structure and complexity of the encoder and on the GOP size and sequence. In broadcasting, the total acceptable delay is less than 1 ms for face-to-face interviews through network links. In transmission applications, it is not important, except when latency on channel switching must be reduced or minimized.

- Editability. Single-frame accuracy is a strong editing requirement in postproduction. Present solutions involve the decoding of several frames (I, B, P) and reencoding after insertion of a new segment. Future solutions are under development to simplify this editing process.

- Random access/channel hoping. Video storage medium requires that any picture be accessed and decoded in a short period of time, using the access point in the bit stream. Similarly, transmission channel selection can only be made at random access points in the selected bit stream. These are provided in the sequence header.

- Complexity and cost. There is a tradeoff between an effective motion estimation technique, which dramatically improves the compression performance, and a reduced complexity and cost of the processing chips.

8.5.2 Data rates and compression ratios

The performance of different compression systems is often judged by comparing their respective compression ratios. This comparison is valid only if the picture format used by different compressions is identical. A meaningful comparison involves the definition of the picture format (number of lines per image and pixels per line, number of frames per second), the sampling structure used (4:2:2, 4:4:4; 4:2:0, or 4:1:1), and the sample resolution (8 or 10 bits).

Once the picture format has been defined, the bit rate of the original image (before compression) can be calculated. Given the bit rate after compression, it is possible to calculate the compression ratio as

$$\text{Compression ratio} = \frac{\text{Original image data rate}}{\text{Compressed image data rate}}$$

Several examples are shown for comparisons.

- The active picture area (APA) of a 4:2:2 format (Y, C_R, C_B) with 8-bit resolution gives an original bit rate of $(720+360+360)\times512\times29.97\times8 = 176.77$ Mbps, where $(720+360+360)$ is the number of Y, C_R, C_B pixels per line, 512 is the number of lines per frame, 29.97 is the exact frame rate per second, and 8 is the sample resolution. This picture format and original frame rate correspond to the Tektronix "Profile" Professional Disk Recorder PDR-100 implementation. If the selected compressed picture quality is 24 Mbps, then the compression ratio is 176.77/24 = 7.4. Some compression scheme implementations consider only 480 lines to reduce the original bit rate to 165.72 Mbps. For the same compressed bit rate of 24 Mbps, the compression ratio is reduced to 6.9. These 480- and 512-line values allow for a macroblock size of 16×16 pixels in 4:2:2 systems and an intrafield compression scheme. Consequently, the total lines per frame has to be a multiple of 32.

- The active picture area (APA) of a 4:2:0 format (Y, C_R, C_B) with 8-bit resolution gives an original bit rate of $(720+360/2+360/2)\times480\times29.97\times8 = 124.29$ Mbps. These picture formats and bit rates correspond to the MPEG MP@ML implementation. In 4:2:0 sampling, a macroblock is formed with

16 pixels by 16 successive lines. Thus, the coded vertical active video area must be a multiple of 16 in progressive scan and a multiple of 32 in interlaced scan. As an example, in the 720×480 scanning format, 720 lines are a multiple of 16 only and cannot be used for interlaced scan. But 1080 lines will be encoded as 1088 lines in progressive systems and as 1080 in interlaced systems.

■ The active picture area (APA) of a 4:1:1 format (Y, C_R, C_B) with 8-bit resolution gives an original bit rate of $(720 + 360/2 + 360/2) \times 480 \times 29.97 \times 8 = 124.29$ Mbps. These picture formats and bit rates correspond to the DVC-Pro recording format implementation. If the selected compressed picture quality is 25 Mbps, then the compression ratio is $124.29/25 = 5$.

It is important to note that some designers consider a number of 704 pixels per line, instead of 720, to only compress the significant pixel area (SPA) of a picture, as shown in Fig. 8.35, where the active pixel area and the significant pixel area for a 720×480 picture format are represented. This SPA is included in the active picture area to take into account the picture-edge transient effects, which are described in Sec. 11.5.7.

Some disk-storage manufacturers use minutes per gigabytes (GB) to express the compression capability of their systems. For example, a 6-min/GB system is capable of 60×6 = 360 s of recording per GB or 1000 MB or 8000 Mbits (Mb) of compressed data stream. Then, the compressed data rate is 8000/360 = 22 Mbps.

Other disk-storage manufacturers use kilobytes per frame to express the compression capability of their systems. The conversion for a 30-fps system is

Figure 8.35 Active and Significant pixel areas of the 720×480 picture format.

easily made in the following example:

$$100 \text{ kB/frame} = 800 \text{ kb/frame} = 800 \times 30 \text{ kbps} = \frac{24{,}000}{1000} \text{ Mbps} = 24 \text{ Mbps}$$

Two other means that can be used to express the effectiveness of video compression algorithms have already been described in Sec. 8.3.5. They are

- Ratio of the number of bits in a pixel block before FDCT to the number of bits in the same block after VLC encoding.

- Ratio of the number of bits used to represent a block of pixels after FDCT to the number of pixels in this block, expressed as bits/pixel.

8.5.3   Video BRR scheme performance

In a constant-bit-rate (CBR) compression system, the quantization noise is related to the accuracy with which the DCT coefficients are quantized. Compared to low-pixel-value-variation blocks, high-pixel-value-variation blocks exhibit more significant DCT coefficients and require an increase of the scaling factor to maintain a CBR. This results in coarser quantization. Consequently, DCT quantization errors are picture-content-dependent and generally appear as random noise, because they are introduced in the spatial frequency domain.

The amplitude of the quantizing noise depends on the picture complexity, for example, it is higher at sharp transitions. The visibility of the quantizing noise depends on the picture brightness and is lower for detailed areas since the HVS is more sensitive to noise in uniform areas than in detailed areas. This degradation is roughly proportional to the compression ratio. However, when this edge-degraded signal is used in postproduction such as in a chroma-keyer, it may cause severe problems.

DPCM coding artifacts generally introduce visible distortions such as granular noise, slope overload, and edge business. Granular noise appears in the quasi-uniform area of a picture, resulting from the quantizer fluctuations between two consecutive quantizing levels in the presence of small-amplitude signals. Slope overload occurs when the quantizer output cannot respond quickly enough to large-amplitude signals, such as with highly contrasted pictures. Edge business refers to a variation in the edge position between successive scan lines as a result of quantizer fluctuations.

8.5.3.1   JPEG performance.   Motion-JPEG is only used for video applications. Its subjective performance at 48 Mbps, which corresponds to a compression ratio of 3.5, shows nearly transparent picture quality. Only critical pictures generate impairments that are detected by experts. At 24 Mbps, a small number of visible impairments can be seen on most of the TV pictures by experts. This quality is not acceptable for broadcast production applications, but it is used for on-air broadcasting.

8.5.3.2 MPEG-1 performance. It is generally said that the MPEG-1 baseline mode has a picture quality performance close to that of the VHS recording format. The maximum video bandwidth is slightly superior to 3 MHz, as 360 pixels give a maximum of 180 cycles during a horizontal line (53.3 μs). However, only odd fields are encoded and the decoded picture may exhibit motion artifacts.

Transmission error visibility is reduced when errors occur in P-and B-frames. When decoded errors are detected, skipped macroblocks replace the data until the beginning of the next slice. However, when an I-frame is affected by errors, the visibility of such errors can be spread over the entire GOP.

8.5.3.3 MPEG-2 performance. The MP@low level (LL) offers the best picture quality for bit rates below 5 Mbps. It is very convenient for off-line viewing applications in broadcasting.

MP@ML is most commonly used today for broadcast transmission. Near-transparent picture quality can be encoded at 9 Mbps with MP@ML and rated as subjectively equivalent to NTSC pictures by viewing experts and nonexperts.

The 4:2:2 profile is better suited to broadcast production applications for the following reasons:

- The active picture area includes several vertical blanking interval (VBI) lines, allowing the encoding and distribution of VBI information.

- With a maximum data rate of 50 Mbps, good quality of pictures is possible, even with the use of I-frames only.

- A 4:2:2 sampling structure facilitates the postproduction work.

- It offers high-quality pictures for multigeneration applications.

It is important to note that the 4:2:2 profile is less efficient when the bit rate decreases. The level of artifacts becomes important and the extra chroma and line information does not improve the picture quality at these bit rates (<15 Mbps), as shown in Fig. 8.36. This figure also shows the variation of the picture quality performance of the MPEG-2 (4:2:2@ML) scheme with the bit rate and the GOP structure.

8.5.4 Video BRR scheme applications

Table 8.7 shows the characteristics of some BRR applications. The motion-JPEG format has limited application because there is no file interchange format standard. It is mostly used in disk recorders where the encoding and decoding processes take place in the same unit.

The use of MPEG-2 is growing rapidly, and a wide range of applications is being developed or already coming to the marketplace, such as nonlinear editing workstations, video servers, disk recorders, VTRs for Electronic News Gathering (ENG), advanced TV, video-on-demand, and interactive TV.

Figure 8.36 Different picture qualities are achieved with MPEG-1 and MPEG-2 standards, which depend on profiles and GOPs (*Courtesy of Hewlett Packard company*).

TABLE 8.7 **Bit-Rate-Reduction Applications**

Applications	Coding standard	Maximum spatial resolution	Temporal resolution, bps	Maximum bit rate
Videophone	H.261	176 × 144 QCIF	7.5	64–128 kbps
Videoconferencing	H.261	352 × 288 CIF	10–30	0.384–1.554 Mbps
Entertainment TV	MPEG-1	352 × 288	30	<1.554 Mbps
Cable	MPEG-2	720 × 576	30	4–9 Mbps
Contribution	DPCM MPEG-2	720 × 576	30	<50 Mbps
Studio/production	Motion-JPEG MPEG-2 (@4:2:2)	720 × 576	30	<50 Mbps
HDTV production	MPEG-2	1920 × 1280	30	100 Mbps
HDTV transmission	MPEG-2	1920 × 1280	30	20 Mbps

8.6   General Concepts of Audio BRR

8.6.1   Need for audio BRR

PCM digital audio signals, as described in Chap. 6, are used in TV, multimedia, and many other applications. TV viewers are now installing multichannel sound systems, also called home theaters, in their homes. These audio signals exhibit a relatively high data rate when disk data storage and transmission channel bandwidth are considered. For example, a 16-bit-resolution audio stereo signal sampled at 48 kHz yields an audio data rate of 1.54 Mbps, whereas a multichannel surround sound system produces a total data rate of about 4.5 Mbps. Therefore, an efficient data compression scheme is required for long-duration material storage and inexpensive narrow-bandwidth channel distribution.

Compressed audio is now integrated in computer-based multimedia applications that allow distribution of program material on CD-ROMs and networks. It is already used in digital broadcast satellite (DBS) transmission applications.

Sound information compression systems rely on the human psychoacoustic characteristics and their limitations to remove unnecessary data information components in audio signals.

8.6.2   Human perceptual system characteristics

The human auditory system (HAS) behaves like a spectrum analyzer and separates the audible sound spectrum in bandpass filters, called *critical bands,* which are 100-Hz wide below 500 Hz, and above 500 Hz increase in proportion to the frequency, as described by Fletcher (1940) and Sharf (1970). A currently used model is made of 25 subbands that physically correspond to 25 cochlear filters. When the audio signal is made of close frequencies, the HAS integrates them as a group of equal energy. Conversely, widely differing frequencies are treated separately and their respective loudness is evaluated. It is worth noting that a plot of the critical bandwidth values versus their center frequency shows a performance close to a constant one-third-octave bandwidth filter, commonly preferred in subjective acoustic measurements.

The HAS sensitivity decreases at low and high frequencies, as shown in Fig. 4.2, which represents the normal equal-loudness level contours versus the audible frequencies. It can be noticed that at low loudness levels, the variation in the HAS sensitivity is very important and decreases at high loudness levels.

Auditory masking of the ear refers to the auditory suppression of one signal in the proximity of another. Experiments with the HAS have detected masking properties in the frequency domain and in the temporal domain.

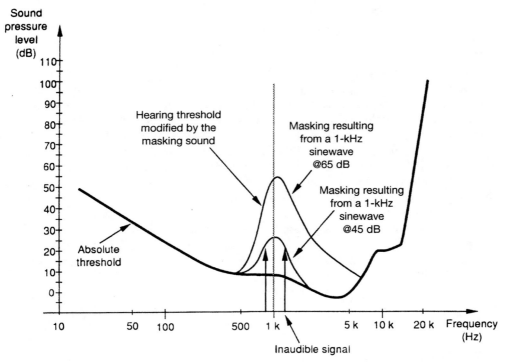

Figure 8.37 Absolute hearing and frequency masking thresholds.

Frequency domain masking is illustrated in Fig. 8.37. The subjectively determined minimum audible pressure for pure tones, also called *absolute hearing threshold,* is also shown. A 1-kHz tone at a sound pressure level of 45 dB raises the hearing threshold to 27 dB. This means that noise below 27 dB is not audible. Therefore, the 1-kHz tone can be quantized such that the quantizing noise does not exceed 27 dB. Using the well-known 6 dB-per-bit rule of thumb to express SNRs in a digitized audio signal, only 3 bits are necessary to encode this tone, because the difference ($45 - 27 = 18$ dB) matches the quantizing range of these three bits. If the 1-kHz tone level is increased to 65 dB, the masking level becomes 55 dB and the difference (10 dB) can be encoded with 2 bits only. The sampling resolution is selectively reduced, and so is the data rate.

It is important to note the presence of pre- and postfrequency masking around the 1-kHz tone shown in the Fig. 8.37. The postfrequency masking is more important and increases with the sound level. This allows the reduction of the encoding precision for frequency signals above the masking tone. The steep slope of the low-frequency side of the 1-kHz masking curve requires sharp filters in audio systems that try to simulate a psychoacoustic model.

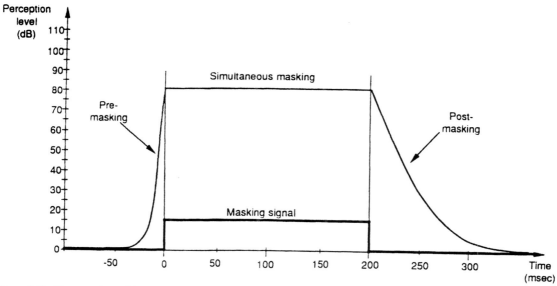

Figure 8.38 Temporal masking.

Tones around 1 kHz at levels below the corresponding masking curve are not audible and do not need to be encoded, thus reducing data rate requirements. This means that the perceptual quality of the reproduced sound is not affected. Compression systems based on this property are called *perceptual encoders.*

Experiments have shown that a temporal masking effect starts before a masking signal tone is applied to the HAS, and after the tone is stopped. These effects are called pre- and postmasking and are shown in Fig. 8.38. Postmasking lasts longer than premasking; 1-kHz audio tone frequency-adjacent signals are masked before and after the tone occurs, if their amplitudes are below the masking curve shown in Fig. 8.37. This effect makes possible the use of high-frequency-resolution filter banks in high-quality audio coding, because the temporal resolution of these filters is poor.

8.7 Audio Data Reduction Techniques

Source coding techniques are used to remove the audio signal redundancy (when the sample-to-sample difference is equal to zero) and psychoacoustic masking techniques are used to identify and remove the irrelevant content (inaudible samples).

The two main data compression techniques are:

■ Time domain prediction coding. This uses differential encoding to retrieve the difference between consecutive samples. A reduced bit rate can then be used to encode and transmit the audio signal information.

- Frequency domain transform coding. This technique uses blocks of linear PCM audio samples transformed from the time domain to a certain number of different bands in the frequency domain.

8.7.1 Lossless data reduction

Lossless compression schemes allow the bit-for-bit recovery of the original data information after decompression. They remove statistical redundancies that exist in audio signals by predicting values from past samples. Small compression ratios, around 2:1 at best, can be achieved and are dependent on the original audio signal complexity.

Lossless compression is possible thanks to time domain prediction coding techniques. They are

- Differential algorithms. Audio signals contain repetitive sounds as well as large amounts of redundant and perceptually irrelevant sounds. Repetitive data information are removed in the encoding process and reintroduced at the decoding end. DPCM techniques are described in Sec. 8.2.2 for video signals, but they also apply to audio signals. The audio signals are first decomposed into a number of subbands that contain a certain number of discrete tones. DPCM is then applied using a predictor suitable for short-term periodic signals. This encoding is made adaptive by looking at the input signal energy to modify the quantizing step size. This leads to the so-called adaptive DPCM (ADPCM).

- Entropy coders exploit the redundancy in the representation of the quantized subband coefficients to improve the coding efficiency, as described in Sec. 8.2.1 for video signals. These coefficients are sent in an increasing frequency order that produces large values at low frequencies and long runs of small or near-zero values for high frequencies. VLCs are taken from different Huffman tables to best match the statistics of low- and high-frequency values.

- Block floating-point systems. Binary values coming from an A/D conversion process are grouped in blocks of data, either in the temporal domain, by taking adjacent samples at the A/D converter output, or in the frequency domain, by taking adjacent frequency coefficients at the FDCT output. Binary values in a block of data are then scaled upward so that the largest value is just below the full-scale value. This scaling factor, called the exponent, is common to all values in the block. Consequently, each value can be represented by a mantissa, which is the sample value, and by an exponent indicating the proper amplitude of the sample. This is a nonuniform requantizing process where the quantizing step size is determined by the number of bits allocated per block. The bit allocation calculation is derived

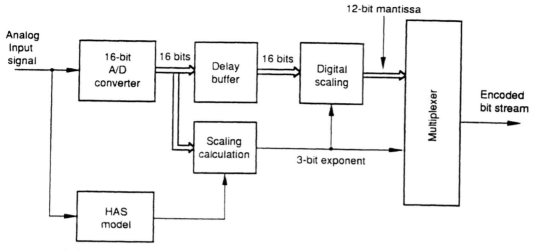

Figure 8.39 Audio data block floating-point encoding system.

from the HAS model, as shown in Fig. 8.39. Data rate reduction is achieved by sending the exponent value once per block of data. The encoding performance is good, although the noise is signal-content-dependent. Masking techniques help in reducing this audible noise.

8.7.2 Lossy data reduction

Lossy compression is obtained by combining two or more processing techniques to take advantage of the HAS's inability to detect certain spectral components among others of higher amplitudes. High-performance data reduction schemes and much higher compression from 2:1 to 20:1 ratios are then possible, depending on the complexity of the encoding/decoding process and the audio quality requirements.

Lossy data reduction systems use a perceptual coding technique. The basic principle is to remove perceptual redundancy in the audio signal by discarding any signal that falls below the threshold curve. For this reason, these lossy data reduction systems are also called perceptually lossless.

Perceptually lossless compression is possible thanks to a combination of techniques, such as

- Temporal and frequency domain masking of signal components, as described earlier in this section.

- Quantizing noise masking for each audible tone by allocating enough bits to make sure that the quantizing noise level is always below the masking curve. At frequencies close to the audible signal, an SNR of 20 to 30 dB is acceptable (4 to 5 bits of resolution).

- Joint coding: This technique exploits the redundancy in multichannel audio systems. It has been found that a significant amount of identical data exists in all channels. Therefore, data reduction can be obtained by coding this identical data once and indicate to the decoder that it must be repeated in other channels.

8.7.3 Audio coding process and implementation

The most important masking effects occur in the frequency domain. To exploit this property, the audio signal spectrum is decomposed in multiple subbands with the time and frequency resolution matching the critical bandwidths of the HAS.

The basic structure of a perceptual encoder is shown in Fig. 8.40 and is made up of

- A multiband filter, generally called filter bank, which is used to decompose the spectrum in subbands.

- A bit allocator, which estimates the mask threshold and allocates bits on the basis of the audio signal spectrum energy and the psychoacoustic (HAS) model.

- A scaler and quantizer processor.

- A data multiplexer, which receives the quantized data and adds the side information (bit allocation and scale factor information) for the decoding process.

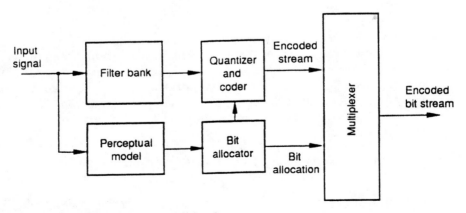

Figure 8.40 Basic perceptual audio encoder.

8.7.3.1 Filter bank. There are three types of filter banks:

- Subband bank. The signal spectrum is divided into equal-width frequency subbands (32 subbands in MPEG Layers I and II). This is similar to the HAS process of frequency analysis, which divides the audio spectrum into critical bands. The width of the critical subbands is variable. Below 500 Hz, the bandwidth is 100 Hz, and it increases to several kilohertz above 10 kHz. Below 500 Hz, a subband contains several critical bands.

  Subband filters, such as polyphase quadrature mirror filters (PQMFs), have a small overlap and are generally used for adjacent temporal samples. For example, in the MPEG Layer II, an audio frame of 1152 original samples is decomposed into 32 equal-width (750 Hz @ 48 kHz sampling) subbands, each containing 36 samples.

  Each subband signal is then uniformly quantized with a bit allocation specific to the subband to maintain a positive mask-to-noise ratio (MNR). This ratio is positive when the masking curve is above the noise level.

- Transform bank. A modified DCT (MDCT) algorithm is generally used to convert the time domain audio signal into a high number of subbands (256 to 1024). Some overlap also exists in this type of filter bank.

- Hybrid filter bank. They are composed of subband filters followed by MDCT filters. This combination provides a finer frequency resolution such as that used in MPEG Layer III. The input signal is first separated in 32 subbands by PQMF and then a MDCT is applied to every 18 samples in each subband, which provides a total of 576 very narrow bands (41.67 Hz @ 48-kHz sampling). This gives a temporal resolution of 3.8 ms.

In the presence of stationary audio signals, a MDCT is performed on every 18 subband output samples providing two transforms per subband per frame. When a transient is detected in the signal, a MDCT is performed on every 6 subband output samples, which results in an improved temporal resolution of 1.27 ms.

Filter resolution characteristics are a consequence of compromises affecting several important factors. A low-resolution (i.e., a wide-frequency-band) filter bank gives a low number of subbands and tonal components of most of audio signal spectrums fall in every subband, hence the necessity to encode every subband. The effectiveness of the masking threshold technique is reduced and many bits are needed to quantize the signal in each subband. However, the low number of subbands in this filter reduces the encoding and decoding complexity while having a good temporal resolution (i.e., a low time duration).

A high-resolution filter bank generates a high number of frequency subbands and tonal components of an audio signal spectrum do not fall into all subbands. Subbands without tonal components do not need to be encoded, thus resulting in a lossless reduction of the encoded data rate. Narrow bandwidths can better simulate the critical bandwidth performance of the HAS. They have, however, a poor temporal resolution, which generates preecho effects during the coding of transient signals. An adaptive filter bank can be implemented by means of a transient detector, which switches the filter frequency resolution from high to low during the transient signal.

It is worth noting that a 256-band transform filter bank is almost similar in complexity with that of a 32-band subband filter bank.

The performance characteristics of the filter bank process are expressed by

- The temporal resolution = Filter bank block length of samples×20.83 µs. A sample duration is equal to 20.83 µs, for a 48-kHz sampling rate. This parameter gives an idea of the encoder behavior in the presence of a signal transient. Example: AC-3 coding: 128×20.83 = 2.66 ms.
- The frequency resolution = Maximum spectrum bandwidth/Total number of frequency subbands. The maximum spectrum bandwidth is equal to 24 kHz for a 48-kHz sampling rate. Example: AC-3 coding: 24,000/256 = 93.75 Hz.
- The frame length = Number of frequency subbands × Number of samples in a block × 20.83 µs. This value is calculated here for a 48-kHz sampling rate. Example: MPEG Layer I coding: 32 × 12 × 20.83 = 8 ms.

8.7.3.2 Perceptual model, masking curve, and bit allocation. An accurate psychoacoustic analysis of the input PCM signals is carried out as to its frequency and energy content, using a fast Fourier transform (FFT) algorithm. A masking curve is computed from the hearing threshold and the frequency-masking properties of the HAS, which are represented in Fig. 8.37. The shape and level of this masking curve is signal-content-dependent. In an example shown in Fig. 8.41, it can be seen that the difference between the spectral signal envelope and the masking curve has been reduced to a maximum of 40 dB. This difference determines the maximum number of bits (on the basis of 6 dB per bit) necessary to encode all spectral components of the audio signal. This bit allocation process ensures that the quantizing noise is below the audible threshold. In Fig. 8.41, the masking curve level is higher than that of the spectral signal envelope above 12 kHz and no bit needs to be allocated in this area.

From the masking curve, masking thresholds are derived for each subband. Each of them determines the maximum quantization noise energy acceptable in each subband at which the noise starts to be audible for perceptually lossless compression systems.

MPEG encoders use a forward-adaptive bit allocation process, because the bit allocation calculation is made on the input signal in the encoder only. In addition, side information needs a significant transmission bandwidth.

Dolby AC-3 encoders use a forward- and backward-adaptive bit allocation process. Complex calculation of the spectral envelope and bit allocation infor-

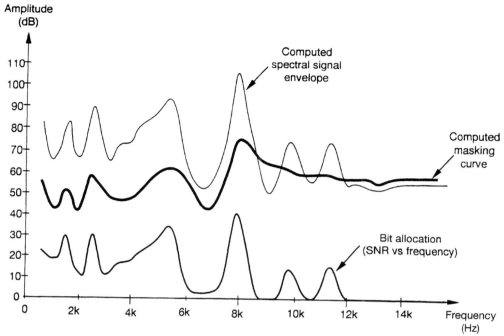

Figure 8.41   Masking curve and bit allocation.

mation is made in the encoder and transmitted within the encoded data stream to the decoder, which carries out a simplified calculation. This allows some changes in the encoding algorithm parameters while maintaining decoder compatibility. Furthermore, the bit allocation information is reduced. For multiple-channel encoding, the bit allocation is made from a common pool of bits.

8.7.3.3 Scaler and quantizer.

Samples from the output of each subband filter are scaled and quantized by means of two methods:

- Block floating-point system (or block companding). As described earlier in Sec. 8.7.1, this system normalizes the highest value in a block of data to the full scale. This block scaling factor is transmitted within the data stream and is used by the decoder for the downscaling of all data values in the block. In MPEG Layer I, a block of data is made of 12 consecutive samples and an audio time frame is constituted of 384 samples (32 subbands × 12 samples). All block data values are then quantized, with a quantizing step size determined by the bit allocator.

- Noise allocation and scalar quantization. In the previous method (layers I and II), each subband has a different scale (or normalization) factor. This second method employs the same scale factor for several bands that have approximately critical bandwidths. This scale factor value does not result from a normalization process, but instead is part of a noise-allocation process. No bit allocation is performed.

After estimation of the masking threshold in each subband, scale factors are used to modify the quantization step size for all values within a scale factor band in order to modify the quantization noise structure to better match the frequency contours of the masking threshold.

A nonuniform quantization process, by means of a 3/4th exponent companding law, is used to adapt the quantizing noise to the signal amplitude in an optimized way. It is followed by a Huffman coding to encode the spectrum values and get better data compression.

8.7.3.4 Data multiplexer. Blocks (also called groups) of 12 data samples from each quantizer output are multiplexed with their corresponding scale factor and bit allocation information to form an audio frame in the encoded bit stream. Optional ancillary (user-definable) data can also be inserted in this bit stream. MPEG standards do not specify what types of data can be transmitted and how they are formatted in the bit stream.

8.8 Audio Compression Standards

8.8.1 MPEG-1 audio subsystem

MPEG-1 standard (ISO/IEC 11172) was primarily developed (in 1991) for the "Coding of moving pictures and associated audio for storage media at up to about 1.5 Mbps." The video part is described in Sec. 8.4.3. The audio part of this standard defines three layers for the coding of PCM signals having a sampling rate of 32, 44.1, or 48 kHz. A basic representation of the MPEG audio encoder and decoder are shown in Figs. 8.42 and 8.43.

There are three distinct layers in the MPEG audio standard that support the following modes of operation:

- Monophonic mode (single channel)

- Dual monophonic mode (two independent audio channels)

- A stereo mode (stereo channels)

- A joint stereo mode (to take advantage of the correlation between stereo channels)

Layer I is made of the basic algorithm, whereas Layers II and III incorporate several enhancements to Layer I. This standard defines two psychoacoustic

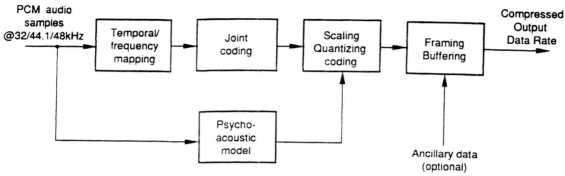

Figure 8.42 Basic structure of MPEG audio encoder.

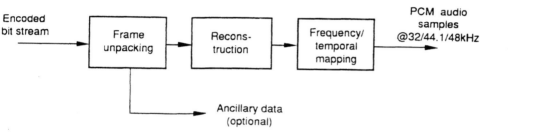

Figure 8.43 Basic structure of MPEG audio decoder.

models applicable to all layers of compression. Model 1 makes compromises to simplify the encoder implementation. In MPEG Layer I, a 512-sample FFT is used to analyze accurately the frequency and energy content of the incoming audio signal. This window is large enough to cover a frame of 384 samples. However, in MPEG Layers II and III, a 1024-sample FFT analysis is used but does not completely cover the audio frame of 1152 samples. This is an acceptable compromise between performance and complexity. In the psychoacoustic model 2, a 1024-sample FFT analysis window is used for all layers.

MPEG Layer I and II frames are of different but constant length to simplify the decoding of the bit stream. Layers II and III process the audio data in frames of 1152 samples. Figure 8.44 shows an MPEG audio encoder.

8.8.1.1 Layer I characteristics. The audio MPEG Layer I frame structure is shown in Fig. 8.45. The most important characteristics are

- Data rates from 32 to 448 kbps (total).
- Input signal divided into frames that contain 384 samples (per channel).
- 8-ms framing @48 kHz ($12 \times 32 \times 20.83$ µs = 8 ms).

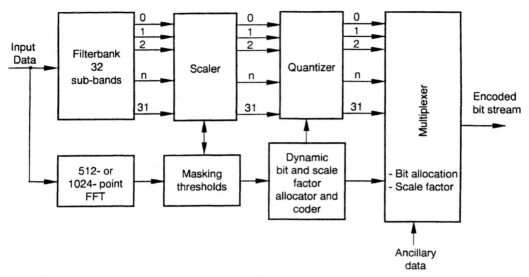

Figure 8.44 Block diagram of an MPEG audio encoder.

- 32 equal-size subbands, generating blocks of 12 samples ($32 \times 12 = 384$ samples).
- 6-bit scale factor per band (120-dB dynamic range), different for each band.
- Forward-adaptive bit allocation.
- Each subband sample is quantized to the precision indicated by the bit allocation calculator.
- Single- or dual-channel, stereo or joint stereo (combined coding of left and right channels of a stereophonic audio signal).
- Most appropriate for consumer applications such as recording and for studio use because the frame size is only 8 ms long. The Philips Digital Compact Cassette (DCC) system uses a compatible Layer I format at 192 kbps per channel.

8.8.1.2 Layer II characteristics. The audio MPEG Layer II improves the Layer I performance, allowing further compression. Target bit rates are around 128 kbps. The frame structure is shown in Fig. 8.46. The most important characteristics are

- Data rates from 32 to 384 kbps (total).
- Input signal divided into frames that contain 1152 samples (per channel).
- 32 equal-sized subbands, generating blocks of 36 samples ($32 \times 36 = 1152$ samples).

Figure 8.45 Audio MPEG Layer I frame structure.

- 24-ms framing@48 kHz ($384 \times 3 \times 20.83$ μs = 24 ms).

- 6-bit scale factor per band (120-dB dynamic range), different for each band, which may be shared by groups of 12, 24, and 36 samples (8, 16, or 24 ms of time) to avoid audible distortion.

- Forward-adaptive bit allocation, which is constant across 24-ms frame and uses fractional bit quantization.

- Single- or dual-channel, stereo or joint stereo (combined coding of left and right channels of a stereophonic audio signal).

- Most widely used of MPEG audio standards in CD-ROM, DVB, DAB, DBS, multimedia, etc.

8.8.1.3 Layer III characteristics.

Layer III provides the best quality at low-bit rates. The temporal-to-frequency filter bank provides high spectral resolution and adaptive temporal and frequency resolution to better cope with different

CRC = Cyclic Redundancy Check
SCFSI = Scale Factor Selection Information

Figure 8.46 Audio MPEG Layer I, II, and III bit-stream formats.

input signal contents. Target bit rates are around 64 kbps. The frame structure is shown in Fig. 8.46. The most important characteristics are

■ Data rates from 32 to 320 kbps (total).

■ Input signal divided into frames that contain 1152 samples (per channel).

■ 24-ms framing@48 kHz ($384 \times 3 \times 20.83 \ \mu s = 24$ ms).

■ 32 equal-sized subbands further divided into 18 MDCT bands (total of 576 bands), with capability of transient block switching down to 192 bands.

■ Scale factors are used to modify quantized noise structure and levels.

■ Forward-adaptive bit allocation.

■ VLC (Huffman) of quantized values.

■ Single- or dual-channel, stereo or joint stereo (combined coding of left and right channels of a stereophonic audio signal).

■ Used for low-bit-rate applications such as narrow ISDN, telecommunications, satellite links, and high-quality audio through internet.

Figure 8.46 gives the structure of MPEG Layers I, II, and III bit-stream formats.

8.8.2 MPEG-2 audio subsystem

In a second standard established in 1994, the MPEG-2 (ISO/IEC 13818), extensions of the MPEG-1 standard have been defined to meet the needs for new applications such as

■ A wide range of audio quality from low to high bit rates, from 32 to 1066 kbps. This wide range is realized by splitting the MPEG-2 audio frame into two parts, a MPEG-1–compatible primary bit stream (384 kbps for Layer II), and an extension bit stream. With Layer III at 64 kbps per channel, five full-bandwidth audio channels can be encoded within 320 kbps.

■ Coding of up to six audio channels, including an optional low-frequency enhancement channel, to support multichannel surround sound requirements.

These extensions are made possible by adding to each layer

■ Half sampling rates (16, 22.05, and 24 kHz)

■ Multichannel capability (multichannel bit rates extended up to 1 Mbps to allow high quality). These data are inserted in the ancillary data space of the MPEG-1 audio frame structure as shown in Fig. 8.47.

The MPEG-2 audio standard is downward-compatible with the MPEG-1 standard. But MPEG-1 decoders can only decode the left and right channels of the MPEG-2 audio data streams. All MPEG-1 and MPEG-2 layers are identical. At low sampling frequencies, the frequency resolution is about 21 Hz at 24-kHz sampling. This allows better matching of the scale factor bands to the HAS critical bandwidths and gives better audio quality at low bit rates, although the audio signal bandwidth is reduced to a maximum of 12 kHz.

Figure 8.47   Audio MPEG-2 extensions in MPEG-1 bit stream format.

Figure 8.48 The MPEG audio family.

A summary of all MPEG audio compression encoding standards is shown in Fig. 8.48.

8.8.3 Other compression schemes

8.8.3.1 Apt-X characteristics. The apt-X audio data compression algorithm is not a perceptual coding system. Input signals are separated into four bands by a pseudo-quadrature mirror filter (PQMF) filter bank that is followed by an ADPCM algorithm. An energy analysis of the previous input signal samples determines the present quantizer step size, which is a characteristic of the backward-adaptive process. The most important characteristics of this compression scheme are

- Bit rates from 56 to 384 kbps
- Fixed compression ratio of 4:1
- Uses a predictive (ADPCM) compression algorithm
- Four time-domain subband algorithms
- User-defined audio bandwidths from 7 to 22 kHz
- Sampling frequencies from 16 to 48 kHz
- Backward-adaptive process
- Mono, stereo, or dual channels
- Short processing delay (2.7 ms@48-kHz sampling and 7.6 ms@16 kHz)
- Used in recording, ISDN, and transmission applications

8.8.3.2 Apt-Q characteristics. The apt-Q is a perceptual algorithm using a MDCT filter bank to transform the time-domain input samples to the frequency domain. Input signal energy and content are used in conjunction with a psychoacoustic model to calculate masking thresholds. A quantization of each of the subband outputs is made according to a bit allocation process similar to that of the MPEG Layers I and II. Then, an adaptive VLC encoding takes place using Huffman code tables. The most important characteristics of this compression scheme are

- Data rates from 56 to 128 kbps

- Variable compression ratio, nominally 12:1 and 18:1

- Uses a MDCT algorithm and a perceptual model to adjust the masking threshold

- 1024 equally spaced frequency subbands

- Adaptive switching to short transform window (128 frequency subbands) to reduce preecho problems in case of rapid changes in the audio content, which yields a temporal resolution of 5.24 ms (256 samples×20.83)

- Adaptive VLC coding (Huffman code words) to further reduce the bit rate by exploiting the redundancy in the quantized coefficients for each of the frequency subbands

- Audio bandwidths from 15 to 22 kHz

- Sampling frequencies from 16 to 48 kHz

- Processing delay typical of similar transform algorithms

8.8.3.3 AC-3 characteristics. The AC-3 audio coding scheme was developed by Dolby for multichannel audio applications. It encodes five full-bandwidth audio channels (3 Hz to 20 kHz), including left, center, right, left surround, right surround, and one reduced-bandwidth, low-frequency enhancement channel (3 Hz to 120 Hz) into a 384-kbps data signal. This audio coding scheme has been selected for ATSC broadcasting, as explained in Chap. 11.

A simplified block diagram of the AC-3 encoding scheme is shown in Fig. 8.49. The input audio samples are transformed in the frequency domain using a 512-point MDCT filter bank (block length of 256 samples), providing 93.75 Hz of frequency resolution at a 48-kHz sampling rate. Adaptive frequency-to-temporal resolution is achieved by switching to a smaller 256-point transform in presence of rapid changes in the input audio signal.

A block floating-point system converts each transform coefficient into an exponent and mantissa pair. The mantissas are then quantized with a variable number of bits, based on a parametric bit allocation model. This model uses psychoacoustic masking principles to determine the number of bits for each mantissa in a given frequency band. This adaptive bit allocation process results in an acceptable SNR for each DCT coefficient. The encoded spectral envelope data and the quantized mantissa data corresponding to the six

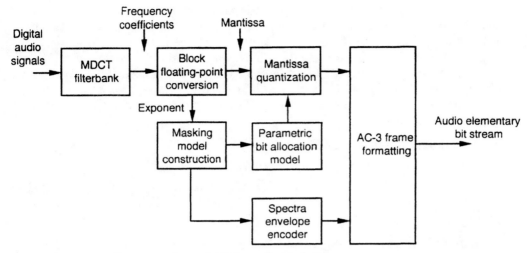

Figure 8.49   Basic Dolby surround sound AC-3 encoder block diagram.

audio blocks are formatted into an AC-3 sync frame. An efficient processing algorithm is used to extract similarities that exist between audio channels, such as between right and left channels or between left surround and right surround channels, to simply encode the identical information once and distribute it between similar channels at the decompression stage.

The AC-3 bit stream is composed of repetitive, independent sync frames, as shown in Fig. 8.50. A sync frame is made of six audio blocks corresponding to the six audio channels and representing $256 \times 6 = 1536$ audio samples, and one auxiliary block, located at the end of the frame, reserved for control or status information of system transmission.

The most important characteristics of this compression scheme are

- Data rates from 32 to 640 kbps
- 32-ms framing@48 kHz ($384 \times 4 \times 20.83 \ \mu s = 32$ ms)
- 256/128 subbands
- 4.5-bit scale factor per band (144 dB range)
- Forward/backward fully adaptive bit allocation
- Mono up to 5.1ch coding
- Extensive consumer features
- Applications in HDTV, CATV, DVB, DBS, multimedia, internet, etc.
- Down-mixing capability to meet the listeners' needs for stereo or mono only.
- Loudness uniformity to reduce level variations when switching between TV channels, programs, and stereo and multichannel programs.

SI = Sync Information
BSI = Bit Stream Information
AB 0 = Audio Block #0

	Purpose
BSF	Block Switch Flags
DF	Dither flags
DRC	Dynamic range control
CS	Coupling strategy
CC	Coupling coordinates
ES	Exponent stategy

Figure 8.50 AC-3 sync frame and audio block structures.

- Dynamic range control, because all coding systems can easily deliver extreme amounts of dynamic range. This capability will permit the user to meet his or her requirements such as full dynamic range for home theater and limited dynamic range for portable TV receivers.

Table 8.8 summarizes the characteristics of MPEG layers, apt-X and apt-Q, and AC-3 audio encoding schemes.

8.9 Audio BRR Scheme Performance

Audio BRR is generally not used in a production environment because it conflicts with the need for high-quality original material in the production process. In addition, compressed audio is difficult to edit and postprocess, because the edit accuracy is limited to frame boundaries (24 or 32 ms apart). However, when a moderate compression (2:1 or 3:1) is applied to PCM audio signals, several generations can be made.

In transmission, low-bit-rate audio coding is very often used. To provide for quality headroom in cascading of codecs and postprocessing, the distribution bit rate must be 1.5 to 2 times the decoded bit rate in the receiver.

TABLE 8.8   Main Audio Compression Coding Scheme Comparison

Audio schemes	Total bit rates, kbps	Filter bank	Frequency resolution @48 kHz	Temporal resolution @48 kHz, ms	Frame length @48 kHz, msec	Bit rate target, kbps per channel
MPEG Layer I	32–448	PQMF	750 Hz	0.66	8	128
MPEG Layer II	32–384	PQMF	750 Hz	0.66	24	128
MPEG Layer III	32–320	PQMF/ MDCT	41.66 Hz	4	24	64
apt-X	Fixed 4:1 compression*	PQMF	12 kHz	<0.1	2.54	192
apt-Q	Fixed 12:1 & 18:1 compression†	MDCT	23.44 Hz	5.34	42.66	64
AC-3	32–640	MDCT	93.75 Hz	2.66	32	64

*Bit rates are from 56 to 384 kbps when the sampling frequency varies from 16 to 48 kHz.
†Bit rates can vary from 56 kbps for a mono signal sampled at 32 kHz to 128 kbps for a stereo signal sampled at 48 kHz.

8.10   Distribution of Compressed Signals

The MPEG-1 standard addresses the compression, decompression, and synchronization of video and audio signals, which constitute the compression layer. The MPEG-2 standard enhances the MPEG-1 standard by adding a system layer. The MPEG-2 layer architecture is shown in Fig. 8.51.

The compression layer describes the syntax of the audio and video streams based on the video and audio data stream architecture described, respectively, in Secs. 8.4.1 and 8.8.1. Independent audio, video, or data sequences are MPEG-2–encoded to form independent data streams called *elementary streams* (ESs).

The system layer defines the combination of separate audio and video streams into a single stream for storage (program stream) or transmission (transmission stream), as shown in Fig. 8.52. It also includes the timing and other information needed to demultiplex the audio and video streams and to synchronize audio and video at the decoding end. To this end, system clock reference (SCR) and presentation time stamp (PTS) are included in MPEG bit streams.

The MPEG standard defines a hierarchy of three data streams, a packetized data stream, a program stream, and a transport stream.

8.10.1   Packetized elementary stream

Through a packetizer, elementary streams are separated into packets whose size can be of any length. The packet content can originate from an MPEG-2–encoded audio or video or data stream. The packetized elementary stream (PES) packet structure is shown in Fig. 8.53.

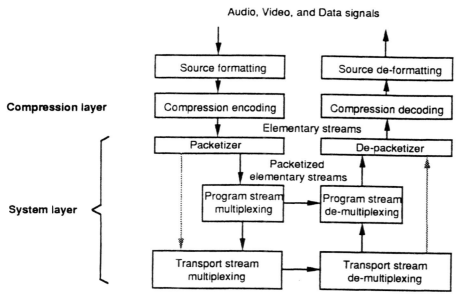

Figure 8.51 The MPEG-2 layered architecture.

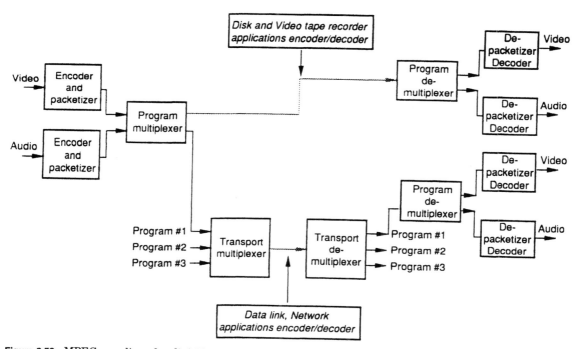

Figure 8.52 MPEG encoding of audio/video signals in program and transport data streams.

Figure 8.53 Packetized Elementary Stream (PES) structure in the MPEG-2 standard.

8.10.2 Program stream

PES packets originating from one or more elementary streams sharing the same time base, such as audio, video, and data, are multiplexed into a single program stream (PS) by means of repetitive packs, as shown in Fig. 8.54. In the pack header, an SCR ensures that audio and video packets are timed. It is a real-time clock that indicates the time the pack is transmitted.

The PS packs can be of any length. The number and the sequence of packets per pack are not defined, but packets from individual streams are sent in chronological order. A PS can carry up to 32 audio streams, 16 video streams, 16 data streams, all having the same time base. Program streams are sensitive to errors and are used in multimedia recording and local network distribution of synchronous multiplexed elementary streams.

8.10.3 Transport stream

A transport stream (TS) can be formed from a combination of one or more program streams with independent time bases or from a combination of PES, as shown in Fig. 8.55. However, a program stream is not a subset of a transport stream, because all the program field information is not contained in the transport stream. Some information must be derived when extracting a PS from a TS. Conversion can be made through a common interchange format of PES packets.

PES packets originating from one or more elementary streams sharing the same or different time base, such as audio, video, and data, are multiplexed into a single transport stream by means of repetitive, small-size transport packets, as shown in Fig. 8.56. One or more PS, having different reference

Figure 8.54 Program stream structure in the MPEG-2 standard.

Figure 8.55 Program and Transport Streams formatting from Packetized Elementary Streams.

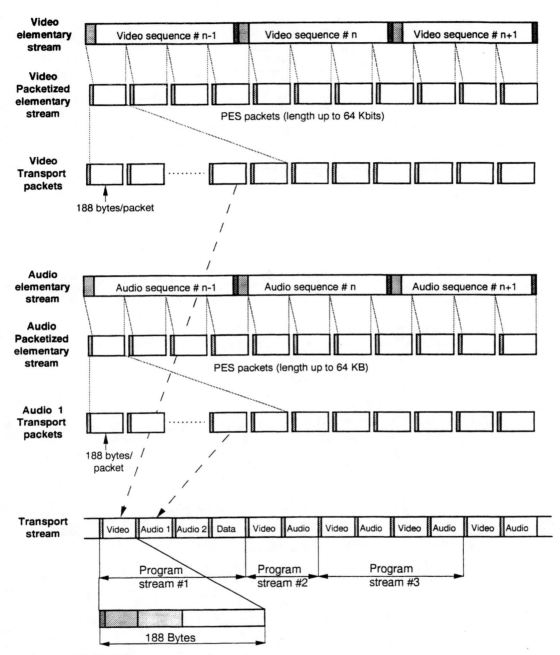

Figure 8.56 MPEG-2 Transport stream formatting.

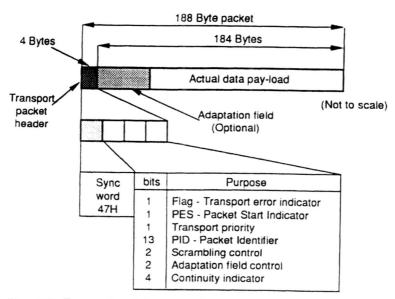

Figure 8.57   Transport packet structure in the MPEG-2 standard.

clocks, can also be multiplexed into a single TS through a conversion in PES packets. The capability of multiplexing programs with different bit rates into one TS is used in the advanced television system, which is detailed in Chap. 11.

TS packets are fixed at 188 bytes in length and their data content is described in Fig. 8.57. They carry timing and synchronization information and jitter-correction mechanisms to ensure reliable long-distance transmissions. In addition, the fixed data packet size facilitates the conversion of TS into ATM (asynchronous transfer mode) network cells. This stream is more resistant to errors, such as those encountered in digital transmission applications.

9

Computers and Television

Previous chapters have explained analog and digital signal generation, processing, and display as presently used in the TV domain. This chapter explains mechanisms for generating video signals, displaying them on monitors designed for computers, and providing these signals to standard-compatible outputs. It gives the user the tools he or she needs to make an informed computer choice based on performance and compatibility with existing production system facilities. It is intended for

- Graphics specialists involved with video imaging

- Computer-oriented specialists moving into the new computer-based television production domain to understand and fulfill the requirements and needs of the broadcast video world

- Video specialists attempting to incorporate computer devices, systems, and concepts in their video environment

Personal computers were widely used in the past for generating texts, spreadsheets, graphics, and games. The result could be displayed and printed. Now, these computers are used for generating pictures and processing the corresponding video signals in real time. Personal computers are invading the video and audio production world. New applications are growing rapidly such as editing, servers, graphic systems, on-air systems, network interface, and access systems. All dedicated video production boxes that we were accustomed to using for broadcast production can now be replaced by a computer equipped with the proper processing boards and production software. These activities result in growing interdependence of the computer graphics industry with broadcast television and other video industries.

The image production process has been slowed in the past because of the high amount of data to process in real time to produce acceptable picture quality. This is now possible thanks to the digital compression techniques allowing very complex data picture processing to be performed in real time at an affordable cost. However, this would not be possible without the important improvements realized in computer architecture, main processor and memory speeds, high-performance expansion cards, and display monitors.

9.1 Computer Architecture

Since its origins, personal computer architecture contained computer components connected through buses, such as the address bus, data bus, and control bus. The address bus carries signals specifying the memory location where data must be stored or retrieved. The data bus carries data itself, and its speed is one of the most important limiting factors in computer performance. The control bus is used to control the data and the address buses, such as memory read or write or peripheral read or write. A minimum configuration of a computer consists of a central processing unit (CPU), a random access memory (RAM), and a read only memory (ROM) exchanging data with external components, such as keyboard, mouse, disk drives, and display through the main system bus. Any device connected to the main system bus becomes an integral part of the computer with direct access to the microprocessor, the memory, and existing external components.

Personal computer performance has improved tremendously over the last 10 years. More and more complex calculations and operations are requested, involving very high data transfers between processors, memories, data storage, and displays. These needs have been fulfilled by making constant improvements to existing computer architecture and more specifically by

- Increasing the speed of the processors, from 8 to 66 MHz and more recently to 250 MHz
- Increasing the bus width, from 16 to 64 bits
- Implementing additional buses such as local bus, over-the-top bus, and switched bus, to alleviate the amount of data processed by the main processor through the main system bus.
- Developing intelligent and rapid display adapters and storage device controllers.

The architecture evolution on Mac, IBM, and Silicon Graphics Incorporated (SGI) computers will be reviewed to show features and performance capabilities of communications buses, expansion cards, and display monitors.

9.2 Internal Computer Communication Buses

9.2.1 Main system buses

The main system bus was originally used for all input and output (I/O) between the system board and peripheral devices, including memory, disks, and display devices.

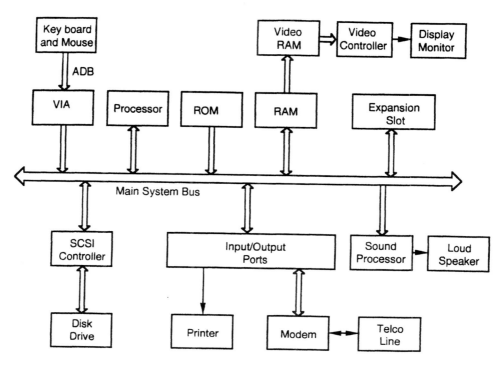

Note: Disk controller could be part of I/O ports
ADB = Apple Desktop Bus
SCSI = Small Computer Interface
VIA = Versatile Interface Adapter

Figure 9.1   Mac-type computer architecture.

To connect various optional devices and provide for open architecture needs, general-purpose expansion buses were designed to provide a buffered interface to devices located either on the motherboard or on expansion cards plugged into expansion connectors.

Manufacturer's proprietary expansion slot designs for various needs and applications created incompatibility between computers and the appearance of a wide range of products on the market at various prices. Figures 9.1 and 9.2 show typical Mac-type and IBM-type architecture designed around main system buses.

9.2.1.1 IBM-type computers.   IBM-type computers currently use main systems buses (also called expansion buses) based on Industry Standard Architecture (ISA), Micro Channel Architecture (MCA), and Extended Industry Standard Architecture (EISA).

- ISA. The original IBM AT open architecture design was never documented.

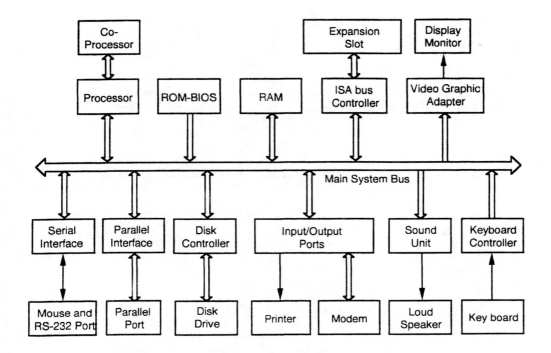

Note: Data Bus only shown.

Figure 9.2   IBM-type computer architecture.

Other IBM-type computer manufacturers and IBM agreed subsequently on an official standard known as the ISA. The ISA standard defines an open architecture allowing third parties to develop adapters. It defines a 16-bit data bus with an 8-MHz clock, based on a 49-pin double-sided expansion connector design. Each expansion card is connected to address lines, data lines, and other control signals. The ISA specification has no standard hardware or software interface for the configuration of resources such as memory and address space and interrupts for each card. Consequently, the ISA card configuration is done manually with jumpers. Many ISA cards are in use today, but this bus standard is no longer used by developers because of its limited performance as compared to other more recent bus architecture.

- MCA. This is a proprietary standard developed for IBM PS/2 computers and is not compatible with the ISA architecture. It defines an asynchronous bus (i.e., events can happen anytime, with no synchronism with the clock) and additional requirements such as data transfer protocols over the bus as used in the IBM PS/2. The 16-bit bus is suitable for the 286 and 386SX processors and is designed with 58-pin double-sided connectors. The 32-bit bus design can be implemented with 386 and 486 processors and uses 93-

pin double-sided connectors. Additional processors can be implemented on adapters to carry work independently to the main processors by sharing control of the bus. A typical application is network interfacing. Important new capabilities are

- Capability to identify each adapter in each computer within a network.
- Capability of remote control of a particular adapter.
- Better grounding, leading to less interference and better data integrity.
- Capability of hardware request interrupts leading to better operational reliability and minimizing loss of data.
- A high-speed clock signal at 14.318 MHz (four times the NTSC subcarrier) is present on this bus.

This bus standard is not widely used because of other competitive bus architectures such as EISA and peripheral component interconnect (PCI).

- EISA. This is an alternative to the MCA bus standard and was developed by a consortium of computer manufacturing companies. It defines 32-bit data and address buses allowing an access to the 4 GB addressing requirements of the 386 and 486 processors. The EISA bus maintains a compatibility with 16-bit ISA expansion cards.

The clock driving the expansion bus remains identical to that of the ISA standard, that is, 8 MHz. The EISA bus operates nominally in synchronous mode, locked to an 8.3-MHz clock derived from a 33-MHz processor (approximately one-fourth). However, two others modes, compressed cycle transfer mode and burst transfer mode, which require less bus cycle per transfer, are respectively 50% and 100% faster than the nominal synchronous mode. The effective transfer rate in the case of burst mode is 33 MBps, using an 8.33-MHz bus and a 32-bit data bus.

The EISA bus has hardware and software mechanisms to identify resources requested by a card and resolve system resource conflicts. It has been implemented in many computer products, although the coming of local buses (such as PCI) has made this standard less attractive.

9.2.1.2 Mac-type computers.

Mac-type computers currently use main systems buses based on processor direct slot (PDS) and Nu-bus standard architecture. Most Mac computers, from the SE series to the more recent Quadra series, have a single PDS that enables external devices to have direct access to data, address, and control buses connected to the main processor.

Starting from the beginning of the Mac II series, an expansion bus called Nu-bus was implemented and allows for several expansion slots. The number of slots depends on the Mac computer model and varies from 1 to 6. The Nu-bus standard, which is an extension of the Texas Instruments Nu-bus specifications, defines a 32-bit-wide multiplexed address or data bus that connects to the expansion slots. Figure 9.3 shows the Mac Nu-bus implementation.

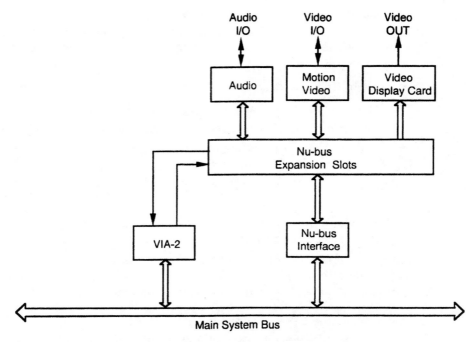

Note: Nu-bus expansion slots, interface and control lines only shown. For other motherboard components, see Figure 9-1.
VIA = Versatile Interface Adapter.

Figure 9.3 Mac-type computer Nu-bus Implementation.

The system configuration is very simple thanks to identification addressing (ID lines) enabling the interface system to be free of DIP switches and jumpers. The Nu-bus is a synchronous bus, but variable clock period length can be used for transferring data. This design emulates an asynchronous bus with a simple synchronous bus implementation.

Recent Mac computer models such as Quadra 840AV and Centris 660AV are equipped with a newer version of the Nu-bus standard, the Nu-bus-90, which provides faster communications with the advantages of the main system bus.

9.2.2 Local buses

Computer-intensive applications generate large amounts of data that must be handled by processors operating at speeds of up to 200 MHz and beyond. Since traditional main system buses are unable to transfer the information quickly enough, new buses were designed to speed up display response, enhance processor interaction with main memory, and improve disk storage performance.

A first step was to connect the main system RAM directly to the processor through a private path, also called the processor bus. Further development led to new designs capable of high-speed communication (using a 25- or 33-MHz clock) from the processor and system RAM to display devices and hard disks, through a local bus. Consequently, local buses allow high-speed transfer interfaces, such as memory, display, disk drives, and audio/video I/O boards, to be removed from the slower (8-MHz) main bus.

The introduction of local bus standards, such as the Video Electronic Standard Association (VESA) VL bus, the PCI bus for IBM-compatible computers, and the Nu-bus for Mac-compatible computers, have permitted the development of multibus-architecture personal computers.

9.2.2.1 VL bus. The VESA local bus (VL bus) has been designed to improve the data transfer rate of the PC/AT architecture. In 1992 VESA created an expansion bus standard that allows up to 10 peripheral components to attach directly to the CPU and thus operate at high CPU speeds through a local bus. Other devices such as sound cards, mouse interfaces, and modems can share the main system bus.

The VL-bus is an open design and its specifications are based on the Intel 486 processor bus performance. Consequently, not every VL bus card is compatible with every VL bus–based systems. The VL bus can operate from 16 to 66 MHz across a 32-bit bus, for motherboard-based components. However, the standard specifies signals at a maximum of 40 MHz on any expansion slot to minimize signal radiation. A revised standard specifies 50-MHz maximum signals for a 64-bit bus. Sustained data rates of 130 and 260 MBps can be achieved, respectively.

9.2.2.2 PCI bus. To provide better expansion system compatibility and a more robust bus for Pentium-based systems, the PCI local bus was designed by major manufacturers and industry partners. As a result, the PCI is a high-performance processor-independent 32- or 64-bit bus designed to be used with devices that have high-bandwidth requirements.

Figure 9.4 shows the PCI bus architecture. It features a 32-bit data bus running at a clock speed of 33 MHz and generating a peak bandwidth of 132 MBps (33 MHz×32 bit/8 = 132 MBps). However, it must be noted that the effective bus bandwidth of the PCI bus is reduced to 80 to 100 MBps because of the multiple control operations taking place on the bus. The PCI standard allows for a transparent upgrade path to a 64-bit bus design, having a theoretical capability of transferring information at up to 264 MBp.

It is important to note that the PCI bus architecture is designed to supplement, not replace, traditional expansion buses such as ISA, EISA, and MCA in personal computers. On these expansion buses, we can find a mouse interface, a printer, modems, and other I/O devices that do not require a fast data exchange with the CPU and peripherals. A separate attachment (PCI bridge) is used to connect these expansion buses to the PCI bus.

The PCI provides increased performance for graphics, full-motion video, hard disk drives, network adapters, and high-speed peripherals, while main-

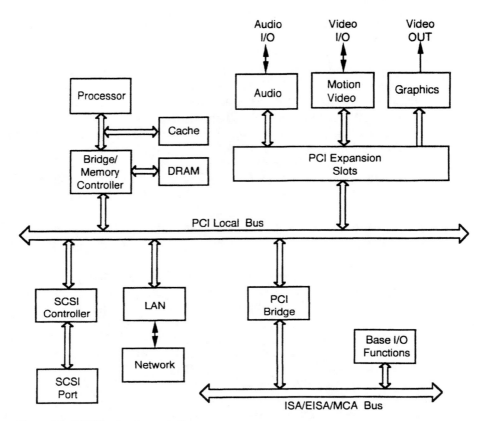

Figure 9.4   PCI Bus Implementation.

taining compatibility with existing ISA, EISA, and MCA expansion buses. It supports the autoconfiguration of compatible Plug-and-Play expansion cards by using device information stored in expansion cards.

Video production applications create situations (e.g., video servers) where several processors and peripherals must be connected using the PCI bus in a scalable high-performance architecture. However, the PCI standard imposes certain electrical and physical constraints to ensure signal integrity. When the bus is running at 33 MHz, a motherboard or passive backplane is typically limited to four plug-in boards on a single PCI bus segment. PCI-to-PCI bridges can then be used to add up to 255 PCI bus segments that are connected in a hierarchical tree, as shown in Fig. 9.5. This interconnect topology generates conflicts when multiple simultaneous transactions take place. In addition, latency and bandwidth problems occur in large systems.

A new standard, called compact PCI, is a superset of the desktop PCI standard. This standard is intended for PCI-based industrial computing products. A new high-density connector provides improved shielding and impedance

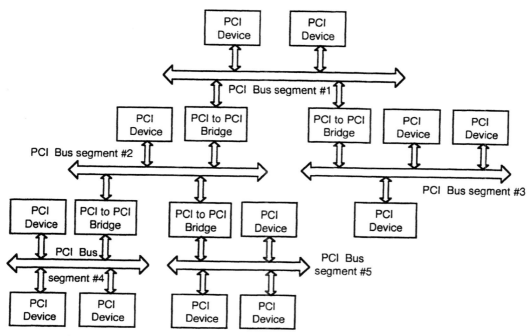

Figure 9.5 PCI hierarchical tree structure for multiple PCI devices.

matching. Compact PCI systems can have up to eight slots, as compared to four in the desktop PCI standard. Using a PCI bridge chip, eight extra slots can be added to a compact PCI system.

In 1995, Apple announced that the PCI standard will be implemented in its new Power PC–based machines. It is expected that the PCI will improve by a factor of 2 to 3 the performance of Nu-bus in graphics, video, and networking applications.

9.2.3 Over-the-top buses

An alternative to the commercial general-purpose buses and their limitations is a dedicated over-the-top expansion bus. This additional bus is used to link high-performance adapters, such as 10-bit quality video processor and CD-quality digital audio boards, within a personal computer. Several over-the-top architectures, including VGA FC, VAFC, Movie-2, and HSV standards, are available.

9.2.3.1 VGA FC. The VGA Feature Connector (VGA FC) standard (March 1994) defines an open hardware interface. It became the first over-the-top expansion bus used in IBM PC-type computers. An enhanced version is specified in the VESA Advance Feature Connector (VAFC) standard. Both VGA FC and VAFC feature a single video channel, no audio support, and their limited performance make them unsuitable for high-performance video adapters.

9.2.3.2 Movie-2. Originally designed by Matrox in November 1995, the Movie-2 digital audio/video expansion bus is designed around an open standard for developers of PC-based broadcast and professional digital video equipment. It is a dedicated over-the-top bus that links specialized adapters within a PC, and offers an increased throughput bandwidth. This is required to support multiple real-time streams of uncompressed 10-bit CCIR 601-quality digital video, key signals, and CD-quality digital audio. The Movie-2 bus is shown in Fig. 9.6. It carries eight continuous uncompressed digital video buses, six digital key buses, and four serial digital audio buses for a constant data rate capability in excess of 242 MBps (or 1936 Mbps).

Examples of some Movie-2 bus adapters under development by Matrox include a digital video mixer, motion-JPEG codecs, MPEG-2 encoders and decoders, 2D and 3D DVE units, multichannel analog video I/O boards, multichannel serial digital interfaces, and DSP-based application accelerators.

The Movie-2 bus uses two high-density right-angle connectors on the top edge of adapter boards. Movie-2 bus–compliant audio/video adapters interface to each other through an over-the-top passive printed circuit board that connects to the two top edge connectors. PCI, EISA, and ISA video adapters can use the Movie-2 bus provided they respect the standard height for PCI adapters.

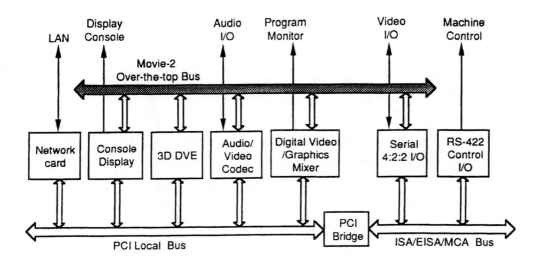

Note: Data bus shown only.

Figure 9.6 Movie-2 Over-the-Top Bus Implementation (*after Matrox.*)

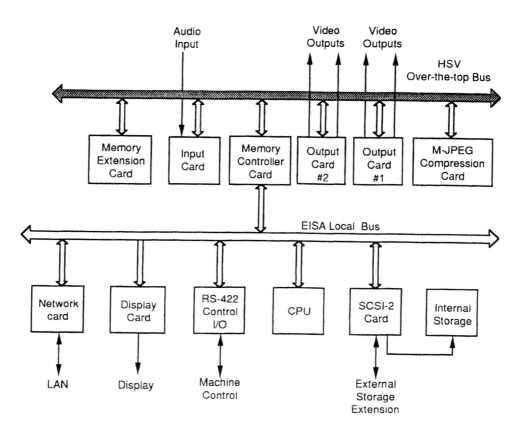

Note: Data bus shown only.

Figure 9.7   HSV Over-the-Top Bus Implementation.

9.2.3.3   HSV.   The high-speed video (HSV) bus, developed by Thomson Broadcast Systems, in France, is very similar to the Movie-2 bus, as shown in Fig. 9.7. It establishes a high-speed communication between processing boards, allowing up to eight simultaneous real-time 4:4:4:4 (10 bits) video signals to be transferred.

9.2.4   Switched buses

Multicomputing systems need to interconnect a considerable number of high-performance peripherals and processors together. A broadcast server configuration requires that up to 20 work stations be interconnected to it, so high connectivity (high number of adapters) is a very important issue.

As described in the previous section, PCI bus segment interconnection capabilities are limited because of the electrical and physical limitations of the PCI standard. A switched-bus architecture, such as the RACEway Interlink

Standard (ANSI/VITA 5-1994) developed by Mercury Computer Systems, Inc., can accommodate up to 20 PCI slots in a small enclosure with high scalability and bandwidth, without stressing the server CPU.

RACEway Interlink* is a relatively recent expansion of the VME and PCI bus architectures. It is an open "bus-less" interconnect technology that allows multiprocessing applications with VME and PCI bus systems. In contrast to a central bus interconnect architecture, RACEway Interlink is constructed with an integrated circuit crossbar switch (switching fabric) that allows simultaneous real-time communications between processing and I/O cards, while keeping latencies very low. Figure 9.8 shows the crossbar switch allowing a maximum of three independent I/O ports to be connected to the other three. For interconnection of a larger number of devices, Mercury has developed the RACEway Interlink family of modules, as shown in Fig. 9.9.

Switching fabrics provide redundant interconnection resources, allowing multiple transactions to proceed simultaneously. This approach is very useful in applications requiring intensive I/O data transfers. Mercury has developed a RACEway-to-PCI bridge chip to adapt the RACEway technology to standard PCI bus–based systems, thus taking advantage of the high-performance PCI adapters sold in large quantities. A 20-slot RACEway Interlink configuration

---

*Complete details about this technology can be found in a document number DS-42-23 available from Mercury Computer Systems, Inc.

Figure 9.8 Basic crossbar switch interconnections (*Courtesy of Mercury Computer Systems, Inc.*).

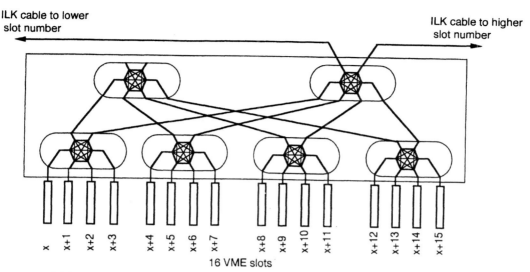

ILK cable to lower
slot number

ILK cable to higher
slot number

x x+1 x+2 x+3 x+4 x+5 x+6 x+7 x+8 x+9 x+10 x+11 x+12 x+13 x+14 x+15

16 VME slots

Figure 9.9 Multiple peripheral interconnections using the ILK16 RACEway Interlink module crossbar switch (*Courtesy of Mercury Computer Systems, Inc.*).

for a PCI-based system can deliver up to 1 GB of I/O in a real-time, low-latency, deterministic, and scalable manner. Each part of the switched bus consists of an intelligent multiport switching device that recognizes a data stream destination message and routes data transfers to the appropriate processing cards.

9.2.5 Data bus router

Figure 9.10 shows a data bus router implementation, as used by Tektronix, in their professional disk recorder, the Profile PDR-100. In this design, the EISA bus is controlled by the application processor and sends data to the various subsystems such as the router, the high-speed data processor, the video JPEG codecs,* and the mix-effects. It also interconnects digital audio signals to the disk recorder subsystem. External reference, time code, and control signal data are carried by the EISA bus.

A high-speed processor bus is used to interconnect the i960 processor with JPEG video data codecs and disk storage units. Multiple digital video data I/O signals are routed through a 32×32 crosspoint matrix to simultaneously interconnect these signals with real-time video processing applications, such as Mix-effects, JPEG codecs, and disk storage units. In this design, bandwidth requirements are limited to a maximum of 27 MBps (using 8-bit digital component video signals) for all routed links.

---

*The generic word codec is the short form of coder-decoder which implies any form of encoding-decoding. Here, it represents a compression encoder-decoder system.

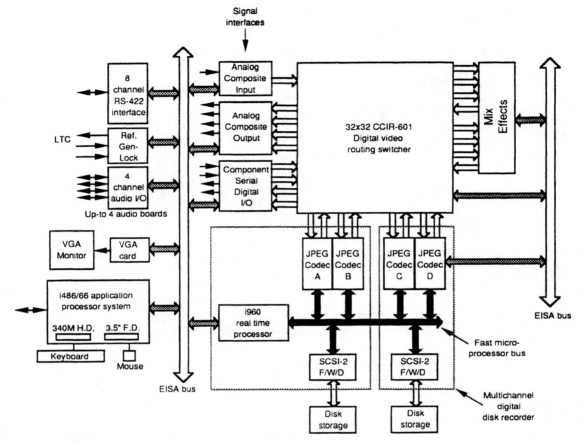

Figure 9.10 Data bus router Implementation (*Courtesy of Tektronix*).

In this section, we have successively described main system buses, local buses, over-the-top buses, switched buses, and finally a data bus router. Their performance and characteristics are summarized in Table 9.1.

9.3 Display Monitors for Computers

With the exception of particular applications, personal computers were developed as stand-alone systems. The final software product, be it a spreadsheet, a graphic, text, or a game, was displayed on a dedicated monitor. Each manufacturer developed a proprietary monitoring format to provide acceptable display performance to the customer. Computer systems remained isolated, with practically no need for exchange of video information between systems from different manufacturers. Consequently, monitor formats were specifically developed to suit the computer industry needs, which are quite different from those of the broadcast industry. A variety of display formats exists, resulting

TABLE 9.1 Computer Bus Data Speed Performance and Characteristics

	Theoretical bus speed, MBps	Bus format, bits	Bus frequency, MHz	CPU frequency, MHz
Main system buses				
▪ ISA	5	8/16	8	8–66
▪ EISA	33	32	8	8–130
▪ MCA	45	16/32	10	10–40
▪ Nu-bus	20	32	20–40	20–130
Local buses				
▪ VL	130	32	40	40–130
	260	64	50	50–130
▪ PCI	132	32	33	33–130
	264	64	33	33–130
	568	64	66	66–130
Over-the-top buses				
▪ Movie-2	242*	32/64	27	40–166
▪ HDV	432*	32/64	27	40–166
Switched bus				
▪ RACEway Interlink	1000	32/64	33–66	33–130

*8-bit and 10-bit video data words can be used.

in increased manufacturing costs, especially to produce multisync monitors. Now, most of the current applications require a color picture monitor to display complex pictures accurately.

9.3.1 The CRT construction

Before going into details about display computer formats, we will make a short review of the cathode-ray tube (CRT) internal construction, as shown in Fig. 9.11. CRT scanning and characteristics are detailed in Sec. 1.3. Three GBR electron beams, whose intensities are modulated by GBR video signals coming from the computer video controller, are deviated in both the horizontal and vertical direction.

A shadow mask, which contains more than 1 million holes in a 17-inch monitor, ensures that each G, B, or R beam strikes only the corresponding dot of the phosphor triad on the monitor screen. Each triad displays a variable mix of green, blue, and red lights generating a wide selection of colors, including gray and white, to form a single point on the monitor screen. Figure 9.12 shows the shadow mask hole distribution, and the horizontal dot and vertical row spacings. Alternate CRT designs use a guard band stripe screen with in-line GBR guns [precision in-line (PIL) or single-gun Sony picture tube].

9.3.2 General considerations

Figure 9.13 shows phosphor triads and an example of a square pixel illumination. Depending on the resolution of the display format and the CRT size, a

Figure 9.11 Color monitor CRT construction based on a delta gun configuration.

pixel can be composed of one or several triads and is defined as the smallest screen area that can be independently controlled by the computer video controller. A pixel is square if on-screen distances between the center of two consecutive pixels, in both the horizontal and the vertical directions, are identical. Computer monitors generally use a 4:3 raster format and must have a horizontal to vertical pixel number ratio of 4:3 to display square pixels.

As an example, a 17-inch monitor having horizontal and vertical screen dimensions of 12.598×9.449 inches (320×240 mm) can be configured as a 640×480 pixel format. Both screen dimensions and pixel ratios are equal to 4/3. This means that a drawn circle will look like a real circle and not an ellipse. A maximum of 640 picture points needs to be displayed in the screen horizontal direction. With a dot pitch of 0.25 mm, the horizontal dot spacing is 0.2165 mm ($0.25 \times \cos 30°$) and the vertical row spacing is 0.125 mm ($0.25 \times \sin 30°$). Consequently, there are 320 mm/0.2165 mm = 1478 dots per row available to generate 640 pixels per line (in this display format example), and several screen dots will be used to generate a single pixel. In the vertical direction, the same CRT exhibits 240 mm/0.125 mm = 1920 rows. Some triads can be partially illuminated, as shown in Fig. 9.13.

The horizontal size of the square pixel is defined by the video pulse width representing one pixel duration, whereas the vertical size depends on the beam spot dimension. Display light interpretation by the human visual system (HVS) will be a square pixel. However, when the same monitor is configured for a high-

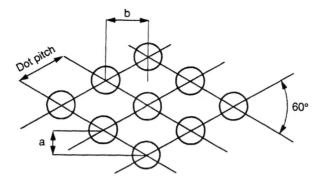

a = Vertical row spacing distance
b = Horizontal dot spacing

Figure 9.12 Shadow mask hole repartition and dot pitch.

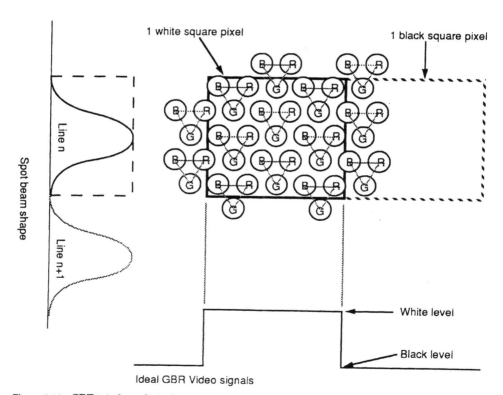

Figure 9.13 CRT triads and pixel generation.

er-resolution format, such as 1024×768 pixels, a smaller number of triads will be used per pixel and a very small beam spot dimension is required. The round-hole shadow mask design must carefully consider the number of dot rows so as to avoid moiré fringes caused by a "beat" with horizontal scan lines. When a monitor is configurable for several display formats, it is important to look for the possible occurrence of this disturbing pattern on the monitor screen.

A monitor screen display is made of successive lines of pixels. If all pixels of a specific line are illuminated, the display shows a straight, continuous line. The illumination control of each pixel allows for the creation of variable pixel brightnesses, from black to white, and colors. As previously explained in Sec. 1.3, lines are drawn by sweeping GBR electron beams over the surface of the CRT screen, from left to right and top to bottom. This process is continuously repeated to update the monitor screen. It is called *refresh rate* by the computer industry and varies from 56 to 75 Hz, and *vertical scan frequency* (or frame rate) by the video industry, which has used, up to now, a single 60-Hz value.

All computers use a progressive vertical scanning, whereas television equipment presently use an interlaced scanning. In 60-Hz progressive scanning, the picture is scanned 60 times per second from top to bottom, whereas in interlaced scanning, the picture frame is scanned in two interleaved halves called *fields*. The first scan generates field 1 by producing the odd-numbered lines from top to bottom, and the second scan generates field 2 and produces the even-numbered lines. This interleaved process is repeated 30 times per second, making the appearance of 60 interleaved fields per second and reducing the display flicker. It must be noted that the ATSC standard defining the future digital television system incorporates several scanning rates, both progressive and interlaced, such as 24, 30, and 60 Hz.

9.3.3 Display monitor performance characteristics

When connecting a display monitor to a personal computer, it is important to verify the compatibility with the video controller card and the performance characteristics to ensure the best on-screen picture quality. Some important picture display performance parameters and features are listed below.

- Video bandwidth and pixel clock. The video bandwidth defines the ability of the internal G, B, and R video amplifiers to drive the corresponding electron beams correctly and reproduce picture details. The picture tube ability to display the video signals is also tied to the high voltage, which moves and accelerates electrons from the cathode to the display screen.

  The highest-resolution format of a picture tube dictates the maximum video bandwidth required to represent all pixels. As an example, if 1024 pixels are to be displayed on the active portion of a horizontal screen line, a maximum of 512 cycles can be generated, because it takes two pixels (one white and one black) to make a video signal cycle. One cycle duration is calculated by dividing the active line duration (e.g., 13.653 μs as used in the SVGA mode) by the maximum number of cycles on this line, which gives a cycle

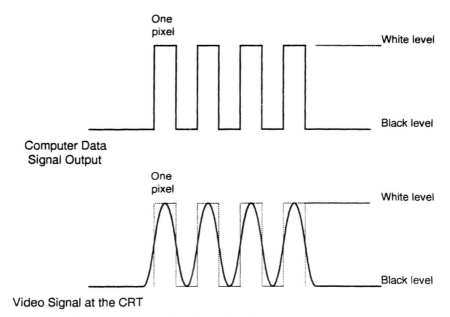

Figure 9.14 Computer data signal and actual video signals.

duration of 26.66 ns. The inverse of this number, 37.5 MHz, is the minimum video frequency that the video amplifier must handle without attenuation to reproduce correctly the sine wave (fundamental) of the pixel square pulse signal, as shown in Fig. 9.14. In this case, each video amplifier is designed for a bandwidth equal to or greater than 37.5 MHz. The pixel clock signal is twice the minimum video bandwidth, because each pixel requires a clock cycle.

- Refresh rate. A "multisync" monitor automatically adjusts the refresh rate to the selected display standard, in a 60- to 75-Hz range. The 60-Hz value may result in flicker, and higher values are now recommended (70 Hz and higher).

- Spatial uniformity. This parameter is a measure of the brightness anywhere on the screen. Low-quality picture displays tend to have a higher brightness in the central area of the screen.

- Color tracking. When the three RGB electron beams are simultaneously driven from a low current to a high current, the light intensity of each color beam (G, B, or R) must keep the same relative value to the others on the screen.

- Maximum usable brightness. Some monitors provide controls for a wide adjustment range of brightness levels to suit user preferences. The brightness level may affect the beam focus and consequently the contrast performance.

- Contrast. This parameter expresses the difference between the brightness of white picture areas and that of the black picture areas, relative to the background illumination [Contrast = $(I_{max} - I_{min})/(I_{max} + I_{min})$]. The contrast calculation shows that an increase in the background illumination at the receiver modifies the displayed image contrast. Values exceed 50 for current monitors.

- Sharpness. The eye is sensitive to the presence of edges in the picture. This parameter depends on the image contrast and the number of picture elements that can be reproduced (i.e., picture resolution).

- Geometry. This parameter characterizes the perceived picture geometry distortion. It is generally defined by the horizontal and vertical scan linearity performances. Geometry controls include horizontal and vertical scan linearity adjustment and also parallelism adjustments of the picture sides with the monitor edges (rotation).

- Horizontal and vertical resolutions. The computer industry has always used this terminology to define the maximum number of pixels in both directions. However, in the television industry, subjective measurements have shown that the effective resolution must be reduced by a factor of about 0.7 (often called the Kell factor) to reflect the aperture loss caused by horizontal and vertical sampling processes. Five resolution computer formats are in common use today, 640×480, 800×600, 832×624, 1024×768, and 1280×1024.

- Dot pitch (or aperture grid pitch). This is another way of expressing the monitor display resolution performance. The dot pitch is the spacing between adjacent holes, taken on a diagonal line, on the shadow mask. Small dot pitches give sharp images. Current dot pitch values are between 0.23 and 0.41 mm (0.009 and 0.016 in).

- Monitor size. Current display specifications indicate the diagonal length of the CRT. Part of the tube is covered by a casing and the displayed picture can be 1 to 3 in smaller.

- Display monitor improvements. New display monitors have desktop on-screen control menus. Power consumption is reduced by adding power-saving features, such as a screen saver, to the monitor. Radiation, such as extremely low frequency (ELF) and very low frequency (VLF) emissions are reduced to comply with the old MPR-II (Swedish initials of Swedish Board for Measurement and Testing) and the more severe TCO (Swedish initials of Swedish Office-Worker's Union) standards, developed by Swedish organizations. No similar standards exist in North America.

9.3.4 Computer monitor formats

Early computer designs used low-resolution display formats to reproduce text and graphics because the data generation and processing speed was very low.

The need to improve graphic and text representations has resulted in numerous display formats with increasingly higher picture resolutions. This evolution has generated a wide variety of formats and incompatible video controllers.

The need for common display formats and scanning parameters (also referred to as timings in the computer industry) has been recognized and has considerably changed the situation. Standardization efforts have been made on the issues of interoperability among computer systems, graphical user interfaces, and display accelerator implementation. Most of the new monitors can support several display formats, configurable from software drivers. Table 9.2 shows several examples of IBM-type computer display formats that have been developed for different application levels. Table 9.3 summarizes Mac-type computer display formats in use today.

TABLE 9.2 Examples of IBM-Type Computer Graphics Display Format Specifications

Standard	VGA	SVGA	SVGA	SVGA	SVGA	SVGA
Resolution, H×V	640×480	640×480	800×600	800×600	1024×768	1280×1024
Color depth	16/256K	16/256K	16/256K	16/256K	16/256K	16
V-active, lines	480	480	600	600	768	1024
V-total, lines	525	525	666	628	806	1068
Hor. freq., kHz	31.469	37.8	48.077	37.879	56.476	76.02
H-active, μs	25.422	20.317	16	20.00	13.653	10.119
Vert. freq., Hz	59.94	72.2	72.188	60.316	70.069	71.18
Pixel clock, MHz	25.2	31.5	50	40	75	126.5
Video bandwidth, MHz	12.6	15.75	25	20	37.5	63.24

TABLE 9.3 Examples of Mac-Type Computer Graphics Display Format Specifications

Standard	Mac II, 13", 15"	Mac II, 16", 17"	Mac II, 19"	Mac II, two-page
Resolution, H×V	640×480	832×624	1024×768	1152×870
Color depth	16/256/16M	16/256/16M	16/256	16/256
V-active, lines	480	624	768	870
V-total, lines	525	667	813	91
Hor. freq., kHz	35.000	49.724	60.24	68.681
H-active, μs	21.164	14.524	12.76	11.520
Vert. freq., Hz	66.67	74.55	75.00	75.06
Pixel clock, MHz	30.240	57.283	80	100.000
Video bandwidth, MHz	15.12	28.7	40	50

9.4   Expansion Cards

Expansion cards have been used extensively since open-bus architecture appeared on the marketplace. Different cards could be plugged into expansion slots to interface the computer with peripherals such as display monitors, modems, and disk storage devices. These expansion cards allow the computer user to choose an expansion interface with the intended level of performance required by a specific application. New computer applications in multimedia and television have stimulated the design of high-performance, rapid, and intelligent expansion cards, and the creation of new interfaces with audio, video, and CD-ROM devices. In the following sections, we will review existing high-performance interfaces that can be used in multimedia applications.

9.4.1   Video controller cards

Video controller cards are responsible for generating GBR video and synchronization signals required to display computer data onto a monochrome or color picture monitor.

CPU-processed data on the bus can be displayed on a picture monitor for user interaction thanks to a display memory located in the computer's memory space and a video controller plugged into the expansion bus. The display memory is dual-port memory, because the processor writes data into it using one port and the video controller reads existing data for on-screen display using another port. The video controller is also called a video graphics adapter (VGA), but this can lead to confusion with the video graphics array (VGA) display standard used in IBM PCs. Through the years, numerous display standards appeared on the market, with an increasing number of available active pixels to meet growing needs for more precise representation of texts and graphics.

The most commonly used display format standards are the VGA, the Super-VGA (SVGA), and the Extended Graphics Array (XGA). The VGA standard was introduced in IBM's PS/2 systems in 1987. It defines 15 modes for text and graphics applications. The 640×480 resolution mode is the most popular in graphics applications. The SVGA standard, also called VGA+ or VESA, offers a superset of the VGA specifications with a capability of 28 modes. The 800×600 resolution is used as the reference of this standard.

The XGA standard was introduced in 1990 by IBM to better suit graphics-based interface requirements. It defines 18 modes that include all VGA modes and adds a text mode with a resolution of 1056×400; 16 colors, having a capability of 132×25 characters for large spreadsheets; and two graphics modes. Table 9.4 summarizes some of the most important modes presently used. However, additional memory is required when greater resolution or color depth is contemplated. This may generate problems with video memory space allocations.

Two different modes are used to represent computer-processed data, the text mode, also called alphanumeric mode, and the graphics mode, also called all-points-addressable (APA) mode or bit-map mode.

TABLE 9.4 Summary of Some Graphic Display Format Standards and Modes

Standards	Resolutions	Color depth
VGA	640×480	16
SVGA	640×480	256
	800×600	16
	800×600	256
	1024×768	16
	1024×768	256
	1280×1024	16
	1280×1024	256
XGA	640×480	16
	640×480	256/65,535
	1024×768	16/256

9.4.1.1 The graphic mode display. An example of a graphics mode display is shown in Fig. 9.15. It represents a monitor screen made of a rectangular array of 640×480 pixels. All these points can be independently addressed to control their brightness and colors. Anything appearing on the screen is created using pixels. Graphics and texts can be generated in various resolution modes. Each pixel has only one datum of information, the color to which it is set.

The possible number of colors for each pixel, also called the color depth, is determined by the number of bits assigned to represent color values. With 4 bits, 16 colors are generated, whereas 8, 16, and 24 bits gives, respectively, 256, 32K, and 16M possible colors. The display controller stores a value (4 to 24 bits) in its video-RAM (VRAM) memory for every pixel on the screen. The amount of VRAM storage needed is determined by the display format resolution and the pixel color depth. Some video controllers specify a 32-bit color depth, where 24 bits are reserved for GBR signals and the remaining 8 bits are used to specify an additional channel called the alpha channel.

When a graphic image is made, the video controller maps the color data for each pixel into the VRAM. With the use of high-resolution formats and graphics-based interfaces, the high amount of processed data becomes difficult to handle. Graphics accelerators and coprocessors are now used on video controller cards. The main CPU sends commands to draw a graphic and the coprocessor calculates the color data information for all pixels involved in the representation of this graphic and then updates the VRAM. Such video controllers can improve the display processing speed by a factor of 10 to 20.

9.4.1.2 The text mode display. The text mode is used to represent characters on a screen, divided in columns and rows. Figure 9.16 shows an example of a text mode format standard where the screen is divided in 80 columns by 30 rows of characters. Each character will be written in a rectangular box of 8 pixels horizontally (640/80) by 16 lines vertically (480/30). An ASCII and extended charac-

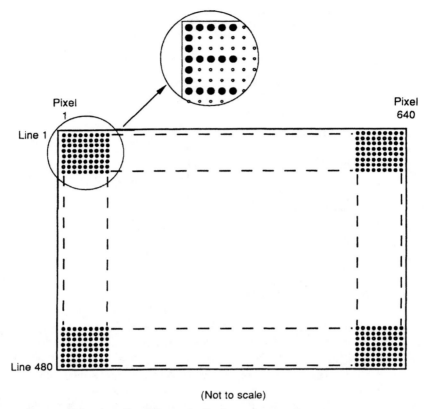

(Not to scale)

Figure 9.15 Computer Graphics mode display representation.

ter set is written in a ROM located into the PC motherboard. Each character datum is composed of 2 bytes, a data component byte corresponding to the character selected and an attribute byte defining the color appearance of character background and foreground. These data are sent directly to the display memory of the video controller, which periodically reads it to refresh the screen display. Several text mode formats are defined in various display format standards and use a different number of pixels to reproduce characters.

A typical video controller design is shown in Fig. 9.17. The 256K VRAM is generally made of fast RAM chips, sold as VRAM chips. The frame buffer controller is a register-controlled array that uses the parameters stored in the control registers to generate and control video data and timing signal outputs. The color look-up table (CLUT) is made of 256 registers (memory locations). Each register contains a 16-bit pattern specifying one of the 16 million possible colors. Therefore, the CLUT converts the digital data representing index values to color values by table look-up in RAM. Then, three 8-bit D/A converters are used to generate analog color signals, which are RS-343A-com-

(Not to scale)

Figure 9.16 Computer Text mode display representation.

patible GBR video signals. The RAM within the CLUT is initialized with default color transformation values. By software, it is possible to read and modify the color palette representing a set of color possibilities.

On Fig. 9.17, an optional video output circuit has been added for interconnection with a video recorder, an NTSC color television receiver, or video production equipment. A frame rate converter is necessary to adjust the computer signal refresh rate to the vertical frequency of 59.99 Hz used in television production. An interlace converter ensures that progressive scan images are properly interleaved. Analog signal encoders are then used to format the video signals to the corresponding video standard.

Many PCs do not have such analog video outputs. A video scan converter can be used to convert video controller formats such as VGA, SGVA, or XGA formats into analog NTSC or component formats.

Figure 9.17   Video controller card implementation.

9.4.2   Video/audio interface cards

Other important interface cards have been recently developed to enable computers to communicate with television production systems: the video and audio I/O cards. These cards allow any user to input video data for manipulation through various processing peripherals and then output the result to storage, video connectors, or networks in formats compatible with the production and broadcast industries. However, some form of audio and video compression may be needed to handle high data rates of original materials and store them on a disk. Typically, such an interface attaches to the high-performance PC bus. On the back of the PC bus card, a series of connectors enables the user to attach a camera, VCR, or any television device in the desired video signal format.

A large number of manufacturers have developed video and audio interface cards, also called *capture cards*. They exist at various audio and video quality levels, which depend on the amount of compression involved, and at different picture sizes, for multimedia applications.

An example of video card products is the True Vision Targa-2000 video graphics adapter for the personal computer. It uses the EISA platform to real-time process an adjustable JPEG video compression and decompression scheme. Independent video input and output (NTSC and PAL) are provided and 16-bit stereo audio can be recorded and played back at a 48-kHz sampling frequency. Additional features include video-in-a-window, a video output from a window with a full-resolution preview on the desktop, a high-resolution desktop at 24 bits/pixel (1152×870) and a gen-lock capability to a sync input.

9.4.3 PCMCIA expansion cards

Computer expansion cards are too large to fit in small computers such as laptops and notebooks. An expansion card designed by the Personal Computer Memory Card International Association (PCMCIA) allows for use in these smaller and cheaper computers. Laptop peripheral interfaces now replace proprietary expansion slots. They are compatible with all existing bus designs, from IBM, Mac, and other computers.

PCMCIA cards have the approximate size of a credit card (85.6×54 mm) and are of three types. Type 1 is a very thin card (3.3 mm thick) that can be used for memory expansion, for example. This card is generally powered by an internal battery. Type 2 specifies I/O circuitry and connectors to allow for computers and electronic device interconnections through networks. This card is 5-mm-thick and uses power from inside the computer via the standard 68-pin connector. Type 2 card features are

- Capability to run programs inside the card without having to download them to the computer memory.
- Capability to advise its host about its configuration and capabilities.
- File formats and data structures defined in a standard.
- Capability to operate independent software links.
- Backward compatibility with type 1 slot to ensure that both card levels can be used. However, a type 1 card may not work in the type 2 slot depending on the intended application of this card.

Type 3 is a 10.5-mm-thick card that may contain a hard disk drive. The PCMCIA-adapter interface has a data bus of 16 bits and 26 address bits.

Multimedia and Television

In the context of this book, a medium is a communication support. The best-known communication medium is the book. Other information media have been introduced over the time as technological advances permitted, such as movie theaters, radio, telephone, television, cable networks, satellites, and computers. New communication supports are currently being developed such as the Internet and information highways.

Presently it is possible to combine all these communication technologies into one single support, called *multimedia,* which is also called *technology convergence.* Our relationship with communication media is changing, since users are now able to select subjects, create their own information content, and their own viewing schedules.

Multimedia communication uses combinations of pictures (text, graphics, animation, video, etc.) and sounds (music, voice, sound effects, etc.) to distribute and communicate information to users. These services can be transmitted over the air, via a network, or through local interfaces.

Multimedia applications are changing the ways in which we communicate and acquire and exploit all types of information. In the home, they provide unlimited access to information and services. For the entrepreneur, multimedia applications are an invaluable means of advertising, distribution, and communication with customers.

10.1 The Multimedia Concept

Until now, television involved the production and transmission of audio and video programs. Digital technology has permitted the real-time processing of these signals in dedicated and expensive digital processors.

Personal computers were designed to generate text, graphics, games, and do other repetitive tasks. The marriage of television and computers has led to the introduction of computer-based systems in the production environment. New requirements for high-speed processing created a strong demand for new hardware architectures and computer software to fulfill these new applications.

Multimedia started with the introduction of audio data into desktop computing. Audio boards were built around tone-synthesis chips. Subsequently, digital video interface chips appeared on the market, making it economical and simple to design video interface boards for analog component or composite format signals. The use of such boards in computers allowed the capture and storage on disk of single video frames from full-motion video material for off-line processing and insertion into text or other types of documents.

The digitization of audio and video signals involves a large amount of data. The processing and integration of full-motion video and audio data in desktop computers require compression and decompression techniques that are capable of reducing the total bit rate by a factor of between 2 and 100. Recently developed and currently available compression techniques permit real-time processing of good-quality audio and very complex picture processing at an affordable cost.

The first-generation compression chips, made by C-cube and Thomson-SGS, were based on the JPEG compression algorithm and have permitted the development of low-end multimedia applications, such as compact disk interactive (CD-I) and digital video interactive (DVI). The development and the introduction of the more advanced MPEG compression algorithm, which includes software programming, has made possible the design of audio and video boards working in real time with processor speed capabilities presently found in personal computers.

Personal computers are presently used for the generation of pictures and their manipulation in real time. Widely used in the past for text, spreadsheet, printed graphics, and games, PCs are invading the video and audio production world. New applications, such as disk storage, editing, servers, graphic systems, on-air systems, network interface and access systems, and interactive video, are growing rapidly. All of these applications create a new emerging multimedia environment. The computer market is increasingly focusing on multimedia applications. This trend will result in increased availability at competitive costs. Figure 10.1 shows a typical multimedia home station.

There are three areas of concern that need to be considered by multimedia system developers. These involve the development of suitable

- Workstations
- Networking concepts
- Related software

10.2 Multimedia Technologies

Multimedia has been made possible by the introduction of new technologies such as

- Digitization of pictures and sounds. Audio and video analog signals are digitized and formatted to transmit the amount of information required for the

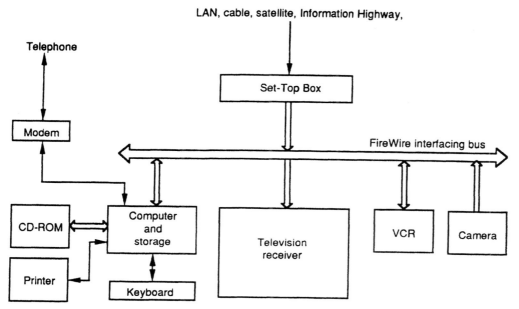

LAN, cable, satellite, Information Highway,

Figure 10.1 Multimedia home station.

intended application. These conversion processes are described in Chaps. 3 and 6. Audio and video acquisition cards, also called capture cards, are available from many manufacturers, with different levels of performance.

- Data compression of digitized audio and video signals. Data reduction has made possible affordable storage and real-time processing hardware, as described in Chap. 8. It helps match the signal content to the channel bandwidth of the processing, storage, and communication systems. Audio and video compression and decompression cards exist on the market for different application requirements.

- Fast real-time processors. Computer performance has evolved by improving processor architecture and the development of high-speed processors, buses, and interfaces. These improvements are described in Chap. 9. The Enhanced Integrated Device Electronics (EIDE) interface is now replacing the old IDE PC connectivity standard for hard disk drives, which was developed in 1988. EIDE allows a maximum burst transfer rate of 16.7 MBps, a maximum disk capacity up to 10 GB, and the possibility to connect new devices such as CD-ROMs and tape drives.

- High-capacity storage systems. Recent years have witnessed an increase and change in the disk storage requirements for computing, broadcasting, multimedia, and audio/video production applications. The size of software is increasing each time a new version appears, requiring increased storage

TABLE 10.1   Examples of Various Interface and Network Line Data Rates

Interfaces and networks	Line data rates
Videoconferencing lines	56 kbps–1.5 Mbps
ISDN	144 kbps
ADSL	6.144 Mbps
DS-1	1.544 Mbps
DS-3	45 Mbps
Ethernet	10 Mbps
Fast Ethernet	100 Mbps
SCSI-2	40 Mbps
SCSI-3	12.5–100 Mbps
CD-ROM	1.2 Mbps
ATM	25, 155, 622
Fiber Channel	133 Mbps–1 Gbps
FDDI	100 Mbps–1 Gbps

space and transfer speed. High-capacity and high-performance disk drives are now more reliable and available at lower prices. Significant new technologies, such as magnetoresistive recording heads, advanced read channels, and fast networking interfaces (Ultra SCSI, SSA, FC-AL) have contributed to their development. These technologies are detailed in Sec. 10.3.3. Table 10.1 shows data space requirements for compressed and uncompressed audio and video data signals.

- High-bandwidth and low-cost data interfaces and networks. Multimedia systems require large data files to be rapidly transferred between system components such as cameras, computers, VCRs, and storage and display units, and between workstations located in the same room or different buildings. The convergence of computers, television, and telecommunications have brought the development of new fast, wide-bandwidth, and affordable interfaces and networks. These technologies are detailed later in Sec. 10.4.

- Compatible architecture. An Open Media Framework Initiative (OMFI) architecture was first introduced by Avid in 1992. A consortium of multimedia workstation manufacturers developed this open-file format that facilitates the exchange of digital information across platforms and between applications.

10.3   Multimedia Hardware and Systems

A multimedia workstation is made of a large number of audio and video processing components, such as a multimedia-version computer, cameras, VCRs, additional large storage, servers, and CD-ROM drives. Each new marketed version of these components brings improved performance, operability, and compatibility with a wide variety of multimedia products.

10.3.1   PC workstations

Multimedia computer platforms are designed with high-speed processors; memory caching; fast local data buses such as PCI and EISA; and separate, large-bandwidth, fast buses such as Movie-2 and VGA-FC. These platforms must handle multiple independent and synchronized data streams using a master video timing controller that can also be used for gen-locking to an external sync reference.

New computer architectures include

- Pentium- and PowerPC-type processors (up to 166 MHz). The new processor architecture allows the execution of more than one instruction per clock cycle, and contains a high-performance 64-bit data bus. Multiple reduced instruction set computing (RISC) processors (up to 250 MHz) can also be used for faster performance but at a higher cost.

- Processor cache (typically 256 kB size). This cache increases memory transaction performance by coordinating the speed of the processor with the slower dynamic random access memory (DRAM). This cache acts as a very fast, temporary storage for commonly used instructions and data between the processor and the main memory. The addition of a processor cache can improve the processing speed by 25%.

- A 64-bit processor bus to increase the data transfer speed between the processor and the system's RAM and memory cache.

- A reduced number of processing cards to eliminate synchronization problems between video, audio, and RS-422 machine remote control.

10.3.2   Audio and video signal-processing systems

The capture of audio and video signals from a camera or a VCR requires the addition of audio and video capture (or grabber) cards. These cards may incorporate analog-to-digital (A/D) converters if the original source signals are analog. Similarly, the distribution of the multimedia production from the computer workstation necessitates interfacing cards to the storage media or distribution networks. Digital-to-analog (D/A) converters may also be required.

Digital audio and video signals must be compressed to make possible and cost-effective their processing, storage, and distribution. JPEG, MPEG, and other compression schemes are available on the market. Figure 10.2 shows a basic block diagram of a video acquisition and distribution card.

Digital audio and video signal processing cards are also needed to manipulate the data and perform special effects, editing, graphics insertions, etc., as explained in Sec. 9.2.3 and shown in Figs. 9.6 and 9.7.

Multimedia video processor chips, such as Texas Instruments TMS 320C80, are used to perform real-time data compression calculations. The Texas Instruments chip includes four digital signal processors connected by means of

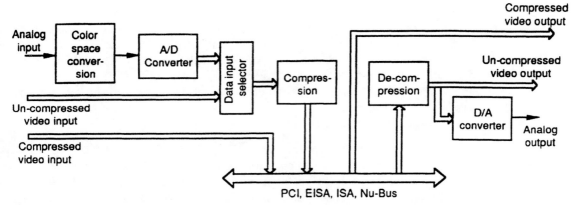

Figure 10.2   Basic implementation of a video acquisition and distribution card.

an internal crossbar switch to five data-RAM and Instruction caches. This switch also allows data transfer to the internal RISC central processing unit (CPU) and to an external memory through a transfer controller.

All processing cards exist in different versions for plugging into PCI, EISA, Nu-bus, or ISA buses. Additional over-the-top high-speed buses such as Movie-2 and VGA-FC may be added to these cards.

10.3.3   Disk and tape storage

Disk storage systems record and play back encoded data signals in a manner similar to that of digital recorders. Data storage applications, transfer speed, and duration capabilities have increased dramatically in recent years.

Performance characteristics of disk storage systems include

- Seek time. This is the average time required by a drive to move its read/write heads from one track position to another on the disk.

- Spindle speed. This is the rotational rate of the disk. High speeds reduce the latency time and improve the transfer rate.

- External data rate. This is a measure of the maximum data rate at the disk drive input/output.

- Latency. This is the amount of time it takes a disk drive to position a specific sector on the disk under the seek plane of the read/write heads. The average latency is defined as the time required for half a rotation of the disk.

- Data caching. A buffer (from 32 kB to 2 MB) is added to the disk system to store frequently used data values for quick access. Cache management provides the fastest way of reading drives and improves their transfer rates.

Disk drives necessitate a periodical thermal calibration of read/write heads to ensure their correct positioning over the proper track. This process takes a few milliseconds and removes the operational thermal effects on disk drive mechanisms.

Some interfaces, such as Fast SCSI-2, incorporate an error detection and correction scheme, guaranteeing a bit error rate of less than 1 error per 10^{13} data bits. Consequently, the major concern in disk array systems is the total failure of one of the drives.

Current hard disk drives exceed 500,000 hours mean time between failures (MTBF), but a single disk failure in an on-air system can generate long interruptions of programs and losses of program data and commercial revenues. For these reasons, the use of a single disk in broadcast applications is not recommended.

Redundant array of inexpensive disks (RAID) is a method of maintaining system operation and providing various levels of data integrity in the event of a disk drive failure. With the most sophisticated levels of RAID (level 3 and 5), a failed disk drive can be hot-swapped and the data rebuilt on the new drive while the system remains on air.

Many disk array configurations are possible, but only a few are practical for data storage from workstations and processing systems. They are shown in Figs. 10.3 and 10.4, and described below.

- RAID-0. The data for the interface is spread across all disks. Several small transfers, each on a separate disk, can take place simultaneously. This configuration provides fast data input/output but no data protection. A disk failure results in a system crash.

- RAID-1. A duplicated copy of the incoming data is stored on a separate disk. This results in very good protection against disk failure at the expense of a reduced storage capability and high acquisition costs.

- RAID-3. A extra drive is added and used only for parity-bit storage. Transferred data are spread across all data disks working in parallel, using a byte-striping technique. These disks, including the parity disk, must be synchronized to allow the recovery of the same parallel sectors on all disks and the error checking and correction if one disk fails. The disk array controller reconstructs the missing information using the remaining good data and parity information. Costs increase by 25%, but a protection has been added to the storage system while maintaining high transfer and storage capabilities. RAID-3 is mostly used in video-based applications where large sequential data files are stored.

- RAID-5. Transferred data can be read from or written on the disks either in parallel, using a block-striping technique if the data file is large, or independently of and concurrently in the case of small files. The parity information is distributed across all the drives in such a way that each disk stores parity information of data blocks recorded on the other disks.

RAID-5 has a redundancy similar to that of RAID-3, but it provides a higher data transfer capability, higher data availability, and higher lev-

RAID Level 0

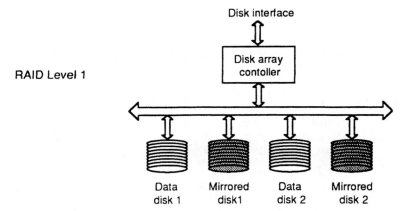

RAID Level 1

Figure 10.3 RAID levels 0 and 1 structure for disk drive implementations.

els of reliability when compared to RAID-3. RAID-3 is better suited to applications in which input/output performance is more important than transfer rate, such as in real-time video servers. When a drive fails, the performance of RAID-3 is maintained, whereas that of RAID-5 drops significantly.

- RAID-6. This level is a further development of level 5 and allows for two drives to fail either simultaneously or a second drive to fail during a reconstruction period.

- RAID-7. This is a proprietary system developed in 1991 by Storage Computer Corporation. It uses an embedded operating system (OS), independent control, and data paths, and supports multiple host interfaces.

RAID Level 3

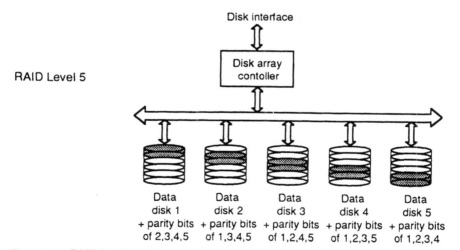

RAID Level 5

Figure 10.4 RAID levels 3 and 5 structure for disk drive implementations.

Other components of the storage system may also fail. Some systems include a hot-swapping of the power supply and redundant disk controller capability.

Digital video tape storage systems still provide the lowest media cost and the highest affordable storage duration, but they also have the highest maintenance requirements and cost for reliable operation.

An important disk storage application is the use of disk-based caching devices in multicassette systems (carts). This combination, often called hybrid storage, cumulates the advantages of both systems, such as more flexibility for last-minute changes in a playlist, reduced maintenance cost (no head wear and periodic cleaning), quick access to the material segment (no preroll), fewer

VTR transports in the cart, huge storage capability, and a 5-year warranty for most of the drives.

10.3.4   Servers

A basic server architecture consists of three parts

- A multiple hard disk drive system capable of fast and simultaneous data access, with enough capacity and redundancy for the contemplated application. A disk array controller manages data distribution and communications between all drives.

- Fast data communication interfaces between disk drives and networks. Several approaches can be used to perform fast data transfer such as very fast CPUs, multiple CPUs and buses, and routing switchers as implemented in the Tektronix Profile PDR-100 professional disk recorder. Interfaces may include data compression encoding and decoding.

- A software-based OS capable of handling multiple digital audio/video data streams in record and playback modes while ensuring correct file management and access.

Server designs and performance are trade-offs between quality of compressed signals, storage capacity, data speed, play time, number of channels, access speed, and reliability. The vertical interval data of incoming video signals inserted by the customer, such as close captioning signals (CCS), vertical interval time code (VITC), video index information, and user data line, should be preserved by stripping line contents, storing them on disks, and reinserting them in playback at the server output. Table 10.2 shows the storage space requirements for different audio and video data signals.

All video servers have in common a large storage capacity and multiple channel capability, but they also have differences in architecture and performance

TABLE 10.2   Storage Space Requirements for Audio and Video Data Signals

Media signals	Specifications	Data rate
Quality-audio voice	1 ch; 8-bit@8 kHz	64 kbps
MPEG audio Layer II	1 ch; 16-bit@48 kHz	128 kbps
MPEG audio Layer III	1 ch; 16-bit@48 kHz	64 kbps
AC-3	6 ch; 16-bit@48 kHz	384 kbps
CD	2 ch; 16-bit@44.1 kHz	1.4 Mbps
AES/EBU	2 ch; 24-bit@48 kHz	3.07 Mbps
MPEG-1 (Video)	352×288, 30 fps, 8-bit	1.5 Mbps
MPEG-2 (MP@ML)	720×576, 30 fps, 8-bit	Max. 15 Mbps
MPEG-2 (4:2:2 P@ML)	720×608, 30 fps, 8-bit	Max. 50 Mbps
CCIR-601	720×480, 30 fps, 8-bit	216 Mbps
HDTV	1920×1080, 30 fps, 8-bit	995 Mbps

because they are designed to meet specific requirements of one of the five main TV applications.

- News. Moderate quality of news materials is acceptable and video compression can be used to reduce the data file size and transfer rate. Multiple access to multiple segments must be possible and a moderate bandwidth is necessary. Guaranteed availability of output ports might be required for direct on-air programming of news materials.

- Production. This is the most demanding application for servers. A wide bandwidth must be provided to meet simultaneous transfer demands for large uncompressed video files with true real-time and random accesses. In a production facility, a server stores all compressed or uncompressed audio and video data files for use in postproduction and distribution. They typically require large storage and multiple fast data communication capabilities through networks with production systems and facilities. These servers are central to the whole production operation and thus determine the overall performance of the entire facility. Consequently, care must be taken in selecting a production server that must have the right features and specifications. Table 10.3 shows production server bandwidth requirements for 4:2:2 video signals.

- Transmission. High compression ratios can be used to broadcast programs. Resulting bit rates and number of channels are low and necessitate low bandwidth requirements.

- Video on demand (VOD). This server application must deliver a very large number of channels, each of low video quality such as MPEG-1 (1.5 Mbps). A high bandwidth may be required to satisfy all demands. Short access times are necessary for VOD whereas long access times permit near-VOD only.

- Data cache. A caching system is basically a temporary random access disk buffer. It is presently used with tape library systems in on-air applications. Caches are very well suited to commercial on-air applications where materials are repeated several times a day and last-moment changes occur fre-

TABLE 10.3   Production and Broadcast Server Bandwidth Requirements for 4:2:2 Video Signals

Production applications	Bandwidth, Mbps	Sample resolution	Compression ratio
High-end postproduction	270	10	1
Usual postproduction	90	10	2.3:1
Light postproduction	25–50	8	6.6:1–3.3:1
News (compressed data)	18–25	8	9:1–6.6:1
Future HDTV broadcast	20	8	10:1–50:1
Good-quality SDTV broadcast	8	8	20:1
Medium-quality SDTV broadcast	3	8	55:1
Low-quality SDTV broadcast	1.5	8	110:1

quently. Bandwidth and number of channel requirements are not critical. Another important application is their use in remote caching. A central main server distributes everyday commercials through networks at local distribution points where caches are used for program insertions of commercials and local programs. In the near future caches will be used at the receiving end and users will access various facility main servers via local networks. The main server could be accessed during low traffic hours. Alternately, the data transfer could be performed at an increased speed (several times the normal speed).

10.3.5   Cameras

Analog camera sources must be digitized by means of a digitizer, or frame grabber in a real-time process, for use in multimedia digital systems.

Digital cameras are appearing on the market. Examples of such cameras are Sony DVCAM DXC-D30 and DCR-VX1000, Panasonic DX1, and JVC GR-DV1. Some of the new digital cameras are equipped with a FireWire connector (IEEE-1394), which allows the direct transfer of digital audio and compressed video data as well as time code, control messages, and error correction bits. FireWire-type digital interfaces are used for interconnections with future advanced television receivers, PC workstations, and already existing digital VCRs.

Users have found that video quality is far superior to that of the S-VHS and Hi-8 formats, and comes with very high quality audio when compared to analog audio provided by these formats.

10.3.6   Videocassette Recorders (VCRs)

Analog VCRs (VHS-type) deliver analog audio and video signals, which need to be digitized before being used in a multimedia system. The playback video quality is poor because of the noise and aliasing in video signals, which might cause problems during the compression encoding session.

Digital VCRs provide high-quality playback of audio and video signals. They are based on the Digital Video (DV) format, which was developed and adopted in 1995 by a consortium of all VCR manufacturers. The DV format includes a 5:1 pseudo-MPEG-2 video compression scheme of original 4:1:1 sampled pictures and two uncompressed digital audio channels. Two audio recording modes are possible:

- A two-channel mode with a sampling resolution of 16 bits at 32, 44.1, or 48 kHz.

- A four-channel mode with a sampling resolution of 12 bits at 32 kHz, with dynamic audio compression and expansion to improve the recording and playback performance.

DV-format machine characteristics are summarized in Table 10.4.

It must be noted that the DV format allows the recording on tape of various types of information related to the recorded data content. This is useful for

TABLE 10.4   Summary of DV Recording Format Characteristics

Parameters	Specifications	
	Video	Audio
Sampling format or frequency	4:1:1 (525/60) 4:2:0 (625/50)	48 kHz
Sample resolution	8 bits	16 bits
Compression	Yes	No
Bit rates, Mbps	24.948	1.5
Data channel	1	2 (16-bit, 32–48 kHz) 4 (12-bit, 32 kHz)
Time code	Yes	
Cassette duration	1 h/4.5 h	
Magnetic tape	Metal evaporated (ME)	
Tape width	6.35 mm	
Track width	10 μm	

adding information at the time of production to keep track of camera settings and specific scenes, and ID numbers for quick search, data compression scheme used, time code, etc. Furthermore, a DV-format VCR could be used as a huge data tape storage device for archiving purposes.

These DVCRs can be equipped with FireWire interfacing connectors. Sony's cassettes include a built-in read/write 4-Mb memory chip to store the first image of all material segments, along with time and date.

10.3.7 CD-ROM and magnetooptical disks

Since their introduction in 1983, CD-ROMs are primarily used for one-time storing of digital audio materials. They can also store up to 650 MB of data information. This is generally insufficient for good-quality digital video programs.

A new format called digital video disk (DVD) has been created to answer the needs for higher data storage capability on small disks. Although their size is identical to that of CD-ROMs, these disks can contain up to 10 GB of MPEG-2 video signals. In addition, the disk player provides two-channel audio outputs as well as an optical digital audio output for connection to an external Dolby digital surround sound (AC-3) decoder. Two competitive disk formats were developed, the MultiMedia Compact Disk (MM-CD) by Sony and Philips, and the Super Density (SD) disk by Hitachi and Panasonic. Lately, the two formats have been combined to create the DVD format.

Magnetooptical disk drives offer high reliability, high media bit density, and few moving parts. The recording medium has slow access time for digital video. In addition, it is typically a write once, read many (WORM) disk, which precludes its use in audio/video postproduction applications where frequent material content changes must be made. However, they can be used for near video on demand (NVOD) server applications where random access capability and media removability are important operational factors. Furthermore, they offer large media storage (3.2 GB), long life, and reliable operation.

10.3.8 Digital Video Interactive (DVI)

DVI is a registered trade name of a multimedia technology developed by Intel Corporation. This technology includes high-speed specialized video processors and software that are available from this company to manufacturers of multimedia computers and to developers of multimedia applications.

10.4 Multimedia Interconnections

Existing data transmission networks are designed around packetized data streams that are switched by packets. New multimedia applications have imposed an evolution of networking requirements to ensure a quality level and a variety of services to customers, broadcasters, and the production industry. Video networks require continuous data streams (real-time) where switching must occur during the vertical blanking interval. The most important multimedia network requirements are

- Bandwidth. Bandwidth depends on the image resolution, the frame rate, and expected picture quality. In addition, faster-than-normal speed capability becomes an important factor in news, file transfer, and dubbing applications. Maximum network bandwidth includes the payload and the file system structure, so the actual throughput is lower and depends on the network standard.

- Error-free communications. Data transmission protocols generally include a means of error detection and correction as they occur. Specialized video networks rely on error concealment and forward error correction (FEC).

- Transmission delay. For interactive multimedia applications, the transmission delay must be considered. Face-to-face conversations, such as in teleconferencing, require that the one-way delay be less than 150 ms.

- Interoperability. This is a critical requirement and guarantees data transfers between multimedia devices through various networks and interfaces.

- Ability to deliver uninterrupted video streams.

- Ability to switch (or remultiplex) from one compressed stream to another in a seamless fashion.

- Cascading capability of compression and decompression schemes. As discussed in Chap. 8, the cascading of different compression schemes may reveal incompatibilities and generate visible artifacts.

- Channel synchronization. Audio, video, and corresponding data must be synchronized at the receiving end, especially when different routes are used for each data file. The MPEG standard provides specific synchronization data to this end, as well as buffer size requirements.

Three modes of data transmission are used depending on the data file content and multimedia application requirements.

- Synchronous. Data files or packets recur exactly with the same period. Thus, transmission delays are defined for each data file or packets.

- Asynchronous. Data files or packets are sent when the network is available for traffic. Transmission delays are variable with this type of network. Data integrity is the most important factor, and erroneous data packets can be resent.

- Isochronous. Data files or packets recur exactly with the same period and with an equal duration over time. The transmission delays are thus constant for each data file. The allocated packet transmission time is fixed and data correction must be applied in the presence of transmission errors. A good example of an isochronous data file is the transmission of digital video data, where packets must be sent continuously with the same delay to avoid intermittent video data delay within pictures. Consequently, the real-time transmission of audio and video data over networks requires an isochronous transport with a guaranteed bandwidth and latency. Furthermore, isochronous data transmission requires less header information in each data packet combined with more data payload, maximizing the network data throughput.

There are basically two types of physical interconnection situations in data transmission systems:

- Between two or more multimedia devices in the same room. In this case, communication is made through an interface.

- Between multimedia workstations in the same building or in different buildings. Communication is made through a network.

Several standard networks (such as Ethernet, FDDI, ATM) have been developed for small-message communication rather than for the large digital files that are common in video applications.

10.4.1 Interfaces

A multimedia system with all components located in the same room can be interconnected through interfaces such as Fiber Channel and Ethernet. However, several interfacing standards may be used in building area networks (BANs) and local area networks (LANs) by means of digital-to-optical converters that allow the removal of some limitations of the original standard characteristics. The most used interfacing standards are described in the following sections.

10.4.1.1   FireWire.   Originally developed and named by Apple Computer, this interfacing format has been standardized as the IEEE 1394-1995 High-Speed Serial Bus (also called P1394). It is intended for connecting digital devices through a short-length cable system, and for interconnecting different parts of a device such as in backplane environments.

Up to 63 devices can be daisy-chained on a single bus with a maximum cable length of 4.5 m between them. High-quality cables can increase this distance to 14 m. The FireWire standard specifications allow for different buses to be bridged together and thousands of devices to be interconnected. Node* IDs are automatically assigned when devices are added or removed from the bus. During transmission, a fairness protocol for arbitration is used to ensure that only one node has access to the bus. Data transfers are made through isochronous channels that are automatically created on the bus with a guaranteed bandwidth and latency. One device may require more than one virtual channel such as one input and two different outputs. Each device is assigned a bus bandwidth, which depends on its bit-rate requirement. A summary of the IEEE 1394 format characteristics is shown in Table 10.5.

The IEEE 1394 interface is aimed at local multimedia device interconnections such as PCs, hard disk drives, cameras, VCRs, printers, scanners, CD-ROMs, cable boxes, ATV receivers, and in industrial use for control, laboratory automation, and measurements on manufactured products. An IEEE 1394-to-PCI interface has been developed by Adaptec Inc. to facilitate the implementation of FireWire connectors and processing capability into computers. Cable and connectors are represented in Fig. 10.5.

---

*A node is a connected bus or network device.

TABLE 10.5   **Summary of FireWire (IEEE 1394) High-Speed Serial Bus Characteristics**

Parameters	Specifications
Maximum bit rate	400 Mbps (50 MBps)
Maximum cable length	4.5 m (low-cost cable) 10 m (thick cable) 14 m (bit rates <200 Mbps)
Cable type	Six conductors (two twisted pairs for signals, two wires for power)
Data transfer mode	Isochronous
Maximum number of virtual channels	63

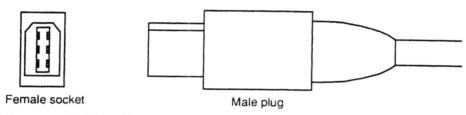

Female socket   Male plug

Figure 10.5 FireWire cable and interconnections.

This interface includes a DV chip set to remotely control a camcorder from a PC using the embedded time code, and compress/decompress digital video data to and from 24-bit RGB data, for computer processing.

10.4.1.2 Serial Storage Architecture (SSA). SSA was originally developed by IBM as a high-speed interface to storage devices. It is now defined in the ANSI X3T10.1 standard.

Several multimedia devices such as media processors and disk storage can be connected in a loop (up to 128 devices) to establish a full-duplex data communication in real time or faster-than-real time. SSA allows multiple simultaneous reads and writes, with a maximum speed of 80 MBps. Each node has two ports and the passthrough delay is about 5 to 10 bytes, which means 0.5 μs for the data to get through. Each link has error detection and recovery capability. Figure 10.6 shows an SSA loop combining two initiators and several disk storage devices. In case of failure in the loop, the initiator can select the correct path to transfer data to a target. Two or more initiators can access the loop simultaneously, provided they communicate with their adjacent targets. Figure 10.7 shows the functional architecture of a target.

10.4.1.3 Fiber Channel–Arbitrated Loop (FC-AL). Fiber Channel has been standardized in an ANSI X3.230-1994 document. It allows high-speed data transfers between multimedia devices and can be implemented by means of twisted

Figure 10.6 SSA loop topology.

Figure 10.7 Target architecture in SSA standard networks.

pairs, coaxial cable, and fiber circuits. Among multiple possible topologies, studio devices can be connected to each other using FC switchers or in a loop, as shown in Figs. 10.8 and 10.9. The loop configuration is less expensive (no data switcher required) and has several advantages.

Fiber Channel provides good interoperability because it is protocol-independent. In studio applications, compressed and uncompressed data can be transferred from servers to workstation or vice versa, at normal speed or faster-than-normal speed.

Devices communicate to each other to determine, through an arbitration process, which one can transmit data on the bus at a given time. The receiving

Figure 10.8 Fiber Channel-Arbitrated Loop and switcher topologies.

Figure 10.9 Implementation of Fiber Channel-Arbitrated Loop with a hub.

device then controls the rate of the data flow by indicating the amount of available buffer memory. Table 10.6 summarizes FC characteristics.

10.4.1.4 Small Computer System Interface (SCSI). This interface was designed in 1986 as a parallel bus to enable quick data transfer (up to 5 MBps) between Mac computers and disk drives, image scanners, CD-ROM drives, laser printers, and tape backup devices. This SCSI-1 standard is documented in ANSI X3.131-1986.

Up to seven devices can be daisy-chained using 50-pin Centronics connectors. The first and the last device in the chain need a termination to reduce electrical reflections and interference with transmission data. Each device has an address number for bus access priority assignment (high number = high priority), and correct data transfer between the computer and the selected peripheral in the chain. Cable length must be as short as possible and below 2 m between devices, with a total chain length below 6 m.

An improved version, the SCSI-2, has three levels:

- The Fast SCSI, which doubles the clock speed and thus the data rate to 10 MBps. A new high-density 50-pin (HD-50 or micro-D) connector is required.

- The Wide SCSI, which expands the number of parallel data lines from 8 to 16 or 32 and increases the transfer rate to 20 MBps.

TABLE 10.6   **Summary of Fiber Channel Specifications**

Parameters	Specifications
Maximum line bit rate	From 133–1062 Mbps (Future: 2 and 4 Gbps)
Maximum data throughput	450–800 Mbps
Frame size	2112-byte payload
Maximum cable length	10 m (miniature coax) >30 m (video coaxial) 500 m (50-μm multimode fiber) 10,000 m (9-μm single-mode fiber)
Supported protocols	SCSI, ATM, IPI, HIPPI, TCP/IP, Ethernet
Maximum number of devices	126

- The Ultra SCSI, which combines the two previous levels and allows a maximum transfer rate of 40 MBps. However, cable quality and length requirements are critical and existing solutions lead to either very short cable lengths or expensive and incompatible implementations.

A recent version, the SCSI-3, uses a very high speed serial transmission data transfer architecture. This architecture removes the distance limitations (no crosstalk between data cable pairs) and increases the number of connected devices. In addition, the maximum transfer rate can range from 12.5 MBps, using a FireWire format protocol, to 100 MBps using a Fiber Channel format protocol. SCSI-3 uses a layered approach and supports existing SCSI protocols as well as several interfacing protocols such as

- FireWire, which is primarily intended for multimedia device interconnections.

- SSA, which as its name implies, was designed for high-speed data transfer to storage systems.

- Fiber Channel, which is suited for network interconnections of digital devices.

This layered approach allows the manufacturer to implement the most appropriate bus for the specific multimedia applications, be they networking, storage, or easy and inexpensive multimedia device interconnections.

A summary of SCSI specifications is shown in Table 10.7.

10.4.1.5 Ethernet.   Ethernet is a LAN standard originally developed by Xerox in 1973 to interconnect computers. It has been later standardized in an IEEE 802.3 document.

The Ethernet cabling system, or topology, conforms to a linear bus consisting of a thick or thin coaxial cable segment with a 50-ohm termination resistor at each end and a ground at one end. This bus is limited in distance and in the number of devices that can be attached to it.

TABLE 10.7   SCSI Standard Specifications and Attributes

Features	SCSI-1	SCSI-2 Fast	SCSI-2 Wide	SCSI-2 Ultra	SCSI-3		
					FireWire	SSA	FC-AL
Maximum data rate, Mbps	1.5–5	10	20	40	12.5–80	80	100
Maximum cable length, m	<6 <25*	<6 <25*	<6 <25*	<6 <25*	4.5	20/1000	Up to 10,000
Maximum number of devices	7	8	8	16	63	127	126
Self-configuring	No	No	No	No	Yes	Yes	Yes
Self-termination	No	No	No	No	Yes	Yes	Yes
Hot-pluggable	No	No	No	No	Yes	Yes	Yes
Connector type	50-pin Centronics	50-pin Micro-D	68-pin	80-pin	6-pin Game-type	9-pin	Optical Coaxial
Cable type	Ribbon, 50 wires	Ribbon, 50 wires	Ribbon, 2×50 wires	Ribbon, 2×50 wires	6 wires	4/6 wires	Optical Coaxial
Minimum number of ports	2	2	2	2	2	2	2

*With differential drivers.

A less-expensive cabling alternative is made of twisted-pair wiring (UTP cables) in a star topology, as shown in Fig. 10.10. A concentrator or multiport repeater is used as a central component to allow multiple workstations to be attached to a network. Maximum data transfer is 10 Mbps for standard Ethernet and 100 Mbps for Fast Ethernet. However, the maximum data throughput is only 50% of the maximum bandwidth.

Ethernet is a shared bus, because any device connected to it can use the same link, provided that specific rules are followed. A device, or node, can access the link if no other devices are transmitting. If two nodes are transmitting simultaneously, a collision is detected and both devices stop for a random pause before making a second attempt. This process is called carrier sense multiple access with collision detection (CSMA-CD). The probability of collision increases with the number of connected devices.

A solution to large multimedia systems is to separate devices in groups and use bridges or routers for intercommunications between these groups. Only data files destined to be received by devices connected to another group go across the bridge or router, which reduces the network occupancy. This bus topology is shown in Fig. 10.11.

There is no acknowledgment of messages at this level of network communications. Each message (also called frames) carries addressing information to identify its source and destination as shown in Fig. 10.12. It must be noted

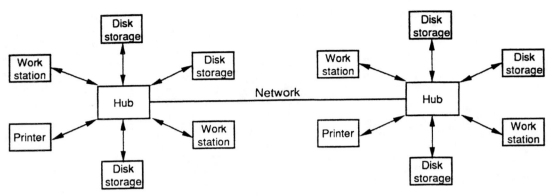

Figure 10.10 Ethernet Star topology, using a hub to attach the network.

Figure 10.11 Ethernet Bus topology, using a bridge for inter-network communications.

Figure 10.12 Ethernet frame structure.

that there is a difference in the frame definition between IEEE 802.3 and Ethernet. The latter uses a type field, whereas the former uses a length field. The Ethernet standard is far more widely used than the IEEE standard. Confusion can be avoided by assigning a number greater than the maximum length (>1500 bytes) to all type field values.

A repeater can be used to connect network segments and allows extension of network distance by regenerating electrical signals passing through it. In designing a system, a repeater must be considered as one device attached to

TABLE 10.8   Summary of Ethernet Cabling Specifications and Performance

Parameters	Thin cable, 10 Base 2	Thick cable, 10 Base 5	UTP, 10 Base T
Maximum number of devices per unrepeated network segment	30	100	30
Maximum cable length per unrepeated network segment	185 m	500 m	100–150 m
Minimum cable length between devices	0.5 m	2.5 m	No min.
Maximum number of segments	5	5	No max.
Connector type	BNC	15-pin AUI	RJ-45
Cable type	RG-58c/u	Belden 9898	UTP

the network. A maximum of two repeaters can be used, in order to limit the accumulated propagation delays generated by each repeater.

A transceiver acts as an isolated interface to the Ethernet link. It is generally included in the Ethernet interfacing board inside the host computer or peripheral device. However, an external box may be required if a device is not compatible with the Ethernet link.

The attachment unit interface (AUI) is defined as the connection between a device and a transceiver. In thick-cable cabling, it also detects data collision on the network. The AUI may be directly connected to the Ethernet interface in the computer via a 15-pin connector. Each device is assigned, by the manufacturer, a unique hexadecimal number (address) for its identification on the network.

Table 10.8 summarizes Ethernet specifications for the three most used cabling implementations, thick (also called 10 Base 5 or Backbone Ethernet), thin (also called 10 Base 2 or Cheapernet) and twisted pairs (also called 10 Base T) cablings. In these names, 10 means a maximum bit rate of 10 Mbps, Base means a baseband-type network, and the last character identifies the cabling. Figure 10.13 shows a basic example of a thin Ethernet coaxial cabling.

Ethernet is not suitable for multimedia distribution applications because the latency is not fixed. It is mostly used in BANs for control, management, and local transfer of non-real-time video and real-time audio files between workstations and storage systems. Audio files can also be transmitted over wide area networks (WANs).

10.4.2 Networks

Multimedia systems located in different buildings or cities can be interconnected through networks, such as LANs or WANs. The most-used networks are described in the following sections.

A baseband network provides a single data transmission channel that occupies the total network bandwidth. Thus, only one device can transmit at a time.

A broadband network provides several different virtual channels by means of frequency-division multiplexing of the total bandwidth.

Figure 10.13   Basic example of Thin Ethernet coaxial cabling.

10.4.2.1 Asymmetric digital subscriber line (ADSL).   To take advantage of the many existing twisted-copper-pair telephone lines connected to homes, telephone company centers (telcos) are standardizing a new modulation and transmission technique to offer digital data services at high-bit rates (up to 6.176 Mbps). This technique provides a good interim solution before the major telco trunks change from copper to fiber optic cables.

The transmission channel bandwidth can be organized in multiple channels of 1.544 Mbps (1.536 Mbps in Europe) with a maximum of four channels or 2.058 Mbps (maximum of 3 Mbps), allowing the simultaneous delivery of several broadband services, among which are digital television, computer application software, fax delivery, video games, home shopping, and VOD. Figure 10.14 shows a functional block diagram of the ADSL distribution to customers.

The ADSL transmission format (ANSI standard T1.E1.4/94) allows the multiplexing of channels that can combine a forward data transmission of several digital services with a high capacity (up to 6.176 Mbps) and a return data transmission of control and user commands with a low capacity (9.6, 16 or 64 kbps), as shown in Fig. 10.15. This asymmetric data transmission frequency content reduces crosstalk between forward and return channels.

ADSL uses a new Discrete Multitone (DMT) modulation technique that divides the channel spectrum into a large number of subchannels. Each subchannel behaves as a Quadrature Amplitude Modulation (QAM) transmitter, using a single carrier to modulate groups of data bits. A variant of QAM has been developed by AT&T, called Carrierless Amplitude/Phase Modulation (CAPM).

In a noisy environment, a maximum transmission distance of 4 km can be achieved, at a bit rate of 2 Mbps, with a regular 0.5-mm copper cable. When the bit rate increases, this maximum distance decreases.

Figure 10.14 Block diagram of ADSL transmission and interconnection.

Figure 10.15 Distribution of ADSL transmission channel and existing telephone signal spectra.

As fiber optic links replace major telco trunks, the remaining short distance to home can deliver high-speed data services (between 10 and 50 Mbps) using Very High Speed Digital Subscriber Line (VDSL) technology.

10.4.2.2 Integrated Services Data Network (ISDN). ISDN is a public-switched digital circuit using standard unshielded twisted pairs (UTPs) of copper telephone wires. There are two ISDN services to help in solving communication bandwidth requirements:

- The Basic Rate Interface (BRI) line, which includes two B (bearer) synchronous 56-kbps channels and one D (dialing) 16-kbps channel for call signaling. Thus, a basic ISDN connection adds up to 128 kbps. B-channels can be used either as a bidirectional 56-kbps data transmission or as a unidirectional 128-kbps channel. Multiple BRI circuits can be grouped to allow high-quality transmission of digital audio signals such as AC-3 @ 384 kbps. BRI is aimed at home and small-company users.

- The Primary Rate Interface (PRI) groups 23 channels in North America and 30 B-channels in Europe and provides a maximum bandwidth of 1.536 and 1.920 Mbps, respectively. An additional 64-kbps D-channel is reserved for signaling. This bandwidth capability can be used for videoconferencing, MPEG-1 video/audio transmission (1.544 Mbps or T1 line), and Internet connections. However, each B-channel may have different path delays and require channel buffering at the receiving end. The ISDN-H11 service guarantees an identical delay on all channels, but at increased cost. PRI is aimed at business users. Figure 10.16 shows two examples of interconnections to ISDN networks.

ISDN services are based on the splitting of the data stream into a number of synchronous digital phone lines with bandwidth capabilities from 64 to 1920 kbps. Interactive wide-area multimedia communications can be established to perform isochronous data transfer.

The Broadband-ISDN (B-ISDN) is an emerging multimedia network that will provide a wide range of combined services such as voice, data, and video. Future B-ISDN networks will use a protocol based on ATM technology.

10.4.2.3 Asynchronous Transfer Mode (ATM). ATM was originally designed for long-distance, high-speed transmission of data from multiple users at differing data rates and quality of service characteristics. It is a packet-switching protocol, not a transmission system, which routes data packets between nodes. Although the transmission is asynchronous, ATM can optionally carry isochronous data files.

A physical layer based on the synchronous data hierarchy (SDH) is used to carry the ATM cells into SDH containers. SDH is a telephone hierarchy of standards that allows synchronous multiplexing of data streams on high-speed data interconnections. The synchronous optical network (SONET) is an optical implementation of SDH. Data are packaged in small, fix-sized, 53-byte packets, of which 48 bytes constitute the actual payload and 5 bytes are reserved for packet header information. Data packets are given a destination header or address and can be easily mixed with various application packets and added or dropped when required at a device location.

ATM uses a switched-base topology with variable bandwidth allocation. It

a) Example #1: A two-way analog audio communication channel using MEPG Layer III encoding and ISDN network

b) Example #2: A two-way digital surround sound communication channel using AC-3I encoding and ISDN network

Figure 10.16 Block-diagram of two typical interconnections to ISDN networks.

can provide switched and permanent circuits to meet requirement of broad-casters and telecommunication companies. ATM data cells can be transmitted over fiber optic, twisted-pair, or coaxial cables. Table 10.9 summarizes ATM characteristics.

It must be noted that the 188-byte MPEG-2 transport stream packets can be easily converted to ATM cells through an ATM adaptation layer (AAL). This capability has been seriously considered in the advanced television standardization process.

10.4.2.4 Fiber Distributed Data Interface (FDDI). FDDI is suited to high-speed digital networks thanks to its wide bandwidth, low signal attenuation over long links, and insensitivity to interference. Multimedia devices are interconnected in a ring configuration (also called loop) running at 100 Mbps, with future capability of 1 Gbps. A data token circulates from device to device, enabling them to transmit their data at the time of the token reception and to transmit the token onto the network. The next device may either pass the received data and add its own data if necessary, or remove the data from the network. Thus, data can be

TABLE 10.9   **Summary of ATM Specifications**

Parameters	Specifications
Maximum bit rate	From 25–155 Mbps (future: 622 Mbps and up to 2.4 Gbps)
Maximum cable length between ATM-based devices	With fiber: unlimited With copper: few hundred feet
Data transfer mode	Burst, CBR, isochronous
Physical transport mode	Synchronous, (SDH/SONET)
Maximum number of devices	Unlimited
Network types	LAN and WAN

transmitted to any device connected in the ring and, after reception, any other device in possession of the token can initiate another data transfer.

FDDI is not widely implemented at this time because of its complexity and compatibility with other protocols; 100-Mbps FDDI has an actual throughput of 30 Mbps and a maximum transfer distance of 200 km of asynchronous, synchronous, and dedicated data with bandwidth allocation capability.

10.5   Multimedia Software

Since 1991, Macintosh and Windows-based computer operating systems have included a multimedia capability, allowing the development of platform-dependent multimedia application software. The market for multimedia applications is rapidly developing because of technology advances and specification standardization around both computer platforms. User-friendly multimedia software facilitates the development of multimedia products, thus reducing production time and costs.

A great deal of multimedia development software exists on the market. These object-oriented and dynamic tools are

- Multimedia ToolBox. This Windows software allows the user to develop multimedia applications thanks to a graphical user interface (GUI), an object-oriented programming language (Openscript), and various tools for editing, creation, and animation.

- QuickTime. This Apple software is an extension of the Macintosh operating system and combines dynamic video and audio to text and graphics documents. It is not an application software. QuickTime performs many multimedia functions such as compression, editing, storage, and display of audio and video files. A version of QuickTime can operate on both Macintosh and Windows platforms using the same data files.

- Authoring software. Authoring software enables the application developer to create, build, and prepare a multimedia production, which will be later recorded on CD-ROM or other media supports, or sent through networks. A variety of authoring software is available and adapted to today's computer platforms. The developer can create interactive productions that have a wide range of applications in education, technical training, information kiosks, product advertising, etc.

■ Animation software. Sophisticated animation software provides high-end animation functions, which include three-dimensional drawing, rotating and scrolling titles, and special video effects similar to those encountered in specialized video effect processors.

10.6 Multimedia Systems and Applications

Multimedia systems combine all possible forms of media supports, such as sound, pictures, graphics, text, full-motion video, and animation, into a wide range of multimedia applications. Such applications and services are possible thanks to the convergence of broadcasting, computers, and telecommunications. Multimedia workstations compute and process in real time continuous synchronized presentations of media supports that can be transmitted and delivered to the customer. Broadcast and delivery systems include satellite, TV cable, TV transmitters, CD-ROMs, and telephone lines.

Today, more and more customers are equipped with a television receiver and a personal computer and can benefit from these new multimedia services. Until recently, the whole family used to watch the same program following a consensus or an arbitrary decision. In the near future, family members will simultaneously access the service of their choice, be it games, home shopping, TV programs, VOD, etc. Multiple display terminals will be required to meet all the needs.

Among new services, interactive multimedia is developing rapidly. Interactivity brings another dimension to the usually passive way of watching TV, namely the possibility of reacting, or choosing, the outcomes of events being displayed.

Interactive multimedia technology provides tools and commands for users, giving them the ability to influence the service they are using by controlling the information flow as well as the direction of the media. Customers can even create their own multimedia documents using personal computers and suitable multimedia services.

Interactive multimedia workstations and components are available, as described in Sec. 10.3, and authoring software is getting simpler and friendlier. Interactive users receive services from a program channel and send their requests through a return channel, which can have a narrow bandwidth for VOD, home shopping, etc., but must have a wider bandwidth for interactive games, videoconferencing, etc.

Multimedia services are developing rapidly because they offer efficient means of training, advertising, providing information, etc., as described in the three following examples:

- Interactive video training is a more effective method than using only text and graphics. The interactivity forces the student to concentrate on the subject, which can be replayed in case of difficulty.

- Interactive CD-based information and promotional kiosks provide a less expensive solution than PC-based systems. These kiosks are spreading in stores, commercial centers, airports, and tourist places. The media support is robust and operation-free. Interactivity allows customers to select the product they need and learn how to use it from an experienced demonstrator.

- The present Internet bandwidth is not practical for downloading 1.4-Mbps video from a Web site with a 28-kbps modem. Future network improvements will change this situation. CDs are an alternative to the Internet because they can be easily shipped with hundreds of pages of text, graphics, video, and audio materials.

Several important multimedia services are described in the following sections.

10.6.1 Video on demand (VOD)

VOD is a new entertainment and programming distribution service that allows users to order programming from a centralized location via phone lines or network connections, for viewing on TV receivers at the time of their choice. Additional functions are included to simulate VCR-like control capabilities such as pause, fast forward, and rewind.

Such a system is designed around a VOD server capable of multiple real-time access of the same or different program materials and distributing them through program channels. A return channel is needed by the user to communicate his or her program selection requests and additionally for polling responses, ordering from shopping catalogs, program browsing, etc.

Different solutions can be used to interconnect VOD servers to customers.

- Telephone line with ADSL technology. This is a point-to-point delivery system and only one subscriber receives the program. Program encryption, which consists of a program-scrambling process to limit the reception to the requestor, is not necessary. ADSL provides a low-bandwidth return channel for subscriber requests.

- Conventional TV cable. This is a typical point-to-multipoint distribution system. Encryption is required, and authorized subscribers can access the program. Most of the TV cable systems are still unidirectional, and the program selection return channel can be the telephone line and, when possible, the TV cable.

- Satellite. This distribution system has identical characteristics to that of conventional TV cable systems.

VOD services include movie programming, interactive games, educational programming, media information, software distribution to computers, and home shopping.

10.6.2   Near video on demand (NVOD)

NVOD is the transmission of a program, such as a movie, at frequent intervals, thus minimizing the wait time. The availability simulates VOD, although true VCR functionality, such as pause, fast forward, and rewind, is not available. A pseudo-pause may be offered by allowing viewers to switch to the next-closest scheduled playing time if they wish to pause their viewing.

10.6.3   PhotoCD

This technique has been developed by Kodak and Philips to digitize standard 35-mm color negative photos and record them on a writable CD-ROM. Over 100 high-quality images can be recorded on a disk, and then randomly accessed and displayed on a customer television receiver.

The high-quality digitization of negatives is carried out by a photoprocessor that can perform color balance correction, picture cropping, zooming, titling, etc. The PhotoCD system can record a hierarchy of five spatial resolutions of each digitized picture (pyramidal coding). The highest resolution, 3000×2000 pixels, has been selected to make possible good and enlarged hard copies. The three lowest resolutions, from 768×512 to 192×128 pixels, use a 4:2:0 structure format and no data compression to store pictures. The 768×512 resolution provides TV-quality pictures, whereas 192×128 resolution is devoted to picture browsing.

The 1500×1000 resolution is suitable for HDTV displays. The compression encoding is made on the difference between this resolution format and the interpolated 768×512 format. Quantization and run-length coding are used to perform a mild compression of about 2.5:1.

A similar process is realized for the coding of the 3000×2000 resolution format using a higher data compression ratio. This pyramidal coding of all picture resolutions allows a fast transfer to the computer and TV receiver of the selected picture quality. PhotoCD images can be played on a PhotoCD player. They are also compatible with CD-I video systems.

10.6.4   Compact Disk Interactive (CD-I)

This is a new multimedia system standard developed by Philips. It allows the recording of full-motion pictures at 1.2 Mbps and audio at 200 kbps on CDs. Applications may combine audio, video, text, graphics, animation, and interactivity. CD-I is intended for home use, training, publishing, interactive kiosks, etc.

MPEG audio and video compression schemes are used to process a maximum picture area of 352 pixels×240 lines (for 60-Hz system) or 352 pixels×288 lines (for 50-Hz systems). However, the picture area can be modified as long as the total number of pixels in the picture remains equal to 84,840 (for 60-Hz systems).

The CD-I system was designed as a consumer electronic product. A low-cost internal computer provides the processing power for signal encoding and decoding, and the production tools required to produce CD-I applications. Full-motion CD-I disks have a capability of playing 72 minutes of good-quality audiovisual material directly on TV receivers.

10.6.5   Computer-Telephony Integration (CTI)

This technology can provide new and effective ways of doing business in the home and office through existing telephone and future ATM networks. Typical applications include multimedia teleworking, telemedicine, VOD, long-distance learning from universities, and multimedia information kiosks.

10.7   Multimedia Standardization Activities

Several groups participate actively in the development of multimedia standards to ensure a certain commonality and interoperability between systems for all multimedia applications. The most important groups are

- Interactive Multimedia Association (IMA). This group of computer industry manufacturers develops multimedia service and format definitions for recommended use throughout the industry.

- Digital Audio Visual Council (DAVIC). This association of over 120 manufacturers works toward achieving a coherent introduction of multimedia applications for interactive broadcasting and services to the home. It also addresses the problems of end-to-end interoperability of interactive communication systems at the production, distribution, and receiving levels.

- Video Electronics Standards Association (VESA). VESA is in the process of standardizing an enhanced video connector (EVC), which will connect a PC to a monitor. It will allow the use of the standard analog RGB signals and a wide array of multimedia signals such as audio and composite video. There is also an option for a digital serial bus such as the P1394 to allow for a bidirectional digital interface between computers and multimedia components (storage, cameras, etc.).

- Multimedia and Hypermedia Information Coding Experts Group (MHEG). This group concentrates on a multimedia document architecture that would be common to all major computer platforms. Existing JPEG and MPEG standards define interchange formats for components of multimedia applications, such as audio and video. They neither define interactions between these components nor the multimedia document structure that should be used for storing, exchanging, and publishing multimedia presentations. The mandate of MHEG is to define and standardize a universal and interoperable multimedia document structure as a Hypertext tool where all components would be internally cross-referenced in order to enable users to navigate from one topic to another.

Advanced Television (ATV) Concepts

The North American Standard Television System (SDTV), better known as NTSC, will in the near future be replaced by a digital television system (DTV) based on the Advanced Television System Committee (ATSC) standard. The term *high-definition television* (HDTV) was introduced several years ago to define a high-quality analog production standard (described in the SMPTE 240M document) and later its digital representation (SMPTE 260M document).

The DTV transmission standard uses digital processing and compression to achieve a simultaneous transmission of several different programs and/or a reception of the same single program at a picture quality level depending on the receiver complexity. In any event the received quality is equivalent to that present at the studio output. It represents a dramatic change for the production and the broadcast industry as well as the users. New technologies have recently brought tremendous flexibility in the use of different picture formats using digital compression systems. Moreover, because of the digital nature of the picture information in this new television system and the emergence of powerful and high-speed processors, the computer industry is directing its main business to the television world. These converging technologies are modifying completely the TV world.

In North America, this new television system has been standardized by the ATSC. In Europe, a digital video broadcasting (DVB) consortium has defined several DVB standards.

11.1 Why the Industry Is Moving to DTV

Although TV programs are aired in the NTSC composite format, many programs are created in an analog component format and converted to the digital domain using 4:2:2 serial digital equipment working at 525 lines per second. Pictures look wonderful in production studios, and postproduction processes such as chroma keying are quite simple. But NTSC encoding and subsequent transmission processes degrade the studio-quality picture delivered to the home.

The NTSC system was developed as a transmission system and was subsequently introduced in television studios for program production. Analog NTSC and associated audio signals have noted deficiencies. In studio operations, chroma keying is rendered difficult by the reduced chrominance signal bandwidth of NTSC signals, and cross-color and chrominance subcarrier crawling effects are clearly visible. Ghosts, interference, and noise affect home reception.

Since its origin, the TV production world has had to deal with varying customer needs, quality requirements, budgets, and multigeneration performance criteria. Any TV production work is tied to job-specific requirements. In designing the new DTV system, multiple scanning formats were considered, aiming at ensuring the same flexibility and providing an optimal integration in the existing user-demand market. Different formats and standards will emerge from different market segments and video applications. No single system will be optimum for all applications.

With more and more TV programs being produced in the digital domain, there is an obvious desire to broadcast these programs directly to the viewer's home with their original quality. As these original pictures involve high bit rates (270 Mbps), various degrees of video compression are being used to produce different picture quality levels in line with the intended applications (news, video on demand, high-quality programs, dramas, etc.).

The digital production and transmission of television programs will have an impact on the following:

- The video and audio performance—in terms of better quality pictures and 16:9 aspect ratio, with a user-selectable choice of picture quality levels, and CD-quality sound system. This performance is the result of the generalization of digital image production and better signal distribution system integrity with the use of error correction.

- The introduction of new auxiliary data services which are embedded into each program data stream for transmission to the home.

- Operational costs due to improved performance stability and reliability afforded by digital systems. Additional cost savings are possible by using automatic monitoring and user-friendly graphical interfaces.

- Program distribution costs, as the present business of satellite, cable, and transmitter program delivery will be extended to the wide telecommunication networks market. The addition of auxiliary data channel services will make this digital broadcasting more attractive for new home business applications.

- Equipment costs, as digital television becomes a common market for television, computer, and telecommunication product manufacturers. In addition, numerous multimedia applications will extend the production program market.

- The use of the RF spectrum as digital transmitters require less power for equal coverage, resulting in reduced interference.

A comparison between present analog NTSC and future digital ATSC broadcasting system attributes is shown in Table 11.1.

The digital transmission technology also provides new business opportunities for broadcasters, thanks to the addition of interactive capabilities at the receiving end. Interactivity can be used for educational services, electronic information services, advertising, banking, audience statistics, video on demand, games, etc. The return channel can take various forms such as cable network, fiber channel, telephone lines, terrestrial links, and satellite. Each has a different performance capability and cost, and each must be selected to meet the requirements of the considered interactive services.

11.2   Standardization Efforts toward a Single Standard

Efforts to achieve a single worldwide production standard began in 1982. The CCIR decided that a production standard should be agreed upon as a first step and that transmission standards should be studied later. These conflicting factors were to be considered:

TABLE 11.1   Comparison between Present Analog NTSC and Future Digital ATSC System Attributes

Features	NTSC	ATSC
Active picture, pixels × lines	525 lines (~720 × 483)	From 1920 × 1080 to 640 pixels × 480 lines
Picture rates, fps	29.97	60, 30, 24 59.94, 29.97, 23.98
Framing	Interlaced	Interlaced and progressive
Image aspect ratio (IAR)	4:3	4:3–16:9
Pixel aspect ratio (PAR)	0.99	0.9–1
Audio transmission	Analog FM, mono/ stereo 80Hz–15 kHz	Digital surround 5+1 20Hz–20 kHz
Interoperability with computers	Difficult	Possible
Scalability	Difficult	Possible
Data compression	None	Yes
Auxiliary data capability	None	Yes
Transmission impairments	High	Very low

- The strong European Community interest in a 50-Hz field rate while NTSC countries insisted on a field rate of not less than 59.94 Hz

- Ongoing research and development activities concerning transmission and alternative production standards

- Opinions that production and transmission standards should be coherent

To further complicate the single worldwide standardization process, two different HDTV specifications for production of programs were already documented in CCIR Report 801, as shown in Table 11.2:

- The 1125/60 standard supported by the United States, Canada, and Japan

- The 1250/50 standard established by the European countries

In the absence of a common standard, efforts were deployed toward achieving parameter commonality where possible, as follows:

- A common data rate (CDR) as specified in the CCIR-601* Recommendation based on a common luminance sampling frequency of 13.5 MHz and an identical number of 720 samples per active line. This approach was extended to cover HDTV signals by doubling the spatial resolution of CCIR-601 and adapting to a 16:9 aspect ratio. This resulted in $720 \times 2 \times 16/9 \times 3/4 = 1920$ samples per active line. Examples of CDR scanning parameters are shown in Table 11.3. A CDR disadvantage concerns the need for a two- or three-dimensional interpolation in systems conversions.

- A common image format (CIF) which specifies spatial characteristics for the active picture area systems. Common elements include the image aspect ratio, the number of samples per active line, and the number of active lines. But the total number of lines, sampling rate, and picture rate may vary. The CIF concept is shown in Fig. 11.1. TV frame formats, from different image scanning dimensions, are made of a common CIF and different inactive horizontal and vertical blanking intervals. As an example, this CIF can be repeated at a rate of 30 fps (frames per second) in the 525 scanning standard and 25 fps in the 625 scanning standard. This approach is commonly used in image compression and computer processing. A CIF advantage is the simple temporal conversion between two

---

*The CCIR-601 standard is now known as the ITU.R BT-601.

TABLE 11.2 CCIR Report 801 HDTV Systems

	1125/60 System	1250/50 System
Total number of lines per frame	1125	1250
Number of active lines per frame	1035	1152
Interlace ratio	2:1	Progressive 1:1
Aspect ratio	16:9	16:9
Field frequency (Hz)	60	50
Line frequency (Hz)	33,750	62,500

TABLE 11.3   Examples of Common Data Rate Scanning Parameters

	29.97-Hz frame rate		25.00-Hz frame rate	
Common parameters				
Active pixels per line	720	1920	720	1920
Interlace	2:1	2:1	2:1	2:1
Luminance sampling rate (MHz)	13.5	74.25	13.5	74.25
Total data rate (Mbps)*	216	1188	216	1188
Uncommon parameters				
Active lines per frame	487	1080	576	1152
Horizontal line rate (kHz)	15,374	33,750	15,625	31,250
Total samples per line†	858	2200	864	2376
Total lines per frame	525	1125	625	1250

*The total data rates are calculated for 8-bit resolution luminance and chrominance format systems.

†The total samples per line represent the digitization of the luminance signal only.

Figure 11.1   Common image format (CIF) concept.

progressive format systems. But the storage of CIF pictures on DVTRs implies different data rates and recording formats.

Another important issue is the selection of a unique colorimetry standard. Colorimetry means the combination of the three-color primaries, transfer characteristics, and matrix components. Agreement on colorimetry, optoelectronic transfer characteristics at the camera, and an assumed electrooptical transfer characteristic at the display end has not been reached. The ATSC standard specifies the SMPTE 274M colorimetry as the default and preferred colorimetry. Several colorimetry standards are possible and are shown in Table 11.4. This colorimetry information is embedded in the MPEG-2 data stream at the production encoding phase and will help the receiver display to create a proper reproduction of the original image. This capability is a necessary requirement for future use of non-CRT-based displays with various archival materials.

With the production standards defined, it became obvious that a future HDTV transmission standard is strongly interrelated to them. It should, therefore, not be dissociated from them and defined separately. Table 11.5 summarizes current production and transmission formats.

11.3   The ATV Emergence

In 1987 at the request of U.S. broadcasters, the Federal Communications Commission (FCC) created an Advisory Committee on Advanced Television Service (ACATS) to investigate the feasibility of terrestrial broadcast standards for an advanced television (ATV) system. By the end of 1988, a total of 23 advanced television proposals were sent to the ACATS. Testing of all these proposals took place at the Advanced Television Test Center (ATTC) in Alexandria, Virginia, and at the Advanced Television Evaluation Laboratory (ATEL) in Ottawa, Canada.

In 1990, the FCC made several decisive key spectrum-related decisions:

- Broadcasters of ATV programs will deliver an identical and simultaneous NTSC program to allow for a transition period before the NTSC system is phased out.
- All TV formats of the future ATV system will use a 6-MHz transmission channel.
- Unused UHF channels will be assigned to all TV stations.

In January 1991, four all-digital ATV systems and two analog systems were proposed for testing at the ATTC and ATEL. After two years of intensive testing, a special panel set up by the ACATS could not select a satisfactory system. The requirement for a digital dual-channel stereo audio system was upgraded to a five-channel surround sound system. It was decided to retain all-digital technology and to ask manufacturers for improvements.

The manufacturers started discussions on merging their technologies. The ACATS strongly supported a Grand Alliance of all remaining companies to merge the four all-digital systems into a single DTV system. By February 1994, most of the key elements of the DTV system were approved by the Committee Technical Subgroup, including the use of the vestigial sideband (VSB) modulation subsystem.

TABLE 11.4 Differing Colorimetry Standards Used in TV Production

Standards	Color primaries Primary	X	Y	Transfer characteristics (gamma)	Matrix coefficients
ITU-R BT.709 SMPTE 295M	Red	640	330	$V = 1.099Lc^{0.45} - 0.099$ for $1 \geq Lc \geq 0.018$; $V = 4.5Lc$ for $0.018 > Lc \geq 0$	$E'_Y = 0.7152E'_G + 0.0722E'_B + 0.2126E'_R$ $E'_{Pb} = -0.386E'_G + 0.500E'_B - 0.115E'_R$ $E'_{Pr} = 0.454E'_G - 0.046E'_B + 0.500E'_R$
	Green	300	600		
	Blue	150	060		
	White (D65)	312.7	329		
ITU-R BT.470-4 system B, G		640	330	Gamma = 2.8	$E'_Y = 0.587E'_G + 0.114E'_B + 0.299E'_R$ $E'_{Pb} = -0.331E'_G + 0.500E'_B - 0.169E'_R$ $E'_{Pr} = -0.419E'_G - 0.081E'_B + 0.500E'_R$
		290	600		
		150	060		
	White (D65)	313	329		
SMPTE-170M (NTSC-1995)		630	340	$V = 1.099L^{0.45} - 0.099$ for $1 \geq Lc \geq 0.018$; $V = 4.5Lc$ for $0.018 > Lc \geq 0$	$E'_Y = 0.587E'_G + 0.114E'_B + 0.299E'_R$ $E'_{Pb} = -0.331E'_G + 0.500E'_B - 0.169E'_R$ $E'_{Pr} = -0.419E'_G - 0.081E'_B + 0.500E'_R$
		310	595		
		155	070		
	White (D65)	312.7	329		
ITU-R BT.470-4 system M NTSC-1953 (FCC)		670	330	Gamma = 2.2	$E'_Y = 0.59E'_G + 0.11E'_B + 0.30E'_R$ $E'_{Pb} = -0.331E'_G + 0.500E'_B - 0.169E'_R$ $E'_{Pr} = -0.421E'_G - 0.079E'_B + 0.500E'_R$
		210	710		
		140	080		
	White (C)	310	316		
SMPTE 240M SMPTE 260M		630	340	$V = 1.1115Lc^{0.45} - 0.1115$ for $Lc \geq 0.0228$; $V = 4.0Lc$ for $0.0228 > Lc$	$E'_Y = 0.701E'_G + 0.087E'_B + 0.212E'_R$ $E'_{Pb} = 0.384E'_G + 0.500E'_B - 0.116E'_R$ $E'_{Pr} = -0.445E'_G - 0.055E'_B + 0.500E'_R$
		310	595		
		155	070		
	White (D65)	312.7	329		

Note: The SMPTE 274M standard (1920 × 1080, scanning and interface for multiple picture rates) specifies two reference primary chromaticities and white values. The ITU-BT-709 is recommended for future generation of signal generation equipment, and the SMPTE 240M is reserved for interim implementations of currently used equipment.

TABLE 11.5   Production and Transmission Formats

Standards	Use	Active picture dimensions	Signal	Format	Aspect ratio
NTSC	Production and transmission	Equivalent to 680 × 483	Analog	Composite	4:3
Component digital	Production	720 × 483	Digital	Component	4:3 16:9
HDTV	Production and transmission	1920 × 1035	Digital	Component	16:9
ATV	Production and transmission	1920 × 1080 1280 × 720 704 × 480 640 × 480	Digital	Component	16:9 4:3

11.4   The Digital Solution

A new TV service based on digital technology creates new opportunities for media productions and services that can be provided by other industries, such as computers and telecommunications. During the period of ATV discussions and experimentation (1987 to 1994), many important issues concerning the development of a new ATV broadcasting system were identified. Since these issues have a considerable impact on its design and costs, and consequently on its success, the ACATS adopted several basic criteria aimed at meeting the needs of consumers and facilitating the implementation of a new DTV system. Several important criteria are discussed in the following sections.

11.4.1   Interoperability

The purpose of interchangeability is to make the coexistence of various new applications such as DTV, computers, and multimedia user-friendly; to facilitate direct exchange of compressed materials between all proponents if the same audio and video compression scheme is used; to expand the marketplace and encourage the electronics industry to adopt DTV technology; to develop a common platform to be used for DTV and other computer and multimedia applications; and to resolve conflicting considerations related to various broadcasting interests (cable TV and consumer electronics, as well as computing, telephone companies, and the national information infrastructure).

Three important principles aimed at achieving some degree of interoperability have been considered:

■ The ATSC system architecture is based on layers, compatible with the open-system interconnection (OSI) model used in many data network standards, as shown in Fig. 11.2. A reference model for data broadcasting is described in CCIR Report 1207 to facilitate the standardization and evolution of future systems.

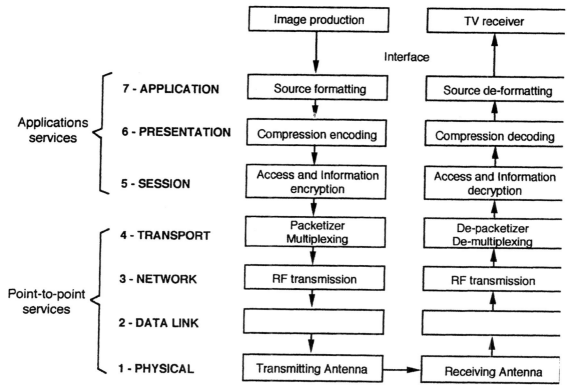

Figure 11.2 The OSI layered architecture and the ATSC system.

- Each ATSC layer is interoperable with other applications at corresponding layers. This facilitates exchanges of program materials by means of various layers of the DTV architecture. For example, a digital VCR can be designed to have input/output ports compatible with the MPEG-2 stream to make possible a direct dubbing at the compressed bit stream level, as specified in a recent SMPTE Working Group document on recording format standardization. This approach also facilitates the conversion of ATSC data packets into asynchronous transport mode (ATM) network data cells, and vice versa.

- Implement adequate operating flexibility in the ATSC system by using a header/descriptor approach.

All possible image formats were studied with the aim of achieving interoperability. Several formats were adopted to resolve the differences between numerous existing pixel (pixel aspect ratio) and scanning formats (picture dimensions, picture aspect ratios, interlaced and progressive, frame rates) such as those used in film, television, and computer standards. Table 11.6 shows all scanning formats and the corresponding pixel aspect ratios that can be found in the ATSC standard documents.

TABLE 11.6   ATSC Image Formats and Associated SMPTE Production Standards

Format name	SDTV formats		HDTV formats	
	VGA 480	CCIR-601	720	1080
Active image formats* (pixels × lines)	640 × 480	720 × 483	1280 × 720	1920 × 1080
Image rates†‡ (fps)	60P, 30P 30I, 24P	60P, 30P 30I, 24P	60P, 30P 24P	30P 30I, 24P
Image aspect ratio (IAR)	4:3	4:3 16:9	16:9	16:9
Pixel aspect ratio (PAR)	1	0.9 1.2	1	1
Signal representation standards		SMPTE 293M CCIR-601	SMPTE 296M	SMPTE 274M
Video interfacing standards		SMPTE 125M SMPTE 259M SMPTE 294M	SMPTE 296M	SMPTE 274M
Maximum bit rates @ 30 fps§ (Mbps)	147	162	442	995
Maximum compression ratios¶	7.8	8.6	23	53

*The CCIR-601 format yields an active picture format of 720 × 483 pixels. It is also used as a 704 × 480 format to ensure computer and compression compatibilities with the CSIF (352 × 240) format.

†These picture rates can be the stated integer value of 1000/1001 times the integer value(say, 59.94 Hz instead of 60 Hz) for ease of conversion of NTSC original pictures.

‡In this table, I means interlace and P means progressive scanning.

§Bit rates are calculated for an 8-bit sampling resolution, a 4:2:2 sampling structure, and the ATSC format-related active image area.

¶These compression ratios are calculated to generate the nominal ATSC video channel bandwidth of 18.9 Mbps with one television program.

11.4.2   Flexibility

The important issue in designing the future ATSC system is that by implementing enough flexibility and expandability, it is not necessary to develop all the ATSC formats simultaneously. A gradual conversion to digital television can be made with more reasonable conversion costs.

The DTV system is flexible enough to be operated over cables, fiber-optic networks, transmitters, and satellites with various program contents (sports, news, movies, and high-end postproduction materials) and display formats. Several approaches add flexibility in the future DTV system:

- Use a header*/descriptor in each data packet rather than a specified data format for each program stream component (audio, video, and auxiliary).

- Use an MPEG-2 packet format transport, allowing for the delivery of a wide variety of picture, sound, and data services.

- Define the coding and the transmission format independently of the program source format and the receiver display format.

---

*A header is included at the beginning of each data packet that constitutes the program data stream.

11.4.3 Compression

The 6-MHz bandwidth constraint for terrestrial TV broadcasting offers no other alternative than video data compression to transmit all the ATSC formats. As an example, the 1920 × 1080 @ 30I format requires a compression ratio of 50 to fit into one transport channel. Table 11.7 shows the bit rates of all ATSC corresponding production scanning formats before compression is applied. Contributing elements to these bit rates are the maximum number of pixels per line, the maximum number of lines per image, the image rate, and the number of bits per pixel for *Y, Pr,* and *Pb* signals.

The ATSC standard is based on the MPEG-2 compression scheme, a main profile (MP) implemented from main level (ML) to high level (HL), and uses a motion-compensated DCT algorithm, as explained in Chap. 8. However, some constraints are applied to this compression standard. This means that an MPEG-2 decoder can decode all ATSC formats, but an ATSC decoder cannot decode all MPEG-2 formats due to ATSC standard constraints.

The MPEG-2 compression scheme allows for the possibility of using a base layer and an enhancement layer adapted to the HDTV scanning formats. These two layers could be sent on separate channels, as more channels become available when NTSC transmissions stop. This will facilitate the introduction of the ATSC transmission of the 1920 × 1080 @ 60P scanning format.

Compression scheme requirements depend on the nature of produced TV pictures. Typical programs such as sports may use progressive scan formats for superior motion fidelity, compared to the interlaced version. Dramas, variety shows, and documentaries require an emphasis on spatial resolution.

Improvements in compression algorithms will provide a superior picture quality while being compatible with the already installed ATSC compression decoders.

TABLE 11.7 Active and total Production Format Bit Rates Used at Input of ATSC System

Active/total video formats (pixels × lines)*	Frame rates	Bit rates (Mbits/s)†			
		8-bit 4:2:2	10-bit 4:2:2	8-bit 4:2:0	10-bit 4:2:0
640 × 480/ (720 × 525)	30I	147/(181)	184/(227)	110/(136)	138/(170)
	30P	147/(181)	184/(227)	110/(136)	138/(170)
	60P	294/(362)	368/(454)	220/(272)	276/(340)
704 × 480/ (858 × 525)	30I	162/(216)	203/(270)	121/(162)	152/(202)
	30P	162/(216)	203/(270)	121/(162)	152/(202)
	60P	324/(432)	406/(540)	242/(324)	304/(404)
1280 × 720/ (1650 × 750)	30P	442/(594)	553/(742)	331/(445)	415/(556)
	60P	884/(1188)	1106/(1485)	662/(890)	830/(1113)
1920 × 1080/ (2200 × 1125)	30I	995/(1188)	1244/(1485)	746/(891)	933/(1113)
	30P	995/(1188)	1244/(1485)	746/(891)	933/(1113)

*Values in parentheses are total scanning formats corresponding to the active video scanning formats defined in the ATSC standard documents.

†The 10-bit 6:3:3 sampling structure, as produced by an 18-MHz sampling frequency, for a 704 × 480 active video format and not shown in this table, will generate a bit rate of 360 Mbps. This production format is not included in the ATSC video input format standards.

11.4.4 Progressive versus interlaced scanning

In analog NTSC distribution systems, interlacing was preferred instead of increasing the frame rate to 60 fps owing to bandwidth constraints. Interlacing is still motivated by the desire for bandwidth reduction. Conventional TV with 30 fps is therefore 2:1 interlaced, resulting in 60 fields per second, thus effectively reducing the visual perception of large area flicker.

Progressive scanning has been primarily chosen, for most of the ATSC formats, to facilitate the interoperability of television production equipment with computers. But it also provides some advantages over interlaced scanning in TV production systems:

- It reduces the processing complexity in motion estimation, scanning conversion, and editing and compression techniques.

- Superior picture performance with fewer motion artifacts can be achieved as the vertical resolution is improved and the interline flicker does not exist.

- Progressively scanned video results in good-quality freeze frames.

Interlaced formats still exist in the ATSC standard. This fact takes into account the existence of 30-year archives of interlaced TV programs as well as huge investments made by TV production houses and broadcasters in interlace-based equipment. The 1080-line progressive version is not included in the ATSC standard since it was not a realistic design at the time.

11.4.5 Image aspect ratio and pixel aspect ratio

The *image aspect ratio* (IAR) is defined as the ratio of the horizontal image dimension (expressed in millimeters or inches) to the vertical image dimension (expressed in millimeters or inches). It takes on the value of 4:3 or 16:9 for television and 2.4:1 for film applications. The 16:9 aspect ratio was established by an SMPTE Working Group on High-Definition Electronic Production in 1985. Any program material or motion picture film can be accommodated within the 16:9 format for production and postproduction, distribution, or display.

The *pixel aspect ratio* (PAR) is defined as the ratio of the horizontal pixel spacing to the vertical pixel spacing dimension. It does not refer to the actual shape of the pixel. Both the IAR and the number of horizontal pixels (Hp) and vertical lines (Vl) determine this pixel aspect (or dot spacing) in the reproduced picture. The relationship between the PAR and the IAR is given by

$$PAR = IAR \times \frac{Vl}{Hp}$$

If IAR and Hp/Vl are identical (or if PAR $= 1$), the image has square pixels (or samples). As an example, the VGA/SVGA standard image format has 640×480 pixels for a 4:3 picture dimension format. So PAR $= 4/3 \times 480/640 = 1$.

This confirms that square pixels are used to generate and display pictures on computers. In video applications, the number of horizontal pixels is determined by the video sampling frequency, which is derived from a multiple of the horizontal scanning or subcarrier frequency. This results in nonsquare pixel displays of digitized NTSC and CCIR-601 signals. A good operational practice would be to create the original material in the IAR and PAR in which it will be displayed. Some pictures and pixel aspect ratios are shown in Fig. 11.3.

11.4.6 Production aperture and clean aperture

The active picture area is defined by the number of active pixels in the horizontal direction (or on a line) and the number of active lines in the vertical direction. This active picture area is also called the *production aperture* because it is the actual format to be considered for use in signal origination and postproduction.

Figure 11.3 Image aspect ratio (IAR) and pixel aspect ratio (PAR).

The SMPTE Standard 274M, which defines a 1920 × 1080 scanning and interface for multiple picture rates, mentions that "improper digital processing of the edges of the picture may introduce transients near the edges, both horizontally and vertically." Such effects are to be minimized or avoided by use of appropriate techniques, such as repetition or mirroring of pixels at the production aperture edges, prior to digital filtering. The following factors contribute to the occurrence of analog transients:

- Bandwidth limitation of component analog signals (most noticeably, the ringing on color-difference signals)

- Analog filter implementation

- Amplitude clipping of analog signals due to the finite dynamic range imposed by the quantization process

- Use of digital blanking in repeated analog-digital-analog conversion

- Tolerance in analog blanking specifications

Hence, recognizing the reality of those picture edge transient effects, the definition of a system design guideline is introduced in the form of a subjectively artifact-free area, called the *clean aperture*. The clean aperture defines an area within which picture information is subjectively unaffected by all edge transient distortions. For example, the SMPTE 296M standard for television, 1280 × 720 scanning and interface, includes dimensions of the affected area as being 16 pixels on both the left and right sides and 8 lines on both the top and the bottom parts of the production aperture edges. The minimum clean aperture is 1248 horizontal active pixels by 704 active lines whose quality is guaranteed for final release. Figure 11.4 shows the clean aperture lying within the production aperture.

11.4.7 Audio system considerations

Consumers are accustomed to CD-quality audio reproduction systems. Film sound tracks on stereo VHS tapes are enjoyed in the home and have created a new high-fidelity industry called home theater. A significant market would

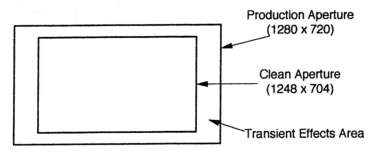

Figure 11.4 Example of production and clean apertures for the 1280 × 720 picture format.

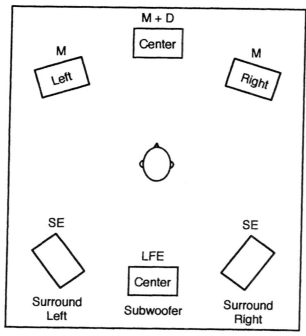

M = Music
D = dialogue
SE = Special Effect
LFE = Low Frequency
Enhancement

Figure 11.5 Surround sound system configuration.

benefit from a new digital television system that incorporated such a multiple-audio-channel capability. However, this multichannel audio system must be flexible enough in providing basic mono or stereo configurations to accommodate a wide variety of consumer needs.

After a performance comparison of various systems, the ATSC selected the Dolby AC-3 surround sound system that was previously established by the motion picture industry for digital motion picture sound tracks. Six discrete channels are used to transport music (M) in the front three loudspeakers, dialogue (D) in the center front channel, and special effects (SE) in the left and right surround channels, as shown in Fig. 11.5. An optional low-frequency enhancement (LFE) bandwidth-limited channel provides high-level low-frequency signals to avoid digital-to-analog converter and main speaker overloads.

11.4.8 DTV compatibility with film originating programs

The importance of film source origination materials cannot be ignored in TV applications. Such materials have a 24-fps temporal resolution. To accommodate existing conventional television systems running at 30 fps (or 60 fields per second), an up-conversion process, called a 3:2 pull-down process, is used to convert the film rate to the CTV rate, as described in Fig. 11.6. All film frames are repeated five times, and then every alternate frame is removed

Figure 11.6   Up-conversion of 24-fps motion pictures to 60-fps (3:2 pull-down process).

from the frame stream. The result is that odd film frames are repeated three times while even film frames are repeated twice.

When compression encoding of TV materials derived from film is performed, the additional and redundant frames brought by the 3:2 pull-down process must be removed from the video signal. This removal process, called the *inverse telecine process,* will eliminate the redundant frames while maximizing the bit allotment for each remaining frame. In addition, the 3:2 pull-down process may generate motion artifacts if compression encoding is performed on such sequences. Source identification is important for the source adaptive capability of devices included in the DTV production and delivery chain (DVTR or storage devices, display, home DVTRs and receivers, etc.). In TV production, the SMPTE RP-186 video index standard allows for data identification of each field or frame of source material to be inserted on line 14 of the vertical blanking interval. In ATSC program delivery, film originating materials are sent with an identifier included in the program data stream.

The postprocessor at the receiver end of the video system may perform interpolation (up-conversion from 24 to 60 fps) to establish the original frame rate. Standard methods, such as the 3:2 pull-down process described earlier, as well as sophisticated motion-compensated interpolation methods, can be utilized. In the latter case, the increased frame repetition resulting from a highly accurate motion field estimation may produce a better picture quality compared to that projected in a movie theater.

A large percentage of TV programming is produced on film that has a frame rate of 24 fps. Past practice was to perform postproduction by editing the film to produce a complete program on film. For North American transmission purposes, the film is transferred to 60-Hz video by employing the 3:2 pull-down technique. For transmission in 50-Hz countries, the film is run at the slightly higher speed of 25 fps.

In North America it is common practice to transfer film to 60-Hz video for postproduction. Once the video is transferred, editing decisions are made based on 60-Hz video practices, not on 24-Hz film practices. This introduces technical problems in international program exchange. Either the film must be cut to conform to the 60-Hz video and then transferred to 50-Hz video, or the video standards must be converted to 50 Hz with a loss of quality.

Equipment is now introduced which allows editing to be performed electronically at the 24-fps rate. The resulting program can be easily converted to 60-Hz video (3:2 pull-down technique) or to 50-Hz video (speeding it up to 25 fps). This new equipment embodies the original concept of the common image format. The images are captured according to the CIF at one picture rate and are presented at a different picture rate. The new 24-Hz equipment comprises cameras, telecines, VTRs, production switchers, and digital effects units. This new equipment permits the creation of an electronic master equivalent of the cut film of the past. It will operate using a picture format of 1920 × 1080 at 24 Hz with progressive capture.

The new format is called *progressive segmented frame* (PsF) and has been included in the recent reissue of ITU-R BT.709 international HDTV standard. Essentially the picture is captured as a 24-fps progressively scanned format (24P). The 24P video signal can be preserved in its original format (progressively scanned frame or 24P) or can be transformed to segmented frames (24PsF). Each progressive frame contains 1080 active lines. Each progressive frame is transformed to two parts, named *segments*: Segment A of each original frame contains the odd lines (1 to 1079), and segment B of each frame contains the even lines (2 to 1080). There is a certain similarity between 24PsF and interlace which may lead to confusion. It is for this reason that the divided parts of the original progressive frames are called *segments*. This term emphasizes that they are not fields. The time represented by the pair of segments equals the time of the original frame ($\frac{1}{24}$ s). The two segments can be joined together to recreate the original frame. Figure 11.7 shows the manner in which a sequence of 24 progressive frames per second (24P) is restructured into a sequence of 48 segments per second (24PsF).

11.5 DTV Picture Formats

The DTV standards do not specify the scanning format to be used by broadcasters. In effect, any scanning format may be used as long as the total transmitted bit rate—a multiplex of video, audio, and ancillary data—is equal to

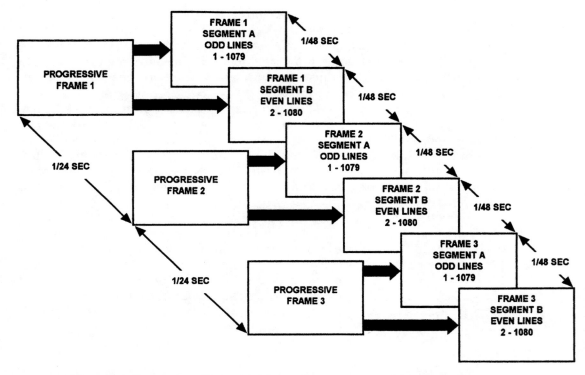

Figure 11.7   Formation of a sequence of segmented frames from a sequence of progressive frames.

19.38 Mbps. This bit rate is specified for terrestrial transmissions in a standard 6-MHz channel, using 8-VSB modulation. Initially 16-VSB modulation was suggested for use in cable networks. At the time of this writing, October 1999, there is no agreement on the modulation type or types that will be used by cable systems. Even though the scanning standards are neither specified nor enforced, a de facto family of scanning formats has emerged. These formats have evolved from the original ATSC recommendations.

11.5.1   The HDTV Formats

Three formats are discussed:

- The 1920 × 1035I format
- The 1920 × 1080 family of formats
- The 1280 × 720P format

11.5.1.1   General considerations

- Signal characteristics. All HDTV scanning formats originate as component analog gamma-corrected E'_G, E'_B, and E'_R signals. These signals may be encoded into component analog E'_Y, E'_{CB}, and E'_{CR} signals, or the original primary color signals may be used. The analog signal shapes are similar to those described in Chap. 2. All analog signals have a nominal p-p analog signal amplitude of 700 mV corresponding to a 100/0/100/0 color bars signal.

▩ Colorimetry standards. The colorimetry of the earlier, legacy, 1920 × 1035I HDTV format, as per SMPTE 240 M, uses a specific set of E'_G, E'_B, E'_R, $E'_B - E'_Y$, and $E'_R - E'_Y$ multiplication coefficients. These coefficients differ from the ITU-R BT.601 coefficients as well as the currently agreed upon coefficients specified by ITU-R BT.709. The discrete signal levels, corresponding to the 100/0/100/0 color bars, as shown in Chap. 2, reflect the coefficients specified by ITU-R BT.601. The same types of signals have different discrete values when ITU-R BT.709 or SMPTE 240M coefficients are used. Table 11.8 recalls the three sets of equations.

Table 11.9 shows the calculated signal amplitudes for individual steps corresponding to the three primary colors (green, blue, and red) and their complements (yellow, cyan, and magenta) of the 100/0/100/0 color bars signals with reference to Figs. 11.8, 11.9, and 11.10. Format conversions from ITU-R BT.601 or SMPTE 240M signals require a recalculation of the coefficients to avoid significant colorimetry errors.

▩ Analog horizontal synchronizing signals. Analog horizontal synchronizing signals have a specific trilevel bipolar shape and an amplitude of ±300 mV and are thus different from those used by the standard definition systems. This type of signal reduces the noise susceptibility of horizontal synchronization.

▩ Timing reference. All timing figures are expressed in multiples of T, where T is the reciprocal of the digital sampling frequency f_S of the luminance E'_Y component signal.

▩ The digital representation. The most common digital representation comprises two separate 10-bit channels operating side by side: one for luminance

TABLE 11.8 Coded Signals in Three Standards

Standard	Coded Signals
ITU-R BT.709	$E'_Y = 0.7152E'_G + 0.0722E'_B + 0.2126E'_R$ $E'_{CB} = 0.5389(E'_B - E'_Y)$ $E'_{CR} = 0.635(E'_R - E'_Y)$
SMPTE 240M	$E'_Y = 0.701E'_G + 0.087E'_B + 0.212E'_R$ $E'_{CB} = 0.5476(E'_B - E'_Y)$ $E'_{CR} = 0.6345(E'_R - E'_Y)$
ITU-R BT.601	$E'_Y = 0.587E'_G + 0.114E'_B + 0.299E'_R$ $E'_{CB} = 0.564(E'_B - E'_Y)$ $E'_{CR} = 0.713(E'_R - E'_Y)$

TABLE 11.9 **100 Percent Color Bars Component Signal Levels (in mV) for Three Colorimetry Standards**

	Signal	White	Yellow	Cyan	Green	Magenta	Red	Blue	Black
Primary colors	E'_G	700	700	700	700	0	0	0	0
	E'_B	700	0	700	0	700	0	700	0
	E'_R	700	700	0	0	700	700	0	0
ITU-R BT709	E'_Y	700	649.46	551.18	500.64	199.36	148.82	50.54	0
	E'_{CB}	0	−344.4	79.94	−263.9	263.9	−79.94	344.4	0
	E'_{CR}	0	38.64	−350.0	−311.57	311.57	350.0	−38.64	0
SMPTE 240M	E'_Y	700	639.10	551.60	490.70	209.30	148.40	−60.90	0
	E'_{CB}	0	−349.93	81.20	−268.66	268.66	−81.20	349.93	0
	E'_{CR}	0	38.64	−349.93	−311.29	311.29	349.93	−38.64	0
ITU-R BT601	E'_Y	700	620.20	490.70	410.90	289.10	209.30	0.114	0
	E'_{CB}	0	−360.43	83.65	−276.71	276.71	−83.65	360.43	0
	E'_{CR}	0	43.40	−393.26	−349.80	349.80	393.26	−43.40	0

(Y) and one for time-division-multiplexed color-difference component signals (C_B/C_R). The two data streams have their individual TRS (EAV and SAV), line number, line CRC information, as well as ancillary data, if used. This contrasts with ITU-R BT.601 signals where the three data streams, respectively, Y, C_B, and C_R, are time-division multiplexed into a single 27 Mwords/s digital data stream with a single set of TRS and ancillary data packet.

11.5.1.2 The 1920 × 1035I format The 1920 3 1035I format is a legacy interlace scanned 16:9 aspect ratio HDTV format initially issued as SMPTE 240M, Signal Parameters—1125-Line High Definition Production Systems, and also detailed in the ITU-R BT.1120 document. The 1035 figure refers to the active number of lines per frame. This standard defines the basic characteristics of the analog video signals associated with origination equipment operating in the 1125-line interlaced high-definition television production systems with 60.00-Hz or 59.94-Hz field rates. The specifications cover

- Scanning parameters
- Colorimetry
- Video signal waveforms

The component signals may be E'_R, E'_G, E'_B or E'_Y, E'_{CB}, and E'_{CR}. Table 11.10 details the picture scanning characteristics of the 1920 × 1035I format. As with all DTV scanning standards, there is a clear relationship to the digital representation of these signals. In this table this trend is reflected by the mention of the number of digital samples per full line and the digital sampling frequencies. There are two slightly different variations related to the field frequency, which may be 60 Hz or the NTSC-friendly 59.94 Hz. The line frequency, as well as the sampling frequency, reflects these two variations.

Figure 11.8 Relationship between analog signal levels of the Y component of a 100% color bars signal and 8-bit and 10-bit digital sample values. The digital levels 00 and FF for 8 bit and 000, 001, 002, 003, and 3FC, 3FD, 3FE, 3FF for 10 bit are not permitted in the data stream. For analog signal amplitude values see Table 11.9.

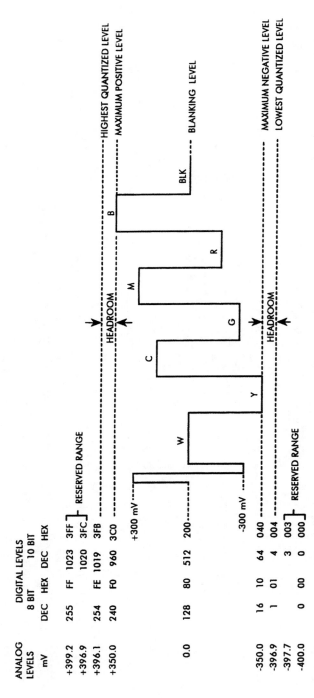

Figure 11.9 Relationship between analog signal levels of the C_B component of a 100% color bars signal and 8-bit and 10-bit digital sample values. The digital levels 00 and FF for 8 bit and 000, 001, 002, 003 and 3FC, 3FD, 3FE, 3FF for 10 bit are not permitted in the data stream. For analog signal amplitude values see Table 11.9.

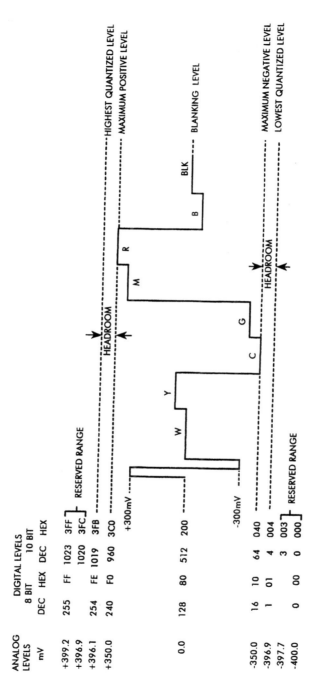

Figure 11.10 Relationship between analog signal levels of the C_R component of a 100% color bars signal and 8-bit and 10-bit digital sample values. The digital levels 00 and FF for 8 bit and 000, 001, 002, 003, and 3FC, 3FD, 3FE, 3FF for 10 bit are not permitted in the data stream. For analog signal amplitude values see Table 11.9.

TABLE 11.10   Picture Scanning Characteristics for 1920 × 1035I Picture Format

Item	Parameter	Value
1	Order of sampling presentation in a scanned system	Left to right, top to bottom
2	Field frequency (Hz)	60 (60/1.001)*
3	Interlace ratio	2:1
4	Line frequency (Hz) (33,750/1.001)	33,750
5	Samples per full line R, G, B, Y C_B, C_R	2200 1100
6	Nominal signal bandwidth† (MHz)	30
7	Sampling frequency R, G, B, Y (MHz)	74.25 (74.25/1.001)
8	Sampling frequency C_B, C_R (MHz)	37.125 (37.125/1.001)
9	Total number of lines Per frame Field 1 Field 2	1125 563 (1–563) 562 (564–1125)
10	Active number of lines Per frame Field 1 Field 2	1035 517(41–557) 518 (603–1120)

*Value in parentheses is for NTSC-friendly systems where the frame divisor is 1.001.
†Bandwidth is for all components before antialiasing and low-pass filtering.

Table 11.11 details the analog representation of the 1920 × 1035I picture format. It refers to the component signals shown in Figs. 11.8, 11.9, and 11.10. It also refers to the details of the horizontal blanking interval and the trilevel horizontal synchronizing pulse detailed in Fig. 11.11. Other specifications such as color primaries, transfer characteristics, and matrix coefficients can be found in Table 11.4 referenced to the SMPTE 240M standard.

Figure 11.12 shows the details of the field blanking period. The vertical synchronizing signal is organized as follows: A five-line duration block contains the vertical synchronizing pulse divided into 10 wide pulses, two per line. Each wide pulse is flanked by trilevel horizontal synchronizing pulses. Vertical synchronization is obtained by differentiating the first leading edge following the first wide vertical pulse. Alternatively an integrator can be used as in the standard television systems. Figure 11.13 shows details of the vertical synchronizing pulse.

TABLE 11.11   Analog Representation of 1920 × 1035I Picture Format

Item	Parameter	Value
1	Nominal level (mV) E'_R, E'_G, E'_B, E'_Y	Reference black: 0 Reference white: 700 See Fig. 11.8.
2	Nominal level (mV) E'_{CB}, E'_{CR}	±350 mV See Figs. 11.9 and 11.10.
3	Form of synchronizing signal	Trilevel bipolar. See Fig. 11.11.
4	Line sync timing reference	0_H. See Fig. 11.11.
5	Sync level (mV)	±300 ± 2%
6	Sync signal timing	Sync on all components. See Table 11.12 and Figs. 11.11, 11.12, and 11.13.
7	Blanking interval	See Table 11.12 and Figs. 11.11, 11.12, and 11.13.

Figure 11.11   1920 × 1035I scanning format details of horizontal blanking interval and trilevel sync. For symbol values, see Table 11.12.

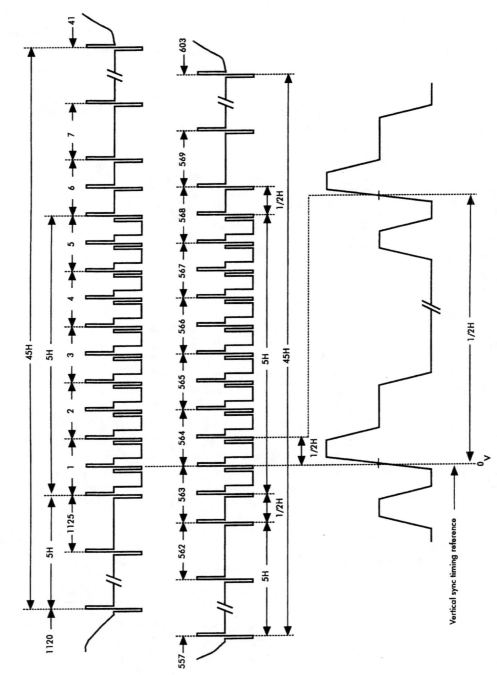

Figure 11.12 1920 × 1035I scanning format details of analog field blanking period and of field synchronizing pulse.

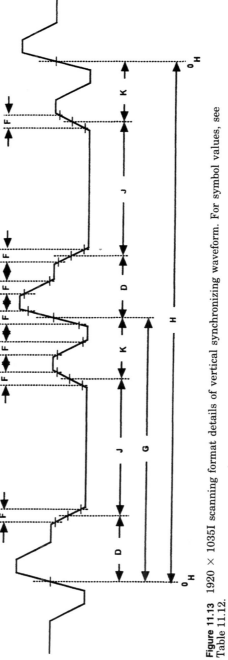

Figure 11.13 1920 × 1035I scanning format details of vertical synchronizing waveform. For symbol values, see Table 11.12.

**TABLE 11.12   Level and Line Timing Specifications of 1920 ×
1035I Picture Format**
(Refer to Figs. 11.11, 11.12, and 11.13.)

Symbol	Parameter	Value
$f_s{}^*$	$R, G, B,$ or Y sampling frequency (MHz)	74.25 (74.25/1.001)
T	Reference clock interval (μs)	$1/f_s$
A	Negative line sync width (T)	44 ± 3
B	End of active line to 0_H (T)	$88 + 6, -0$
C	Positive line sync width (T)	44 ± 3
D	Clamp period (T)	132
E	0_H to start of active video (T)	$192 + 6, -0$
F	Rise/fall time (T)	4 ± 1.5
G	Half-line interval (T)	1100
H	Total line interval (T)	2200
J	Vertical sync width (T)	880
K	End of vertical sync width (T)	88
S_n	Negative pulse amplitude (mV)	300 ± 6
S_p	Positive pulse amplitude (mV)	300 ± 6
S	P-p sync pulse amplitude (mV)	600 (nom.)
V	Video signal amplitude (mV)	700
	Active line interval (T)	$1920 + 0, -12$

*Value in parentheses is for NTSC-friendly systems where the
frame divisor is 1.001.

Table 11.12 describes the symbols used in Figs. 11.11, 11.12, and 11.13 and
details the level and timing specifications.

The SMPTE Standard 260M, *Digital Representation and Bit-Parallel
Interface—1125/60 High Definition Production System,* specifies the digital representation of the signal parameters of the high definition production system as described by the SMPTE Standard 240M. The digital coding
is based on one luminance E'_Y and two color-difference signals E'_{CB} and E'_{CR}
or on the use of the three primary signals E'_G, E'_B, and E'_R. The E'_Y, E'_{CB},
and E'_{CR} are transmitted at the 22:11:11 level of the CCIR digital hierarchy
for digital television with a nominal sampling frequency of 74.25 MHz (22
× 3.375 MHz) for the luminance signal and 37.125 MHz (11 × 3.375 MHz)
for the color-difference signals. The E'_G, E'_B, and E'_R signals would be transmitted at the 22:22:22 hierarchy level. It is interesting to note that this
document, published in 1992, follows the original CCIR 601 nomenclature
and philosophy, scaled up and adapted to higher sampling strategies. Table
11.13 details the digital representation of the format. The antialias filter
characteristics can be found in Figs. 11.14 and 11.15.

The most common digital representation assumes a dual-channel data
stream consisting of

■ A digital data stream conveying digitized luminance signals (Y) sampled at
74.25 MHz (or 74.25/1.001 MHz) with 10-bit accuracy represented by a
parallel data stream of, nominally, 74.25 Mwords/s

- A digital data stream conveying time-division-multiplexed C_B and C_R signals, each sampled at 37.125 MHz (or 37.125/1.001 MHz) with 10-bit accuracy, or a parallel data stream of, nominally, 74.25 Mwords/s

Each data stream (Y and C_B/C_R) carries the active video information, the timing reference signals (EAV and SAV), and the ancillary data, if present. Figure 11.16 shows the analog and digital timing relationships along a scanning line for the Y component. A similar situation exists in the C_B/C_R scanning line, not shown here. The 0_H datum represents the analog line start. A digital sampling instant in a line is defined by a number from 0 (the first active digital sample) through a number representing the total number of samples in a line less one. The digital line starts with sample 0. The total Y line has a duration of 2200 samples, of which 1920 samples convey video data. The total C_B/C_R multiplexed color-difference line has a duration of 2200 samples (1100 C_B and 1100 C_R), of which 1920 convey video data (960 C_B and 960 C_R). The horizontal blanking interval of each of the two types of lines has a duration of 280 samples, of which the EAV (end of active video) and SAV (start of active

TABLE 11.13   Digital Representation of 1920 × 1035I Picture Format

Item	Parameter	Value
1	Coded signal	R, G, B or Y, C_B, C_R
2	Sampling lattice R, G, B, Y	Orthogonal, line and picture repetitive
3	Sampling lattice C_B, C_R	Orthogonal, line and picture repetitive co-sited with each other and alternative Y samples
4	Samples per full line R, G, B, Y C_B, C_R	 2200 1100
5	Number of active samples per line R, G, B, Y C_B, C_R	 1920 960
6	Total number of lines	1125
7	Number of active lines	1035
8	Coding format	Linear 8 or 10 bits per component
9	Quantization levels Black level: R, G, B, Y Achromatic: C_B, C_R Nominal peak: R, G, B, Y C_B, C_R	8-bit coding 10-bit coding 16 64 128 512 235 940 16 and 240 64 and 1019
10	Quantization level assignment Video data Timing reference	8-bit coding 10-bit coding 1 through 254 4 through 1019 0 and 255 0–3 and 1020–1023
11	Filter characteristics R, G, B, Y C_B, C_R	 See Fig. 11.14. See Fig. 11.15.

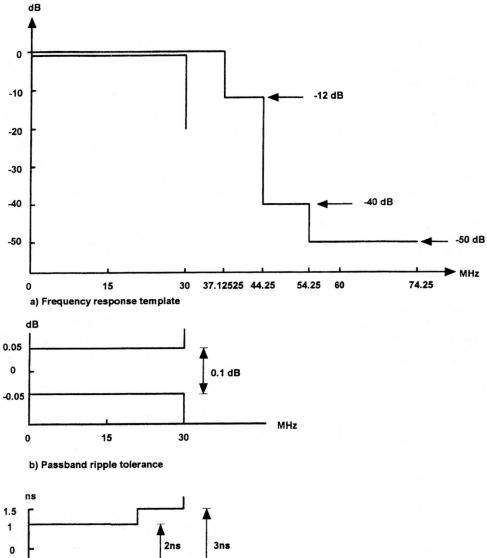

a) Frequency response template

b) Passband ripple tolerance

c) Passband group-delay tolerance

Figure 11.14   1920 × 1035I scanning format specification of *R, G, B, Y* low-pass filter.

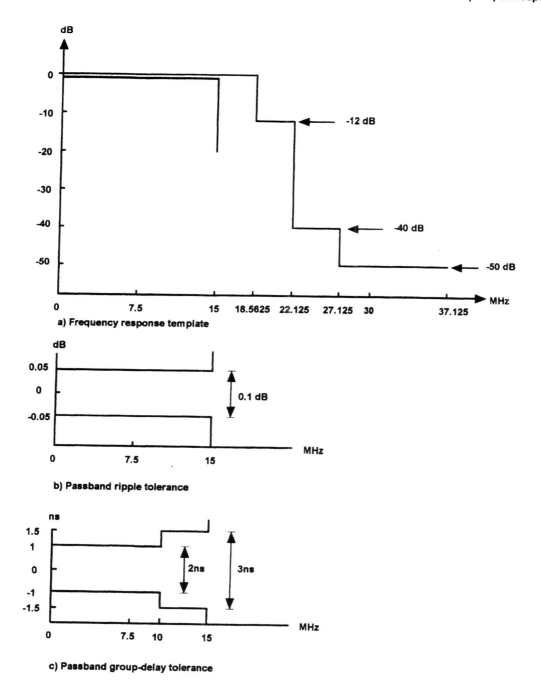

Figure 11.15   1920 × 1035I scanning format specification of color-difference low-pass filter.

Figure 11.16 1920 × 1035I scanning format analog and digital relationship. Note: Horizontal axis is not to scale.

video) signals have a duration of four samples each. This leaves a total of 272 Y samples and 272 multiplexed C_B/C_R samples for other potential uses.

Each timing reference signal, EAV and SAV, consists of a four-word sequence:

- Three synchronizing words with hexadecimal values of, respectively, 3FF, 000, and 000.

- The XYZ word which carries the V bit, the F bit, and the H bit. These bits define the vertical and horizontal blanking. In addition, bits P0, P1, P2, and P3, which assume values depending on the status of the V, F, and H bits, provide a limited error correction (single errors) and detection (2 bit errors) of these bits. Table 11.14 details the information carried by the EAV and SAV signals.

Figure 11.17 is a spatial representation of the timing reference signals during a television frame. Table 11.15 details significant instants shown in Fig. 11.17.

11.5.1.3 The 1920 × 1080 format. The 1920 × 1080 scanning format family comprises several HDTV scanning systems using a 16:9 aspect ratio and a

TABLE 11.14   Timing Reference Signals for 1920 × 1035I Picture Format

Bit	First word	Second word	Third word	Fourth word
9	1	0	0	1
8	1	0	0	F
7	1	0	0	V
6	1	0	0	H
5	1	0	0	P3
4	1	0	0	P2
3	1	0	0	P1
2	1	0	0	P0
1	1	0	0	0
0	1	0	0	0

F = 0 during field 1	V = 0 elsewhere	H = 0 in SAV
F = 1 during field 2	V = 1 during field blanking	H = 1 in EAV

Note: P0, P1, P2, P3 protection bits based on value of F, V, H.

TABLE 11.15   Field Interval Timing Specifications for 1920 × 1035I Picture Format

(Refer to Fig. 11.17.)

Symbol	Definition	Digital line number
	Total number of lines	1125
	Total number of lines in field No. 1	563
	Total number of lines in field No. 2	562
	Number of active lines	1035
	Number of active lines in field No. 1	517
	Number of active lines in field No. 2	518
L1	First line of field No. 1	1
L2	Last line of digital field blanking No. 1	40
L3	First line of field No. 1 active video	41
L4	Last line of field No. 1 active video	557
L5	First line of digital field blanking No. 2	558
L6	Last line of field No. 1	563
L7	First line of field No. 2	564
L8	Last line of digital field blanking No. 2	602
L9	First line of field No. 2 active video	603
L10	Last line of field No. 2 active video	1120
L11	First line of digital field blanking No. 1	1121
L12	Last line of field No. 2	1125

Note: Digital field blanking No. 1 denotes the field blanking period that is prior to the active video of field No. 1, and digital field blanking No. 2 denotes that prior to the active video of field No. 2.

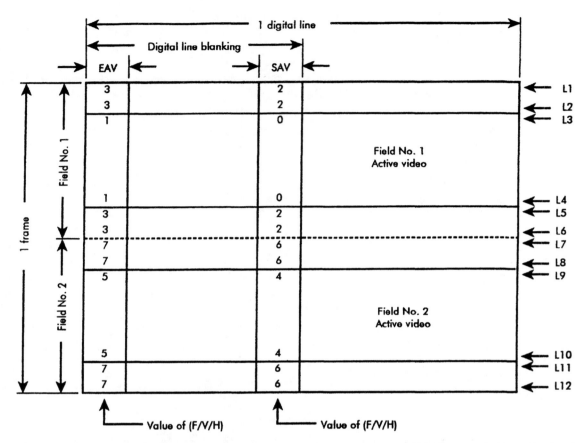

Figure 11.17 Field timing relationship of 1920 × 1035I picture format. Note 1: The values (*F/V/H*) for EAV and SAV represent the status of bits for *F, V,* and *H* in a way that the 3-bit word composed of *F, V, H* represents a binary number expressed in decimal notation. (*F* corresponding to MSB and *H* corresponding to LSB.) For example, the value 3 represents the bits *F* = 0, *V* = 1, and *H* = 1. Note 2: Drawing not to scale. For symbol values see Table 11.15.

total of 1125 lines, of which 1080 are active. The latest revision of the SMPTE Standard 274M. *1920 × 1080 Scanning and Analog and Parallel Digital Interfaces for Multiple Picture Rates*, still under study, was issued for comments in 1998. It specifies 11 scanning systems sharing a common image format (CIF). Of these, eight are progressive and three are interlaced. The novelty here is the introduction of 50-Hz, 25-Hz, and 24-Hz frame rate options.

In June 1999, the International Telecommunications Union issued the third revision of the ITU-R BT.709 Standard, *Parameter Values for the HDTV Standards for Production and International Program Exchange*. This new version of the standard reaches out to encompass the 50-Hz- and 60-Hz-based systems in both their progressive and interlaced aspects. In addition it introduces the recently developed progressive sequential frame (PsF) concept as applied to 24-Hz, 25-Hz, and 30-Hz frame systems. Most of the data in this section are borrowed from ITU-R BT.709-3.

Table 11.16 details the picture scanning characteristics of the various 1920 × 1080 formats. There are three distinct categories:

- 60-Hz-related: 60P, 60I, 30PsF, and 30P with a choice of nominal or NTSC-friendly vertical scanning, numbered 1 through 8. All formats in this category have 2200 Y samples per full line and 1100 each of C_B, C_R samples per full line. The H and V scanning rates and the related Y and C_B, C_R digital sampling frequencies depend on the scanning system. The highest-quality format, 60P, has a Y digital sampling frequency of 148.5 MHz and a C_B, C_R digital sampling frequency of 74.25 MHz each. At 10 bits per sample, this results in a Y, C_B, C_R time-division-multiplexed serial distribution bit rate of 2.97 Gbps. The other formats in this category have an equivalent serial bit rate of 1.485 Gbps.

- 50-Hz-related: 50P, 50I, 25PsF, and 25P numbered 9 through 12. All formats in this category have 2640 Y samples per full line and 1320 each of C_B, C_R samples per full line. The H and V scanning rates and the related Y and C_B/C_R digital sampling frequencies depend on the scanning system. The highest-quality format, 50P, has a Y digital sampling frequency of 148.5 MHz and a C_B, C_R sampling frequency of 74.25 MHz each. At 10 bits per sample, this results in a Y, C_B, C_R time-division-multiplexed serial distribution bit rate of 2.97 Gbps. The other formats in this category have an equivalent serial bit rate of 1.485 Gbps.

- 24-Hz-related: 24PsF and 24P, with a choice of nominal or NTSC-friendly vertical scanning, numbered 13 through 16. All formats in this category have 2640 Y samples per full line and 1320 each C_B, C_R samples per full line. The H and V scanning rates and the related Y and C_B/C_R digital sampling frequencies are the same for all formats. All formats have a Y digital sampling frequency of 74.25 MHz and a C_B, C_R sampling frequency of 37.125 MHz each. At 10 bits per sample, this results in a Y, C_B, C_R time-division multiplexed serial bit rate of 1.485 Gbps.

Table 11.17 details the analog representation of the 1920 × 1080 picture format. It refers to the component signals shown in Figs. 11.8, 11.9, and 11.10. It also refers to the details of the horizontal blanking interval and the trilevel horizontal synchronizing pulse shown in Fig. 11.18. Other specifications such as color primaries, transfer characteristics, and matrix coefficients can be found in Table 11.4 referenced to the ITU-R BT.709 standard.

Figure 11.19 shows the details of the analog field blanking period of the 1920 × 1080I interlaced scanning format. All the scanning systems of this format use a total of 1125 lines. In an analog-only system, this would imply 562.5 lines in each field. However, because a digital interface must also be supported, only whole numbers of lines in each field are allowed, in order to

Table 11.16 Picture Scanning Characteristics for 1920 × 1080 Picture Format

		System									
		1 (2)	3 (4)	5 (6)	7 (8)	9	10	11	12	13 (14)	15 (16)
Item	Parameter	60P	60I	30PsF	30P	50P	50I	25PsF	25P	24PsF	24P
1	Order of sample presentation in a scanned system	Left to right, top to bottom. For interlaced and segmented frame systems, first active line of field 1 at top of picture									
2	Field/frame/segment frequency (Hz)*	60 (60/1.001)		30 (30/1.001)		50		25		48 (48/1.001)	24 (24/1.001)3
3	Interlace ratio	1:1	2:1	1:1sF	1:1	1:1	2:1	1:1sF	1:1	1:1sF	1:1
4	Picture rate (Hz)	60 (60/1.001)		30 (30/1.001)		50		25		24 (24/1.001)	
5	Line frequency (Hz)†	67500 (67500/1.001)		33750 (30/1.001)		56250		28125		27000 (27000/1.001)	
6	Samples per full line R, G, B, Y; C_B, C_R	2200 1100				2640 1320				2750 1375	
7	Nominal signal bandwidth (MHz)‡	60		30		60			30		
8	Sampling frequency R, G, B, Y (MHz)§	148.5 (148.5/1.001)		74.25 (74.25/1.001)		148.5		74.25		74.25 (74.25/1.001)	
9	Sampling frequency C_B, C_R (MHz)¶	74.25 (74.25/1.001)		37.125 (37.125/1.001)		74.25		37.125		37.125 (37.125/1.001)	

*Value in parentheses is for NTSC-friendly systems where the frame divisor is 1.001.
†The tolerance on line frequency is ±0.001 percent.
‡Bandwidth is for all components before antialiasing low-pass filtering.
§The tolerance on sampling frequency is ±0.001 percent.
¶C_B, C_R sampling frequency is half of luminance sampling frequency.
Note: All systems in this table have a total number of 1125 lines per frame.

TABLE 11.17   Analog Representation of 1920 × 1080 Picture Format

Item	Parameter	Value
1	Nominal level (mV) E'_R, E'_G, E'_B, E'_Y	Reference black: 0 Reference white: 700 See Fig. 11.8.
2	Nominal level (mV) E'_{CB}, E'_{CR}	±350 mV See Figs. 11.9 and 11.10.
3	Form of synchronizing signal	Trilevel bipolar See Fig. 11.18.
4	Line sync timing reference	0_H See Fig. 11.18.
5	Sync level (mV)	±300 ± 2%
6	Sync signal timing	Sync on all components See Table 11.18 and Figs. 11.18, 11.19, and 11.20.
7	Blanking interval	See Table 11.18 and Figs. 11.18, 11.19, and 11.20.

Figure 11.18 1920 × 1080 scanning format details of horizontal blanking interval and trilevel sync. For symbol values, see Table 11.18.

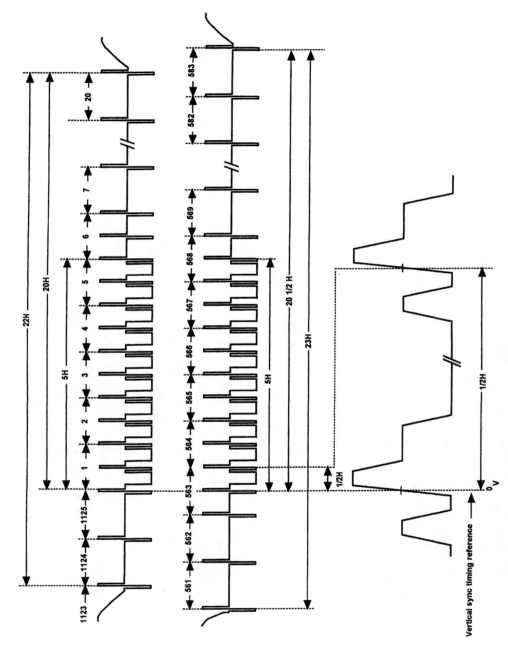

Figure 11.19 1920 × 1080I scanning format details of analog field blanking period and of field synchronizing pulse.

permit unambiguous identification of lines by the digital timing reference sequences. As a consequence, the interlaced versions define integer, and hence unequal, numbers of lines (563 and 562) in each of the two fields comprising a frame. Analog sync sequences, however, must remain equally spaced in time and are therefore not aligned to the fields as defined for the digital interface. This results in the analog vertical sync for the second digital field beginning one half-line before the end of the first digital field.

The first field comprises 563 lines, including

- Vertical blanking: lines 1 through 20 inclusive and lines 1124 through 1125, for a total of 22 lines
- Picture: 540 lines, 21 through 560 inclusive

The second field comprises 562 lines, including

- Vertical blanking: lines 561 to 583 inclusive, for a total of 23 lines
- Picture: 540 lines, 584 through 1123 inclusive

Interlaced analog vertical sync is located in lines 1 through 5 for the first field and from halfway through line 563 to halfway through line 568 for the second field.

Figure 11.20 shows the analog frame blanking period of the 1920 × 1080P progressive scanning format. The assignment of the lines within the frame is as follows:

- Vertical blanking period: lines 1 through 41 inclusive (including vertical sync, lines 1 through 5 inclusive) and lines 1122 through 1125, for a total of 45 lines
- Picture: 1080 lines, 42 through 1121 inclusive

Figure 11.21 shows details of the vertical synchronizing waveform for progressive scanning and interlaced scanning. Table 11.18 describes the symbols used in Figs. 11.18 and 11.21 and shows level and line timing specifications referenced to the 0_H datum. As shown, the number of samples per active line is equal to 1920 for all 16 systems, thus meeting HDTV CIF requirements.

The production aperture defines a region of 1920 samples by 1080 lines. The clean aperture of the picture is defined as a region 1888 samples in width by 1062 lines high symmetrically located in the production aperture and clean of transient effects due to blanking and signal processing. The aspect ratio represented by the production aperture is 16:9. The sample aspect ratio is 1:1 (square pixels).

The digital coding is based on one luminance E'_Y, and two color-difference signals E'_{CB} and E'_{CR}, or on the use of three primary signals E'_G, E'_B, and E'_R. Table 11.19 details the digital representation of the 1920 × 1080 format.

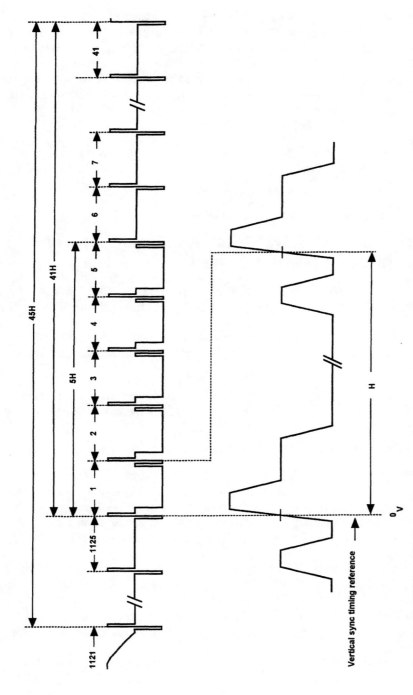

Figure 11.20 1920 × 1080P scanning format details of analog frame blanking period and of frame synchronizing pulse.

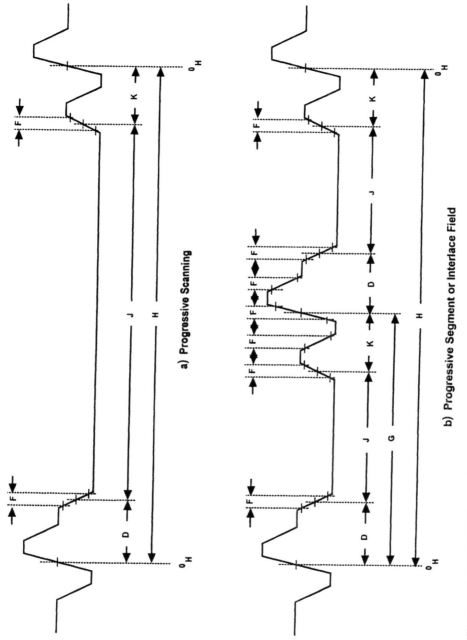

a) Progressive Scanning

b) Progressive Segment or Interlace Field

Figure 11.21 1920 × 1080 scanning format details of vertical synchronizing waveform. For symbol values, see Table 11.18.

529

TABLE 11.18 Level and Line Timing Specifications for 1920 × 1080 Picture Format

(Refer to Figs. 11.18, 11.19, 11.20, and 11.21.)

Item	Parameter	System									
		1 (2)	3 (4)	5 (6)	7 (8)	9	10	11	12	13 (14)	15 (16)
		60P	60I	30PsF	30P	50P	50I	25PsF	25P	24PsF	24P
T	Reference clock interval (μs)*	1/148.5† (1.001/148.5)		1/74.25 (1.001/74.25)		1/148.5	1/74.25			1/74.25 (1.001/74.25)	
A	Negative line sync width (T)					44 ± 3					
B	End of active video (T)‡		88 + 6, − 0			528 + 6, − 0				638 + 6, − 0	
C	Positive line sync width (T)					44 ± 3					
D	Clamp period (T)					132 ± 3					
E	Start of active video (T)					192 + 6, − 0					
F	Rise/fall time (T)					4 ± 1.5					
	Active line interval (T)					1920 + 0, − 12					
S_p	Positive pulse amplitude (mV)					300 ± 6					
S_n	Negative pulse amplitude (mV)					300 ± 6					
S	P-p sync pulse amplitude (mV)					600					
V	Video signal amplitude (mV)					700					
G	Half-line interval (T)		1100					1320		1375	
H	Total line interval (T)		2200					2640		2750	
J	Vertical sync width (T)	1980 ± 3		880 ± 3		1980 ±3		880 ±3	1980 ± 3	880 ± 3	1980 ± 3
K	End of vertical sync pulse (T)		88 ± 3			528 ± 3		308 ± 3	528 ± 3	363 ± 3	638 ± 3

*T denotes the duration of a reference clock or the reciprocal of the Y clock frequency (see Table 11.16).
†Value in parentheses is for NTSC-friendly systems where the frame divisor is 1.001.
‡An analog "line" starts at the line sync timing reference 0_H (inclusive) and ends just before the subsequent 0_H (exclusive).

TABLE 11.19   Digital Representation of 1920 × 1080 Picture Format

Item	Parameter	Value		
1	Coded signal	R, G, B, or Y, C_B, C_R		
2	Sampling lattice R, G, B, Y	Orthogonal, line and picture repetitive		
3	Sampling lattice C_B, C_R	Orthogonal, line and picture repetitive cosited with each other and alternative Y samples		
4	Number of active samples per line R, G, B, Y C_B, C_R	System 1–8 2200 1100	System 9–12 2640 1320	System 13–16 2750 1375
5	Number of active samples per line R, G, B, Y C_B, C_R	1920 960		
6	Total number of lines	1125		
7	Number of active lines	1080		
8	Coding format	Linear 8 or 10 bits per component		
9	Quantization levels Black level: R, G, B, Y Achromatic: C_B, C_R Nominal peak: R, G, B, Y C_B, C_R	8-bit coding 16 128 235 16 and 240	10-bit coding 64 512 940 64 and 1019	
10	Quantization level assignment Video data Timing reference	8-bit coding 1 through 254 0 and 255	10-bit coding 4 through 1019 0–3 and 1020–1023	
11	Filter characteristics R, G, B, Y C_B, C_R	See Fig. 11.22. See Fig. 11.23.		

The antialias filter characteristics can be found in Figs. 11.22 and 11.23.

The most common digital representation assumes a dual-channel data stream consisting of

- A digital data stream conveying digitized luminance signals (Y) sampled at one of the frequencies listed in Table 11.16 with 10-bit accuracy

- A digital data stream conveying time-division multiplexed signals C_B and C_R sampled at one of the frequencies listed in Table 11.16 with 10-bit accuracy

Each data stream carries the active video information, the TRS information, and the ancillary data, if present.

Figure 11.24 shows the analog and digital timing relationships along a scanning line for the Y component. A similar situation exists in the C_B/C_R scanning line, not shown here. The 0_H datum represents the analog line start. A digital sampling instant in a line is defined by a number from 0 (the first active digital sample) through a number representing the total number of samples in a line less one. In Fig. 11.24 a number of significant sampling instants are identified with alphabetical symbols a through and including p.

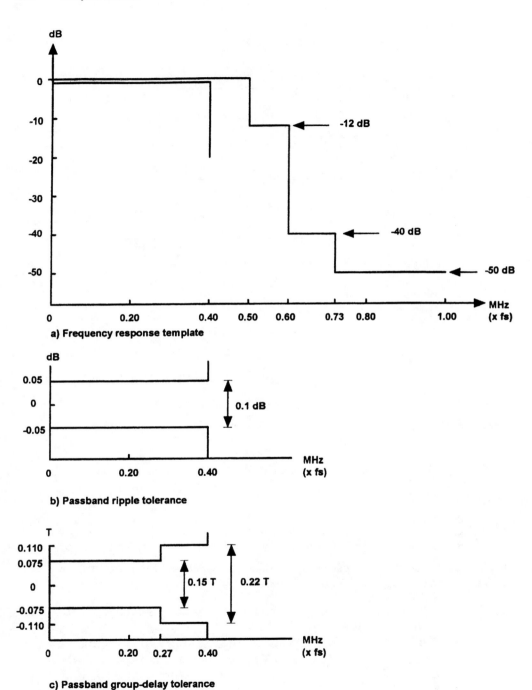

Figure 11.22 1920 × 1080 scanning format specification for *R, G, B,* and *Y* low-pass filter.

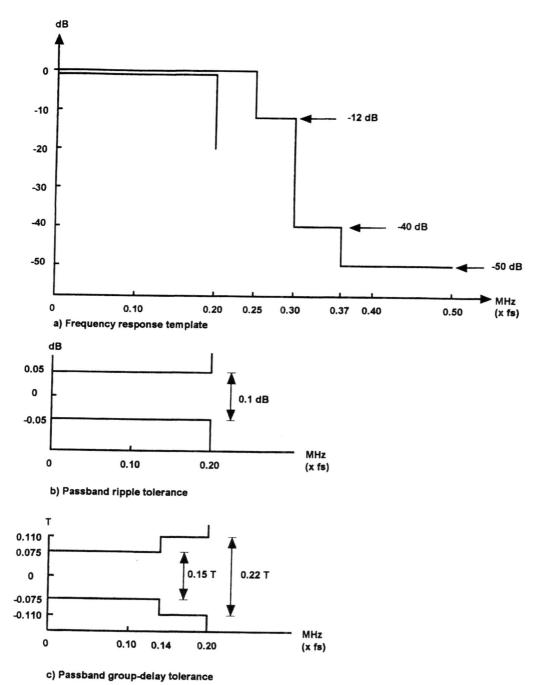

Figure 11.23 1920×1080 scanning format specification for color-difference low-pass filter.

Figure 11.24 1920 × 1080 scanning format analog and digital timing relationship. For symbol values, see Table 11.20. Note: Horizontal axis not to scale.

The values assumed by the symbols in Fig. 11.24 represent the number of the digital sample. They depend on the scanning system and are listed in Table 11.20.

The digital line starts with sample o, which assumes the value 0 in all scanning systems. The last active video sample p assumes the value 1919 in all scanning systems. This indicates that all scanning systems have the same number (1920) of active samples per line. Samples a, b, c, and d identify the four EAV words and assume identical values (1920, 1921, 1922, and 1923) in all scanning systems. Sample e, the first sample in the horizontal digital blanking interval, assumes the value 1924 in all scanning systems. The values of samples f through and including n assume different values depending on the scanning system. As a result, symbol B, representing the duration of the horizontal blanking interval, assumes a value of $272T$ (the same as in

Table 11.20 Values for Symbols in Fig. 11.24 for Different Scanning Systems
(For scanning systems see Tables 11.16 and 11.18.)

Scanning system	Sample numbering															
	a	b	c	d	e	f	g	h	i	j	k	l	m	n	o	p
1, 2, 3, 4, 5, 6, 7, 8	1920	1921	1922	1923	1924	1964	2007	2008	2009	2052	2196	2197	2198	2199	0	1919
9, 10, 11, 12	1920	1921	1922	1923	1924	2404	2447	2448	2449	2492	2636	2637	2638	2639	0	1919
13, 14, 15, 16	1920	1921	1922	1923	1924	2514	2557	2558	2559	2602	2746	2747	2748	2749	0	1919

Scanning system	Duration of reference clock periods (T)		
	A	B	C
1, 2, 3, 4, 5, 6, 7, 8	44	272	2200
9, 10, 11, 12	484	712	2640
13, 14, 15, 16	594	822	2750

SMPTE 240M) for systems 1, 2, 3, 4, 5, 6, 7, 8. For systems 9, 10, 11, and 12, B assumes a value of $712T$; and for systems 13, 14, 15, and 16, B assumes a value of $822T$. This results in an equal number of 1920 active samples per line in all scanning standards, a requirement of the HDTV CIF concept.

The durations of EAV and SAV are equal in all scanning standards. They each consist of a four-word sequence:

▨ Three synchronizing words with hexadecimal values of, respectively, 3FF, 000, and 000.

▨ The XYZ word which carries the V bit, the F bit and the H bit. These bits define the vertical and horizontal blanking. In addition, bits P0, P1, P2, and P3, which assume values depending on the status of the V, H, and F bits, provide a limited error correction (single errors) and detection (2 bit errors) of these bits.

Table 11.21 details the information carried by the EAV and the SAV. Note that in the 1920 × 1080P format F = 0 as there are no fields needing identification.

Figure 11.25 is a spatial representation of the timing reference signals during a television frame of the 1920 × 1080I interlaced scanning format. The L symbols identify specific lines in the active and blanking intervals. Table 11.22 describes the L symbols shown in Fig. 11.25.

Figure 11.26 is a spatial representation of the timing reference signals during a television frame of the 1920 × 1080P progressive scanning format. The L symbols identify specific lines in the active and blanking intervals. Table 11.23 describes the L symbols shown in Fig. 11.26.

11.5.1.4 The 1280 × 720P format. The 1280 × 720P format is defined by the SMPTE Standard 296M, *1280 × 720 Scanning, Analog and Digital Representation and Analog Interface*. It is a progressive scanning system. The 720 figure refers to the active number of lines per frame.

Table 11.24 details the picture scanning characteristics of the 1280 × 720P format. There are two slightly different variations related to the frame frequency, which may be 60 Hz or the NTSC-friendly 60/1.001 (59.94) Hz. The line frequency, as well as the digital sampling frequency, reflect these two variations.

Table 11.25 details the analog representation of the 1280 × 720P format. It refers to component analog signals shown in Figs. 11.8, 11.9, and 11.10. It also refers to the horizontal blanking interval and the trilevel horizontal synchronizing pulse detailed in Fig. 11.27, the frame blanking interval in Fig. 11.28, and the vertical synchronizing waveform in Fig. 11.29. Other specifications such as color primaries, transfer characteristics, and matrix coefficients can be found in Table 11.4 referenced to the ITU-R BT.709 Standard.

Figure 11.28 shows the details of the analog frame blanking period of the 1280 × 720P progressive scanning format. The assignment of the lines within the frame is as follows:

TABLE 11.21   Timing Reference Signals for 1920 × 1080 Picture Format

Bit	First word	Second word	Third word	Fourth word
9	1	0	0	1
8	1	0	0	F
7	1	0	0	V
6	1	0	0	H
5	1	0	0	P3
4	1	0	0	P2
3	1	0	0	P1
2	1	0	0	P0
1	1	0	0	0
0	1	0	0	0
1080P Format	F = 0	V = 0 elsewhere V = 1 in frame blanking	H = 0 in SAV H = 1 in EAV	
1080I Format	F = 0 in field 1 F = 1 in field 2	V = 0 elsewhere V = 1 in field blanking	H = 0 in SAV H = 1 in EAV	
P0, P1, P2, P3 protection bits are based on values of F, V, H				

TABLE 11.22   Field Interval Timing Specifications for 1920 × 1080I Picture Format
(Refer to Fig. 11.25.)

Symbol	Definition	Digital line number
	Total number of lines	1125
	Total number of lines in field No. 1	563
	Total number of lines in field No. 2	562
	Number of active lines	1080
	Number of active lines in field No. 1	540
	Number of active lines in field No. 2	540
L1	First line of field No. 1	1
L2	Last line of digital field blanking No. 1	20
L3	First line of field No. 1 active video	21
L4	Last line of field No. 1 active video	560
L5	First line of digital field blanking No. 2	561
L6	Last line of field No. 1	563
L7	First line of field No. 2	564
L8	Last line of digital field blanking No. 2	583
L9	First line of field No. 2 active video	584
L10	Last line of field No. 2 active video	1123
L11	First line of digital field blanking No. 1	1124
L12	Last line of field No. 2	1125

Note: Digital field blanking No. 1 denotes the field blanking period that is prior to the active video of field No. 1, and digital field blanking No. 2 denotes that prior to the active video of field No. 2.

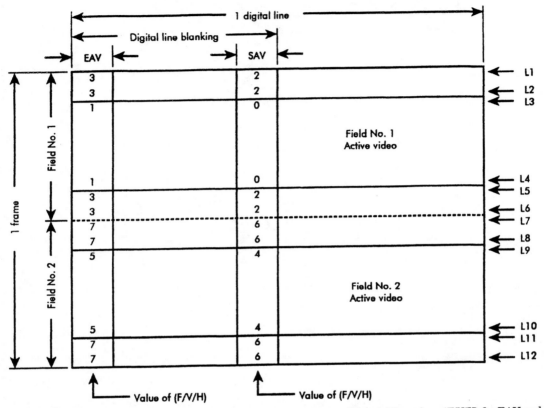

Figure 11.25 Field timing relationship of 1920 × 1080I picture format. Note 1: The values (*F/V/H*) for EAV and SAV represent the status of bits for *F*, *V*, and *H* in a way that the 3-bit word composed of *F*, *V*, *H* represents a binary number expressed in decimal notation (*F* corresponding to MSB and *H* corresponding to LSB). For example, the value 3 represents the bits *F* = 0, *V* = 1, and *H* = 1. Note 2: Drawing not to scale. For symbol values see Table 11.22.

TABLE 11.23 Frame Interval Timing Specifications for 1920 × 1080P Picture Format
(Refer to Fig. 11.26.)

Symbol	Definition	Digital line number
	Total number of lines	1125
	Number of active lines	1080
L1	First line	1
L2	Last line of digital frame blanking	41
L3	First line of active video	42
L4	Last line of active video	1121
L5	First line of digital frame blanking	1122
L6	Last line	1125

Figure 11.26 Frame timing relationship of 1920 × 1080P picture format. Note 1: The values (*F/V/H*) for EAV and SAV represent the status of bits for *F*, *V*, and *H* in a way that the 3-bit word composed of *F*, *V*, *H* represents a binary number expressed in decimal notation (*F* corresponding to MSB and *H* corresponding to LSB). For example, the value 3 represents the bits *F* = 0, *V* = 1, and *H* = 1. Note 2: Drawing not to scale. For symbol values see Table 11.23.

TABLE 11.24   Picture Scanning Characteristics for 1280 × 720P Picture Format

Item	Parameter	Value
1	Order of sampling presentation in a scanned system	Left to right, top to bottom
2	Frame frequency (Hz)*	60 (60/1.001)
3	Interlace ratio	1:1
4	Line frequency (Hz)	45,000 (45,000/1.001)
5	Samples per full line R, G, B, Y C_B, C_R	 1650 825
6	Nominal signal bandwidth (MHz)†	30
7	Sampling frequency R, G, B, Y (MHz)	74.25 (74.25/1.001)
8	Sampling frequency C_B, C_R (MHz)	37.125 (37.125/1.001)
9	Total number of lines	750
10	Active number of lines	720 (26–745)

*Value in parentheses is for NTSC-friendly systems where the frame divisor is 1.001.

†Bandwidth is for all components before antialiasing and low-pass filtering.

TABLE 11.25   Analog representation of 1280 × 720P Picture Format

Item	Parameter	Value
1	Nominal level (mV) E'_R, E'_G, E'_B, E'_Y	Reference black: 0 Reference white: 700 See Fig. 11.8.
2	Nominal level (mV) E'_{CB}, E'_{CR}	±350 mV See Figs. 11.9 and 11.10.
3	Form of synchronizing signal	Trilevel bipolar. See Fig. 11.27.
4	Line sync timing reference	0_H. See Fig. 11.27.
5	Sync level (mV)	±300 ± 2%
6	Sync signal timing	Sync on all components See Table 11.25 and Figs. 11.27, 11.28, and 11.29.
7	Blanking interval	See Table 11.26 and Figs. 11.27, 11.28, and 11.29.

- Vertical blanking period: lines 1 through 25 inclusive (including vertical sync, lines 1 through 5 inclusive) and lines 746 through 750 inclusive, for a total of 30 lines
- Picture: 720 lines, lines 26 through 745 inclusive

Figure 11.29 shows the details of the vertical synchronizing waveform. Table 11.26 describes the symbols used in Figs. 11.27, 11.28, and 11.29 and specifies the values assigned to them. Line timing specifications are referenced to the 0_H datum.

The production aperture defines a region of 1280 samples by 720 lines. The clean aperture of the picture is defined as a region of 1248 samples by 702 lines high, symmetrically located in the production aperture and clean of transient effects due to blanking and signal processing. The aspect ratio of the image represented by the production aperture is 16:9. The sample aspect ratio is 1:1 (square pixels).

The digital coding is based on one luminance E'_Y and two color-difference signals E'_{CB} and E'_{CR}, or on the use of the three primary signals E'_G, E'_B, and E'_R. Table 11.27 details the digital representation of the 1280 × 720P format. The antialiasing filter characteristics can be found in Figs. 11.30 and 11.31.

The most common digital representation assumes a dual-channel data stream consisting of

- A digital data stream conveying digitized luminance signals Y sampled at 74.25 MHz (74.25/1.001 MHz for NTSC-friendly systems) with 10-bit accuracy
- A digital data stream conveying time-division-multiplexed digital C_B and C_R signals each sampled at 37.125 MHz (37.125/1.001 MHz for NTSC-friendly systems) with 10-bit accuracy

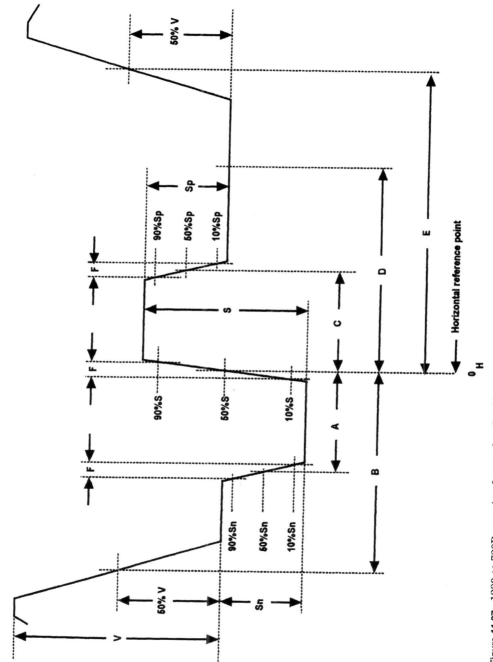

Figure 11.27 1280 × 720P scanning format details of horizontal blanking interval and trilevel sync. For symbol values, see Table 11.26.

541

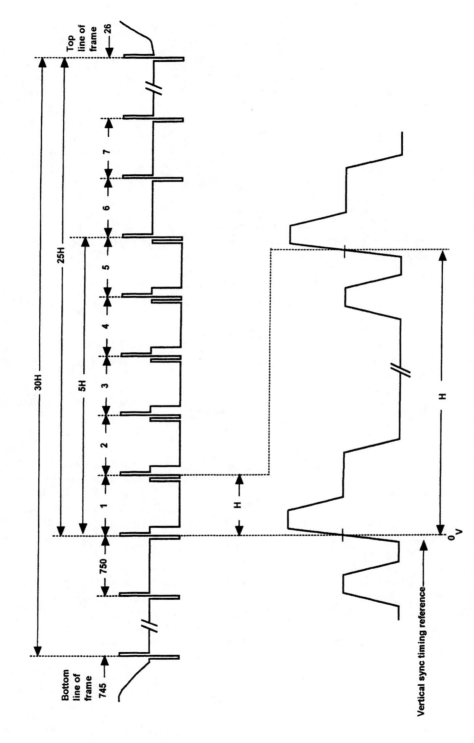

Figure 11.28 1280 × 720P scanning format details of analog frame blanking period and of frame synchronizing pulse.

TABLE 11.26   Level and Line Timing Specifications 1280 × 720P Picture Format
(Refer to Figs. 11.27, 11.28, and 11.29.)

Symbol	Parameter	Value
T	Reference clock interval (μs)	$1/f_s$
A	Negative line sync width (T)	40 ± 3
B	End of active line to 0_H (T)	110
C	Positive line sync width (T)	40 ± 3
D	Clamp period (T)	260
E	0_H to start of active video (T)	$260 - 6, + 0$
F	Rise/fall time (T)	4 ± 1.5
H	Total line interval (T)	1650
J	Vertical sync width (T)	1280
K	End of vertical sync width (T)	110
S_n	Negative pulse amplitude (mV)	300 ± 6
S_p	Positive pulse amplitude (mV)	300 ± 6
S	P-p sync pulse amplitude (mV)	600 (nom.)
V	Video signal amplitude (mV)	700
	Active line interval (T)	$1280 -12, +0$
	0_H to end of active line (T)	$1540 - 6, +0$

Each data stream (Y and C_B/C_R) carries the active video information and the ancillary data, if present. Figure 11.32 shows the analog and digital timing relationships along the scanning line for the Y component. A similar situation exists in the C_B/C_R scanning line, not shown here. The 0_H datum represents the analog line start. A digital sampling instant in a line is defined by a number from 0 (the first active digital sample) through a number representing the total number of samples in a line less one. The digital line starts with sample 0. The total Y line has a duration of 1650 samples of which 1280 convey video data. The total C_B/C_R multiplexed color-difference line has a duration of 1650 samples (825 C_B and 825 C_R) of which 1280 convey video data (640 C_B and 640 C_R). The horizontal blanking interval of each of the two types of lines has a duration of 370 samples of which the EAV and SAV have a duration of four samples each. This leaves a total of 362 Y samples and 362 multiplexed C_B/C_R samples for other potential uses.

The EAV and SAV each consist of a four-word sequence as follows:

- Three synchronizing words with hexadecimal values of, respectively, 3FF, 000, and 000.

- The XYZ word which carries the V bit, the F bit, and the H bit. These bits define the vertical and the horizontal blanking. In addition, bits P0, P1, P2, and P3, which assume values depending on the status of the V, F, and H bits, provide a limited error correction (single errors) and detection (2 bit errors) of these bits.

Table 11.28 details the information carried by the EAV and the SAV. Note that in the 1280 × 720P format F = 0 as there are no fields needing identification.

Figure 11.33 is a spatial representation of the timing reference signals during a television frame of the 1280 × 720P progressive scanning format. The L

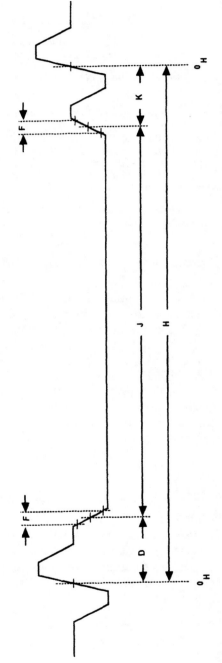

Figure 11.29 $1280 \times 720P$ scanning format details of vertical synchronizing waveform. For symbol values, see Table 11.26.

TABLE 11.27 Digital Representation of 1280 × 720P Picture Format

Item	Parameter	Value	
1	Coded signal	$R, G, B,$ or Y, C_B, C_R	
2	Sampling lattice R, G, B, Y	Orthogonal, line and picture repetitive	
3	Sampling lattice C_B, C_R	Orthogonal, line and picture repetitive cosited with each other and alternative Y samples	
4	Samples per full line R, G, B, Y C_B, C_R	1650 825	
5	Number of active samples per line R, G, B, Y C_B, C_R	1280 640	
6	Total number of lines	750	
7	Number of active lines	720	
8	Coding format	Linear 8 or 10 bits per component	
9	Quantization levels Black level: R, G, B, Y Achromatic: C_B, C_R Nominal peak: R, G, B, Y C_B, C_R	8-bit coding 16 128 235 16 and 240	10-bit coding 64 512 940 64 and 1019
10	Quantization level assignment Video data Timing reference	8-bit coding 1 through 254 0 and 255	10-bit coding 4 through 1019 0–3 and 1020–1023
11	Filter characteristics R, G, B, Y C_B, C_R	See Fig. 11.30. See Fig. 11.31.	

symbols identify specific lines in the active and blanking intervals. Table 11.29 describes the L symbols shown in Fig. 11.33.

11.5.2 The standard definition 720 × 483P format

The 720 × 483P format is defined by the SMPTE Standard 293M, *720 × 483 Active Line at 59.94-Hz Progressive Scan Production—Digital Representation.* This format is also described in ITU-R BT.1358 for 625- and 525-line progressive systems. The format has a 16:9 aspect ratio (optionally 4:3) and uses progressive scanning. The 483 figure refers to the active number of lines per frame. The principal application of this standard is to provide enhanced standard definition television signals for digital television broadcasting.

Table 11.30 details the picture scanning characteristics of the 720 × 483P format. The frame frequency is NTSC-related (60/1.001 = 59.94 Hz). The line frequency at 31,000/1.001 Hz reflects the progressive scan format's higher data rate.

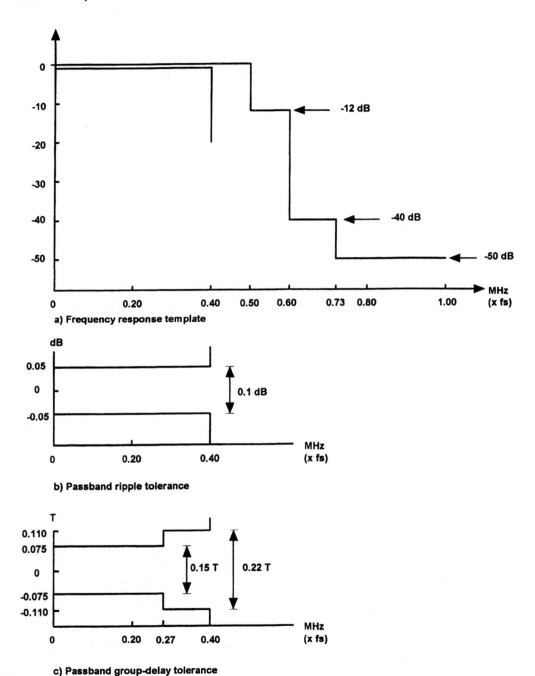

a) Frequency response template

b) Passband ripple tolerance

c) Passband group-delay tolerance

Figure 11.30 $1280 \times 720P$ scanning format specification of R, G, B, and Y low-pass filter.

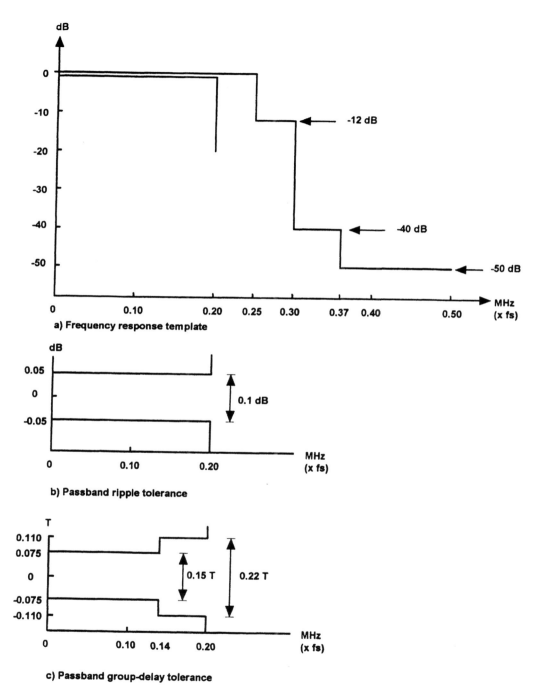

Figure 11.31 1280 × 720P scanning format specification of color-difference low-pass filter.

Figure 11.32   1280 × 720P scanning format analog and digital relationship. Note: Horizontal axis not to scale.

Table 11.31 details the analog representation of the 720 × 483P format. It refers to component analog signals shown in Figs. 3.42, 3.43, and 3.44. The quoted nominal signal bandwidths of 12 MHz (for E'_Y, E'_G, E'_B and E'_R signals) and 6 MHz (for E'_{CB} and E'_{CR} signals) reflect the higher vertical and horizontal scanning rates of the progressive format and are required to maintain the same horizontal resolution as available with interlaced scanning. The resulting horizontal resolution is equivalent to that obtained with the interlaced scanning format ITU-R BT.601, which only requires signal bandwidths of 5.75 and 2.75 MHz, respectively. Other specifications, such as color primaries, transfer characteristics, and matrix coefficients, can be found in Table 11.4 referenced to SMPTE 170M.

Figure 11.34 details the analog horizontal blanking interval and the characteristics of the standard (legacy) bilevel horizontal synchronizing pulse. Table 11.32 describes the symbols used in Fig. 11.34 and specifies the values assigned to them.

TABLE 11.28   Timing Reference Signals for 1280 × 720P Picture Format

Bit	First word	Second word	Third word	Fourth word
9	1	0	0	1
8	1	0	0	F
7	1	0	0	V
6	1	0	0	H
5	1	0	0	P3
4	1	0	0	P2
3	1	0	0	P1
2	1	0	0	P0
1	1	0	0	0
0	1	0	0	0
F=0		V = 0 elsewhere		H = 0 in SAV
		V = 1 in frame blanking		H = 1 in EAV
		P0, P1, P2, P3 protection bits based on values of F, V, H		

TABLE 11.29   Frame Interval Timing Specifications for 1280 × 720P Picture Format

(Refer to Fig. 11.33.)

Symbol	Definition	Digital line number
	Total number of lines	750
	Number of active lines	720
L1	First line	1
L2	Last line of digital frame blanking	25
L3	First line of active video	26
L4	Last line of active video	745
L5	First line of digital frame blanking	746
L6	Last line	750

Figure 11.33 Frame timing relationship of 1280 × 720P picture format. Note 1: The values (*F/V/H*) for EAV and SAV represent the status of bits for *F*, *V*, and *H* in a way that the 3-bit word composed of *F*, *V*, *H* represents a binary number expressed in decimal notation (*F* corresponding to MSB and *H* corresponding to LSB). For example, the value 3 represents the bits *F* = 0, *V* = 1, and *H* = 1. Note 2: Drawing not to scale. For symbol values see Table 11.29.

TABLE 11.30   Picture Scanning Characteristics for 720 × 483P Picture Format

Item	Parameter	Value
1	Order of sampling presentation in a scanned system	Left to right, top to bottom
2	Scanning format	Progressive 1:1
3	Frame frequency (Hz)	60/1.001
4	Total number of lines	525
5	Active number of lines	483 (line 43–525)
6	Aspect ratio	16:9 (4:3)
7	Line frequency (Hz)	31,500/1.001 ± 3 ppm

TABLE 11.31   Analog Representation of 720 × 483P Picture Format

Item	Parameter	Value
1	Nominal level (mV) E'_R, E'_G, E'_B, E'_Y	Reference black: 0 Reference white: 700 See Fig. 3.42.
2	Nominal level (mV) E'_{CB}, E'_{CR}	±350 mV See Figs. 3.43 and 3.44.
3	Nominal signal bandwidth (MHz) E'_R, E'_G, E'_B, E'_Y E'_{CB}, E'_{CR}	12 6
4	Form of synchronizing signal	Bilevel bipolar. See Fig. 11.34.
5	Line sync timing reference	0_H. See Fig. 11.34.
6	Sync level (mV)	-300 ± 7.5
7	Horizontal sync and blanking interval signal timing	Sync on all components See Table 11.32 and Fig. 11.34.
8	Vertical sync and blanking interval signal timing	See Table 11.33 and Fig. 11.35.

TABLE 11.32   Details of Horizontal Synchronizing Pulse

(Refer to Fig. 11.34.)

Symbol	Parameter	Value
H	Nominal line period (µs)	1001/32.5 (31.778)
A	Horizontal blanking interval (µs)	5.35 + 0.15, −0.1
B	Interval between time datum (0_H) and back edge of horizontal blanking pulse (µs)	4.6 + 0.1, −0.05
C	Front porch (µs)	0.75 ± 0.05
D	Synchronizing pulse duration (µs)	2.35 ± 0.05
E	Rise time (10% to 90%) of edges of horizontal blanking pulse (µs)	0.07 ± 0.01
F	Rise time (10% to 90%) of edges of horizontal synchronizing pulse (µs)	0.07 ± 0.01

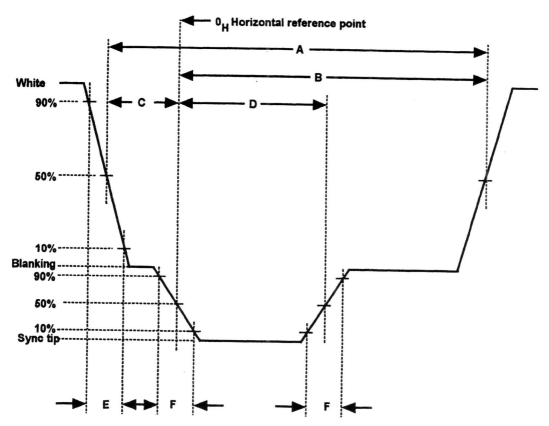

Figure 11.34   720 × 483P scanning format details of horizontal blanking period and bilevel sync. For symbol values, see Table 11.32.

Figure 11.35 shows details of the analog frame blanking period and the frame synchronizing pulse. The assignment of the lines within the frame is as follows:

- Vertical (frame) blanking period: lines 1 through 42 inclusive (including vertical sync, lines 7 through 12 inclusive)
- Picture: 483 lines, lines 34 through 525 inclusive

Table 11.33 describes the symbols used in Fig. 11.35 and specifies the values assigned to them.

The digital coding is based on one luminance E'_Y and two color-difference signals E'_{CB} and E'_{CR}, or on the use of three primary signals E'_G, E'_B, and E'_R. To accommodate the wider analog signal bandwidths of the format, the R, G, B, Y sampling frequency is 27 MHz, and the C_B, C_R sampling frequency is 13.5 MHz. Table 11.34 details the digital representation of the 720 × 483P format.

The antialiasing filter characteristics can be found in Figs. 11.36 and 11.37.

Figure 11.35   720 × 483P scanning format details of analog frame blanking period and frame synchronizing pulse. For symbol values, see Table 11.33.

TABLE 11.33   Details of Vertical Synchronizing Signals

(Refer to Fig. 11.35.)

Symbol	Parameter	Value
V	Nominal frame period (ms)	1001/60 (16.683)
A	Interval between front edge of vertical blanking interval and front edge of first vertical synchronizing pulse	6H
B	Duration of sequence of vertical synchronizing pulses	6H*
C	Interval between back edge of last vertical synchronizing pulse and back edge of vertical blanking interval	30H*
D	Vertical blanking interval	42H + 5.35 μs†
	Rise time (10% to 90%) of edges of vertical blanking pulse (μs)	0.07 ± 0.01
P	Duration of vertical synchronizing pulse (μs)	29.428 ± 0.05
R	Interval between vertical synchronizing pulses (μs)	2.35 ± 0.05
S	Rise time (10% to 90%) of vertical synchronizing pulses (μs)	0.07 ± 0.01

*H = 16.683 μs.
†Tolerances: +0.15, −0.1.

TABLE 11.34 Digital Representation of 720 × 483P Picture Format

Item	Parameter	Value	
1	Coded signal	R, G, B, or Y, C_B, C_R	
2	Sampling lattice R, G, B, Y	Orthogonal, line and picture repetitive	
3	Sampling lattice C_B, C_R	Orthogonal, line and picture repetitive cosited with each other and alternative Y samples	
4	Sampling frequency R, G, B, Y (MHz)	27 ± 3 ppm	
5	Sampling frequency C_B, C_R (MHz)	13.5 ± 3 ppm	
6	Samples per full line R, G, B, Y C_B, C_R	858 429	
7	Number of active samples per line R, G, B, Y C_B, C_R	720 360	
8	Total number of lines	525	
9	Number of active lines	483	
10	Coding format	Linear 8 or 10 bits per component	
11	Quantization levels Black level: R, G, B, Y Achromatic: C_B, C_R Nominal peak: R, G, B, Y C_B, C_R	8-bit coding 16 128 235 16 and 240	10-bit coding 64 512 940 64 and 1019
12	Quantization level assignment Video data Timing reference	8-bit coding 1 through 254 0 and 255	10-bit coding 4 through 1019 0–3 and 1020–1023
13	Filter characteristics R, G, B, Y C_B, C_R	See Fig. 11.36. See Fig. 11.37.	

The most common digital representation assumes a dual-channel data stream consisting of:

■ A digital data stream conveying digitized luminance signals Y sampled at 27 MHz with 10-bit accuracy

■ A digital data stream conveying time-division-multiplexed digitized C_B and C_R signals each sampled at 13.5 MHz with 10-bit accuracy

Each data stream (Y and C_B/C_R) carries the active video information, the timing reference signals (EAV and SAV), and the ancillary data, if present. Figure 11.38 shows the analog and digital relationships along the scanning line for the Y component. A similar situation exists in the C_B/C_R scanning line, not shown here. A digital sampling instant in a line

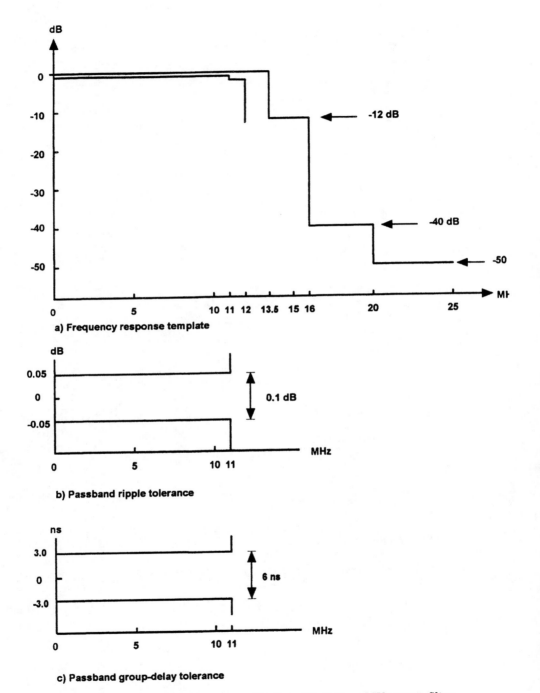

Figure 11.36   720 × 483 scanning format specification of *R*, *G*, *B*, and *Y* low-pass filter.

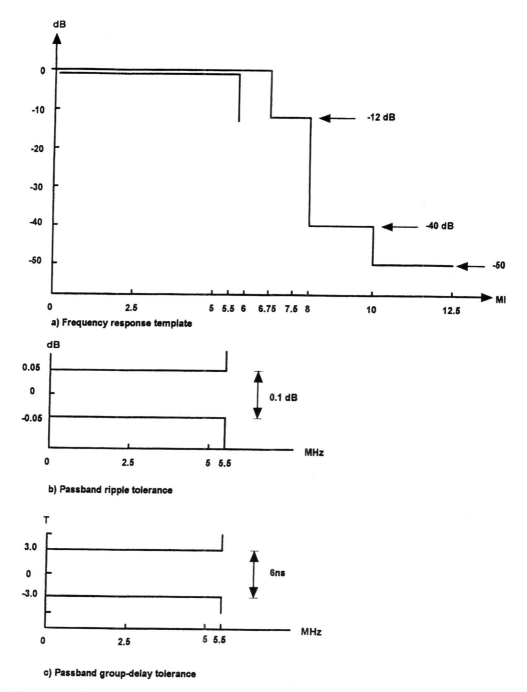

a) Frequency response template

b) Passband ripple tolerance

c) Passband group-delay tolerance

Figure 11.37 720×483 scanning format specification of color-difference low-pass filter.

Figure 11.38   720 × 483 scanning format analog and digital timing relationship. Note: Horizontal axis not to scale.

is defined by a number from 0 (the first active digital sample) through a number representing the total number of samples in a line less one. The total Y line has a duration of 858 luminance samples of which 720 convey video data. The C_B/C_R multiplexed color-difference line has a duration of 858 samples (429 C_B and 429 C_R) of which 720 convey video data (360 C_B and 360 C_R). The horizontal blanking of each of the two types of lines has a duration of 138 samples of which the EAV and SAV have a duration of four samples each. This leaves a total of 130 Y samples and 130 multiplexed C_B/C_R samples for other potential uses.

The EAV and SAV each consist of a four-word sequence as follows:

- Three synchronizing words with hexadecimal values of, respectively, 3FF, 000, and 000.

- The XYZ word which carries the V bit, the F bit, and the H bit. These bits define the vertical and horizontal blanking. In addition, bits P0, P1, P2, and P3, which assume values depending on the status of the V, F, and H bits, provide a limited correction (single errors) and detection (2 bit errors) of these bits.

Table 11.35 details the information carried by the EAV and SAV signals. Note that in the 720 × 483P format F = 0, as there are no fields needing identification.

Figure 11.39 is a spatial representation of the timing reference signals during a television frame of the 720 × 483P progressive scanning format. The L symbols identify specific lines in the active and blanking intervals. Table 11.36 describes the L symbols in Fig. 11.39.

11.6   Serial Transport Interfaces in the Studio

11.6.1   General

Chapter 7 discusses the bit-serial digital distribution of SDTV signals in the composite (NTSC or PAL $4f_{sc}$) or component (4:2:2) digital format, as defined by SMPTE 259M. The concept of the HDTV source format, the bit-parallel

TABLE 11.35   Timing Reference Signals for 720 × 483P Picture Format

Bit	First word	Second word	Third word	Fourth word
9	1	0	0	1
8	1	0	0	F
7	1	0	0	V
6	1	0	0	H
5	1	0	0	P3
4	1	0	0	P2
3	1	0	0	P1
2	1	0	0	P0
1	1	0	0	0
0	1	0	0	0

F = 0	V = 0 elsewhere	H = 0 in SAV
	V = 1 in frame blanking	H = 1 in EAV

P0, P1, P2, P3 protection bits based on values of F, V, H

TABLE 11.36   Frame Interval Timing Specifications for 720 × 483P Picture Format

(Refer to Fig. 11.39.)

Symbol	Definition	Digital line number
	Total number of lines	525
	Number of active lines	483
L1	First line of digital frame blanking	1
L2	Last line of digital frame blanking	42
L3	First line of active video	43
L4	Last line of active video	525

Figure 11.39 Frame timing relationship of 720 × 483 picture format. Note 1: The values (*F/V/H*) for EAV and SAV represent the status of bits for *F*, *V*, and *H* in a way that the 3-bit word composed of *F*, *V*, *H* represents a binary number expressed in decimal notation (*F* corresponding to MSB and *H* corresponding to LSB). For example, the value 3 represents the bits *F* = 0, *V* = 1, and *H* = 1. Note 2: Drawing not to scale. For symbol values see Table 11.36.

digital data stream, differs significantly from that of the SDTV formats. The main difference, apart from the higher sampling rate, is in the manner in which the *Y*, C_B, and C_R digital signals are organized.

The SDTV source format is a 10-bit parallel interface with time-division-multiplexed *Y*, C_B, and C_R signals resulting in a 27-Mwords/s data rate. The 4:2:2 format carries the timing reference signals EAV and SAV in its bit-parallel interface as well as ancillary data, if present. This signal is available on a cable with 10 pairs of conductors conveying the 10 bits and a 27-MHz parallel clock signal. This bit-parallel signal is serialized as detailed in Chap. 7.

The HDTV source format uses two bit-parallel channels as follows:

- A bit-parallel channel carrying *Y* signals
- A bit-parallel channel carrying time-division-multiplexed C_B/C_R signals

Each of the two data channels carries its own EAV and SAV timing signals. In addition to the timing signals the two channels may carry additional data in the space in between the EAV and the SAV as follows:

- Line number data composed of two words indicating the line number
- CRCs (cyclic redundancy codes) used to detect errors in the active digital line and the EAV which follows it

When these data are not available in the original format, they have to be added in a coprocessor prior to multiplexing, ancillary data insertion, and serialization. Figure 11.40 shows details of the horizontal blanking interval and the position of the line number and CRC data.

11.6.2 The 1.485-Gbps HDTV bit-serial distribution interface

SMPTE Standard 292M, *Bit-Serial Digital Interface for High-Definition Television Systems,* defines an interface common to several HDTV source formats operating at 1.485 Gbps and 1.485/1.001 Gbps. This standard deals with 1125-line, 1250-line (European HDTV), and 750-line scanning formats.

ITU-R BT.1120 Standard, *Digital Interfaces for 1125/60/2:1 and 1250/50/2:1 HDTV Studio Signals,* defines an interface common to 1125-line and 1250-line standards. It does not cover the 750-line standard, which is not considered HDTV in Europe, where it is commonly referred to as an ATSC format.

Most of the data in this section are borrowed from the latest version of the SMPTE 292M Standard. Table 11.37 details essential source format parameters. The source formats are identified as A through M, referenced to four SMPTE standards as follows:

- SMPTE 260M: The legacy HDTV format with 1125 lines per total frame, 1035 active lines per frame 2:1 interlaced with a 30-Hz or 30/1.001-Hz (NTSC-friendly) frame rate

- SMPTE 295M: The European HDTV format with 1250 lines per total frame, 1080 active lines per frame 2:1 interlaced with a 25-Hz frame rate.

- SMPTE 274M: A family of 1125 lines per total frame, 1080 active line formats

- SMPTE 296M: An HDTV format with 750 lines per total frame, 720 active lines per frame, progressively scanned with a 60-Hz or 60/1.001-Hz (NTSC-friendly) frame rate

As shown in Table 11.37, all formats are transmitted using the same nominal 1.485-Gbps rate. This is obtained by adjusting the number of total lines

Figure 11.40 Details of the horizontal blanking interval showing the position of the number and the CRC data (luminance channel shown).

TABLE 11.37 Source Format Parameters

Reference SMPTE Standard:	260M		295M	274M									296M	
Format:	A	B	C	D	E	F	G	H	I	J	K	L	M
Total lines per frame	1125	1125	1250	1125	1125	1125	1125	1125	1125	1125	1125	750	750
Y words per total line	2200	2200	2376	2200	2200	2640	2200	2200	2640	2750	2750	1650	1650
C_B/C_R words per total line	2200	2200	2376	2200	2200	2640	2200	2200	2640	2750	2750	1650	1650
Total active lines	1035	1035	1080	1080	1080	1080	1080	1080	1080	1080	1080	720	720
Y words per active line	1920	1920	1920	1920	1920	1920	1920	1920	1920	1920	1920	1280	1280
C_B/C_R words per active line	1920	1920	1920	1920	1920	1920	1920	1920	1920	1920	1920	1280	1280
Frame rate (Hz)	30	30/N	25	30	30/N	25	30	30/N	25	24	24/N	60	60/N
Interlace ratio	2:1	2:1	2:1	2:1	2:1	2:1	1:1	1:1	1:1	1:1	1:1	1:1	1:1
Divisor	1	N	1	1	N	1	1	N	1	1	N	1	N
Serial bit rate (Gbps)	1.485	1.4835	1.485	1.485	1.4835	1.485	1.485	1.4835	1.485	1.485	1.4835	1.485	1.4835

per frame and words per total line while maintaining the appropriate number of total active lines per frame and words per active line. All nominal frame rate formats (A, C, D, F, G, I, J, L) are operating at a bit-serial data rate of 1.485 Gbps. Formats with an NTSC-friendly frame rate (B, E, H, K, M) operate at 1.4835 Gbps (1.485/N). The divisor N has a value of 1.001. The difference between the two data rates is unimportant.

This section focuses on the D format that is typical of the other formats. The D format source data consist of two bit-parallel data streams operating in tandem:

- A bit-parallel Y data stream with a resolution of 10 bits per sample and a data rate of 74.25 Mwords/s

- A bit-parallel time-division-multiplexed C_B/C_R data stream with a resolution of 10 bits per sample and a data rate of 74.25 Mwords/s

Each of the two data streams carries its own EAV and SAV timing reference signals. The details of the timing reference signals are shown in Table 11.38.

Table 11.39 details the values assigned to the fourth EAV and SAV word (XYZ) at specific time instants of the picture including the values of the protection bits (P0, P1, P2, P3).

The line number data are composed of two words, LN0 and LN1, and their assigned values are shown in Table 11.40.

CRCs (cyclic redundancy codes) are used to detect errors in the active digital line and the EAV which follows it. The error detection code consists of two words determined by the polynomial generator equation

$$\mathrm{CRC}(X) = X^{18} + X^5 + X^4 + 1$$

The initial value of the CRC is set to zero. The calculation starts at the first active line word and ends at the final word of line number LN1. Separate CRCs are calculated for luminance data, YCR0 and YCR1, and

TABLE 11.38   Timing Reference Signals for 1920 × 1080I Picture Format

Bit	First word	Second word	Third word	Fourth word
9	1	0	0	1
8	1	0	0	F
7	1	0	0	V
6	1	0	0	H
5	1	0	0	P3
4	1	0	0	P2
3	1	0	0	P1
2	1	0	0	P0
1	1	0	0	0
0	1	0	0	0

F = 1 in field 1	V = 0 elsewhere	H = 0 in SAV
F = 0 in field 2	V = 1 in field blanking	H = 1 in EAV

For protection bits values see Table 11.39.

TABLE 11.39   10-Bit Hexadecimal (XYZ) and Binary Values of HDTV Interlaced Scanning Standard TRS (SAV and EAV) Fourth Word for Specific Instants and Related Protection Bits (P0, P1, P2, P3)

Instant	Hex	9 (MSB)	F	V	H	P3	P2	P1	P0	1	0 (LSB)
Field 1 Active video SAV	200	1	0	0	0	0	0	0	0	0	0
Field 1 Active video EAV	274	1	0	0	1	1	1	0	1	0	0
Field 1 Field blanking SAV	2AC	1	0	1	0	1	0	1	1	0	0
Field 1 Field blanking EAV	2D8	1	0	1	1	0	1	1	0	0	0
Field 2 Active video SAV	31C	1	1	0	0	0	1	1	1	0	0
Field 2 Active video EAV	368	1	1	0	1	1	0	1	0	0	0
Field 2 Field blanking SAV	3B0	1	1	1	0	1	1	0	0	0	0
Field 2 Field blanking EAV	3C4	1	1	1	1	0	0	0	1	0	0

TABLE 11.40   Line Number Data

Word	9 (MSB)	8	7	6	5	4	3	2	1	0 (LSB)
LN0	Not b8	L6	L5	L4	L3	L2	L1	L0	R	R
LN2	Not b8	R	R	R	L10	L9	L8	L7	R	R

Notes: L0 to L10 are line numbers in binary code. R = reserved, set to 0.

for color-difference data, CCR0 and CCR1. Table 11.41 shows the values assigned to the CRC data.

The two 74.25-Mwords/s bit-parallel data streams Y and C_B/C_R are multiplexed word by word into a single 148.5-Mwords/s 10-bit parallel data stream in the order C_B, Y, C_R, Y, C_B, Y, C_R, Y, and so on, as shown in Fig. 11.41.

Figure 11.42 shows the conceptual block diagram of an HDTV serializer. The serializer performs several functions:

▪ Each of the two the bit-parallel data streams is fed to a coprocessor, which inserts the line number and CRC data. An additional coprocessor may be used to insert ancillary data.

▪ The formatted bit-parallel data streams feed a multiplexer whose output is shown in the bottom row of Fig. 11.41. The data rate at the output of the multiplexer is 148.5 Mwords/s.

▪ The output of the multiplexer feeds a parallel-to-serial converter whose output is an NRZ (nonreturn to zero) coded bit-serial data stream with a 1.485-Gbps bit rate.

TABLE 11.41 CRC data

Word	9 (MSB)	8	7	6	5	4	3	2	1	0 (LSB)
YCR0	Not b8	CRC8	CRC7	CRC6	CRC5	CRC4	CRC3	CRC2	CRC1	CRC0
YCR1	Not b8	CRC17	CRC16	CRC15	CRC14	CRC13	CRC12	CRC11	CRC10	CRC9
CCR0	Not b8	CRC8	CRC7	CRC6	CRC5	CRC4	CRC3	CRC2	CRC1	CRC0
CCR1	Not b8	CRC17	CRC16	CRC15	CRC14	CRC13	CRC12	CRC11	CRC10	CRC9

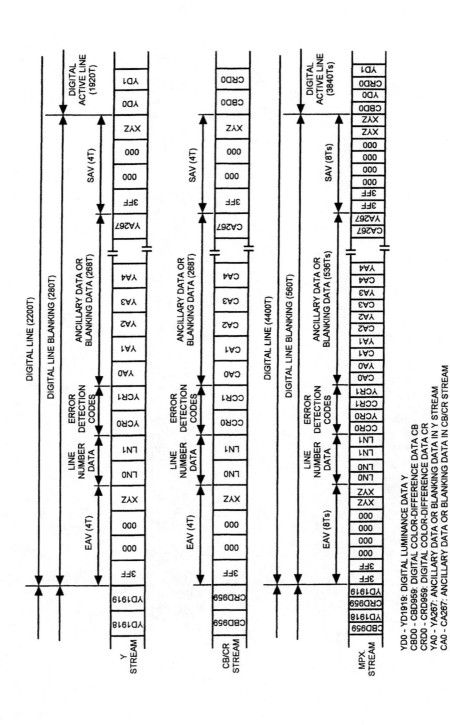

Figure 11.41 Formation of a Y, C_B, C_R multiplexed data stream from separate Y and C_B/C_R data streams.

YD0 - YD1919: DIGITAL LUMINANCE DATA Y
CBD0 - CBD959: DIGITAL COLOR-DIFFERENCE DATA CB
CRD0 - CRD959: DIGITAL COLOR-DIFFERENCE DATA CR
YA0 - YA267: ANCILLARY DATA OR BLANKING DATA IN Y STREAM
CA0 - CA267: ANCILLARY DATA OR BLANKING DATA IN CB/CR STREAM
T:1/74.25 MHz = 13.468 ns
Ts: T/2

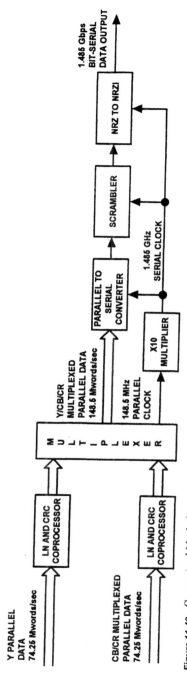

Figure 11.42 Conceptual block diagram of HDTV serializer.

■ The NRZ bit-serial signal feeds a scrambler which randomizes long sequences of 0s and 1s as well as repetitive data patterns that could result in clock regeneration difficulties. It produces a pseudorandom binary sequence (PRBS) which is combined with the NRZ data in order to randomize them. The generator polynomial for the scrambled NRZ is

$$G_1(X) = X^9 + X^4 + 1$$

■ The scrambled NRZ data stream feeds an NRZ-to-NRZI converter, which converts long runs of 1s to transitions, thus further helping the clock recovery process. The generator polynomial for NRZI is

$$G_2(X) = X + 1$$

Table 11.42 summarizes the characteristics of the HDTV bit-serial interface. Of particular importance are the tolerances on the parameters of the waveform (eye diagram) present at the output of a bit-serial digital signal source (the transmitter). Figure 11.43 shows a drawing of an eye diagram and identifies the measurement parameters listed in Table 11.42.

Table 11.43 lists the jitter specifications for bit-serial HDTV signals. The jitter specifications fall in two categories:

■ Timing jitter—jitter measured in a bandpass extending from 10 Hz to one-tenth of the clock rate

■ Alignment jitter—jitter measured in a bandpass extending from 100 kHz to one-tenth of the clock rate

Figure 11.44 shows a comparison of three types of jitter measurement bandpass characteristics.

TABLE 11.42   Characteristics of Bit-Serial Interfaces

Channel coding	Transmitter characteristics (see Fig. 11.43)	Receiver characteristics
Scrambled NRZI Input signal polarity: Positive logic Generator polynomials: 　$G_1(X) = X^9 + X^4 + 1$ 　$G_2(X) = X + 1$ Data word length: 10 bits Transmission order: LSB of any 　data word transmitted first	Unbalanced output Source impedance: 75 Ω nominal Return loss: ≥15 dB 　(5 MHz to 1.485 GHz) Output signal amplitude: 　800 mV$_{\text{p-p}}$ ± 10% DC offset: 0.0 V ± 0.5 V Rise time and fall time: ≤270 ps 　between 20% and 80% of signal 　amplitude. Differences not to exceed 　100 ps. Overshoot of rising and falling edges: 　≤10% of signal amplitude Jitter: See Table 11.43.	Unbalanced input Input impedance: 75 Ω nominal Return loss: ≥15 dB (5 MHz to 　clock frequency of signal) Optional cable loss equalization: 　20 dB at half clock frequency Interference tolerance: 　DC:　　　　　±2.5 V 　<5 MHz:　　　<2 V$_{\text{p-p}}$ 　5 to 27 MHz: <100 mV$_{\text{p-p}}$ 　>27 MHz:　　<40 mV$_{\text{p-p}}$

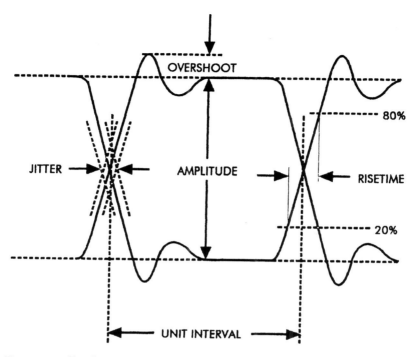

Figure 11.43 Eye diagram measurement dimensions.

TABLE 11.43 **Bit-Serial Digital Jitter Specifications**

(See Fig. 11.44.)

Parameter	Magnitude	Notation
Timing jitter lower band edge	10 Hz	B1
Alignment jitter lower band edge	100 kHz	B2
Upper band edge	$>^1/_{10}$ the clock rate	B3
Timing jitter	1 UI (673 ps p-p)*,†	A1
Alignment jitter	0.2 UI (134.6 ps p-p)	A2
Test signal	100% color bars‡	100/0/100/0
Serial clock divider	≠10 (preferred)§	n

*Some bit-parallel signal clocks may contain jitter of up to 2 ns p-p, which could cause excessive bit-serial jitter.

†UI stands for unit interval and has a value of 673 ps.

‡Color bars are chosen as a nonstressing test signal for jitter measurements. The use of a signal with long runs of zeros may give misleading results.

§This refers to the reference signal for triggering the jitter-measuring oscilloscope. Using a times-10 division is acceptable, but it may mask some jitter components.

See SMPTE RP 184 for definition of terms.

Amplitude

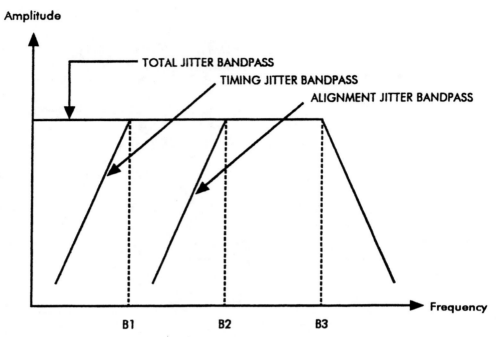

Figure 11.44 Jitter measurement bandpass.

The third column of Table 11.42 summarizes the HDTV receiver (deserializer) characteristics. The deserializer performs several functions:

- Cable-loss equalization. An automatic cable-loss equalizer corrects for high-frequency (>8 MHz) and low-frequency (<8 MHz) losses introduced by the coaxial cable. The equalization capability is a manufacturer's choice. Some circuits are capable of automatically equalizing losses introduced by up to 100 m of Belden 1694A coaxial cable. If a safety margin of 15 m is taken into consideration, the equalization capability is on the order of 85 m of cable length. For longer distances coaxial cables have to be replaced with fiber-optic distribution systems.

- Serial clock recovery. The bit-serial digital signal is self-clocking. The receiver contains a clock regenerator that recreates the clock through a phase-locked loop (PLL) and a voltage-controlled oscillator (VCO). The PLL bandpass is a compromise between noise rejection (narrow bandpass) and pull-in range (wide bandwidth). It is typically 2 MHz. The clock regeneration circuitry uses the received scrambled NRZI signal before the recovery of the original NRZ signal.

- NRZI-to-NRZ conversion. A reverse process recovers the NRZ signals by restoring the original long sequences of 1s.

- Descrambling. A reverse process removes the random transitions introduced by the scrambler to recover the original NRZ signal.

▨ Serial-to-parallel conversion to a 148.5-Mwords/s time-division-multiplexed Y, C_B, C_R data stream.

▨ Demultiplexing of the Y and C_B/C_R data streams.

Figure 11.45 shows the conceptual block diagram of an HDTV deserializer.

1.7 Data Multiplexing

The digital representation of analog video signals results in a specific number of samples per total line and a smaller number of samples per active line. The difference between the total line duration and the active line duration is an analog television legacy related to the necessity of formatting and transmitting an analog horizontal as well as vertical synchronizing signal to ensure a correctly timed reproduction of the picture. Component digital television replaces the analog synchronizing signals with TRS (timing reference signals). The EAV (end of active video) and SAV (start of active video) TRS have a duration of four words each, or a total of eight words, ensuring the proper synchronization of the digital video signals. This leaves a considerable quantity of unused words in the horizontal and vertical blanking intervals for other uses such as the transport of ancillary data. Ancillary data include digital audio and time code, as well as other data.

Source format D of SMPTE 292 standard specifies a digital horizontal blanking duration of 280 words for each of the two basic data streams Y and C_B/C_R. As shown in Fig. 11.41, the combined EAV, line number data, error detection codes, and SAV signals amount to 12 words leaving 268 Y words, numbered YA0 to YA267, and 268 C_B/C_R words, numbered CA0 to CA267, which can be used to transmit horizontal ancillary data (HANC). Source format D specifies a vertical blanking duration of 45 lines for each of the two basic data streams Y and C_B/C_R. Each active line has a duration of 1920 Y words, numbered YD0 to YD1919, and 1920 multiplexed C_B/C_R words, numbered CBD0 to CBD959 and CRD0 to CRD959, respectively, which can be used to transmit vertical ancillary data (VANC).

11.7.1 HANC multiplexing

The most important use of the HANC data space is for the insertion of audio signals accompanying the video signal. The SMPTE Standard 299M, *24-Bit Digital Audio Format for HDTV Bit-Serial Interface,* defines the manner in which 24-bit AES digital audio data and associated control information are embedded into the ancillary data space of bit-serial digital video signals conforming to SMPTE 292M.

11.7.1.1 General considerations. The preferred implementation is the use of 48-kHz sampled audio synchronous to video for intrastudio implementations, but 44.1- and 32.0-kHz can also be used. Audio is defined as being

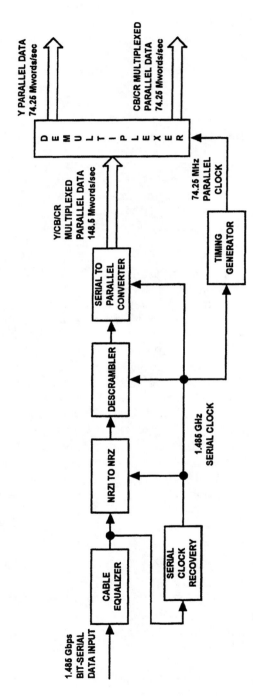

Figure 11.45 Conceptual block diagram of HDTV deserializer.

synchronous with video if the sampling rate is such that the number of audio samples occurring within an integer number of video frames is itself a constant integer number. Table 11.44 shows the relationship between audio sampling frequencies and the number of audio samples per synchronous integer number of video frames.

The standard provides for the embedding of a minimum of 2 audio channels (one audio pair) and a maximum of 16 audio channels (eight audio pairs). The audio channels are transmitted in pairs combined, where appropriate, into groups of four. Each group is identified by a unique ancillary data ID.

Audio data packets are multiplexed into the horizontal ancillary data space of all the lines of the C_B/C_R bit-parallel data stream except the line right after the switching line. Since the switching point for the 1125/60 system exists in lines 7 and 569, the audio data packets are not permitted on lines 8 and 570. Only audio data packets are allowed for use in the horizontal ancillary data space of the C_B/C_R data stream.

Audio control packets are multiplexed into the horizontal ancillary data space of the Y bit-parallel data stream. They are transmitted in the second line after the switching point or on lines 9 and 571.

Figure 11.46 shows details of the horizontal blanking interval of the C_B/C_R bit-parallel data stream. This example is typical of the 1920 × 1080 scanning format D where the total digital line (Y or C_B/C_R) has 2200 samples and the active digital line (Y or C_B/C_R) has 1920 samples. The digital line blanking has a duration of 280 samples. There are 268 data words in the C_B/C_R data stream, numbered CA0 through CA267, available for carrying ancillary data. The structure of the horizontal blanking interval is as follows:

- EAV. This is the end of active video timing reference. It consists of four words: 3FF, 000, 000, and XYZ.

- The line number data. This information consists of two words, LN0 and LN1.

- The error detection codes. This information consists of two words (CCR0 and CCR1).

- Ancillary data or blanking data: 268 words, CA0 to CA267 inclusive, which can be used to carry digital audio information.

- SAV. This is the start of active video timing reference. It consists of four words: 3FF, 000, 000, and XYZ.

TABLE 11.44   Relationship between Audio Sampling Frequencies and Number of Audio Samples per Synchronous Integer Number of Video Frames

Audio sampling rate	30-fps format	29.97-fps format
48.0 kHz	1600/1	8008/5
44.1 kHz	1470/1	147,147/100
21.0 kHz	3200/3	16,016/15

Figure 11.46 Details of the horizontal blanking interval of the C_B/C_R bit-parallel data stream.

11.7.1.2 The audio data packet. Audio data derived from two channel pairs (two AES/EBU data streams or four individual audio channels) are configured as an audio data packet. Figure 11.47 shows the structure of an audio data packet. It consists of

- The ancillary data flag (ADF): Three words: 000, 3FF, and 3FF.
- One data identification (DID) word. Its value depends on the audio group it identifies.
- One data block number (DBN) word.
- One data count (DC) word.
- 24 user data words (UDW), UDW0 through UDW23.
- One checksum word.

The UDW consists of three types of data:

- CLK (audio clock and phase data) is used to regenerate the audio sampling data at the receiving site. Table 11.45 details the bit assignment of the two CLK words UDW0 and UDW1.
- CHn (audio data), where n varies from 1 to 4, identifies the four audio channels making up a group. As shown in Fig. 11.47, each channel is represented by four 10-bit user data words, for a total of 16 words, numbered UDW2 through UDW17. Table 11.46 details the bit assignment of audio data (CHn).
- ECCs (error correction codes) consist of six user data words numbered UDW18 through UDW23. They are used to correct or detect errors in 24 words from the first word of ADF through UDW17. The error correction code is called BCH(31,25) and comes from a family of codes developed by Bose, Chaudhuri, and Hocquenghen. The BCH code is formed for every bit sequence of b0 to b7, respectively. The six user words are determined by the polynomial generator equation

$$\text{ECC}(X) = (X + 1)(X^5 + X^2 + 1) = X^6 + X^5 + X^3 + X^2 + X + 1$$

Table 11.47 shows the bit assignment of ECC.

Figure 11.48 details the manner in which an audio data packet is formatted from two AES/EBU data streams, 1 and 2. The audio of each channel consists of 29 bits as follows:

- AES audio: 24 bits.
- Validity (V): 1 bit.
- User data (U): 1 bit
- Channel status (C): 1 bit.

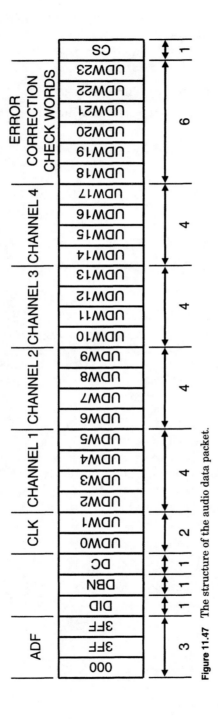

Figure 11.47 The structure of the audio data packet.

TABLE 11.45   Bit Assignment of CLK

Bit number	UDW0	UDW1
b9 (MSB)	Not b8	Not b8
b8	Even parity*	Even parity*
b7	ck7 Audio clock phase data	0
b6	ck6 Audio clock phase data	0
b5	ck5 Audio clock phase data	0
b4	ck4 Audio clock phase data	ck12 Multiplex position flag
b3	ck3 Audio clock phase data	ck11 Audio clock phase data (MSB)
b2	ck2 Audio clock phase data	ck10 Audio clock phase data
b1	ck1 Audio clock phase data	ck9 Audio clock phase data
b0 (LSB)	ck0 Audio clock phase data (LSB)	ck8 Audio clock phase data

*Even parity for b0 through b7.

TABLE 11.46   Bit Assignment of Audio Data (CHn)

	CH1			
Bit number	UDW2	UDW3	UDW4	UDW5
b9 (MSB)	Not b8	Not b8	Not b8	Not b8
b8	Even parity*	Even parity*	Even parity*	Even parity*
b7	aud1 3	aud1 11	aud1 19	P1
b6	aud1 2	aud1 10	aud1 18	C1
b5	aud1 1	aud1 9	aud1 17	U1
b4	aud1 0 (LSB)	aud1 8	aud1 16	V1
b3	Z	aud1 7	aud1 15	aud1 23 (MSB)
b2	0	aud1 6	aud1 14	aud1 22
b1	0	aud1 5	aud1 13	aud1 21
b0 (LSB)	0	aud1 4	aud1 12	aud1 20
	CH2			
Bit number	UDW6	UDW7	UDW8	UDW9
b9 (MSB)	Not b8	Not b8	Not b8	Not b8
b8	Even parity*	Even parity*	Even parity*	Even parity*
b7	aud2 3	aud2 11	aud2 19	P2
b6	aud2 2	aud2 10	aud2 18	C2
b5	aud2 1	aud2 9	aud2 17	U2
b4	aud2 0 (LSB)	aud2 8	aud2 16	V2
b3	Z	aud2 7	aud2 15	aud2 23 (MSB)
b2	0	aud2 6	aud2 14	aud2 22
b1	0	aud2 5	aud2 13	aud2 21
b0 (LSB)	0	aud2 4	aud2 12	aud2 20
	CH3			
Bit number	UDW10	UDW11	UDW12	UDW13
b9 (MSB)	Not b8	Not b8	Not b8	Not b8
b8	Even parity*	Even parity*	Even parity*	Even parity*
b7	aud3 3	aud3 11	aud3 19	P3
b6	aud3 2	aud3 10	aud3 18	C3
b5	aud3 1	aud3 9	aud3 17	U3
b4	aud3 0 (LSB)	aud3 8	aud3 16	V3
b3	Z	aud3 7	aud3 15	aud3 23 (MSB)
b2	0	aud3 6	aud3 14	aud3 22
b1	0	aud3 5	aud3 13	aud3 21
b0 (LSB)	0	aud3 4	aud3 12	aud3 20

TABLE 11.46 Bit Assignment of Audio Data (CHn) *(Continued)*

Bit number	CH4			
	UDW14	UDW15	UDW16	UDW17
b9 (MSB)	Not b8	Not b8	Not b8	Not b8
b8	Even parity*	Even parity*	Even parity*	Even parity*
b7	aud4 3	aud4 11	aud4 19	P4
b6	aud4 2	aud4 10	aud4 18	C4
b5	aud4 1	aud4 9	aud4 17	U4
b4	aud4 0 (LSB)	aud4 8	aud4 16	V4
b3	Z	aud4 7	aud4 15	aud4 23 (MSB)
b2	0	aud4 6	aud4 14	aud4 22
b1	0	aud4 5	aud4 13	aud4 21
b0 (LSB)	0	aud4 4	aud4 12	aud4 20

*Even parity for b0 through b7.
Z = AES block sync.
Un = AES user bit of CHn.
Pn = AES parity bit of CHn.
aud(0-23) = 24-bit AES audio data of CHn.
Vn = AES channel status bit of CHn.
The values of Vn, Un, Cn, and Pn are equal to those of the AES subframe, respectively.

TABLE 11.47 Bit Assignment of ECC

	UDW18	UDW19	UDW20	UDW21	UDW22	UDW23
Bit number	ECC0	ECC1	ECC2	ECC3	ECC4	ECC5
b9 (MSB)	Not b8	Not b8	Not b8	Not b8	Not b8	Not b8
b8	Even parity*	Even parity*	Even parity*	Even parity*	Even parity*	Even parity*
b7	ecc0 7	ecc1 7	ecc2 7	ecc3 7	ecc4 7	ecc5 7
b6	ecc0 6	ecc1 6	ecc2 6	ecc3 6	ecc4 6	ecc5 6
b5	ecc0 5	ecc1 5	ecc2 5	ecc3 5	ecc4 5	ecc5 5
b4	ecc0 4	ecc1 4	ecc2 4	ecc3 4	ecc4 4	ecc5 4
b3	ecc0 3	ecc1 3	ecc2 3	ecc3 3	ecc4 3	ecc5 3
b2	ecc0 2	ecc1 2	ecc2 2	ecc3 2	ecc4 2	ecc5 2
b1	ecc0 1	ecc1 1	ecc2 1	ecc3 1	ecc4 1	ecc5 1
b0 (LSB)	ecc0 0	ecc1 0	ecc2 0	ecc3 0	ecc4 0	ecc5 0

*Even parity for b0 through b7.

- Parity (P): 1 bit.

- Z flag: 1 bit derived from the preamble of the AES-3 stream. The Z bit is common to the two channels of an AES channel pair.

This representation is for one audio data packet carrying four audio channels from AES/EBU data streams 1 and 2. Carrying 16 audio channels requires an additional three audio data packets containing, respectively, data from AES/EBU data streams 3 and 4 (four channels), 5 and 6 (four channels), and 7 and 8 (four channels). Each group of four channels is identified by the DID word which assumes the following values:

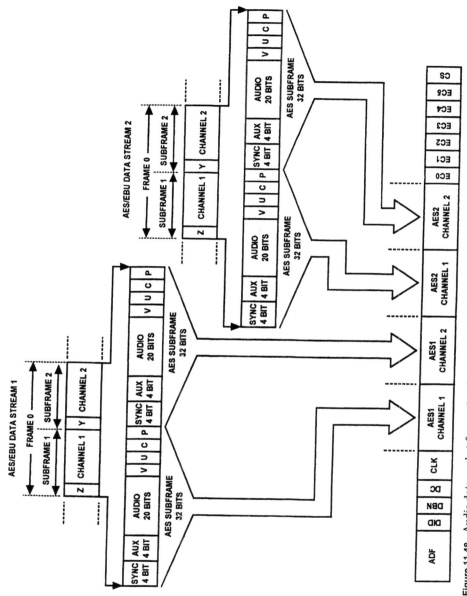

Figure 11.48 Audio data packet formatting from two AES/EBU data streams.

577

■ 2E7h for audio group 1 (channels 1 to 4)

■ 1E6h for audio group 2 (channels 5 to 8)

■ 1E5h for audio group 3 (channels 9 to 12)

■ 2E4h for audio group 4 (channels 13 to 16)

11.7.1.3 The audio control packet. The audio control packet is an ancillary data packet occurring once every field and containing data used in the process of decoding the audio data stream. Figure 11.49 shows the structure of the audio control packet. The audio control packet consists of.

■ The ancillary data flag (ADF). Three words: 000, 3FF, and 3FF.

■ One data identification (DID) word. Its value depends on the audio group it identifies.

■ One data block number (DBN) word.

■ One data count word.

■ 11 user data words, UDW0 through UDW10.

■ One checksum (CS) word.

The UDW consists of five types of data:

■ AF (audio frame) number data provide a sequential numbering of video frames to indicate where they fall in the progression of a noninteger number of samples per video frame (audio frame sequence). The first number of the sequence is always 1, and the final number is equal to the length of the audio frame sequence. A value of AF equal to all zeros indicates that the frame number is not available. The AF is common to all channels in a

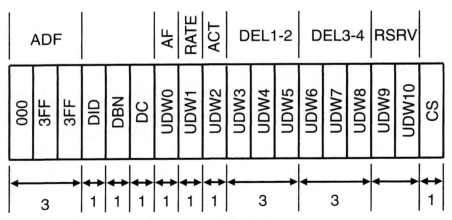

Figure 11.49 The structure of an audio control packet.

given audio group. Table 11.48 shows the bit assignment of AF. The correct use of the audio frame number requires that the audio sequence be defined. Three synchronous audio sampling rates are defined in this standard: 48.0, 44.1, and 32 kHz. All audio frame sequences are based on two integer numbers of samples per frame: m and $m + 1$. Odd-numbered audio frames (1, 3, 5, etc.) have the larger integer number of samples, and even-numbered audio frames (2, 4, 6, etc.) have the smaller integer number of samples with the exception shown in Table 11.49.

▣ RATE (sampling rate). The sampling rate of all channel pairs is defined by the word RATE. Table 11.50 shows the bit assignment of RATE. The sync mode bit, asx, when set to 1, indicates that the channel pairs in a given audio group are operating asynchronously. The rate code is defined in Table 11.51.

▣ ACT indicates active channels. Bits a1 to a4 are set to 1 for each active channel in a given audio group; otherwise they are set to 0. Table 11.52 shows the bit assignment of ACT.

▣ DEL m-n: These words indicate the amount of accumulated audio processing delay relative to video, measured in sampling intervals (T) of each channel pair of CH m and CH n. Table 11.53 shows the bit assignment of DEL m-n. The delay points are referenced to the point where AES/EBU data are input to the formatter. The delay words represent the average, inherent, delay value over a period no less than the duration of an audio frame sequence plus any of the preexisting audio delay. Audio delay data (del0 to del25) are represented in the format of 26-bit two's complement. Positive values indicate that the video leads the audio.

▣ RSRV: The words marked RSRV are reserved for future use. Their bit assignment is shown in Table 11.54.

The audio control packets are transmitted once every field.

11.7.2 VANC multiplexing

The manner in which various types of data are multiplexed in VANC is currently under study in various SMPTE committees. At the time of writing this book, October 1999, there is no definite standard available.

11.8 The Grand Alliance System

The ATSC standard is a terrestrial transmission standard which defines the bit stream content and transport, and its digital transmission, in a 6-MHz bandwidth TV RF channel. Production, display, or user-interface standards are not defined. The ATSC system employs multiple display transmission formats, digital audio and video compression, data packetization, and new RF signal modulation techniques.

TABLE 11.48   Bit Assignment of AF

	UDW0
Bit number	AF
b9 (MSB)	Not b8
b8	f8 Audio frame number (MSB)
b7	f7 Audio frame number
b6	f6 Audio frame number
b5	f5 Audio frame number
b4	f4 Audio frame number
b3	f3 Audio frame number
b2	f2 Audio frame number
b1	f1 Audio frame number
b0 (LSB)	f0 Audio frame number (LSB)

TABLE 11.49   Exception to Audio Frame Sequence

Frame rate (Hz)	Sampling rate (kHz)	Frame sequence	Basic numbering		Exceptions	
			Samples per odd audio frame (m)	Samples per even audio frame ($m+1$)	Frame number	Number of samples
30	48.0	1	1600		None	
	44.1	1	1470		None	
	32.0	3	1067	1066	None	
30/1.001	48.0	5	1602	1601	None	
	44.1	100	1472	1471	23, 47, 71	1471
	32.0	15	1068	1067	4, 8, 12	1068

TABLE 11.50   Bit Assignment of RATE

	UDW1
Bit number	RATE
b9 (MSB)	Not b8
b8	0
b7	0
b6	0
b5	0
b4	0
b3	X2 (MSB)
b2	X1 Rate code
b1	X0 (LSB)
b0 (LSB)	asx Synchronous audio: 0
	Asynchronous audio: 1

The packetization allows audio, video, and auxiliary data to be separated into fixed-size units, suitable for forward error correction, program stream multiplexing and switching, time synchronization, flexibility and extendibility, and compatibility with ATM format. The transport bit rate of 20 Mbps can accommodate either a single-channel high-definition TV service or multiple-program standard TV services.

TABLE 11.51   Assignment of Rate Code

X2	X1	X0	Sample rate
0	0	0	48.0 kHz
0	0	1	44.1 kHz
0	1	0	32.0 kHz
1	1	1	Free-running
0	1	1	Reserved
⋮	⋮	⋮	⋮
1	1	0	Reserved

TABLE 11.52   Bit Assignment of ACT

Bit number	UDW2 ACT
b9 (MSB)	Not b8
b8	Even parity*
b7	0
b6	0
b5	0
b4	0
b3	a4 Active: 1, inactive: 0 (CH4)
b2	a3 Active: 1, inactive: 0 (CH3)
b1	a2 Active: 1, inactive: 0 (CH2)
b0 (LSB)	a1 Active: 1, inactive: 0 (CH1)

*Even parity for b0 through b7.

TABLE 11.53   Bit Assignment of DEL*m-n*

Bit number	UDW3	UDW4	UDW5	UDW6	UDW7	UW8
	DEL 1-2			DEL 3-4		
b9 (MSB)	Not b8	Not b8	Not b8	Not b8	Not b8	Not b8
b8	del 7	del 16	del 25 (±)	del 7	del 16	del 25 (±)
b7	del 6	del 15	del 24 (MSB)	del 6	del 15	del 24 (MSB)
b6	del 5	del 14	del 23	del 5	del 14	del 23
b5	del 4	del 13	del 22	del 4	del 13	del 22
b4	del 3	del 12	del 21	del 3	del 12	del 21
b3	del 2	del 11	del 20	del 2	del 11	del 20
b2	del 1	del 10	del 19	del 1	del 10	del 19
b1	del 0 (LSB)	del 9	del 18	del 0(LSB)	del 9	del 18
b0 (LSB)	e	del 8	del 17	e	del 8	del 17

11.8.1   System overview

The layered ATSC transmission system architecture is discussed in Sec. 11.4.1. It is implemented by means of five subsystems that are the video and audio source coding and compression, the ancillary data services, the program multiplex and transport, the RF/transmission, and the receiver. A digital terrestrial television broadcasting (DTTB) model is shown in Fig. 11.50.

TABLE 11.54   bit Assignment of RSVR

Bit number	UDW9 RSRV	UDW10 RSRV
b9 (MSB)	Not b8	Not b8
b8	Reserved (set to 0)	Reserved (set to 0)
b7	Reserved (set to 0)	Reserved (set to 0)
b6	Reserved (set to 0)	Reserved (set to 0)
b5	Reserved (set to 0)	Reserved (set to 0)
b4	Reserved (set to 0)	Reserved (set to 0)
b3	Reserved (set to 0)	Reserved (set to 0)
b2	Reserved (set to 0)	Reserved (set to 0)
b1	Reserved (set to 0)	Reserved (set to 0)
b0 (LSB)	Reserved (set to 0)	Reserved (set to 0)

Figure 11.50   Digital terrestrial television broadcasting and cable network models.

11.8.2   Video system characteristics

The ATSC standard supports a range of program materials originating in different scanning formats. ATSC picture formats and associated SMPTE production standards are shown in Table 11.6, discussed in Sec. 11.5. Two program format levels, HDTV and SDTV (standard definition television), are represented.

Format variations include the active pixel and line numbers, and frame rates that affect the picture resolution. The choice of these pixel and line numbers was made to facilitate the interformat conversion. The 640 × 480 format, which is twice the CSIF, simplifies the interoperability with computer systems. In addition, when both the image aspect ratio and the active horizontal pixels to active vertical lines ratio are equal, this choice results in square pixels for most of the ATSC formats. The ATSC table takes into account the existing CCIR-601 format that can be displayed in both 16:9 and 4:3 aspect ratios. Decoded NTSC and computer-generated pictures will be transmitted in 4:3 aspect ratios as well.

Frame rates include 60 Hz to improve the picture quality and allow interoperability with computer systems. Frame rates of 59.94 and 29.97 Hz are included to simplify interworking with NTSC originating materials and simulcast transmissions. Frames rates of 24 and 23.98 Hz allow film originating materials to be transmitted in a film mode. The ATSC video encoder is designed to identify and remove the extra pictures that were added by the 3:2 pull-down process in film-to-video conversion.

Interlaced and progressive-scanning formats will accommodate the huge existing TV production investment in interlaced systems and more than 40 years of video program archives, and computer-oriented TV production requiring progressive scanning. In addition, the 1920 × 1080 format is specified with an interlaced version only, to reflect the current technology limitations. But the encoder architecture can support a 60-Hz frame rate version when technologies and costs are appropriate.

Accounting for the all-picture scanning formats and frames rates, there are 18 video transmission formats allowed by the ATSC standard. If we take into consideration the NTSC-friendly scanning formats (59.94 Hz instead of 60 Hz etc.), we end up with 36 choices. This wide variety of formats will allow program producers, application developers, and broadcasters to select the best format for the intended application. Materials in any other formats can be converted to one of the allowed scanning formats.

Video data compression is used to reduce the original high bit rate to a nominal compressed data rate of approximately 18.9 Mbps. Tables 11.6 and 11.7 show the original and maximum compressed bit rates for the ATSC image formats. The video compression scheme is based on the main profile syntax of the MPEG-2 video standard. The allowable parameters are bounded by the upper limits specified for the Main Profile @ High Level. However, some constraints* are applied to the MPEG-2 source coding specifications.

The compression scheme uses a motion-compensated DCT algorithm and B frame prediction. The video encoder supports the wide-motion estimation range needed for tracking fast motion pictures. In addition, it employs source-adaptive coding, field and frame motion vectors, and other techniques to improve compression efficiency. The compressed video data are then MPEG-2 formatted in an elementary bit stream.

---

*Details of these constraints can be found in annex A of the ATSC document A-53.

The video system design comprises two OSI layers, the video source formatting and the compression encoding, as shown in Fig. 11.51. The source formatting is a necessary step since today most of the program sources are produced in various analog component formats, using *G, B, R* or *Y, Pr, Pb* signals. Existing NTSC program materials can be decoded to *Y, Pr,* and *Pb* signals. A component signal digitization is then performed using a 13.5-MHz sampling frequency for ATSC SDTV signals and a 75-MHz frequency for ATSC HDTV signals. Different colorimetry standards are used in picture production. To ensure an identical colorimetry rendition on the receiver screen, colorimetry picture data parameters are specified by the color primaries, transfer characteristics (gamma), and matrix coefficients.

If *G, B, R* input signals are provided, a color space conversion to *Y, Pr, Pb* is made using matrix coefficient values indicated in Table 11.4. The *Y, Pr, Pb* component format is preferred before applying data compression, as a first data reduction is made during the color space conversion, compared to the *G, B, R* data signal format (see Sec. 2.3.2). Nonlinearity gamma correction is made by means of *Y, Pr, Pb* data lookup tables. A film detection circuit calculates the interfield correlation to look for the 3:2 pull-down sequence and remove the extra fields that were added in the TV data stream at the time of the film-to-tape conversion. Then a vertical chroma decimation (4:2:2 to 4:2:0) is performed before MPEG-2 compression encoding takes place.

The compression layer uses an MPEG-2-compatible encoder. To process HDTV bit rates, the original data stream can be separated on a block basis into several panels. Each panel compresses data to the MPEG "slice" layer, using B frames, wide-motion estimation to track fast motion, and adaptive field and frame DCT coding. These compressed data are multiplexed in the time domain and formatted into packets with the corresponding slice, picture, GOP, and sequence layers. These packets are then sent to the transport system.

11.8.3 Audio system characteristics

Audio system characteristics are defined in the ATSC standard document A-52.* The digital audio compression system is a constrained subset of the AC-3 system, developed by Dolby Labs. It encodes five full-bandwidth audio channels (3 Hz to 20 kHz) including left, center, right, left and right surround, and one reduced-bandwidth low-frequency enhancement (LFE) channel (3 Hz to 120 Hz) into a 384-kbps data stream. This LFE channel carries about one-tenth the bandwidth of the other channels, so the AC-3 system is frequently mentioned as 5.1 or 5+1 channels.

The ATSC A-52 Standard document specifies the encoding of a complete main audio service as well as additional audio services. Table 11.55 lists the audio service types contained in this standard. Main audio services may con-

---

*The reader can also consult annex B of the ATSC document A-53 and the Guide to the use of ATSC Digital Television Standard A-54.

Figure 11.51 Video source formatting and encoding.

tain from 1 to 5.1 channels while associated services are typically a single-channel service. The complete main (CM) audio service is the normal mode of operation and is complete with dialogue, effects, and music. The music and effects (ME) audio service does not include dialogue, which can be provided by an associated dialogue service (D). This makes possible the use of a different language or multiple languages at the expense of main service channel availability.

Associated services are for the visually impaired (VI) or hearing-impaired (HI), for emergency (E) message announcements, commentary (C) (optional program content), voiceover (V) (similar to the emergency service), and dialogue (D) to be used in the ME main audio service. The audio system also provides features to the user to control fluctuations in audio level between programs and to adjust the dynamic range of the original audio to the reproduction environment.

After digitization of the six analog audio channels, digital signals are sent to the AC-3 encoder. Figure 11.52 shows a simplified representation of an AC-3 encoder block diagram. These signals are converted from the time to the frequency domain using a modified DCT compression algorithm. A block floating-point process converts the transform coefficient to exponent/mantissa pairs. The mantissas are then quantized with a variable number of bits, based on a parametric bit allocation model. This model uses psychoacoustic masking principles to determine the number of bits for each mantissa in a given frequency

TABLE 11.55 Audio Service Types Contained in an AC-3 Elementary Stream

ATSC designation	Type of service	Number of channels	Bit rates
Complete main (CM)	Main audio	1 to 5 + 1	64 to 384
Music and effects (ME)	Main audio	1 to 5 + 1	64 to 384
Visually impaired (VI)	Associated	1	128
Hearing impaired (HI)	Associated	1	128
Dialogue (D)	Associated	1	128
Commentary (C)	Associated	1	128
Emergency (E)	Associated	1	128
Voiceover (VO)	Associated	1	128

band. This adaptive bit allocation process results in an acceptable S/N ratio for each DCT coefficient. The encoded spectral envelope and the quantized mantissa data corresponding to the six audio blocks are formatted into an AC-3 frame. An efficient processing algorithm is used to extract similarities that exist between audio channels, such as between right and left channels or between left surround and right surround channels. The identical information is then encoded and distributed between similar channels after decompression.

The AC-3 bit stream is composed of repetitive, independent, sync frames, as shown in Fig. 11.53. A sync frame is made of 6 audio blocks, representing $256 \times 6 = 1536$ audio samples, and one auxiliary block, located at the end of the frame, reserved for control or status information of system transmission. At the beginning of each frame, there is a header consisting of sync information (SI) (sync code, frame size, etc.) and bit stream information (BSI) (number of coded channels, etc.). Error detection codes (CRCs) are also included in each audio frame to allow the decoder to check its transmission transparency.

This elementary encoded bit stream is multiplexed with elementary video and auxiliary streams to obtain a stream as shown in Fig. 11.54. A typical 16-bit audio signal (one channel) has a bit rate of 768 kbps before compression. A 12:1 compression ratio is applied to reduce the bit rate to 64 kbps. After compression, the six audio channels are multiplexed into a single 384-kbps data stream to obtain an audio elementary bit stream.

The most important characteristics of the ATSC AC-3 data stream are as follows:

- There are 5 + 1 audio channels (5 main and 1 LFE).
- Each audio channel is sampled with a 48-kHz sampling frequency locked to the 27-MHz main system clock ($27 \times 2/1125$).
- There is an audio sample resolution of 16 up to 24 bits.
- The noise level is 95 dB below clipping.
- There is a maximum bit rate of 384 kbps for a main audio service.

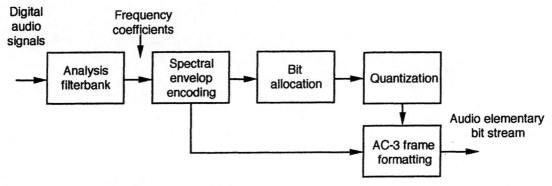

Figure 11.52 Simplified AC-3 encoder block diagram.

SI = Sync Information
BSI = Bit Stream Information
AB 0 = Audio Block #0

Figure 11.53 The AC-3 bit stream is made of repetitive sync frames.

Figure 11.54 AC-3 encoded bit stream formatting and multiplexing into a program stream.

- The maximum bit rate is 128 kbps for a single-channel associated audio service.

- The maximum bit rate is 512 kbps for the simultaneous transmission of a main and an associated audio service.

- There is interoperability with the digital video disk (DVD) format.

11.8.4 Ancillary data services

In addition to the audio and video services, new digital services can be included in the MPEG-2-based transport layer defined in the ATSC standard. The ancillary data services are as follows:

- Text-based ancillary services such as program subtitles (closed captions) and emergency messages.

- Program guide information. This optional capability allows a receiver to build an on-screen grid of program information. Additional control information will facilitate navigation through the multiplicity of programs and services. Program ratings and theme category and subcategory information are provided for each programming event through seven types of default override record data, each rated with 16 levels, such as category, subcategory, content advisory, violence, sexual content, language, and ratings. Other type values are reserved for future extension. Program rating specifications, defined in the ANSI/EIA-608-94 Standard, and under review in 1996 Section Extended Data Service (XDS), are used for program content rating. Daylight savings time control is also provided and will be used to convert program guide time (GMT) to local time.

- System control information. This optional capability can allow interoperability with other broadcast applications such as cable, satellite, multichannel multipoint distribution service (MMDS), and satellite master-antenna television (SMATV). Information on available transport streams and their constituents will ease network operator control of multiple transport streams.

- Future additional services. Also referred to as private data, this optional capability provides the necessary extensibility to envision new business opportunities and new services. TV distributors can better compete by offering a unique set of services at any time in the future.

11.8.5 Program multiplex and transport system characteristics

The program multiplex and transport system employs a fixed-length transport stream packetization approach based on a subset of the MPEG-2 transport system specification. Each data packet contains only one type of data and can be identified by a packet ID header. This design makes possible the transmission of new services, from broadcast distribution of computer software to the transmission of high-quality pictures to computers.

Elementary audio, video, and auxiliary data streams are packetized at the application-encoding layer or at the transport layer. This packetization is application-dependent and results in elementary stream packets of variable length.

Packetized elementary streams (PESs) are then converted to fixed-length 188-byte packets, each with a 4-byte header, before being fed to the transport layer. Audio, video, and auxiliary PES packets are multiplexed together to constitute a program stream (see Figs. 11.50 and 11.54). The mix of these different packets can be dynamically allocated to meet the available channel bandwidth.

The transport layer provides the multiplexing and synchronization of services (audio, video, auxiliary) that comprise a program. The transport syntax is flexible enough to allow a program to be comprised of a large number of elementary bit streams. Each transport packet header contains 1 sync byte and several flags for data stream content and program identification, as shown in Fig. 11.55. An optional adaptation header of variable length is included in the data payload and contains additional information such as time and media synchronization, bit stream splice point flag, etc.

A conditional access mode capability, by means of data packet scrambling, can be implemented in the transport stream with bits defined in the packet header. The transport syntax supports all key encryption and descrambling functions, applicable to program streams or elementary streams.

The transport system output is a continuous MPEG-2 transport stream at a constant rate of 19.39 Mbps, which is intended for the primary application of an 8-VSB terrestrial broadcasting.

11.8.6 RF/transmission system characteristics

The ATSC standard recommends the use of vestigial sideband (8-VSB) digital transmission to meet the needs of a variety of media. A transmitter performs the transport-to-transmission layer conversion by implementing several processing functions, which are described hereafter. A typical transmitter block diagram is shown in Fig. 11.56.

The incoming transport data feed a frame synchronizer that locates the transport layer sync bytes and aligns the serial data stream into bytes. The transmission system is fed with a 19.39-Mbps bit-serial data stream made of 188-byte MPEG-compatible data packets. These 188 bytes include 1 sync byte and 187 bytes of data which represent a useful payload data rate of 19.39 $\times$ 187/188 = 19.28 Mbps.

Figure 11.55 Multiplexing of elementary streams into a program stream and program streams into a transport stream.

Figure 11.56 Typical implementation of a VSB transmitter.

A data randomizer ensures that constant data values do not exist in the serial bit stream by breaking the data value uniformity. Such data sequences would generate a nonuniform spectrum that could produce interference effects on other services such as existing NTSC transmitters. Random data also improve the receiver recovery loop performance.

Reed-Solomon check bytes (20 R-S parity bytes) for forward error correction (FEC) are added to the 187 data bytes of each packet to provide for an efficient correction of transmission errors. The sync byte is excluded in this process.

A data interleaver spreads the data bytes from the same R-S block over time and provides further error protection by disseminating bursts of errors.

A trellis coding process is used to increase the input signal redundancy by doubling the data levels. Each 208-byte data block is converted in 832 two-bit words that are trellis-encoded (four-state trellis encoder) and mapped into a one-dimensional constellation, defined as eight-level VSB.

A multiplexer receives the 832 three-bit symbols from the trellis encoder and inserts data segment sync (four symbols per segment) and data field sync (one segment per field) signals. This helps the receiver in packet and symbol clock acquisition and phase locking during adverse receiving conditions.

A parallel-to-serial converter allows data bits to be processed 3 bits at a time, at the 10.76-MHz symbol rate, and D/A-converted to eight-level pulses. A linear phase raised-cosine Nyquist filter is used to minimize intersymbol interference. A small pilot (-11.3 dB below the average data signal power) is then added to the VSB signal at the suppressed-carrier frequency (310 kHz from the lower band edge). This facilitates signal acquisition and increases the pull-in range in the receiver.

A VSB modulator provides a filtered VSB IF signal at a nominal 46.69-MHz frequency (44 MHz + 10.76/4 MHz offset). Finally, the RF up-converter translates the IF data signal spectrum to the desired RF channel. Terrestrial field tests have shown that the TV signal reception quality is significantly improved with the VSB transmission system compared to that of the NTSC.

The high data mode has a payload data rate of 38.57 Mbps (twice the terrestrial mode) and is devoted to less demanding transmission environments, such as cable networks. The transmission system uses a 16-level VSB modulation and looks identical to the terrestrial system. However, a mapper that creates multilevel data symbols replaces the trellis encoding.

11.8.7 Receiver characteristics

The ATSC standard does not specify the requirements for a compliant receiver. However, the FCC has draft a recommendation specifying that all TV receivers be capable of decoding the audio, video, and auxiliary signals specified in the ATSC standard documents, regardless of how the signals are displayed or broadcast. The functionality of receiving multiple services could be implemented with either receivers or set-top adapters to convert digital ATSC to analog NTSC, S-video, or component analog signals.

The ATSC receiver reverses the functions of the RF transmission and transport layers and, after decompression and decoding, generates video and audio suitable for the display format and listening conditions chosen.

11.8.7.1 Video receiver characteristics.
The display format is independent of the transmission formats. To simplify designs, TV receivers may not display different formats. They may be built according to a "native" display format standard, be it a 1920 × 1080, a 1280 × 720, or a 720 × 480 picture format. A display format conversion involving a built-in transcoding capability of picture format dimensions, frame rates (for example, 24 to 30 fps), scanning processes, and image aspect ratios may take place in the TV receiver to convert whatever is received to the native display format. The TV receiver or set-top box will receive coded bit streams, made of a set of data and computer instructions, which it will execute to recreate the original picture and sound in the receiver's native display format.

An important consideration is the screen type and size used for HDTV displays. A common SDTV 4:3 picture format screen has a 27-in diagonal and a picture height of 18 in. The normal viewing distance is about 6 times the picture height, or 9 ft in this case. At this distance the scanning lines are barely visible, and the analog horizontal resolution of 333 LPH (for an ideal 4.2-MHz luminance video bandwidth and a Kell factor of 0.7) is comfortably perceived by the viewer. HDTV pictures with a total of 1080 active lines per frame would result in a horizontal resolution of about 756 LPH (for an ideal 30-MHz luminance video bandwidth). To perceive this amount of detail, and assuming a screen with a 16:9 format but the same 18-in picture height, would require that the viewer be seated at a distance 3 ft from the screen. Alternately the screen size would have to be doubled, resulting in a screen height of 36 in. Contemporary CRT technologies cannot produce a CRT meeting these size requirements at either an affordable price and weight or an acceptable neck dimension. Alternate back-projection technologies offer limited screen brightness and viewing angle. More recent technologies such flat-panel displays do not meet the vertical resolution requirements or size and are very high-priced. Using smaller screens would, by necessity, result in a loss of perceived HDTV picture detail, and all the advantages of high definition would be lost. This situation definitely requires either different viewing conditions or the development of large-dimension flat screens of adequate resolution at an affordable cost.

11.8.7.2 Audio receiver characteristics. The AC-3 standard defines a very flexible system that can provide eight levels of audio decoding configurations, from mono, stereo, to multichannel surround sound with six discrete channels. Specific instructions are encoded at the production stage in the AC-3 data stream header in order to deliver audio signals simultaneously optimized for many different types of listeners. These specific instructions comprise

- One or two channel mixing coefficients of the original multichannel program to be used for mono or stereo playback.
- Speech-level information to allow programs and channels with different operating levels to be adjusted at the reproduction level chosen by the listener.
- Compression control signals to adapt the dynamic range of the decoded signals to the user reproduction environment. A listener can decide to have loud effects brought down in level while the quiet ambiance is brought up and the dialogue is left unaffected.

Three types of receivers could appear on the marketplace:

- High-end receiver (1920 × 1080). SDTV and 1280 × 720 format programs are up-converted to 1920 × 1080. High-end video converters will include frame rate conversion (say, 24 to 30P), progressive to interlaced, etc. Multichannel surround sound capability would be included.
- Medium-end receiver (1280 × 720). The 1920 × 1080 format programs are down-converted to 1280 × 720, and SDTV format programs are up-converted to 1280 × 720. Multichannel surround sound capability would be included as well.
- Low-end receiver (720 × 480). HDTV and 1280 × 720 format programs are down-converted to 720 × 480, in a 16:9 aspect ratio. It may have frame rate conversion depending on the converter chip cost. NTSC receivers with

the addition of set-top boxes will enable the customer to receive NTSC and SDTV/HDTV programs on an interlaced 4:3 receiver. An audio stereo signal mix could be provided or a multichannel surround sound capability if the AC-3 decoding chip cost is inexpensive.

Other receiver functionalities include the following:

▥ The detection of closed-captioning and time code signals.

▥ The handling of data or graphics-related services.

▥ The provision for an interface with a computer, camera, and videotape recorder.

▥ The implementation of a conditional access security system. A joint subcommittee of the Electronic Industries Association (EIA) and the National Cable Television Association (NCTA) has proposed a plug-in security module technique to implement such a capability in consumer equipment.

Figure 11.57 shows a conceptual block diagram of an ATSC receiver. An RF tuner followed by a transmission and a transport layer performs the reverse process previously described in the respective sections. Decoded audio, video, and auxiliary signals can be selected by the user for on-screen monitoring or for videotape recording through an IEEE P-1394 standard high-performance serial bus interface, which could be used to interconnect with interactive games, computers, cameras, and other devices. However, the implementation of this interface, in ATSC receivers, is still being discussed.

11.9   The European Digital Video Broadcasting (DVB) System

The DVB project started in September 1993 when several industry leading companies met together under the European Broadcasting Union (EBU) to develop specifications for transmitting MPEG-2-based digital video by

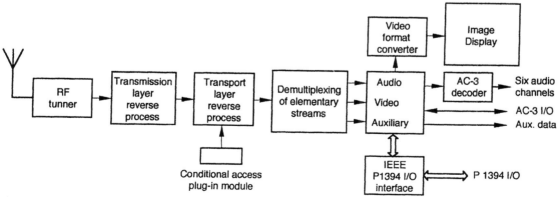

Figure 11.57 Typical design of an ATSC receiver.

cable, satellite, or terrestrial means in Europe. Since then, a significant number of participants from the Far East and Australia have joined the initial group, which is now composed of broadcasters, satellite and cable operators, regulatory organizations, and representatives of the consumer electronics industry.

These participants shared identical concerns to their North American counterparts in designing a future TV system that would combine an open architecture with full interoperability, and result in market-oriented systems. So it is not surprising that a lot of commonalties exist between the ATSC standard and the DVB project specifications.

The DVB project has developed standard specifications for the harmonization and commonality of future digital television systems to be used by all delivery environments and to realize an economy of scale in system design, component development and manufacturing, and user acceptance. The delivery systems, shown in Fig. 11.58, are made of common components and application-specific components such as the modulation scheme that must be adapted to the delivery medium. The DVB family of delivery system standards includes the following:

Figure 11.58 A typical DVB transmission chain model.

- **DVB-S.** This is a satellite transmission system, configurable to meet a range of transponder bandwidths and power requirements, in the 11 to 12-GHz band. The DVB-S system uses a quadrature phase-shift keying (QPSK) modulation and is optimized for a single carrier per transponder and satellite power efficiency. The maximum data rate from the MPEG-2 transport layer is about 38.1 Mbps.

- **DVB-C.** This is a cable delivery system to be used with 7- to 8-MHz cable channels. The DVB-C uses a 64 quadrature amplitude modulation (QAM) system, optimized for high levels of S/N ratios and a low level of intermodulation products. The maximum data rate from the MPEG-2 transport layer is about 38.1 Mbps.

- **DVB-CS.** This is a community or localized cable system for single-building distribution, called the satellite master antenna TV (SMATV) system, using 8-MHz cable channels.

- **DVB-T.** This digital terrestrial broadcasting TV system is used with 7- to 8-MHz channels. The maximum data rate from the MPEG-2 transport layer is about 24 Mbps.

- **DVB-MS.** This is identical to DVB-S and is for a multichannel multipoint distribution service (MMDS) broadcasting system (>10 GHz).

- **DVB-MC.** This is identical to DVB-S and is for an MMDS broadcasting system (<10 GHz).

The DVB family also includes universal applications for the DVB delivery systems:

- **DVB-SI.** This service information system is similar to that of the ATSC standard.

- **DVB-CI.** These are specifications of a common interface for conditional access implementations.

- **DVB-TXT.** This teletext transport specification transmits data information in elementary streams, independently of the vertical blanking interval.

As in the ATSC standard, DVB guidelines define the minimum requirements for the first generation of satellite, cable, and transmitter receivers of conventional quality TV programs. When there is a market for HDTV, the receiver guidelines can be updated. To access digital services, the viewer needs an extra decoder that connects to the conventional receiver. This low-cost set-top box or integrated receiver decoder (IRD) will not bring the benefits of wide-screen pictures. An affordable integrated wide-screen digital television receiver will allow the customer to benefit from the improved picture quality, multichannel sound, and format attractiveness.

All DVB systems are compatible with one another. However, several important differences exist with the ATSC system specifications:

- The audio encoding uses the MPEG-2 layer II (Musicam) audio standard.

- The video encoding uses the MPEG-2 video standard, Main Profile @ Main Level (MP @ ML), with a maximum resolution of 720 × 576 pixels. Lower format resolution down to 352 × 288 can be transmitted, and the receiver will provide appropriate up-conversion to produce a full-screen luminance display of 720 × 576. The DVB project does not standardize HDTV formats, but the transport and transmission systems have the capability to handle future compressed HDTV data. The current MPEG-2 LSI implementations at MP @ ML allow a variable bit rate operation from 1.5 to 15 Mbps.

- The transmission system must deliver 4:3, 16:9, and 20:9 (2.21:1) image aspect ratios at a 50-Hz frame rate.

- The terrestrial broadcasting standard specifies the use of a transmission scheme based on the coded orthogonal frequency-division-multiplexing (COFDM) principle already used in digital audio broadcasting. Either 1704 ("2k") or 6816 ("8k") individual carriers may be used to transmit digital data to receivers. The 8k system is more robust but increases the receiver complexity and cost. Some countries have already adopted the 2k system even though it is known that 2k receivers will never be able to decode 8k transmissions when the need and the technology arrive.

12

Transition to DTV

The world of broadcasting is changing rapidly because of the emergence of several new technologies. The introduction of DTV has been facilitated by ATSC and DVB standardization activities and has generated the coexistence of uncompressed and compressed SDTV/HDTV signal formats in production, contribution, and distribution areas. Computer-based audio and video editors are now widely used and have introduced the concept of file transfer techniques whereby program materials are stored as data files and sent to other computers using standard computer networks.

The access of business data networks such as local-area networks (LANs) and wide-area networks (WANs) with high speed and low delays has been facilitated by the Internet success. Audio and video compression techniques have made possible the transfer of TV program materials as data through contribution and distribution communication networks. The implication of data networks in studio designs brings the needs of basic technical knowledge for broadcast people. This chapter reviews the transport technologies needed in intra- and interfacility applications to transfer audio/video streams and media files.

Multiformat video and audio program materials are unavoidable in production and distribution. Converters are required to generate programs in specific formats suitable for distribution networks and broadcasting requirements. Characteristics and performance of these standard converters are briefly explained.

MPEG-2 improvements and requirements for production applications are described. The MPEG-2 introduction in studio applications is spreading toward the ultimate goal of a complete MPEG studio. Several transition scenarios from current analog composite and digital component to DTV architecture are then developed.

12.1 Interfaces, Networks, and Transmission Protocols

The convergence of digital television, computer, and telecommunication technologies is enabling the distribution of TV program materials as bit streams between studios and facilities. New network infrastructures are now made possible by combining very fast digital video processors, compression-based storage systems, and high-speed networks.

Multimedia archives have been gathered for the past 40 years and constitute a very important source of materials available for reuse in new programs. Digital storage and retrieval systems now constitute the basic elements of any broadcast architecture. This has created the need for an accurate description of each material using a metadata descriptor and for fast browsing of large quantities of media materials that can easily be achieved through computers and networks.

The SMPTE and the EBU early on recognized the need for a common approach to design and implementation of new television systems. A joint Task Force for "Harmonized Standards for the Exchange of Program Materials as Bit Streams" (TF-HS) has been established by the EBU and the SMPTE. An impressive amount of work has been carried out by this task force of more than 200 worldwide experts, and a final report was issued in July 1998, "Analyses and Results."*

The task force work and report are structured in layers almost similar to the OSI model, where each layer performs a specific task. However, some applications used in broadcast and described in this report are considered part of the presentation layer of the OSI model, e.g., compression data formats. The layered approach facilitates the standardization of data broadcasting systems and their evolution. Additions or changes within a layer will not affect the others.

The task force report is divided into four sections: systems, compression, wrappers and metadata, and networks and transfer protocols. The SMPTE has recently adopted a similar structure for its standards committee activities, as shown in Fig. 12.1.

12.1.1 Reference architecture

The TF-HD has developed a reference architecture (RA) that allows interoperability and flexibility between existing and future applications, and between mixed software and hardware systems. This RA recommends interfaces; push and pull file transfer protocols; streaming protocols for video, audio, and data across channels and networks; and a file system access de facto standard.

An object-oriented structure has been adopted by the TF-HS which is based on a preliminary work on object-based architecture made by DAVIC. Each object contains its own data and performance specifications corresponding to specific processes. This approach facilitates the division of complex studio operational problems into manageable blocks and subblocks, and the development of solutions by means of software design, computer, and network technologies.

---

*This task force report is available from the SMPTE Web site (www.smpte.com) or directly from the EBU and the SMPTE.

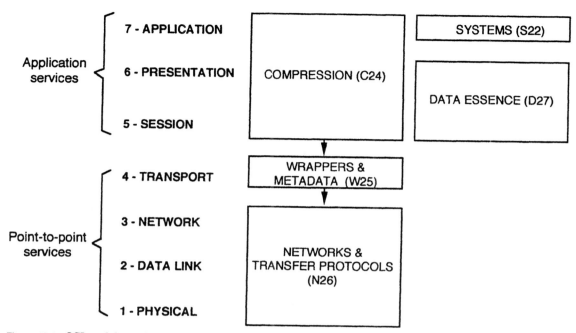

Figure 12.1 OSI model and SMPTE standards committee structure.

All standardization activities now use new concepts, and their knowledge is essential for the understanding of all new standards and related developments. The packaging of audio, video, and related data materials together has made necessary the definition of a data architecture (RA components) to facilitate the exchange of program materials as bit streams.

12.1.2 Reference architecture components

■ Essence. Audio or video essence is constituted by the actual raw audio or video sample data that can be compressed or not. Data essence refers to text information such as in HTML, program guides, and closed captioning.

■ Metadata. Metadata are information related to essence. They are generated at the time of the production of programs as information that fully describes the program material content in a very short and concise manner. The program content is stored in a separate area and easily retrieved by adding a unique material identifier (UMID). Metadata examples are time code, source ID, program ID, edit decision list (EDL), version number, etc. It allows a fast browsing of program materials, fast access through networked connections to archives and on-line servers, and asset management during the production, postproduction, distribution, and delivery of programs. SMPTE is standardizing several classes of metadata such as essential metadata, access metadata, and compositional metadata that are described in an SMPTE dictionary.

- Content. A content unit is any small element of essence data that has a duration of a frame, a shot, or any specific duration.

- Content element. A content element is a type of essence data added to its corresponding metadata. A content element can be made of audio, video, or associated data essence.

- Content item. A content item is a collection of up to 255 content elements of the same essence type. Additional metadata is added to associate content elements together.

- Content package. A content package is a combination of one or several content items used to generate a program. Additional metadata is added to associate content items together. Figure 12.2 shows an example of program materials and related information structured in a content unit, content element, content item, and content package. A more complex content package would include any combination of system, video, audio, and text items. Content multiplexing is very efficient in preserving the timing between all items.

- Wrapper. The wrapper (or container) concept is directly related to file formats and links together essence and metadata to form a content. Different wrappers are used depending on the application such as streaming wrappers and file storage wrappers. Figure 12.3 shows an example of a wrapper and its content. The content may also be a complex content package composed of one or several content packages and content items.

- Unique material identifier (UMID). It allows the unique identification (or labeling) of all media material items as they pass in processing systems involving streaming and storage, be it locally or remotely, and archiving. A UMID is composed of a content unit, which can be of variable duration depending on the type of materials, and clip information defining the number of contiguous content units. SMPTE is developing several standard documents related to the UMID specifications and the metadata formats that are used in conjunction with a UMID.

- Unique program identifier (UPID). A complete program is generally made of several audio, video, and text materials, each having its UMIDs. As soon as this program is completed, it receives a UPID and some metadata is removed.

12.1.3 Networks

A network interconnects many data sources to receiving devices (also called sinks). They are classified in two main categories: local-area networks for the interconnection of users in one geographical area and wide-area networks used to interconnect several geographic areas. With the rapid evolution of digital networks and increase in available bandwidth, new ways of sending data can be used in broadcast applications.

1) Content Element = Essence + Metadata

2) Content Package = Audio + Video Content Items

Figure 12.2 Example of audio and video materials, and related information structured in content packages.

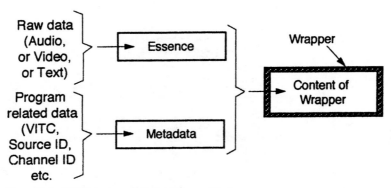

Figure 12.3 Wrappers and content formatting.

The TF-HS has recognized the fact that no single network can accommodate user requirements in intra- and interstudios used in the broadcast industry. Several interfaces, networks, and transport mechanisms have been recommended. They are described in Secs. 12.3 and 12.4.

12.1.4 Transmission protocols

A transmission protocol organizes the transport of data from a source to a given destination. It must provide an efficient access mechanism to the communication channel. Due to the complexity of the communication system requirements, protocols are divided into layers that are stacked on top of one another. The OSI model is generally used for the specification of protocols for all open networks, i.e., for networks that are defined by a set of international standards. Such a model is defined in Sec. 11.4.1 of this book. At the link and network layers, data streams are formatted into long packets of bytes called frames or small packets called cells which are added with frame or cell information to be used by the receiving devices. Different transmission protocols have been developed to suit different applications and still coexist. Figure 12.4 summarizes the data stream formatting into cells or frames sent over these four transmission protocols.

Asynchronous protocol. This communication mode (cell relay and ATM) was the first used to transmit characters. It is characterized by a start and a stop bit for each character and a fixed bit sequence for each cell. The timing is then derived from these start and stop bits, and the duration between each sequence is variable. Data files or packets are sent when the network is available for traffic. Cells are made of small packets of bytes and are suitable for efficiently multiplexing different types of data into the same channel.

Isochronous protocol. Similar to the asynchronous protocol, it uses a start and a stop bit, but the time between each sequence is kept constant. This protocol was developed to improve the synchronization of the first generation of receiver modems, by sending bits at a regular rate. It brings the advantages

Cell header   Fixed payload length (short packets)

1) Asynchronous protocol (ATM)

1 cell

Variable intercell space

2) Isochronous protocol

Constant intercell space

Frame Header   Frame Trailer   Fixed or variable payload length (long packets)

3) Plesiochronous protocol (PDH)

Frame Header   Frame Trailer   Fixed or variable payload length (long packets)

4) Synchronous protocol (SDH)

Figure 12.4 Data stream representation of current transmission protocols.

of predictable transmission delays and the possibility of establishing a guaranteed bandwidth value. This protocol is well suited to real-time transmission of audio and video data files or packets.

Plesiochronous protocol. A nearly synchronous network has a well-controlled nominal data rate but is not fully synchronized to a common data clock (frame relay and PDH networks). Some variation in clock frequency and phase of the various links and nodes is allowed. This type of network was used as the first digital communication system and has now been superseded by synchronous networks. The DS hierarchy in North America (E hierarchy in Europe) is a well-known application of the plesiochronous digital hierarchy (PDH). However, major operational problems exist with PDH networks. A complete demultiplexing and remultiplexing on the whole hierarchy must be performed to realize data stream add/drops. In addition, measurement of network performance and rerouting after network failure are difficult due to the multiplexer structure.

Synchronous protocol.   This communication mode uses a common clock to synchronize two communicating devices. It allows several digital signals to be directly multiplexed or demultiplexed into a higher- or lower-rate synchronous signal. This is called synchronous multiplexing. Transmission delays are well defined for each data file or packet. Synchronous digital hierarchy (SDH) network interfaces are based on this standardized ITU protocol and use a synchronous transport module hierarchy (STM-0 to STM-64).* Due to the arrival of inexpensive fiber-optic transmission technologies, the synchronous optical network (SONET) interface standard has also been developed, thus creating the synchronous transport signal hierarchy (STS-1 to STS-192) or optical carrier hierarchy (OC-1 to OC-192).

12.2   Data Streaming and Data File Transfers

Audio/video program materials are currently transmitted from studio to studio or between locations by means of a continuous data stream. New technologies now permit the transfer of streams of compressed materials in a faster-than-real-time manner between equipment or between studios. But the recent availability of low-delay and high-bandwidth networks is changing broadcast studio distribution and delivery methods. Any audio/video program materials can now be transmitted as media files in a manner similar to the exchange of files between computers.

During investigations of asset sharing, the TF-HS has defined a reference architecture for both content file and streaming transfers. Interfaces, networks, and applicable transfer protocols have been considered for recommendation. Each of these two transfer methods has advantages and disadvantages that are described in the following sections.

12.2.1   Streaming transfers

- Links are built to provide a high quality of service (QOS) because they do not offer a guaranteed delivery† of information. BER, delay, jitter, and rate are the main components of the QOS. Most of the existing digital TV production systems are based on stream interfaces and pipes.

- Link cost is high.

- Delivery speed is in real time for most of the transfers and is requested for live events. Compressed materials can be transmitted using faster-than-real-time transfers.

- Delivery is synchronous or isochronous.

- Streaming transfers are dependent on format attribute differences.

- There is guaranteed latency.

- Communication is unidirectional.

---

*STM-0, OC-1, and STS-1 have a common basic level of 51.84 Mbps. Higher levels are n times the basic rate (OC-192 = STS-192 = STM-64 = 9.953 Gbps, with STM-1 = OC-3 = 155.52 Mbps).

†A guaranteed delivery is realized when the information is delivered uncorrupted.

12.2.2   File transfers

- Transport layer ensures guaranteed reliable delivery of a file from one location to another, even with the use of existing low-quality networks.
- Cost of network infrastructure is low.
- Delivery speed can be faster, normal, or slower.
- Delivery is asynchronous.
- File transfers are not dependent on format attribute differences.
- There is unknown latency as errored packets must be resent.
- Communication is bidirectional.
- Interconnected devices do not need to be equipped with video/audio input and output ports.

12.2.3   Transmission control protocols

During the development of the reference architecture concept, the TF-HS has investigated transfer protocols for the transmission of data content and has issued several recommendations:

- For file transfers, the file transfer protocol (FTP) is the industry standard protocol. Fast and point-to-multipoint transfer requirements can be answered by improving FTP to become FTP+ and adopting the express transport protocol (XTP). This will also make possible the selection of maximum transfer rates and transfers of partial files.
- For file sharing, the network file system (NFS) enables clients to have full read/write access to portions or the entirety of the file located on the server computer. A complete control over file permissions is provided.

12.3   Transport Technologies for Intrafacility Applications

The TV production and broadcast industry is no longer limited to the analog NTSC and PAL video formats. Materials are now generated in analog composite and component formats, digital composite and component formats, and several different compressed video formats at different frame rates, image aspect ratios, pixel aspect ratios, sampling structures, color spaces, and picture sizes. These materials originate from cameras, film scanners, slides, VTRs, servers, etc., and more often from video computer systems. Thus there is the necessity to interconnect them all in order to create a production architecture.

Network and computer technologies are capable of moving digital assets between all TV production processing areas. New video networking systems combined with the most recent technologies make possible the real-time transfer of very large data files of uncompressed video and high resolution of DTV formats over a network. However, broadcast operations put additional requirements such as fixed delay, frame synchronization, real-time and faster-than-real-time transfer, and low bit error rates. The reference architecture recommends several transport technologies (SDTI, FC, IEEE1394, and Ethernet) for intrafacility applications. SDTI is a broadcast infrastructure while the others are primarily computer infrastructures. DVB-ASI is also a possible choice for intrafacility interconnections as it is implemented in most of the MPEG-2 encoders and is used to carry MPEG-2 TS data packets to the multiplexer. DVB-ASI technology is described in Sec. 12.4.2.

12.3.1 Serial data transport interface

The serial data transport interface (SDTI) protocol specifies the mapping of many different data types into the serial digital interface (SDI) defined in the SMPTE 259M document. This protocol is described in the SMPTE 305M document and in an EBU draft proposal. SDTI allows for the exchange of digital audio and video signals in their native compressed formats through existing links designed to carry 270- and 360-Mbps data streams. In addition, it supports addressing which can be used for automatic routing purposes. Other SDTI implementations are being considered by the SMPTE standards organization for 540- and 1485-Mbps interfaces. Figure 12.5 shows an OSI model block diagram of the SDI physical layer and the SDTI transport layer. This interface has been recommended by the TF-HS and is already used for real-time and faster-than-real-time streaming applications.

The SDTI can carry 8- or 9-bit data packet payloads in the SDI active video line area. However, it is possible to extend the payload by inserting data packets in the unused horizontal ancillary (HANC) data space area. The tenth bit B9 (most significant bit) is set as the complement of B8 to avoid the generation of illegal word values (that is, 3FF, 000, 001, etc.). SMPTE 305M does not

Figure 12.5 SDTI structure based on the OSI reference model.

specify the data structure inside the user data block. It is defined in each document particular to a specific application.

Figure 12.6 shows an SDTI implementation in a 270-Mbps data stream. The standard document specifies that user data can be inserted on any line, in the active video line area between SAV and EAV flags. It means that the vertical ancillary (VANC) data space area can also be used. When data stream switching occurs, data may be corrupted at the time of the switch.

Data packets (also called blocks) can be of a fixed block size or variable block size. An SDTI data block includes a data-type flag and a word count when the variable-block-size mode is used. A maximum of 256 types can be registered by the SMPTE. Data types have already been allocated to various applications such as DVC-Pro, DVC-Pro50, DV-Cam, Beta-SX, HD-Cam, Digital-S, and MPEG-2 program streams, MPEG-2 transport streams, generic content packages for system audio AES, video, and auxiliary data (SDTI-CP), encapsulation of various packetized formats (SDTI-PF), etc.

A 53-word header is located in the HANC data space area immediately after the EAV and only for each line that carries payload packets. It allows for

■ Destination and source addressing

■ Line numbering

■ Flag indication of block error correction presence/absence at the end of fixed data block size only

■ Code indication for the size of the payload (in 270- or 360-Mbps streams)

■ Authorized address identification (AAI) which defines the address word format

Presently, two 4-bit words have been allocated, the first one for nonspecified format (00h) and the second one for the Internet protocol v6 addressing format (01h). Other addressing schemes can be registered in the future.

The SDTI has the following features:

■ It uses existing widespread SDI infrastructures to provide a unidirectional interconnection between two devices. Point-to-multipoint applications can also be considered. It provides a simple and inexpensive packetized transport solution having significant advantages over networking technologies in studio applications. The networking overhead is very limited, and the bandwidth and quality of service (QoS) can be guaranteed.

■ It can reuse existing digital routers and distribution amplifiers and can coexist with multiple formats. Intelligent routers could self-route the signal by using the destination address data embedded in each serial data stream. This addressing scheme is identified by the AAI word. Figure 12.7 shows an SDTI implementation in an existing SDI infrastructure and another implementation in a server application using its addressing scheme capability.

Figure 12.6 SDTI embedding in a 270-Mbps SDI data stream.

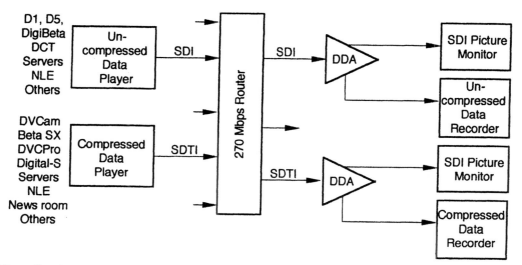

a) SDTI application in an existing studio infrastructure

b) SDTI application using an addressing scheme

Figure 12.7 SDTI implementation in broadcast studios.

■ It transmits data packets of variable lengths. High-performance transfers of compressed data can be achieved in real-time and faster-than-real-time applications. A 270-Mbps SDTI can carry four DV-based video channels in real time or one identical channel at 4 times the normal speed.

■ It eliminates the need for multiple decompression and recompression in the production process, thus maintaining signal integrity.

- SDTI data stream recording could be made possible on component DVTRs having a strong error correction. Several modifications are necessary to disable the error concealment, to allocate a header and extended data recording area on tape, etc. However, VANC payloads cannot be handled by current DVTRs, which cannot record a total of 525 lines of data. Forward error correction could be added prior to recording to allow for increased robustness during the recording/playback process and to minimize the data error sensitivity of compressed streams.

- When SDTI carries synchronous data streams, the buffer requirements are minimal. In asynchronous data stream transfers where, e.g., the SDTI runs at 30 fps and the compressed data originate from 24-fps material, a significant buffer is required to recreate the data streams at the receiving point.

- EDH can be added to SDTI data streams. This is extremely useful when compressed data are transmitted through SDI systems. The EDH error detection is very sensitive and can detect errors before a small amount of errored bits starts significantly degrading the decompressed data program. In addition, multiple EDH data flags can be used to determine if errors originate from the transmitter or the interconnection path.

- However, if SDTI is used in an SDI system, the receiver must be able to recognize and process SDI and various SDTI data rates corresponding to non-real-time and faster-than-real-time operations. In addition, EAV and SAV sync patterns must be protected to allow for a correct detection of the SDTI header data.

The SMPTE is presently defining an HD-compatible version of the SDTI, called HD-SDTI, which is based on the bit-serial digital interface for high-definition interface (SMPTE 292M standard for HD-SDI). The serial interface bit rate is 1.485 Gbps with a maximum payload of about 1 Gbps, inserting 8-bit words in the active line area. Most of the parameters are compatible with the SDTI characteristics. Using the extended mode, payloads for 1125- and 750-line source formats can be accommodated at various frame rates such as 24, 25, 30, and 60 fps. However, *C* and *Y* samples are separated to create two independent HD-SDTI transports on the same data link. Table 12.1 shows the maximum SDTI payload for SDTV and HDTV interfaces.

12.3.2 Fiber channel networking

Fiber channel (FC) was originally developed for the computer industry to improve LAN performance. Further FC developments have enabled its wide application to high-bandwidth servers and storage systems in a LAN environment. FC features scalable gigabit storage and delivery, flow control, self-management, and high reliability. Multiple topologies and protocols can be accommodated with a high efficiency due to a low transmission overhead. FC is already used in fast and large file transfer applications but can also be used in data streaming transfers.

TABLE 12.1   SDTI Maximum Possible Payloads for SDTV and HDTV Interfaces

Transport interface	Bit rate (Mbps)	Max. payload (Mbps)	Max. extended payload (Mbps)
SDTI	270 (8-bit)	181	208
	270 (9-bit)	204	235
	360 (8-bit)	242	281
	360 (9-bit)	272	316
	540 (8-bit)*	362	416
	540 (9-bit)†	408	470
HD-SDTI	1485 (8-bit)†	1037	1167

*Future implementations.
†Being standardized by the SMPTE.

The FC standard is defined in the ANSI X3.320-1994 document. The FC structure is made of five layers based on the OSI model, as shown in Fig. 12.8. The frame structure of the FC data stream is shown in Fig. 12.9. The FC protocol supports asynchronous and synchronous transmission. It uses a dc balanced 8B/10B encode scheme that facilitates clock recovery, byte synchronization, and error detection, and it provides robust transmission characteristics. Each 8-bit word is encoded to a 10-bit word by adding two running disparity bits representing the difference of "1" bits and "0" bits in the transmitted 8-bit data word. This disparity characteristic allows the dc balance of the serial data stream. But this transmission coding reduces the effective payload by 20 percent.

FC provides six classes of service:

▒ Class 1: acknowledged connection, also called dedicated channel

▒ Class 2: acknowledged connectionless, also called packet switch traffic with receipt confirmation, analogous to a multiplex link

▒ Class 3: unacknowledged connectionless, suitable for distribution to single or multiple locations, without receipt confirmation

▒ Class 4: fractional-bandwidth connection-oriented. Suitable for real-time video transfer application with guaranteed bidirectional delivery, gigabit bandwidth, and bounded end-to-end delay

▒ Class 5: not defined

▒ Class 6: unidirectional connection; suitable for real-time video transfer application

Several topologies are supported by FC, such as point-to-point, switched (also called fabric), and arbitrated loop. In the switched topology, a switch receives frames from a source node* and routes them to the destination node

---

*A node is a combination of transmit and receive ports that are interconnected by point-to-point connections alone or point-to-point connections within a fabric.

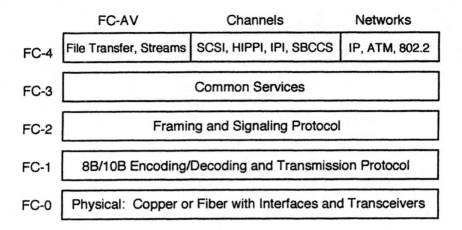

SCSI = Small Computer System Interface
IP = Internet Protocol
IPI = Intelligent Peripheral Interface
HIPPI = High Performance Parallel Interface
ATM = Asynchronous Transfer Mode
SBCCS = Single Byte Command Code Set mapping

Figure 12.8 Layered structure of the fiber channel transport technology.

SOF = Start of Frame
CRC = Cyclic Redundancy Check
EOF = End of Frame

Figure 12.9 Framing structure of the fiber channel data stream.

indicated in the frame header. The arbitrated loop works similarly to the token ring protocol, where at each port data are verified and transmitted to another port if the address does not correspond to the received port address. A hub is connected to all nodes. Up to 126 devices can be connected in a loop, but the available bandwidth decreases with the number of devices. These topologies are shown in Figs. 10.8 and 10.9.

A recent and popular application of FC is the storage area network (SAN). An SAN combines FC networking with high-performance drives and management software to provide a high-speed network dedicated to storage and data sharing. Most of the existing computer workstations (CWSs) have a high-bandwidth connection to their local storage device but are connected together through a low-bandwidth connection by a LAN. Figure 12.10 shows typical peer-to-peer, client-server, and SAN configurations. In a client-server architecture, the server is a powerful, high-bandwidth machine incorporating network security and file system protocols that is connected to a high-capacity storage system. A large number of clients have access to the server via a LAN or can run applications on it. In an SAN, the applications and the file system reside in the same workstation, and storage devices are directly attached to a high-speed network (such as fiber channel), thus eliminating the file server. All workstations are connected to a common storage, allowing fast transfer between workstations and sharing of the same disks and files at the same time.

12.3.3 GigaEthernet networking

Ethernet networks are widely used in computer equipment interconnections owing to their simplicity, scalability, and low cost. The Ethernet family started to interconnect devices, also called data terminal equipment (DTE), with 10-Mbps links and has grown to include Fast Ethernet (100 Mbps) and recently GigaEthernet (actually 1250 Mbps). Section 10.4.1.5 gives a brief introduction of this packet-switching transport mechanism and current topologies used in computer and broadcast interconnections.

Ethernet networks can carry data and control information between audio, video, control, and synchronization equipment between devices utilizing multiple protocols. It thus considerably simplifies interconnection requirements in broadcast studio applications. This allows media file transfers over a high-speed network in a manner similar to the data file process between computers, as well as connectivity to existing Fast Ethernet and ATM networks. Although the TF-HS has recommended its use in low-performance transfers only, the recent GigaEthernet standard has generated interest and development of TV production equipment using this networking solution.

GigaEthernet retains the same Ethernet frame structure, media access control (MAC), and flow control, ensuring backward compatibility between Ethernet and Fast Ethernet frames. It uses a complex TCP/IP to provide guaranteed data delivery, requiring a high CPU bandwidth to process packets at gigabit rates. Upon receiving an IP packet, the interconnected device CPU must respond to the interrupt, verify the address content, calculate the checksum and compare, acknowledge the packet, and then send the data to the appropriate application. This long task must be performed in each client and server device in less than 12 μs for a 1500-byte frame size. This overloads the CPU and considerably reduces the actual throughput rate while

Peer-to-Peer

Client-Server

Storage Area Network

Figure 12.10 Client-server networking and SAN implementation.

compromising other foreground tasks. Figure 12.11 shows the layers and data stream frame structure of the GigaEthernet transport technology.

Improvements to the TCP/IP overhead processing and handling limitations have been developed using new network interface cards (NICs) and switched Ethernet topologies in building GigaEthernet networks. Repeating and switching Ethernet topologies are shown in Fig. 12.12. An Ethernet repeater connects all devices to a shared bus and relies on the CSMA-CD control protocol to give access to the bus for a unique device pair. An Ethernet switcher establishes multiple connections at full bandwidth between device pairs using IP addresses and addressing modes (point-to-point, point-to-multipoint, bridge-to-bridge) found in Ethernet frame streams, thus removing the possibility for data collisions on the bus and the time-consuming handshaking process between interconnected devices.

Figure 12.11 Layers and data stream frame structure of the GigaEthernet transport technology.

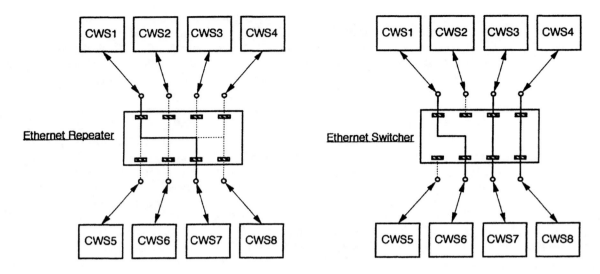

Figure 12.12   Repeating and switching topologies of GigaEthernet.

GigaEthernet connections are too fast for existing servers and network interface cards. The data speed throughout the network interface is 1.25 Gbps, but the 8B/10B encoding reduces the effective payload to about 1 Gbps with full duplex capability. New powerful server platforms and faster NICs are being developed. A new NIC can provide hardware processing of packet addressing and checksum verifications, handle transmitted data packet storage in NIC memory, and improve the management of interrupts, thus alleviating the overall device CPU workload. High real data rates are possible over GigaEthernet, and data rates up to 960 Mbps have been experienced.

The IEEE 802.3z GigaEthernet standard document describes three physical specifications for fiber-optic media (LX and CX with connections up to 5 km and 550 m, respectively) and shielded copper cable (CX with connections up to 25 m). Only laser-based sources are fast enough to create gigabit-modulated pulses. Another IEEE 802.3ab standard defines the use of category 5 unshielded twisted-pair (UTP) cabling for distances up to 100 m.

12.3.4   IEEE 1394 networking

The IEEE 1394-1995 standard for a high-performance serial bus is now widely implemented in both computer peripherals and consumer electronics high-speed data stream interconnections. Section 10.4.1.1 gives a preliminary description of this transport mechanism. IEEE 1394a presently operates at 400 Mbps, but work is under way to define an IEEE 1394b allowing speeds between 800 Mbps and 3.2 Gbps using 8B/10B encoding and long cable lengths using fiber-optic cables. In addition, the SMPTE/EBU task force, in its recommendations for a content transport system, has indicated that this transport protocol has the capability to carry asynchronous file transfers and isochronous audio/video streaming transfers on the same bus and at the same time. The layered structure of the IEEE 1394 transport technology is shown in Fig. 12.13.

Professional video equipment has started implementing IEEE 1394 connections* and transport protocol for the interchange of video, audio, and other data types. It offers a number of advantages:

- It operates with very low-cost connectors, no termination, hot plug-in, dynamic configuration, and 64-bit addressing with automatic address selection.

- It supports up to 63 devices on a single bus and 64,000 devices when buses are combined with bridges. Multiple devices can simultaneously access the same data through the IEEE 1394 serial bus.

- It allows large payload packets with no handshaking, data-type format independence, and scalability.

- Asynchronous transfer mode with data acknowledgment allows guaranteed delivery.

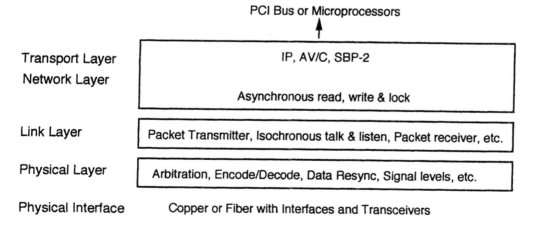

Figure 12.13   Layered structure of IEEE 1394 transport technology.

---

*The ITU-T has recently approved Recommendation J117, *Home Digital Network Interface Specifications*, which is based on the IEEE 1394 standard, for home interconnections of digital VCR, DTV set, computer, camera, etc.

■ Both asynchronous and isochronous packets can be handled by the same transmission link. Isochronous packets use about 80 percent of the cycle (125 μs), and asynchronous traffic data fills the remainder of the cycle. Figure 12.14 shows the framing structure of the IEEE 1394 data stream.

■ Short-cable-length and high-speed issues will be solved by optical interface additions. But a new connector design may be required to allow transmission of high-speed data and low-RF emissions.

IEEE 1394 technology can packetize data of any format of compressed video, audio, or data and can deliver them in that form. The IEC 61883 standard describes the packetization of DV and MPEG-2 data streams into IEEE 1394 transport interfaces. It allows the transportation of multiple video, audio, control, synchronization, and data streams simultaneously on the same cable. By establishing a single connection between servers and storage devices, it facilitates the distribution of IEEE 1394 data streams to each device or user through switchers or routers.

Table 12.2 summarizes the characteristics of the four networking solutions described in this section. Figure 12.15 shows a possible mix of several networking solutions within a typical studio. A summary of stream transfer technologies used in intra- and interfacility applications is shown in Fig. 12.16.

12.4 Transport Technologies for Interfacility Applications and DTV Signal Delivery

Interfacility links (IFLs) have been in existence since the beginning of TV. Transmission techniques are needed to deliver multiple channels of video and audio to production centers for contribution and to interconnect studios in the course of a program, such as live programs, sports, and news. Program distribution to affiliates and program delivery to transmitters, satellites and cable

Figure 12.14 Framing structure of IEEE 1394 data stream.

TABLE 12.2 TV Production Networking Technology Comparison

Architecture	SDTI	Fiber Channel	GigaEthernet	IEEE 1394
Technology application	Studio, LAN, BAN	Studio, SAN, LAN, BAN	Studio, LAN, BAN	Studio, LAN, BAN
Topology	Point-to-point Switched Point-to-multipoint	Point-to-point Switched Loop/hub	Point-to-point Switched Hub	Daisy-chain Switched Loop/hub
Direction mode	Unidirectional	Bidirectional	Bidirectional	Bidirectional
Speed (Gbps)	0.27, 0.36	1.06	1.25	0.4
Transmission mode	Asynchronous Synchronous	Asynchronous Synchronous	Isochronous	Asynchronous Isochronous
File and stream transfer capability	Stream	File, stream	File	File, stream
Latency	Low, constant	Low, constant variable	Variable	Variable constant
Guaranteed delivery	Yes	Yes	No	Yes No
Frame duration (μs)	63.5	20	12	125
Scalability to higher data rates (Gbps)	1.485	2.12, 4.24	No	0.8, 3.2
Physical media	Copper, fiber	Copper, fiber, twisted pair	Copper, fiber, UTP cat. 5	Copper, fiber
Practical length (per standards)	60 km	10 km	5 km	10 km

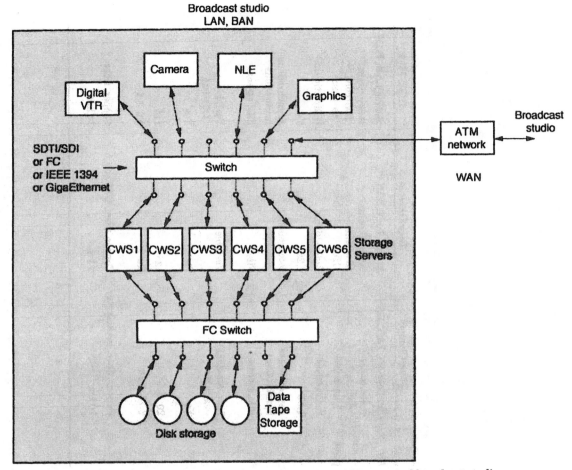

Figure 12.15   Possible combination of several networking solutions within a typical broadcast studio.

operators, and satellite master antenna television (SMATV)* and multichannel multipoint distribution service (MMDS)† broadcast systems are also carried by these transport technologies.

The rapidly increasing demand for content creation to accommodate available new DTV program channels has generated immediate needs for the interconnection of local, national, or international production resources to move multimedia program content between them. This rapid move is driven by the fact that

---

*A typical SMATV example would be the distribution of services to a large number of users living in collective buildings using a satellite receiving antenna connected to a local cable.

†MMDS also means microwave multipoint distribution services and is known as wireless cable. It uses microwave frequencies for direct distribution to users.

Figure 12.16 Data stream transfer technologies in intra- and interstudio applications.

- Programs are now composed of digital streams or files.
- Multiple production locations may already be interconnected by analog (satellites, microwave, etc.) or high-bandwidth and cost-reduced digital (WAN, MAN, etc.) links.
- Production facilities are being implemented with more efficient LAN infrastructures.

Figure 12.17 shows a typical TV broadcast interconnection model. Current IFL applications include

- Contribution feeds from local and affiliate stations, through satellite receive station, microwave radio, and telecommunication company (Telco) network.
- Electronic news gathering (ENG) and field acquisition feeds coming from microwave radio, satellite, and Telco network
- Remote broadcast feeds via local station or satellite carrying important live-event programs
- Studio-to-transmitter (STL) and transmitter-to-studio (TSL) links
- Distribution to local and affiliate stations and to cable and satellite operators

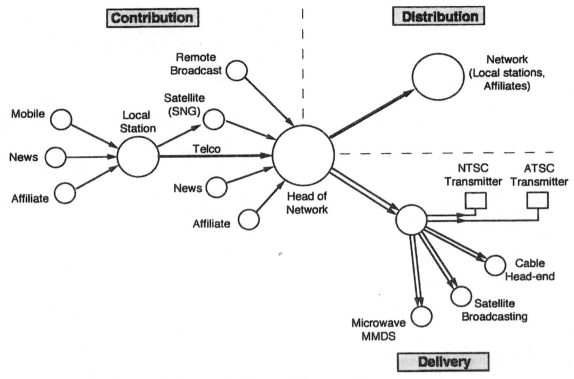

Figure 12.17   TV broadcast production, contribution, and delivery model.

All user applications have common and application-specific system requirements:

■ Transport of both analog NTSC and digital ATSC signals between the same facilities.

■ Additional signal bandwidth to carry intercom, telemetry, transmitter remote control, etc.

■ Unidirectional or bidirectional transport systems.

■ Different bandwidths (bit rates) corresponding to specific applications such as ASTC delivery (19.39 Mbps), contribution/distribution links (45 to 270 Mbps), and ENG (4 to 18 Mbps).

■ Low BER. There is a strong desire for error-free IFLs due to the high bit error intolerance of the MPEG-2 streams and the low acceptable bit error rate requested by users (10^{-14} to 10^{-16}). Current data networks are not designed to perform at this BER level and require an appropriate forward error correction (FEC) to improve BER performance. This is realized at the expense of bandwidth.

■ BER monitoring and handling.

■ Backup link for full redundancy in STL application. A separate link (microwave, fiber cable, Telco, etc.) is provided.

The user can choose several physical transports depending on the application, distances, and costs:

■ Direct-connect copper and fiber-optic cables. Dark (unused) fiber links are increasingly becoming available in metropolitan networks and suitable in high-interference environments. SSI, ASI, SDTI, and DS-3-based data signals can be carried by these cables over short and local distances up to 50 to 70 km.

■ Microwave radio. This is used in STL, TSL, station-to-satellite, remote broadcast, news-to-station, delivery to cable head-end, and station-to-network links. ATSC, ENG, and DS-3-based signals can be carried through these microwave links.

■ Telco networks such as metropolitan-area network (MAN) and WAN. ATM-based Telco networks, from 2 Mbps to several gigabits per second, can accommodate a wide range of data rates.

■ Satellite. SSI, ASI, SDTI, and DS-3-based data signals can be transported through satellite links.

Some LAN technologies such as FC and IEEE 1394 can also be used in interfacility applications when distances are less than 10 km for interchange of high-quality materials at high bit rates.

12.4.1 Synchronous serial interface

Synchronous serial interface (SSI) is a point-to-point interface intended to carry MPEG-2 TS packets at rates up to 40 Mbps (such as for ATSC and DVB data stream applications) between a studio and a transmitter. SSI data are synchronous to the MPEG-2 TS transmitted on the serial link. They must be used in low-noise applications, i.e., so that a noise level would not corrupt more than one MPEG-2 data packet per day at the transport clock rate. In the presence of higher noise levels, an error correction must be added. SMPTE 310M and DVB-A010 standard documents describe the physical interface and modulation characteristics. Table 12.3 shows the SSI electrical characteristics for the transmitter and the receiver. These SSI signals can be distributed by means of existing analog wide-band video distribution amplifiers with removable clamping, such as Leitch VEH-680 ($BW = 100$ MHz at ± 1 dB) and VEH-6810 ($BW = 200$ MHz at $+1$ dB/ -3 dB).

A single MPEG-2 TS (carrying a single program or multiple programs) can be carried by the SSI as no multiplexing capability is defined at the interface level. Figures 12.18 and 12.19 show the MPEG-2 TS packet adaptation into the synchronous serial interface and its signal waveforms for an 8-VSB (19.32-Mbps) ATSC data stream. Byte synchronization is achieved through the use of the sync byte (47H) in each MPEG-2 TS packet. An 8-VSB transport stream

TABLE 12.3   Synchronous Serial Interface (SSI) Electrical Characteristics

General characteristics	■ Channel coding: Biphase mark ■ Interface clock frequency: 38.6 MHz for the 8-VSB mode
Transmitter characteristics	■ Unbalanced coaxial cable with BNC connector ■ Fiber link, multimode, or single-mode with SC-type connector ■ Return loss: >30 dB (0.1 to 39 MHz) ■ Output signal amplitude: 0.8 V ± 10% across 75-Ω load ■ DC offset: 0.0 V ± 0.5 V ■ Rise and fall time: 0.4 to 5.0 ns between 20 and 80% of signal amplitude points ■ Overshoot of rising and falling edges: <80 mV ■ Maximum jitter: 2 ns p-p over a bandwidth of 1 Hz to 0.39 MHz ■ Interface clock frequency drift error: < ±2.8 ppm ■ Interface clock frequency drift rate: < 0.028 ppm/s
Receiver characteristics	■ Cable equalization of up to 3 dB @ 19.4 MHz for 75-Ω ohm coaxial cable ■ Optical receiver for optical reception @ 1310- and 1550-nm nominal wavelength ■ Automatic reception of any bit rates from 20 to 50 Mbps

Notes: Parameter values above are taken from the SMPTE 310M standard. The DVB-0101 standard document specifies the DVB-SSI and ASI interfaces. Several parameters listed above are different in DVB-0101 standard document, such as amplitude and return loss.

Note: BPM = Biphase Mark

Figure 12.18 MPEG-2 to SSI adaptation.

Transport
Clock (19.32 MHz)

MPEG-TS NRZ Data
(19.32 Mbps)

Interface Clock
(38.6 MHz)

BPM Data
(38.6 Mbps)

Figure 12.19 Biphase mark (BPM) encoded signal waveform for a 19.38-Mbps ATSC data stream signal.

yields a bit rate of 19.32 Mbps with a transport clock of 19.32 MHz. After biphase mark encoding, the bit rate becomes 38.8 Mbps and the interface clock frequency is 38.8 MHz. For a 16-VSB transport stream, multiply all these values by 2.

The ATSC DTV standard A/64 requires that the 19.32-MHz transport clock and the 10.76-MHz symbol clock be locked in frequency to each other. In addition, to minimize cochannel and adjacent-channel interferences, a tight jitter tolerance (0.04 UI) is required for the RF pilot carrier clock of the transmitter. A VCO can be used in this RF pilot carrier design to track the variations of the symbol clock against a precise reference. The SMPTE 310M standard specifies a maximum SSI clock frequency drift limit of ±2.8 ppm (or ±108 Hz) with a maximum frequency drift rate of 0.028 ppm/s (1.10 Hz/s). This yields a maximum ATSC transport clock (TC) frequency drift of ±2.8 ppm (or ±54 Hz) and a symbol clock (SC) frequency drift of ±30 Hz (SC = $\frac{1}{2}$ × 208/188 × 313/312 × TC, as explained in Sec. 11.8.6).

12.4.2 Asynchronous serial interface

The asynchronous serial interface (ASI) is a point-to-point interface intended to carry MPEG-2 TS packets at different data rates between MPEG encoders, multiplexers, and modulators. It uses a dc balanced 8B/10B encoding scheme which produces one 10-bit word for each MPEG-2 TS byte at the input. Dummy bytes are inserted to fill up the available payload of 216 Mbps (27 MBps × 8) and generate a constant data rate of 270 Mbps. This 8B/10B encoding allows for self-error-checking capability and byte synchronization of the link. In addition, it provides a marking of the dummy bytes and removes any dc component in the ASI data stream signal.

The DVB-A010 (or EN 50083-9) standard describes the ASI physical interface and modulation characteristics. Figure 12.20 shows the adaptation of a single MPEG-2 transport stream (carrying a single program or multiple programs) to the ASI as no multiplexing capability is defined at the interface level. Two 10-bit sync words (comma character = K28.5) precede each mapped

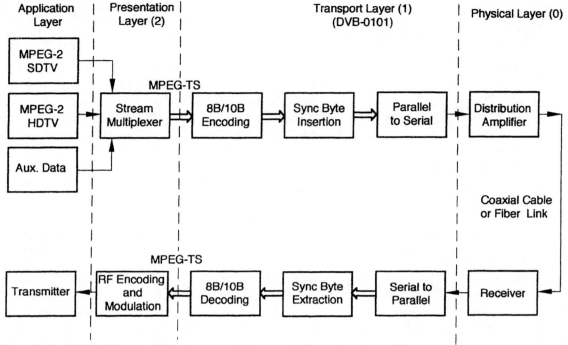

Figure 12.20   MPEG-2 TS data packets to asynchronous serial interface adaptation.

MPEG-2 TS packet. They are inserted to establish byte synchronization in the ASI transmission link. And 188-byte packets are carried through this interface whereas 204-byte packets can optionally be transported. The 204-byte packets are made of 188-byte data and 16 additional bytes that can be either 16 dummy bytes or 16 RS bytes. The dummy bytes can also be used to carry additional data. This allows for a proper selection of the appropriate interface characteristics to match the kind of equipment to be interconnected.

The ASI decoder will properly align a byte stream after the receipt of two sync words (K28.5). ASI sync bytes and stuffing bytes use the same 10-bit word value.

The ASI specifications allow transport packets to be conveyed into this interface through three transmission formats to allow for flexibility in designing such an interface. But this situation creates compatibility issues between encoders and decoders from different manufacturers. These three formats are described in Fig. 12.21 and have the following data structure:

- A burst of contiguous 8B/10B coded bytes originating for an MPEG-2 transport packet. Stuffing data are inserted between bursts using the special character commas (K28.5).

- Individual 8B/10B coded MPEG-2 TS data bytes spread out in time and interspersed with a variable number of stuffing bytes (K28.5).

Figure 12.21 Three possible ways of encoding MPEG-2 data packets into an ASI.

■ A combination of contiguous 8B/10B coded MPEG-2 TS data bytes and stuffing bytes (K28.5).

These transmission formats must be correctly identified and handled by any ASI receiver. MPEG-2 TS must be accurately reproduced at the intended time. Jitter in the ASI transmission may increase PCR jitter in the decoded MPEG-2 TS stream.

The ASI standard document describes the physical interface that provides for both LED-based multimode fiber and coaxial cables. Table 12.4 shows a summary of the ASI characteristics. The commonly used cabling for interconnecting SMPTE 259M signals at 270 Mbps is suitable for ASI distribution. ASI bit streams are polarity-dependent. Noninverting interfacing equipment must be used to ensure correct operation. It also means that only one-half of the existing SDI distribution amplifier and router outputs can be used.

Some SDI equipment may have problems passing ASI signals as no EAV/SAV flags exist within the data stream. These flags are required by some SDI receiver chips relying on EAV/SAV flags to recover the SDI signal. In addition, the DVB-ASI standard requires that the interface be coupled to the coaxial cable via a transformer. A manufacturer has introduced true DVB/ASI reclocking distribution amplifiers and routing switchers that conform to this requirement.

TABLE 12.4   Asynchronous Serial Interface (ASI) Electrical Characteristics

General characteristics	■ Channel coding: 8B/10B ■ Encoded line rate: 270 MHz ± 100 ppm
Transmitter characteristics	■ Unbalanced coaxial cable with BNC connector ■ Fiber link, multimode using LED emitters ■ Output signal amplitude: 0.8 V ± 10% across 75-Ω load ■ Rise and fall time: 1.2 ns (max) between 20 and 80% of signal amplitude points ■ Maximum jitter: 0.3 UI (1.110 ns p-p) ■ Maximum deterministic jitter: 10% p-p (0.2 UI) ■ Maximum random jitter: 8% p-p (0.16 UI)
Receiver characteristics	■ Minimum sensitivity: 200 mV ■ Maximum input voltage: 880 mV p-p ■ Minimum return loss: 15 dB ■ ASI stream polarity sensitivity ■ Bit stream resynchronization in no more than 1 ms ■ Compatibility with three ASI transmission modes

12.4.3 SDTI

The SDTI data format structure is described in Sec. 12.3.1. Generic packets can be multiplexed over 270-Mbps SDI links. The use of hybrid fiber-optic and coaxial (HFC) cables in metropolitan networks can extend its application to long transmission distances up to 60 km. SDTI includes an FEC and an accurate mechanism for packet retiming at the decoder. It can realize intrafacility interconnections of all major production equipment and an interfacility link to the transmitter and local-area distribution networks.

12.4.4 Transporting MPEG-2 data streams in DS-3-based networks

DS-3-based networks and private point-to-point networks are widely available for broadcast-quality IFLs. The mapping of MPEG-2 transport streams (TSs) into the DS hierarchy is addressed in the ANSI T1.519-1999 document. This standard describes the transmission of generic packet data over the DS digital hierarchy, which is defined in the ANSI T1.107-1995 document. Figure 12.22 shows the MPEG-2 to DS adaptation.

Figure 12.23 shows how a descriptor is added to a fixed-length generic packet of 188 bytes to form either a type 1 or a type 2 adaptation packet (AP). This AP is directly mapped into the payload of a DS frame and does not need to be aligned with the DS framing structure. In the DS-1 case, packets are byte-aligned with DS-1 payload, whereas in the DS-3 case, packets are byte-aligned with DS-3 M-frame payload.

Null packets are inserted into the DS payload when generic packets are not available. In this case, the GP_Present flag takes the value 0. The DS-3 interface is configured as a one-time channel slot payload of 44.210 Mbps (unchan-

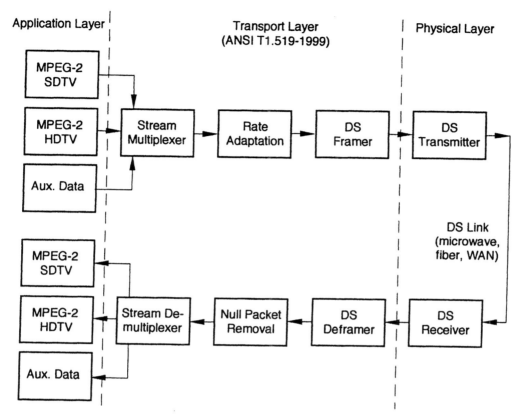

Figure 12.22 Adaptation of MPEG-2 TS packets to DS network.

nelized mode per ANSI T1.107). DS-3 performance and alarm/status monitoring information is supported.

The main features are as follows:

- One or more independent multiplexed generic packet streams are transmitted over a single DS hierarchy network channel.

- A generic packet stream may be composed of an MPEG TS or an MPEG multiprogram TS or other non-MPEG-2 TS data stream formats.

- Type 1 adaptation packet is 189 bytes long without forward error correction (FEC).

- Type 2 adaptation packet is 204 bytes long with 16 bytes of FEC to add protection over a noisy DS link. The FEC_En flag indicates the FEC validity.

- Up to 16 different source payloads multiplexed onto a DS interface can be identified.

1) Generation of Type 1 Adaptation packets

Note*: There is no requirement to have the AP in a fixed position relative to the DS framing.

2) Generation of Type 2 Adaptation packets

Figure 12.23 Generic packet mapping into nonchannelized DS-3 PDH networks.

- Rate adaptation is provided between generic packet stream and DS physical link data rates by means of null packet insertion. This allows for VBR stream transmission over DS channels.

- There is robust receiver packet synchronization using fixed packet length.

- There is a very low byte overhead rate of 1/189 (0.53 percent) or 1/205 (0.49 percent) with FEC. This approach provides the maximum number of bits for distribution of the best compressed picture quality.

- There are DS-1 applications for non-real-time transmission of MPEG-2 streams. In addition, it has been proposed that MPEG-4 (with a bit rate below 1.5 Mbps) be carried by MPEG-2 TS.

- Higher data stream payloads could similarly be embedded into SONET OC-3 and SDTI transport layers.

The DS-3 physical layer characteristics are described in the ITU-T G.703 document, while requirements for central office coaxial cables can be found in the Bellcore GR-139-CORE document. Also, 75-Ω coaxial cables are recommended for cross-connect and interconnect applications. Typical attenuation for a 100-m coaxial cable at 22.368 MHz (DS-3) is -8.2 dB for a 26 AWG size and -3.7 dB for a 20 AWG size.

12.4.5 Transporting MPEG-2 data streams in ATM-based networks

A single ATM infrastructure can be used to interconnect remote studios for communication services (phone, LAN, Internet access) and TV contribution and distribution services. Multiple connections can be multiplexed on a single link, allowing multiple bidirectional transfers. ATM data can be mapped into WAN transmission streams, such as used in PDH (DS and E) networks and SDH and SONET networks. ATM connections may be through a permanent virtual circuit (PVC) or a switched virtual circuit (SVC). The PVC guarantees a permanent end-to-end connection, but costs correlate to this service. SVC guarantees temporary end-to-end connection, complete with signaling and billing intelligence. This offers great flexibility in establishing connections between any subscribers. In addition, ATM is a scalable solution that can handle bursty, CBR, and isochronous data rates; guaranteed QoS; and dynamic switching, as the destination address is in the header.

SMPTE is developing a standard to define a common layer for transporting packetized video, audio, and data over ATM, using the AAL1 adaptation scheme. In addition, another document has been proposed to define an object data format for the exchange of DV-based audio, data, and compressed video using the ATM common layer. This approach will permit real-time and faster-than-real-time transfer, and multiprogram transfer as well, of DV-based 25- and 50-Mbps data structures for 525/60 and 625/50 systems over ATM networks. For payloads above 44 Mbps, the solution is ATM at 155 Mbps into SONET OC-3 or SDH networks.

The DVB organization has issued a standard, ETS 300 814, that defines the transmission of MPEG-2 TS between two DVB interfaces within SDH networks working at 155.5- or 51.8-Mbps rates (ITU-T Rec. G.707) based on the use of ATM cells. The adaptation of RS-coded MPEG-2 packets into ATM cells utilizes the AAL1 scheme per ITU-T Rec. J.82, and ATM cells are then adapted to SDH framing per ITU-T Rec. G.707.

DVB has also issued a standard, ETS 300 813, that defines the transmission of MPEG-2 TS between two DVB interfaces within PDH networks working at DS and E bit rates (per ITU-T Rec. G.702) or any bit rates. The adaptation of RS-coded MPEG-2 packets into ATM cells utilizes the AAL1 scheme per ITU-T Rec. J.82, and ATM cells are then adapted to PDH framing per ITU-T Rec. G.804.

ATM uses different adaptation layer levels. As an example, in the case of mapping an MPEG-2 TS stream onto ATM, the ATM adaptation layer (AAL) mapped 188-byte TS packets onto the payload of the 53-byte ATM cells. The aim of AAL is to enhance the quality of service provided by the ATM layer by specifying several classes of service, which consider parameters such as time relation between sources and destination, constant or variable rate, connection-oriented or connectionless mode, and cell loss correction capabilities. There are five AALs corresponding to different classes of service, but two are more appropriate for streaming applications:

- AAL 1 corresponds to class A specifications which require timing between source and destination, operation at constant rate, and connection-oriented mode. In addition, it provides functions for the MPEG-2 clock transmission transparency and information transparency when FEC is enabled. The overhead is 4 bytes for each MPEG-2 TS packet mapped onto the payload of four ATM cells.

- AAL 5 is defined for high-speed transfer, operating at variable rate, but contains no error correction. The overhead is 8 bytes for two MPEG-2 TS packets mapped onto the payload of eight ATM cells. If the last cell is lost, two AAL5 frames (four MPEG-2 TS packets) are corrupted. Figure 12.24 shows the MPEG-2 data streams to ATM adaptation.

Both ATM-AAL1 and -AAL5 allow for streaming and file transfers and can be efficiently used in local-area networks and wide-area networks. ATM provides the following features:

- Transparent transmission of MPEG-2 TS packets for ATM-AAL1

- MPEG-2 and network clock independence

- Any physical layer medium (copper or fiber-optic cables)

- Multiple MPEG-2 TS that can be routed to separate destinations, using the destination address in each cell header, within the ATM networks without the need to demultiplex at the MPEG-2 layer (virtual circuit switching)

- Extremely robust error correction performance even for most congested network situations for ATM-AAL1

1) Generation of Type AAL1 ATM cells   2) Generation of Type AAL5 ATM cells

Figure 12.24 Mapping of MPEG-2 TS packets onto ATM cells.

Figure 12.25 describes the MPEG-2 TS protocol stack in an ATM environment.

12.4.6 Microwave radio link

Existing analog microwave radio links transmit one analog NTSC signal and several audio channel signals that frequency-modulate different carriers and are multiplexed together in a 17-MHz-wide channel. Typical allocated microwave bandwidth can vary from 17 to 25 MHz. The implementation of DTV services by broadcasters has brought new requirements for carrying ATSC, ENG, and DS-3-based data streams through microwave links. Presently, broadcasters have three STL approaches for the implementation of DTV programming while maintaining existing NTSC transmissions:

■ Add a DTV microwave STL to the existing analog NTSC microwave link when a second RF channel is available. Experiments have shown that 14- to 29-Mbps data streams can be passed transparently through existing analog microwave systems using a 25-MHz channel bandwidth. And 19.39-Mbps ATSC signals easily fit into this bandwidth. This type of link is useful for ENG contribution links to transmit news materials originating from Betacam-SX (18-Mbps) and DVC-Pro (25-Mbps) systems to the station. Carefully designed DTV microwave systems have a wider bandwidth of up to 32 Mbps, allowing the simultaneous transport of ATSC and other data stream signals on the same link.

MPEG-2 Video Elementary Streams	MPEG-2 Audio Elementary Streams	Other Data
MPEG-2 Packetized Elementary Streams		MPEG-2 Specific Information

Figure 12.25   MPEG-2 TS protocol stack in ATM environment.

- Multiplex an ATSC data stream (19.39 Mbps) with an NTSC digitally compressed data stream (15 to 20 Mbps). The NTSC signal is decoded and compressed using an MPEG-2 compression scheme. The resulting data stream easily accommodates a DS-3 microwave link capacity of 44.736 Mbps and can be transmitted with a single RF carrier. But this link approach requires NTSC decoding/recoding, A/D and D/A conversions, and MPEG-2 compression and decompression. It causes increased system costs, differences between audio and video processing delays, and complexity.

- Use a dual-carrier microwave system. The ATSC data stream is digitally modulated by a 16-level quadrature amplitude modulation (16-QAM). The analog NTSC signals frequency-modulate a separate carrier that is converted to a 70-MHz intermediate frequency (IF). They are then combined into a 25-MHz bandwidth channel and up-converted to the microwave channel frequency. Additional channel data (1.544 Mbps) can be added to the ATSC stream and used for data broadcasting, LAN traffic, or multiple voice channels. Microwave modems* use FEC to improve link performance, at the expense of a reduced bandwidth (10 to 15 percent), and dynamic adaptive equalization to reduce degradation due to multipath conditions. This approach offers a flexible architecture that will accommodate future needs. The analog channel can be removed and replaced by a second DTV channel.

When DTV microwave links are implemented, a careful examination of the microwave beam path profile must be made to ensure acceptable system and link characteristics. With highly compressed DTV signals, brief fades may severely interrupt the programming.

---

*A microwave modem is a gateway device interfacing the incoming digital data stream and the microwave.

Figure 12.26 summarizes all possible solutions for studio-to-transmitter links.

12.4.7 Satellite

Transmitting and receiving satellite stations are generally located outside cities and at a significant distance from broadcast transmitter sites to avoid electromagnetic interferences generated by them. So satellite links are generally combined with either a microwave or a Telco network link.

Satellite modems* require a strong FEC to improve system performance at low RF levels. Up to 40 percent of error correction redundancy may be used, thus significantly reducing the effective data bandwidth.

Satellite signal transmission performance is a tradeoff between power, bandwidth, and minimum RF signal level. Satellite transponders have typical bandwidths of 23 to 49 MHz at −1 dB.† As satellite transmissions are

Note 1: The SSI has been standardized by SMPTE for ATSC distribution .

Figure 12.26 Studio-to-transmitter links (STLs).

---

*Satellite modem is a gateway device interfacing the incoming digital data stream and the satellite.

†The DVB ETS 300 241 document describes in Annex C several examples of allowable bit rates versus transponder bandwidths. Annex D describes the satellite link bit rate performance over a 33-MHz transponder.

expensive, bandwidth-efficient modulation schemes are used to maximize the satellite throughput. For a transponder bandwidth of 36 MHz, a maximum transmission data rate of 43 Mbps can be achieved using a QPSK modulation scheme, and a maximum of 49 to 65 Mbps with 8PSK, while 16QAM provides 73 to 86 Mbps. And 16QAM offers the highest bandwidth and noise margin but is more vulnerable to amplitude and phase channel nonlinearities compared to 8PSK. Powerful encoding schemes increase bandwidth but reduce the satellite-radiated signal power.

Figure 12.27 describes several IFL implementations that are suitable to analog NTSC, DTV, and simulcast broadcasting.

12.5 DTV Standard Converters

Current video converter characteristics are designed to interface a multiformat production and distribution environment and to accommodate different operational requirements. They are all described in the following sections and summarized in Fig. 12.28. In addition, a good up-converter must include several processing and system features that are needed in a DTV studio:

- A high-performance noise reduction system to compensate for the increased visibility of noise after up-conversion
- An audio-to-video delay compensation to avoid lip-sync problems
- An advanced motion-adaptive conversion to eliminate motion artifacts

In minimum-cost transition scenarios, the most costly and necessary production equipment is without any doubt the up-converter. A careful examination of features and performance must be made before buying such hardware. Any low-cost solution will inevitably necessitate external equipment to solve performance and operational deficiencies.

12.5.1 Picture aspect ratio conversion

Picture aspect ratio conversions take place at the production plant and in the ATSC receiver. Program material originating in 4:3 format will be converted for insertion in 16:9 program distributions. Similarly, 16:9 originating programs will be converted to 4:3 for simulcast NTSC transmission during the phase-out period. The ATSC receiver will have to convert various picture and scanning formats to its "native" format, as discussed in detail in Sec. 11.8.7. The DVB ETR-154 document includes an additional 2.21:1 aspect ratio which is optional for the 25-Hz HDTV integrated receiver device (IRD).

DTV and DVB production and transmission systems use three different picture format signals to meet production, transmission, and receiver display requirements:

Figure 12.27 Interfacility links.

Note 1: The SSI has been standardized by SMPTE for ATSC signal delivery to the transmitter site. The ASI can also be used with HFC links.

Note 2: MCR = Master Control Room

- The source picture format (or native format). This format is used in TV production areas. The picture aspect ratio can be 4:3, 16:9, or 2.21:1, this latter being part of the DVB format specifications only.

- The encoded picture format. This format has been developed to suit SDTV and HDTV transmission and delivery requirements. It may involve several parameter conversions, including the aspect ratio conversion.

Figure 12.28   Examples of conversion steps between DTV production formats.

■ The display picture format. This format is generated by the IRD. It can be 4:3, 16:9, 14:9, and optionally 2.21:1 for 25-Hz receivers. Note that 14:9 is a display format only, and no picture source generation works in this format at present. Format 14:9 is not included in the DVB standard ERT-154 and the ATSC standard A/53 Table 3.

An application example of these three different picture formats would be a production or network feed in 1080I/16:9 format converted to 720P/16:9 before digital transmission and to 480I/4:3 before analog NTSC transmission. The receiver displays the digital signal in 480P/16:9, 480P/14:9, or 480I/4:3 format.

The most used picture aspect ratio conversions are 4:3 to 16:9 and vice versa:

■ 4:3 to 16:9. This picture aspect ratio conversion is realized by means of two methods that result in two different displayed pictures. The first method is a "vertical crop" of the original picture and is represented in Fig. 12.29. The original 4:3 picture is stretched by a factor of 1.33 (16:9/4:3 = 4:3) in both the horizontal and vertical directions. The 16:9 framing is made by extracting 362 lines (483 × $^3/_4$) of the original 4:3 picture and displaying them on the 16:9 raster as 483 lines. An electronic conversion realizes the expansion of the 362 lines to 483 lines by using a vertical interpolation process. This results in a vertical picture resolution loss of about 25 percent (121/483).

Vertical interpolation can be done by processing every field of the interlaced signal, but degradation in the rendering of diagonal contours reduces the quality of the interpolated signal. This intrafield vertical interpolation process is simple to implement. A more accurate and more complex vertical interpolation process involves several steps such as a conversion from interlaced to progressive scanning, an intraframe vertical interpolation on every frame, and a vertical subsampling of the progressive interpolated signal to restore the interlaced structure.

The second 4:3-to-16:9 picture aspect ratio conversion method inserts the original 4:3 picture inside a 16:9 raster, a shown in Fig. 12.29. This results in two black side panels.* The 720 horizontal pixels of the 4:3 picture are decimated to fit into the 540 pixels (720 × $^3/_4$) of the 16:9 raster. This method does not require a frame memory and is easy to implement, but investigations of viewer preferences showed low interest in this picture display method.

■ 16:9 to 4:3. Two solutions can be used to resolve the 16:9 to 4:3 picture aspect ratio conversion issue, as shown in Fig. 12.30. In the first solution, the original 16:9 picture is cropped on both sides to extract a "central window" that fits into the 4:3 raster. A horizontal pixel interpolation is used to expand the 540 pixels into 720 pixels. Two line memories are necessary to perform this conversion. However, this central-window solution may not be wide enough to reproduce two interesting areas located on both sides of original 16:9 picture materials. An improvement by means of "pan and scan" information allows the positioning of the window inside the wide picture. This information can be provided on a field-per-field basis, in the encoded video data stream, to be used by the TV receiver. However, pan and scan requires additional production operations that make it expensive to use.

The second solution is called *letter box,* as the original 16:9 picture is vertically compressed into 362 lines and results in two horizontal black bars at the top and bottom of the 4:3 raster, as shown in Fig. 12.30. A vertical decimation process, combined with a frame memory, is used to realize this picture aspect ratio conversion. The SMPTE RP-199-1999 recommends the mapping of 4:3 and film-based pictures into a wide-screen (16:9) scanning structure in a manner that retains the original aspect ratio of the work.

Pillar box is the equivalent DVB term for side panels. It describes a picture that does not completely utilize the full width of the raster.

(a) Method 1: "Vertical crop". The original 4:3 image is cut at the top/bottom to horizontally fill a 16:9 raster format.

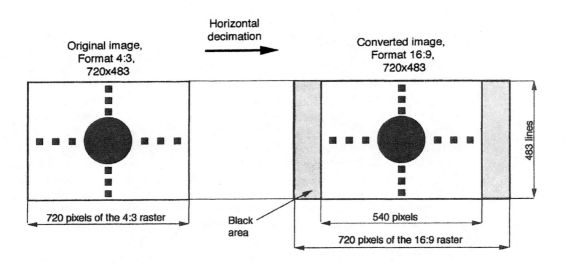

(b) Method 2: "Side panels". The original 4:3 image is inserted in a 16:9 raster format.

Note: In these two examples, for ease of comparison, both display screens, 16:9 and 4:3, have the same vertical dimension. In addition, both display screens have the same active picture array, 720 pixels x 483 lines.

Figure 12.29   A 4:3-to-16:9 picture aspect ratio.

(a) Solution 1: "Central window" (or Edge-cropping). The original 16:9 image is cut on each side to vertically fill a 4:3 raster format.

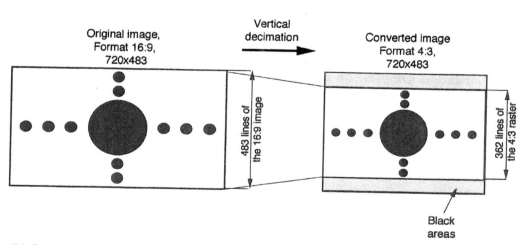

(b) Solution 2: "Letter Box". The original 16:9 image is inserted in a 4:3 raster format.

Note: In these two examples, for ease of comparison, both display screens, 16:9 and 4:3, have the same vertical dimension. In addition, both display screens have the same active picture array, 720 pixels x 483 lines.

Figure 12.30 A 16:9-to-4:3 picture aspect ratio.

■ Anamorphic conversion. The word *anamorphic* is derived from the ancient Greek word *anarmorphoun,* which means transformation, i.e., a picture having an abnormal or unusual aspect ratio. In TV applications, it describes, e.g., the display of a 16:9 native picture on a 4:3 display monitor without the use of an aspect ratio converter. The monitor exhibits a vertically stretched picture. Such a display is often called *full-height anamorphic 16:9*. Conversely, a 4:3 native picture looks horizontally stretched on a 16:9 screen.

The transmission of multiple different DTV formats on the same channel has created a need for an active format description (AFD), information that would help the receiver in optimizing the display of the transmitted material. This AFD flag is being standardized by the DVB consortium (Draft ETR-154 Rev. 3) and identifies the coded active video data only of each DTV format. Its transmission and use are both optional. It is similar to the wide-screen signaling (WSS) concept described in the ETS 300 294 and ITU-R BT.1119-2 standards, and will be carried out in the user data of the video elementary stream in network transmission.

12.5.2 Picture scanning format conversion

DTV picture scanning formats differ by the number of pixels per line and the number of lines per frame. Line rate and pixel number conversions are realized by interpolation or decimation filters. An interpolation process increases the number of video samples in the horizontal and vertical scanning directions by calculating an approximate sample value based on those of two or several adjacent samples. An example of an interpolation process is shown in Fig. 6.19. Decimation is the reverse process of interpolation. An example of a decimation process is shown in Fig. 8.34.

■ 1080- and 720-line formats. The two HDTV scanning formats, 1920 × 1080 and 1280 × 720, are related by a factor of 3:2, and a simple interpolation factor is used to convert one to the other.

■ 720- and 480-line formats. A 2:1 relationship exists between pixel and line specifications of the 1280 × 720 and the VGA 640 × 480 scanning formats, the first one being a 16:9 aspect ratio and the second one a 4:3 [1280/640 = (720/480) × (16/9)/(4/3) = 2] aspect ratio. But the widely used 720 × 480 format requires a more complex processing, as shown in Fig. 12.31. The vertical line ratio is 720/480 = 1.5. It implies that for two lines in the 480 format, three lines must exist in the same time interval in the 720 format and an intermediate line must be created. In the horizontal scanning direction, two different horizontal sample interpolation processes must be used to accommodate 480-line materials produced in 16:9 and 4:3 aspect ratios.

■ 1080- and 480-line formats. The 1920 × 1080 can be realized by doubling the spatial resolution of ITU-R BT.601 and adapting to a 16:9 aspect ratio, which leads to 720 × 2 × (16/9)/(4/3) = 1920 samples per horizon-

Figure 12.31 Conversion between 720 × 480 and 1280 × 720 picture scanning formats.

tal line. As the ITU-R BT.601 format does not have square pixels (PAR = 4/3 × 480/720 = 0.888), the number of lines is calculated as follows: 480 × 2/0.8888 = 1080 lines.

Another possible video format, used in computer picture generation, is 1440 × 1080 (4:3). This format is a subset of the 1920 × 1080 (16:9). Both have square pixels, but the difference lies in the raster aspect ratio. The 1440 × 1080 is part of the MPEG-2 standard and can be up-converted to 1920 × 1080. Down-conversion to 720 × 480 is done by decimating horizontal pixels by 2 and decimating vertical lines as follows: (1080/2) × 4/3 × 480/720 = 480.

- 1035- and 1080-line formats. Digital programs produced in the early days of the HDTV were based on the SMPTE 260M standard and have an active image format of 1920 × 1035 pixels. An interpolator can be used to convert this 1035-line format to an ATSC 1920 × 1080 format.

- Line doubler. When SDTV materials are displayed on large-size (above 32-in) monitors and receivers, visible horizontal scanning lines appear on the picture. Monitors and receivers can eliminate this annoyance by doubling the number of scanning lines of the SDTV signal. An interpolated line is added between two successive lines. Furthermore, the resulting picture appears sharper.

12.5.3 Frame rate conversion

A high number of frame rates is used by the motion picture industry to accommodate various film and television requirements. These frame rates are 23.976, 24, 25, 29.97, 30, 50, 59.94, and 60 fps. In addition, they can be combined with an interlaced or progressive scanning, as shown in Fig. 8.29.

- Interlaced to progressive.* The interlaced fields to progressive scanned frames format conversion performance is affected by the motion content of the picture. When there is no motion between two fields, the progressive frame can be easily obtained by interleaving lines from both fields. The resulting displayed picture exhibits an improved quality performance. But when there is motion, this interleaving combination results in picture judder (moving vertical edges of picture details) due to the skewed superimposition of the two fields. It is then preferable to calculate the "missing" lines from one interlaced field, thus reducing the vertical resolution. Several intrafield processing techniques are used and can be optimized for stationary or moving images. Some techniques require field memories, while others need line memories. Cost, complexity, and desired picture quality influence the choice of a deinterlacing technique.

- Progressive to interlaced. A simple progressive (30P) to interlaced scanning (30I)† conversion is realized by separating odd and even lines of the progressive frame. The odd lines constitute field 1, and the even lines, field 2. In the case of an original progressive picture scanned at 60 Hz and converted to 30I, odd lines of odd frames constitute field 1, and even lines of even frames are field 2.

- 60 to 59.94 Hz. The conversion ratio is also noted as 1/1001 or 30 to 29.97 Hz.‡

---

*An interlaced format is sometimes noted as a 2:1 scanning format while a progressive format is noted as 1:1.

†The 1080I at 30 fps is sometimes noted as 1080 at 60I, which really means 60 fields per second. All progressive formats are expressed in frames per second, and interlaced formats should use the same notation.

‡NTSC signals are broadcast at 29.97 fps. The change from 30 to 29.97 fps was decided after investigations made by the NTSC in 1954. Interfering beat frequencies occurred between the color subcarrier (3.5795 MHz) and the sound intercarrier (4.5 MHz), and between the color subcarrier and the higher frequencies of the video luminance signal. They were minimized by choosing a horizontal frequency (f_H = 15,734.25 Hz) having its 286th harmonic equal to 4.5 MHz and a color subcarrier frequency being equal to 227.5 times the horizontal frequency. Thus, the vertical frequency became $f_H/525$ = 29.97 Hz.

■ 24P to 30 Hz. This conversion is similar to the 3:2 pull-down process used in film-to-DTV transfers, as explained in Sec. 11.4.8.

■ 24P to 24sF. This conversion process is explained in Sec. 11.4.8.

■ 30P to 60P. The 30P signal is split into two fields, where odd lines constitute field 1 and even lines comprise field 2. This approach allows the resulting field-paired signal to be recorded on current storage devices (DVTRs and disk recorders). The recombined frame can be displayed twice to provide a 60P scanning format.

12.5.4 Colorimetry conversion

Colorimetry conversions are necessary when SDTV signals such as ITU-R BT.601 signals at 480 are converted to HDTV SMPTE 274 M/ ITU-R BT.709 HDTV signals. Accuracy in colorimetry coefficient values is needed to maintain color fidelity. An SMPTE Engineering Guideline EG-36 defines two transformation matrices to be used when conversions are made between ITU-R BT.601 and BT 709 standards, and vice versa.

In SDTV and HDTV analog-to-digital conversions, analog difference signals $E'_R - E'_Y$ and $E'_B - E'_Y$ are first normalized to have an excursion range of ±0.5. These new difference signals are called E'_{CR} and E'_{CB}. All analog signals E'_Y, E'_{CR}, and E'_{CB} are then digitized using two different conversion factors:

$$Y = 219E'_Y + 16$$

$$C_R = 224E'_{CR} + 128$$

$$C_B = 224E'_{CB} + 128$$

Two sets of matrices per conversion are necessary to define analog and digital transformations due to differences in scaling of Y luminance and C_R/C_B color-difference signals (219/224).

Table 11.4 summarizes all chromaticity coordinates (x, y) used in SDTV and HDTV systems that are necessary to represent colors in the 2D Plankian locus representation of colors (for reference, see Sec. 2.1.1).

The DVB standard document ETR-154 recommends that the chromaticity coordinates of the source primaries, optoelectronic transfer characteristics of the source picture, and matrix coefficients comply with ITU-R BT.470-3 System B, G, I for 25-Hz SDTV, with ITU-R BT.709 for 25- and 30-Hz HDTV, and with SMPTE 170M for 30-Hz SDTV formats.

12.5.5 Color space conversion

Component video source generators and broadcast equipment need three channels to process and distribute the three-color components of given color image. *GBR* or YC_BC_R color space can be used to represent these three-color signal components. Characteristics of these color spaces are defined in the ITU-R BT.601-5 document for both 525- and 625-line systems. Note that Y, G, B, and R channel signals have 8 percent headroom and 6 percent footroom while C_B and C_R have 7 percent headroom and footroom. Recent experience has shown that software-based color space conversion is a time-consuming process. A hardware implementation is preferred whenever possible.

Illegal and invalid signals must be avoided to maintain accurate color reproduction. A signal is illegal when its values are not within the specified voltage range of 0 to 700 mV. This range is also called the gamut. A legal signal may be invalid if it generates illegal signals when converted to another format.

12.5.5.1 *GBR*-to-*YC_BC_R* conversion. Most of the color picture sources (cameras, computer-based graphic systems, etc.) generate *GBR* signals that are always legal and valid. But more and more production video materials are processed in the YC_BC_R color space (NLE systems, video procamps, compression, etc.) to reduce the total processing bandwidth and provide accurate video processing. Several significant advantages drive this trend:

- A bandwidth reduction in chroma channels facilitates signal processing and storage.

- It is suitable to JPEG, MPEG, or DV image compression schemes that require YC_BC_R input signals.

- Small amplitude errors in *GBR* channels generate visible color temperature variations, affecting white and neutral gray areas. With YC_BC_R, signal amplitude errors may affect chroma saturation, but it is less visible compared to color temperature variations.

- Small timing errors in *GBR* channels will affect both the luminance and the chrominance transitions, thus severely degrading the picture sharpness. With YC_BC_R, only the registration of luminance and chrominance signals is affected.

- Reduction in development and manufacturing costs can be expected with a YC_BC_R architecture as chroma filters, signal timing, and amplitude levels have more relaxed specifications for the same performance at the receiving end.

GBR signals are gamma-precorrected to compensate for the nonlinearity of the phosphors used in CRT-based color TV displays. As a result, when *GBR*-to-YC_BC_R signal conversion is performed, the luminance signal is made of a combination of nonlinear signals and does not represent the true luminance. This signal is often called luma. The difference between the true luminance and the luma signal is carried in the chroma signals. These chroma signals are generally bandwidth-limited and compressed so that the luminance information is lost at high frequencies. This nonconstant luminance error increases with saturated pictures. In addition, some components of the chroma noise can be decoded as luminance signals and added to the displayed luminance.

With the arrival and future general use of plasma displays which do not require gamma-precorrected signals at their input, it is reasonable to think that this gamma correction will be eliminated, thus removing this significant picture degradation from the processing path.

12.5.5.2 YC_BC_R-to-GBR conversion. Some production equipment such as DVTRs, test signal generators, and production mixers generates and processes signals in the YC_BC_R domain. However, color correction, chroma keying, etc., are generally done in the *GBR* domain to enable users to modify black level, gamma precorrection, and white level on each *GBR* signal.

Contrast, brightness, and saturation adjustments can easily be made on YC_BC_R signals while hue adjustment is difficult and may generate illegal values in the *GBR* domain. Figure 12.32 shows the limited color space limits of *GBR* signals compared to YC_BC_R limits. As an example, if the luminance Y signal amplitude is 0 or 100 percent, any nonzero values of C_B and C_R signal amplitudes, which are legal in the YC_BC_R domain, generate illegal *GBR* signals. So, in this case, the original YC_BC_R signals are legal but invalid because they create illegal values in the *GBR* domain.

GBR gamut validity can be verified using a special component analog display called the Tektronix diamond, as shown in Fig. 12.33. Note that the diamond display can also be used to monitor timing errors, black balance, and gamma adjustments.

When YC_BC_R signals are digitally generated or processed and then converted to *GBR* signals, illegal values may appear. An acceptable solution is to equally reduce the C_B and C_R signal amplitudes (rescaling) until valid *GBR*

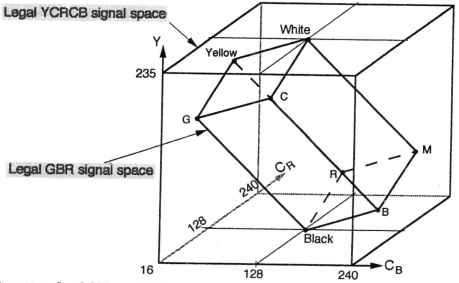

Figure 12.32 Legal *GBR* and *YPBPR* signal color space limits.

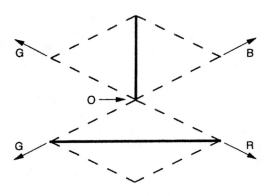

Legal GBR signals are within the diamonds.

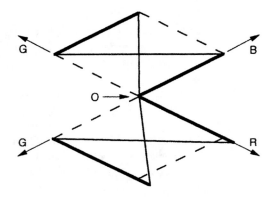

The R channel voltage exceeds the nominal value.

Notes: The center O represents the location for black G,B,R signals. An error in the G channel will affect both the top and bottom halves of the Diamond display.

Figure 12.33 Diamond displays of legal and illegal *GBR* component signals.

values are generated. This ensures that no clipping occurs in the *GBR* domain, thus compromising the color reproduction due to unexpected hue shifts. As a result, only the color saturation is reduced, which is far less objectionable than a hue shift.

Section 2.1.1 mentions that colors outside the Plankian locus create negative and greater-than-100-percent *GBR* values in cameras. These signals cannot be reproduced by negative light emissions of a current display system. Consequently, they are clipped in the subsequent process to conform to the signal format of the system. New display devices being introduced as well as film materials are capable of reproducing a wider color gamut, required to reproduce pure spectral colors. An extended color gamut system allowing greater than 100 percent (129 percent) and negative (-27 percent) *GBR* signal values is described in the ITU-R BT.1361 standard document. It gives detailed explanations on how to calculate YC_BC_R values from *GBR* values obtained from both conventional and extended systems. Compatibility with conventional systems is maintained by using the same ITU-R BT.601 equations to calculate YC_BC_R from E'_Y, E'_{CB}, and E'_{CR} for both systems. Extended gamut *GBR* signals are quantized using different scaling factors and dc offset values. Figure 12.34 shows the 8-bit digital encoding of conventional (ITU-R BT.601) and extended (ITU-R BT.1361) gamut systems. It is important to note that the extended gamut system quantizes *GBR* signals with 160 levels only from 0 to 100 percent, thus reducing the *S/N* ratio (see Sec. 3.1.6 for the calculation of the quantizing noise). To maintain an acceptable *S/N* ratio in *GBR* signal chains, a 12- or 14-bit quantizer is used.

Conventional Gamut

G = 219 (E'G) + 16
B = 219 (E'B) + 16
R = 219 (E'R) + 16

Y = 219 (E'Y) + 16
CB = 224 (E'CB) + 128
CR = 224 (E'CR) + 128

Extended Gamut

G = 160 (E'G) + 48
B = 160 (E'B) + 48
R = 160 (E'R) + 48

Y = 219 (E'Y) + 16
CB = 224 (E'CB) + 128
CR = 224 (E'CR) + 128

Headroom
(19 levels)   — Level 254
               — Level 235 (100%)
                   White

Footroom   Quantizing range
(12 levels)   (219 levels)

    — Level 16 (0%) — Black
    — Level 04

Conventional digitization of GBR signals

Headroom
(46 levels)   — Level 254
               — Level 208 (100%)
                   White

Footroom   Quantizing range
(44 levels)   (160 levels)

    — Level 48 (0%) — Black
    — Level 04

Extended digitization of GBR signals

Figure 12.34   Digital encoding for conventional and extended gamut systems.

12.5.6 Sampling structure conversion

Different sampling structures are used in the production, compression, and distribution of TV materials. Graphics and film-to-video conversions require the best quality, which is provided by a 4:4:4 sampling structure. All digital video facilities are designed around a 4:2:2 sampling structure for video signals, whether uncompressed or MPEG-2-compressed at the 4:2:2 profile. The ATSC standards require a DTV transmission by means of MPEG-2 video signal compression encoding using the main profile (4:2:0). Sampling structure conversions then take place in TV production facilities.

■ 4:2:2 to 4:2:0 conversion. Section 3.3.10.3 describes the 4:2:0 sampling structure and the generation of this structure from 4:2:2 signals. Subsampling is used to derive a 4:2:0 structure by a vertical interpolation filter of four lines and field sequential cosited C_B and C_R samples. Luminance samples remain unchanged during this conversion. The DVB ETR-154 document defines two possible down-sampling processes from 4:2:2 to 4:2:0. An MPEG-2 parameter indicates if the down-sampling process has been applied to the two fields independently or if the two fields have been combined into a single frame before down-sampling.

■ 4:4:4 to 4:2:2 conversion. This is achieved by a subsampling of chrominance signals only.

12.5.7 Film-to-DTV conversion

The film-to-DTV conversion is based on the 3:2 pull-down process that is described in Sec. 11.4.8, which covers DTV compatibility with film originating programs. To reduce flicker, each frame of the 24-fps film is projected two or three times using a two- or three-bladed shutter that provides a video signal at 48 or 72 fps. Most of the film production materials in 16- and 35-mm are converted using this process. This conversion approach is also used by 480P DTV proponents where 480P at 30 fps is used for acquisition and 480P at 60 fps provides an acceptable display format.

12.5.8 Composite and component analog-to-DTV signal conversions

Up-converted NTSC pictures look terribly bad when compared to up-converted 4:2:2 materials. Any residual subcarrier is further compressed as noise and reduces the allocated bit rates for the program. In addition, analog NTSC signals themselves have already been severely bandwidth-compressed.

TV facilities have a tremendous amount of archives in analog tape formats. So care must be taken to use a high-quality NTSC decoder before up-conversion can take place. Similarly, analog NTSC production is going to continue for some time, and a high-quality encoder should be used to later facilitate its decoding and conversion. TV program archives have been recorded on 2-in, 1-in, Betacam SP, MII, D-1, D-2, D-3, D-5, DCT-700, and digital Betacam videotape recorders providing different picture quality performance in playback. Therefore, any up-conversion must be carefully considered to generate an acceptable DTV signal for broadcasting. Users must also remember that not all video component formats are compatible with one another, as explained in Sec. 2.3.

- 2- and 1-in VTR. For almost 40 years, these VTRs have been recording TV program materials in NTSC format on tape with a maximum bandwidth of 4.2 MHz. Chrominance signals are bandwidth-limited to 1.3 and 0.6 MHz, thus providing an analog compression ratio of 4:1 between luminance and chrominance signals. The up-conversion performance can be improved by using noise reduction and signal enhancement processors.

- Analog Betacam and MII VTRs. Although NTSC inputs/outputs are used in most cases, all analog Betacam and MII tape materials are recorded in analog component format on the tape medium. Therefore, it is highly recommended to use analog component output signals for conversion to DTV. In addition, it is known that analog Betacam VTR performance specifications are very wide (for example, ±1 dB at 3 MHz and ±3 dB at 4.2 MHz using a metal tape). It is possible to carefully optimize the component signal channel alignments to get substantial improvements in the overall VTR performance and resulting DTV conversion. Multigeneration performance of Betacam and MII materials degrades very rapidly due to the moiré components that are compounded at every generation.

Therefore, it is recommended to use first-generation Betacam and MII materials and transfer them to a higher-quality storage medium for post-production, if needed, and then up-convert the final program to DTV. Any new recording and playback should be made using analog component input/output.

- D-2 and D-3 DVTRs. These DVTRs have a recording bandwidth up to 6 MHz and provide a superior picture quality after up-conversion. But an analog compression ratio of 4:1 between luminance (6-MHz bandwidth) and chrominance (1.5-MHz) signals still exists. Good-quality digital composite-to-component converters are used before up-conversion.

- D-1, D-5, DCT-700, and digital Betacam DVTRs. These DVTRs have a maximum bandwidth of 5.75 MHz for luminance and 2.85 MHz for chrominance signals. Very good-quality up-conversions are obtained due to the lack of NTSC composite artifacts in the program signal.

12.6 Technical and Practical Considerations about DTV Systems

New DTV systems are bringing new system issues or aggravating already known problems. The multiple choices of new technologies have made possible a high number of combinations of new production architectures in which all system issues must be understood and controlled. Careful system considerations contribute to a well-designed and understood transition scenario to DTV.

The fast-growing use of computers in servers, operational control, on-air systems, etc., has made power interruption a more critical issue. Power distribution reliability in sensitive areas (master control rooms, on-air servers, and storage servers) or even the complete production facility needs to be improved by using uninterruptible power systems.

The new broadcast model will provide more services with fewer technical people. It can be seen as being made of more computer- and disk-based systems, broadcast/computer networks, and more "on-screen" virtual control panels gradually replacing routers, DVTRs, and dedicated control panels. However, the broadcast industry does not have full confidence in the computer hardware/software reliability and will always fear "reboot" warning messages during critical broadcasts, such as commercials. So total redundancy and intelligent mix of known broadcast and computer technologies are important in the design of new DTV facilities. Another consequence is the growing need for better-qualified technical support people in production and postproduction areas.

12.6.1 Existing Infrastructure in production facilities

Existing data "pipes" based on AES, SDI, and HD-SDI standard specifications can be used to carry fixed data rates through current production infrastructures including routers, distribution amplifiers, and storage devices. Two main data pipes already exist in most production facilities (3.072 and 270 Mbps), and two others are being considered in some production installations (540 Mbps and 1.5 Gbps). Numerous standards have been developed to allow the embedding of compressed and uncompressed video and audio data formats into these fixed-data-rate pipes. In addition, the storage of these formats has been made possible thanks to the development of data recording standards such DVC-Pro, Beta-SX, Digital-S, and DV, which allow the interconnection of recorders through an SDTI-based network at normal and 4 times normal speed.

The AES infrastructure offers a maximum data rate of 3.07 Mbps which can be used to carry compressed audio streams such as Dolby-E, AC-3, and Apt-X; metadata; non-PCM audio data; time code; etc. The SDI infrastructure offers a maximum data rate of 270 Mbps which can be used to carry compressed video streams such as MPEG-2, DV, DVC-Pro, Beta-SX, Digital-S, up to 16 channels of audio data, time code, metadata, machine control, etc. The HD-SDI infrastructure will offer a high-bandwidth transport which can be used to carry any combination of uncompressed video scanning formats that does not exceed a maximum payload of 1.2 Gbps. Several standards are being developed by the SMPTE for the specification of an HD-SDTI transport layer, the embedding of 480P into an HD-SDI data stream, and the multiplexing of any combination of video formats with low latency.

12.6.2 Mezzanine compression

In the course of upgrading existing production facilities to DTV, an original idea was to use existing data pipes (cables, routers, and distribution devices) to implement a mezzanine compression scenario. Such mezzanine compression uses a mild compression ratio for any video source format to produce a bit rate in the range of 270 to 360 Mbps. ATSC and SMPTE working groups have discussed mezzanine compression implementation. But it quickly appeared that it would involve the concatenation of too many compression algorithm schemes, too many cascaded compression-decompression cycles that would degrade the picture quality and add delays of several frames, at a very high cost, etc. However, this concept will be used in small islands (ENG and EFP suites) where compressed data are mapped into SDTI and distributed to compressed DVTRs, servers, and storage.

12.6.3 Media asset management

Broadcasters are now faced with a situation in which they can broadcast several simultaneous TV programs on a DTV channel to get additional revenue from a variety of programs and services. All these TV programs must be generated using existing and new media materials, but broadcasters can afford to employ a very limited additional production staff only. The reuse (or repurposing) of existing program materials then becomes an absolute necessity. In addition, a media asset management system is needed when it is important to know the location of program materials on various servers, all postproduction versions of the same media, which native formats, which production or distribution format to convert to, when and where the conversion must be done, what routing must be used to get them where requested, etc.

A new concept has been developed that defines a media asset as being made of the program material itself (the content) and the proper information to identify it, called metadata.* Each asset has also its own right to use it. A fast retrieving and access to media assets requires a proper management infrastructure that would allow storing, searching, sharing, billing, and manipulating audio, video, and other data types. Asset management also allows tracking of media versions and of how to move media through the production process. This new infrastructure also facilitates the handling of the intellectual properties, cuts costs, and increases productivity and asset value. A standardization process is mandatory to define cataloging, languages, media file and format coding, system architecture, storage, and production networks. Unique identifiers such as UMID† and UPID (described in Sec. 12.1.2) are embedded in media files or streams.

12.6.4 Latency requirements in the studio

Latency is one of the most important issues in designing studios and systems operating in real time. Latency describes the processing duration of a signal in a particular device from its input to its output. It is sometimes called propagation delay. Audio and video systems exhibit very different latencies due to signal processing differences between them. Video processing requires frame stores to accomplish signal timing in digital processing equipment while audio equipment generally uses small buffers. This difference between audio and video signal latencies generates audio-video synchronization (often called lip-sync) problems. Predictable delays are essential for on-air applications. It is mandatory to have a rigorous control of latencies in audio and video production to avoid lip-sync problems.

In current TV studio applications, processing devices are gen-locked to a common reference, usually a color black signal. Latency in these devices is generally below one horizontal line duration. Digital production switchers allow for a maximum of $\pm \frac{1}{2}H$ timing differences between all inputs. So the total system latency is a video frame and can cause lip-sync problems if the audio timing is not corrected by the same amount.

When external feeds or processing devices having a latency greater than H are connected to a production switcher, the timing compensation must be made using one of the following:

---

*Metadata have already been described in Sec. 12.1.2.

†UMIDs are defined in the SMPTE Recommended Practice RP-205.

■ A frame synchronizer for each high-latency input. When used for external feeds, it is recommended to implement a frame synchronizer with internal audio processing and delay compensation to maintain the original audio-video timing.

■ A delay device for each low-latency input.

■ Various reference timing signals.

So with large latency or timing differences, the design of a TV studio or station becomes more complicated and difficult. Figure 12.35 shows video signal latencies in a typical SDTV studio system.

12.6.5 Audio-video timing considerations

With the advent of DTV broadcasting, audio-video timing differences can occur in production, postproduction, compression encoding, transmission through networks, delivery to customer homes, and DTV receiver decoding. Video, audio, and auxiliary data may take totally different paths depending on signal formats, interfacing, compression usage, and distribution system choices. EBU Rec.37 and ITU-R BT.1359 documents indicate the maximum accept-

Figure 12.35 Video signal latencies in an SDTV studio system.

able relative audio/video timing at the transmitter site, the reference timing zero being at the master control location, in both advanced (+22.5 ms, sound being in advance with respect to vision) and delayed (−30 ms) situations. It is therefore important that in-plant timings be controlled and known during the production process.

Synchronization can be made easier if time stamping is used both in audio and video data processing. MPEG-2 compressed audio, Dolby-E and new enhanced apt-X compressed audio, and MPEG-2 video stream formats have provisions for the inclusion of time stamps in the production process. Data broadcast services may also need to be synchronized to audio and video. In this case, time stamps must be added to ensure presentation accuracy within one video field or frame. This timing relationship is established during the authoring process.

- Synchronization of compressed audio-video data stream. Presentation time stamp (PTS) information is inserted in video and audio PES headers at the time of compression encoding. This 90-kHz PTS clock is locked to a 27-MHz system time clock (STC) or station master clock (27 MHz/300 = 90 kHz) having an accuracy of at least ±0.002 percent (±540 Hz). Variable-length PES packets are repackaged into fixed-length transport packets to generate a transport stream (TS). Each TS packet header receives 27-MHz program clock reference (PCR) information that is derived from the STC. This PCR information is incorporated in the ATSC data stream at intervals no greater than 100 ms and establishes an accurate time-base information that will be used by the receiver decoder to recreate this STC. Note that the DVB ETR-154 document recommends a value not greater than 40 ms. Accurate audio, video, and data timing at the MPEG-2 decoder output is realized by presenting video and audio frames and data when their PTSs match the running STC. Figure 12.36 shows the PTS and PCR embedding in PES and TS at the encoding stage.

  PCR jitter may be introduced during transmission due to switching between encoders during commercials, thus creating PCR discontinuities and variable delays in networks or demultiplexing/remultiplexing operations. PCR jitter causes frequency variations of the decoded STC, which may generate severe artifacts in the decoded video or audio signal. MPEG-2-compliant decoders include a buffer that can be used to reduce PCR jitter.

- Digital HDTV, SDTV, and audio synchronization. The SMPTE 318M standard describes reference signals for the synchronization of 59.94- or 50-Hz related video and audio systems in analog and digital areas. This document is an improved version of the RP-154. A new reference signal has been created to help in managing buffer problems, jitter, and a defined relationship between video and audio signals by adding VITC signals on lines 14 and 277 only for 525/59.94 (NTSC) systems or on lines 19 and 332 only for 625/50 (PAL) systems. Furthermore, a five-video frame sequence signal is inserted on lines 15 and 278 for 525/59.94 systems to synchronize audio and video frames per AES-11 standard specifications. This synchronization requirement is described in Sec. 6.10.

Figure 12.36 PTS and PCR information generation and embedding in PES and TS.

12.6.6 DTV monitoring

12.6.6.1 Video monitoring. In a multiple-video-format environment, HDTV and SDTV monitors are required. With multichannel broadcasting, there is a need for a larger and more flexible monitoring wall, which must also accommodate a gradual monitor conversion to a 16:9 aspect ratio. However, due to the high cost of broadcast-quality HDTV monitors, broadcasters will take any opportunity to use computer monitors and consumer receivers and will monitor down-converted HDTV signals on existing high-quality SDTV monitors. This is a quite acceptable solution on a monitor wall, in editing areas, where picture quality adjustments are not performed. However, in camera, graphic, and mix/effects areas, high-quality HDTV monitors are required.

Computer monitors such as SVGA monitors can provide a high-quality picture at a reasonable cost. They have refresh rates between 60 and 100 fps and can display a 16:9 aspect ratio, such as the Sony Multiscan GDM-W900 (16:10,

1920 × 1200 pixels). However, users must be aware that their colorimetry does not conform to international broadcast standards (higher color temperature of 9300 K). This type of monitor should not be mixed with broadcast-quality monitors in the same room or production area. In addition, the color uniformity is not very good.

Analog YP_BP_R HDTV signals are supplied to computer monitors through a converter box that provides analog *GBR* signals and generates properly timed SVGA H- and V-drive synchronization signals. This timing is not identical in all makes of SVGA computer monitor formats, and several interfacing parameters must be verified, such as polarity of sync signal, input amplitude (generally TTL level), monitor input impedance (2 kΩ), and H- and V-drive relative timings to the H and V active video position. Generally, computer monitors are equipped with fine H and V adjustments to center the active picture area in the raster. Figure 12.37 shows the relative timing of HDTV video signals and SVGA H- and V-drive synchronization signals.

A low-cost down-converter allows an HDTV 16:9 aspect ratio program to be displayed in a 4:3 raster format as letter box, central window, 14:9 aspect ratio, zoom-in, or anamorphic squeeze, as described in Sec. 12.5.1. Figure 12.38 shows a low-cost down-converter developed for monitoring applications only. It provides NTSC, SDI, and analog component $GBRS/YP_RP_B$ output signals.

DTV consumer receivers are also necessary in a TV station to monitor the transmitter operation and incoming satellite and cable feeds. Silicon providers are planning on VSB (terrestrial delivery), QAM (cable and satellite delivery), QPSK (satellite delivery), and other modulation schemes for consumer home delivery of DTV signals. Section 11.8.7 describes the receiver characteristics as part of the ATSC system.

A picture quality monitoring comparison between 16:9 MPEG-2 4:2:0 525-line and NTSC receivers of a simulcast program will show an improved resolution of the picture (sharpness), a better color reproduction (higher color gamut, an accurate reproduction of hue, saturation, and luminance), and a quasi-absence of noise in the picture. In addition, thanks to recent advances in camera design (12-bit A/D, advanced DSP, etc.), an improved video dynamic range and a better reproduction of the gray scale will be observed on TV materials originating from these cameras.

12.6.6.2 Audio monitoring. The monitoring of multiple channels and multiple audio services requires new methods and equipment for studio surround listening, display of reference levels, and channel balance and phasing. SMPTE is modifying the current Recommended Practice RP-173 to include loudspeaker arrangements for multichannel sound systems used in production and post-production areas. This document also includes recommendations for speaker placement relative to 4:3 and 16:9 aspect ratio screens, relative channel reproduction levels, and bass management. The SMPTE RP-200 specifies the relative and absolute sound pressure levels for multichannel sound systems that

Note: Figure not to scale

Figure 12.37 1080I at 30-fps HDTV sync and SVGA drive signal timing.

Figure 12.38 Low-cost down-converter for monitoring applications.

are used in review rooms. The ITU-R BS.1116 gives recommendations about the monitoring room design (size and shape, reverberation time, etc.) and loudspeaker characteristics.

Dedicated 5.1 speaker systems are already available on the market and provide an optimized position for each speaker. The monitoring of 5.1 channel surround sound (AC-3 or MPEG-2 audio layer 2) creates acoustic issues in the design of production rooms. The high-energy LFE channel generates acoustic waves that can be heard in adjacent rooms. Additional costs must be expected for isolation reinforcement of floors, ceilings, and walls.

The recommended level below clipping level (0 dBFS) of reference tones for transmission is different in North America and Europe. The DVB-ETR 154 document specifies 18 dB while SMPTE recommends 20 dB.

12.6.7 MPEG streams in production studio applications

MPEG-2-based DTV transmission standards adopted by DVB and ATSC are progressively invading all parts of the production chain, from acquisition, distribution, and postproduction to archiving, in a manner reminiscent of the introduction of NTSC and PAL transmission concepts into production facilities. Broadcasters are looking at a complete MPEG-2 infrastructure that would give them the convenience they currently enjoy with analog production facilities. If some basic production processes are now possible on constrained MPEG-2 bit streams, many others are not available, such as mix/effects and zooming. Considerable work has been undertaken by the industry and standards organizations to improve compression scheme performance and develop a set of MPEG-2 specifications that would permit an all-MPEG studio environment.

Recent studies have shown that few improvements can be made to the DCT encoding/decoding process and the entropy encoding (run length and Huffmann). But several important compression functions can be improved and new features added. These compression functions and new features are briefly described below:

- Quantization/unquantization process. This process generates pixel value errors, and current studies are concentrating on reducing their visibility. Several improvements have already been made using an adaptive quantization process. Unavoidable errors could be disseminated in areas where they are less visible, such as close to edges.

- Motion prediction. Intensive studies have been published on this subject, and new improvements will be realized in the motion-prediction domain to extend the range of motion vectors. New motion measurement models require a very large number of computations that will be possible with current or future high-speed microprocessors.

- Digital prefiltering. TV source materials may have a narrow bandwidth, and higher-frequency components are likely to be noise. Noise and unnecessary high frequencies in the picture can be removed before compression encoding to assign all available bits to more important information contained in the picture. During DCT encoding, noise is wrongly identified as useful information of the picture and DCT encoded. This noise encoding brings additional DCT coefficients that occupy a certain bandwidth of the *CBR* compressed channel, thus reducing the number of bits allocated to the useful (visible) information. In addition, encoded noise becomes much more visible after up-conversion or reencoding of the same material.

  Video signals, when represented in the two-dimensional (2D) frequency domain, exhibit more information in the low-frequency components than in the high-frequency components. This video signal characteristic matches that of the HVS, for which horizontal and vertical details (e.g., lines) are very important. Consequently, removing high-frequency coefficients in DCT blocks has no important impact on the perceived picture quality.

  White noise is evenly distributed throughout the entire 2D frequency spectrum while spike noise is mainly located at high vertical and horizontal frequencies. There are obvious benefits to removing these high-frequency components in DCT blocks.

  Simple two-dimensional filters are implemented by cascading a vertical filter with a horizontal filter. This results in a square passband filter where vertical and horizontal cutoff frequencies can be made adjustable. This filter provides the best results in picture resizing applications (e.g., conversion of 720×480 to 360×240). A cross-shape filter has the particularity of having no horizontal cutoff frequency when the vertical frequencies are close to zero, and vice versa. It provides good reduction of noise while maintaining the visual and perceptual quality of the pictures. A diamond-shaped filter will reduce diagonal frequencies, as shown in

Fig. 12.39. When vertical and horizontal frequencies are equal to 1, diagonal frequencies are reduced to 0.707. This diagonal filter will shape the spatial response of a picture to match the spatial response implied by the MPEG run length encoding that is optimized for a zigzag scanning of DCT coefficients. This combined picture processing allows a higher compression ratio. Figure 12.39 shows the 2D frequency spectrum differences between square, diamond, and cross-shape filters.

Edge enhancement is often required for pictures originating from camera systems to compensate for the high-frequency rolloff of optics, the pickup tube spot size, and the amplifier response. This compensation is called aperture correction. Another enhancement is often used to take into account the transfer function of the eye. Such correction provides the subjectively best picture within the bandwidth constraints of the data channel. An edge enhancement unit is then used to restore the response flatness at high frequencies and to provide picture clarity before MPEG encoding takes place.

As an example, the Sony HD-Cam design incorporates a digital prefiltering to reduce the original video data rate by five-eighths, followed by a compression algorithm that further reduces it by 4.4:1 using a sophisticated adaptive field/frame DCT-based algorithm. The combined data rate reduction is 7:1.

▓ Statistical multiplexing. A constant bit rate (CBR) implies a fixed bit rate regardless of the criticality of the TV programs to compress. Difficult TV

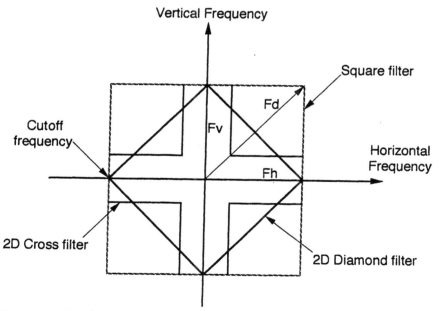

Figure 12.39 Two-dimensional diamond and cross filters used for compression performance improvements.

pictures are generally found in sport materials with high temporal detail, scene cuts, explosions, dissolves, etc. When such critical materials are presented to the compressor, an increase in the allocated bit rate could maintain the same picture quality performance. It is then possible to allocate a higher CBR, for example, for a sports data channel. But broadcasters have a fixed bandwidth (19.39 Mbps) to transmit multiple compressed program data. Occasional critical pictures in one or several CBR channels may generate picture blockiness, resolution losses, etc.

Statistical multiplexing temporarily allocates more bits to a channel compressing a critical sequence by taking these extra bits from one or several channels carried in the same TS. This bit reduction on other channels may introduce several small artifacts on them, but the severe visible artifacts affecting the critical sequence are reduced. This concept assumes that critical sequences do not occur simultaneously on all channels. Newly developed statistical multiplexers are based on recent studies of human sensitivity thresholds of image distortions introduced by MPEG-2 compression algorithms. CBR picture quality can be improved by 10 to 30 percent when statistical multiplexing is used.

- **MPEG-2 encoding alignment.** Multiple concatenations of MPEG-2 encoding/decoding cycles are expected in the production, contribution, and distribution of TV programs. A slight pixel shift in the DCT block alignment, vertically or horizontally, between two successive DCT encoders involves two different DCT calculations and rounding processes that create cumulative video signal degradations. In addition, macroblock boundaries must be identical to reduce artifacts between encoders and decoders from various equipment manufacturers. A proposed SMPTE Recommended Practice is being developed to specify recommended ranges for various video scanning formats. It also recommends that horizontal and vertical active video area reductions be made at the final production stage to ensure macroblock alignment. An example is the reduction from 720 to 704 pixels. Broadcast-quality MPEG-2 encoders are based on the 422 profile @ ML. They use 720 pixels per active line while non-broadcast-quality encoders use 704 pixels only.

- **MPEG-2 TS splicing.** The wide use of video servers in broadcast facilities has created a strong need for their use in distributing compressed video in a format compatible with DTV broadcasting. Switching (also called splicing) and seamless* concatenation of data streams are commonly performed with uncompressed video data. These operations are very complex in the compressed domain due to the different encoding combinations allowed by the MPEG-2 standard. Each picture can be MPEG-2-encoded with different characteristics (I, B, or P frames), resulting in a variable number of bits per frame (see Sec. 8.3.9). Two encoded data streams may have different pro-

---

*A seamless splicing between two bit streams results in a continuous MPEG-2 stream and preserves the MPEG-2 TS structure.

files, levels, GOPs, and bit rates. Some specific points in the bit stream are better suited for adding or removing new picture data. These points are called splicing points. To facilitate seamless splicing, the SMPTE Standard 312M specifies MPEG-2 TS constraints and defines splice points where switching can take place. I-frames and aligned frame boundaries are required at splice points.

Proper buffer management is critical in compressed data stream switching. Decoder overflow and underflow must be controlled at a splice point to ensure a smooth and transparent video transition. However, a controlled underflow is possible at the splice point. This generates a nonseamless splicing where the decoder must wait for data to arrive, thus generating a momentary frozen frame depending on the GOP size and the switching instant within the GOP. Several MPEG-2 data stream splicing products are available on the market for commercials and logo insertions.

■ Recoding data information. Successive encoding-decoding cycles bring picture degradations. If the first-pass MPEG encoding decisions are passed on to the next encoder, significant picture quality performance improvements can be achieved in all subsequent compression-decompression cycles. These levels of improvement increase with low-bit-rate compressed data streams. Any subset of the encoding decisions, also called the recoding data set, will bring visible picture quality improvements. However, if a video processing is applied to the complete picture area between two cycles, the reencoding data set is no longer valid and the following encoder must generate new encoding decisions.

This concept provides independent encoding decisions for each macroblock, as detailed in the proposed SMPTE 327M standard document on the MPEG-2 recoding data set. In case of a keying process, the modified macroblocks must be encoded as a first-pass encoding. All other unchanged macroblocks are recoded using the first-pass encoding decisions. A transport mechanism must be used to feed forward the previous encoding decisions to the next encoder. A serial digital component interface carries the active picture area data along with the transport of the recoding information data, as shown in Fig. 12.40. This ensures both simultaneous processing of data and a correct alignment of the recoding information with the decoded macroblocks in the decoded picture, both spatially and temporarily. This will allow transparent cascading, frame-accurate editing, and logo/caption insertion to be performed.

The transport mechanism should preserve the 8 most significant bits of each active video data word. The recoding data are transmitted in the least significant bit of each 10-bit chrominance data word, as described in the proposed SMPTE 319M standard document on transporting MPEG-2 recoding information through 4:2:2 component digital interfaces.

Several MPEG-related issues must be considered when source materials are assembled in a program for distribution and delivery:

Figure 12.40 SDI transport of recoding data information and corresponding active video area data between two compression-decompression cycles.

- VBI data transparency. Broadcasters have been using the vertical blanking interval (VBI) to deliver data to users or simply to allow external company data to be sent to remote offices. Similarly, the new DTV system has provision for the delivery of multiple services to the end user. An auxiliary data channel is multiplexed to the video and audio PESs, as described in Sec. 11.8.4. As an MPEG-2 encoder has a specified encoding range, its design must provide transparency to the existing VBI data present at the encoder input and multiplex them in the output TS.

- MPEG-2 interoperability ranges. Although the MPEG-2 standard specifies a unique set of tools, there are many possible implementations at different profiles and levels. These implementations must be tested for compliance to ensure a real interoperability in TV studio applications. SMPTE and MPEG Forum are developing a proposed standard to define MPEG-2 interoperability ranges as subsets of the MPEG-2 profiles and levels. The intent is to facilitate the design and the integration of MPEG-based systems in TV studio applications by defining equipment interoperability specifications and application-specific parameter settings. In addition, these new MPEG-2 constraints will lower implementation costs, make editing possible in some cases, and provide predictable latency.

12.6.8   Practical considerations in the interconnection of DTV systems

12.6.8.1   Coaxial cable interconnections.   Most of the transition scenario must take into account the existing TV facility infrastructure, be it analog composite (NTSC or PAL) or digital component production equipment, routing and distribution equipment, and interconnecting cables.

The installation of new cable runs in a broadcast plant necessitates a significant investment. It is then important to consider the possible reuse of existing cable runs, which can offer a smooth financial transition to DTV.

The reuse of existing cables is possible if a proper and clean installation has previously been made. Good cabling practices allow quick cable identification and good interconnection performance. However, in many cases, a complete cable reinstallation will be necessary to make sure that high bit rates (1.5 Gbps) can be distributed with an acceptable performance.

The cable distance limitation must be considered. For example, a Belden 8281B coaxial cable carrying 270-Mbps data can have a maximum cable length of 280 m before receiver failure.* But a reasonable safety margin must also be applied to take into account performance degradation in the transmitter, the interconnection, and the receiver, such as component aging, temperature and power supply variations, interference, connection losses, return losses, patch cord manipulation, signal jitter, and cable stress.

It is then important to consider a reasonable safety margin that can be calculated on the basis of a 6-dB margin used in the past. Calculation shows that a margin of 60 m of Belden 8281B cable is necessary at 270 Mbps, and then installed cable runs must not exceed 220 m to satisfy the required safety margin. A good practice would be to test all cable runs at the end of the equipment installation by adding successively to each cable run an extra cable length to obtain 280 m, and verify if the receiver failure is about to occur. This would certify that the cable installation was properly made.

Another coaxial video cable with low loss, low cost, and low weight is available for digital installation only. The Belden 1694A is an example of such newly designed cable. It allows a maximum cable length of 370 m at 270 Mbps and 100 m at 1.5 Gbps. The practical maximum lengths are reduced to 310 and 85 m, respectively, taking into account a reasonable safety margin. Table 12.5 shows the practical maximum cable length for 143-, 270-, and 540-Mbps and 1.5-Gbps installations.

Another issue to consider in coaxial cable distribution is the BNC connector type. Previous analog and possibly digital interconnections were made with a 50-Ω instead of a 75-Ω version. A large number of BNC connectors must be cascaded before significant return loss degradation appears. The cumulated

TABLE 12.5   Maximum Coaxial Cable Lengths (m) for SDTV and HDTV Systems

Transport interface (Mbps)	143	270	540	1485
Belden 8281B				
Maximum cable length before failure	400	280	190	70
Safety margin length	80	60	40	15
Practical cable length	320	220	150	55
Belden 1694A				
Maximum cable length before failure	480	370	270	100
Safety margin length	80	60	40	15
Practical cable length	400	310	230	85

*And 280 m corresponds to a 30-dB coaxial cable attenuation that must be compensated by the cable equalizer at the receiver input, as specified in the SMPTE Standard 259M.

number of one-quarter wavelengths sets the limit.* Nonetheless, as a given infrastructure is constantly growing, it is a good practice to use true 75-Ω BNC connectors, especially with the introduction of the 1.5-Gbps HDTV serial interface where it becomes an absolute requirement.

12.6.8.2   Fiber cable interconnections.   Optical network technologies seem to be the answer to the rapidly increasing demand for new services such as the Internet, broadcast contribution and distribution over telecommunication networks, bidirectional cable TV systems, computer networks, interactive programs, and multimedia services.

Broadcast facilities are converting from analog to SDTV and HDTV data signal formats. As distances are limited in using coaxial cables, fiber-optic cables can be used as a transport layer with unlimited bandwidth over long distances. In addition, this solution eliminates electromagnetic and noise interferences and grounding problems between buildings. New connectors facilitate the termination of fiber cables, making this work easier, with low cost and installation time. However, a small amount of dirt can bring a significant loss in the transmitted light. Dust caps on male and female connectors must always be used when they are disconnected. Single-mode fiber-optic cables are preferred to multimode fiber cables, the latter having a limited bandwidth capability due to a greater broadening of high-frequency pulses that severely degrades the shape of data pulses and causes interference of adjacent bits. Single-mode fibers have a lower attenuation, but their smaller size renders the coupling between the light and the core more critical.

Improvements in optical network technologies have been made possible by the use of the following:

■ Wavelength division multiplexing (WDM), where several signals of different wavelengths corresponding to independent data streams are combined and transmitted on a single fiber. Bidirectional transmission is also possible. Furthermore, dense WDM (DWDM) enables the transmission of multiple groups of services over a fiber-optic cable and multiple independent data streams per group.

■ Optical amplifiers such as the erbium-doped fiber amplifier (EDFA) to reduce space, power requirements, and costs.

■ Optical add/drop multiplexers (OADMs) to provide system flexibility in collecting and distributing data between sites.

The wide availability of inexpensive optical hardware makes it possible to apply cost-effective solutions to interconnection needs between facilities and even between studios. The conversion to HDTV with 1.5-Gbps serial data interfaces has dramatically reduced the critical distances in a studio. A close cost

---

*$\lambda/4 = 300/(4f) = 300/(4 \times 135)$ MHz $= 0.55$ m at 270 Mbps. But $\lambda/4 = 0.1$ m at 1.5 Gbps.

comparison between a coaxial cable solution and an optical cable solution shows that when cable interconnection length requirements start exceeding 3 times the practical maximum length of available coaxial cables, an optical solution is less expensive.

Because of the need for ever-increasing bandwidths and the spreading use of the Fiber Channel network protocol, it is now very important to include dark fibers* in the design of any new facility or conversion project. In addition, the distribution of SDI and HD-SDI signals over optical fibers provides a distribution of accurately timed video and multiple channels of audio and auxiliary data.

12.7 Transition from NTSC to DTV

The current transition from NTSC to DTV broadcasting is perceived differently among broadcasters depending on their status (network, affiliate, local station, etc.). The format of their choice will then be different, and very often dictated by cash flow. Some want to broadcast multiple SDTV programs in the same allocated digital channel; others think that customers want to see at least one HDTV combined with one or several SDTV programs. DTV broadcasting can even be a combination of several SDTV programs during the day to allow for the addition of educational programming, e.g., and a single HDTV program in the evening. Given the importance of the decision and the different technical and operational experience, a cost-effective transition to DTV must be realized step by step with due consideration given to existing analog facilities and their integration and interfacing with a new DTV production facility. In addition, the possibility for new data services and the implementation cost versus revenues of each step must be considered. Any DTV equipment added at each step must be reusable for the following transitional steps. Broadcasters should anticipate a DTV transition phase of at least 3 to 5 years depending on the size and type of the broadcast station.

In transitioning to DTV, broadcasters must consider many important factors before selecting a new DTV broadcast format:

■ Status of existing production facilities. Their current performance, the presence of analog and/or digital NTSC infrastructures at 29.97 fps, digital 4:2:2, 4:3 to 16:9 switchable aspect ratio production equipment, and the reusability of existing SDTV equipment are important factors. Any scenario will involve up-conversion of existing analog programs, be they analog NTSC- or digital 4:2:2-based materials, for inclusion in a DTV channel. The present infrastructure may be old enough to justify replacing by an all-digital component (4:2:2) SDTV infrastructure. Digital systems increase productivity while reducing maintenance costs.

---

*The term *dark fiber* describes an unused fiber.

- Status of broadcaster. An affiliate (regional and local stations) or a major network do not have the same investment capability and revenue cash flow. Some unavoidable costs (digital transmitter and STL, antenna, tower) may not be affordable for small stations. Production equipment for DTV broadcasting is much less expensive, and a variety of solutions exist from a minimum implementation to a complete HDTV production facility.

- Type of DTV program. A main HDTV program or several additional SDTV programs on the same channel will lead to a different transition scenario. Some broadcasters think that there is not enough good programming to generate multiple SDTV programs. This situation will make advertising revenues too low and not profitable. Starting with one HDTV program will attract new viewers in search of good-quality programs while maintaining the present audience. Several SDTV programs could later be added to provide more programming choices.

- Budgetary constraints. HDTV equipment is more expensive than SDTV equipment at present, and available HD program materials are very limited. Up-conversion of existing SDTV program materials becomes a necessity. Many local stations will rely on their network for the HDTV programming.

- Delivery of new services. Factors such as HDTV attraction, multiple SDTV programs, data broadcasting, pay TV, and competition with the Web, e-mail, and interactive services can help in determining which new services make the most sense for generating added revenues.

- Choice of the HDTV format. At present, no existing digital equipment has a multiformat capability to meet 1080I, 720P, and 480P specifications at various frame rates for production equipment. It does not make sense to design multiformat DTV production systems, so a native format must be selected as the primary format that is used in all production and postproduction activities. A secondary native format can be added if necessary. NTSC can be this secondary format. An affiliate may not necessarily rebroadcast in the same format (e.g., receiving 1080I and rebroadcasting at 720P). The choice of this format has an impact on the efficient use of the available channel bandwidth of 6 MHz or 19.39 Mbps. Choosing 1080I may take up to 80 percent of the bandwidth.

  Although 720P exhibits a slightly lower picture quality performance than 1080I, there is a small advantage for the broadcaster. For the same picture quality, it can be compressed to a slightly lower bit rate (say, 17 Mbps instead of 19 Mbps), giving additional auxiliary data bandwidth that will increase broadcaster revenues.

- Choice of the frame rate. Present analog and/or digital NTSC infrastructures run at 29.97 fps. It would be very difficult to operate a TV facility at 29.97 and 30 fps simultaneously, so 29.97 fps will be the primary choice.

- DTV architecture flexibility. It must be similar to existing analog implementations and must allow adaptability to rapidly changing multimedia technologies and business plan requirements. A large use of media servers in a networking environment combined with a comprehensive asset management application can provide a valuable solution to the requirements of the new DTV architecture and can reduce operational costs.

- Logistic constraints. FCC requirements for simulcast until 2006 or later allow the broadcasters to maintain their current analog broadcast facilities while adding a new digital transmitter and antenna, upgrading the analog STL or adding a new digital STL, and modifying the existing antenna tower. The year 2006 limit makes any purchase of NTSC analog and digital equipment a nonprofitable investment.

- TV viewers. Early investigations both in Japan and Europe showed a strong customer preference for the 16:9 format. The wider picture is more pleasing, with an enhanced sense of realism. Programs based on sport events, nature documentaries, dramas, and feature films benefit from this format. Customers are first looking at a wider aspect ratio, and then a better sound and picture quality. Any good 4:3 SDTV program can be up-converted to a 16:9 HDTV format with an acceptable quality for the average viewer. DTV set sales will not be critically influenced by the choice of broadcasters for an HDTV or an SDTV format. Rather, price and availability of new DTV programs will be the major factors in the acquisition of new DTV sets. Viewers see the highest improvement in picture quality when transitioning from a 480 at 30I to a 480 at 60P that can be realized at the lowest DTV implementation cost, as shown in Fig. 12.41. On the contrary, higher levels of HDTV formats (720P and 1080I) provide lower differential improvements at much higher costs.

- Extracting the most possible picture quality information from existing archived TV materials. Section 12.5.8 explains how to best convert TV program materials recorded on 2-in, 1-in, Betacam SP, D-1, D-2, D-3, D-5, DCT-700, and digital Betacam videotape recorders to DTV formats. A very important picture quality difference can be noticed after DTV up-conversion between the same materials originating in NTSC and 4:2:2 formats. It is thus very important when up-converting TV programs to implement a conversion system that extracts the best quality from existing tape archives.

- DTV audio production and broadcasting. The lack of high-quality audio becomes noticeable in HDTV programs. There is no need to use a very high-quality audio in news and talk and game shows, but all other TV programs will benefit from surround sound. It provides a sound that replicates the cinema experience to capture public attention, so higher ratings and revenues can be expected. Surround sound is a low priority for some broadcasters, but many postproduction facilities are already using it for commercials. The reason is that surround sound brings an extra creative tool in production applications.

It is obvious that no one implementation will meet all the needs of broadcasters, the production industry, and DTV viewers. In the context of the simulcast

Figure 12.41   Differences in perceived picture quality between ATSC DTV formats.

requirement, different approaches will be developed to initiate the DTV transmission and the ultimate complete transition to DTV services. But any transitional scenario to DTV broadcasting necessarily implies the following:

■ The digital transmitter and power system must be purchased and modifications made to the antenna and the tower. A digital STL may be implemented by means of one of the three options described in Sec. 12.4.6.

■ Program and system information protocol (PSIP) data are inserted into the transport streams. These data allow multichannel programs to be transmitted to consumers along with information that can be displayed on the ATSC receiver, such as time, channel name, event name, and extended event description information. Users will be able to build an electronic program guide (EPG) on their DTV receiver screens. Then a PSIP server is requested in any transitional scenario. PSIP is defined in the ATSC Standard A/65 and supersedes the former EPG and SI Standards A/55 and A/56.

■ Local branding is an essential requirement. Insertion of logos (video overlays, chroma keying, screen logo) must be made possible even in a minimum scenario because it generates revenue and recognition of the local TV station.

- Important parts of the DTV infrastructure such as ATSC encoder/multiplexer, digital STL, PSIP, and data services servers should be fully redundant for the most important channels being broadcasted.

- Statistical multiplexing of MPEG-2 data streams is a very important feature that allows the broadcasting of several superior-video-quality programs into the same channel. This option may require a software addition compatible with the ATSC multiplexer installed in the DTV infrastructure.

- Studio engineers must try to eliminate the NSTC signal generation in the production chain by keeping the 4:2:2 original quality delivered by the camera or the source up to the up-converter for DTV transmissions. An NTSC encoder is then used to provide for simulcasting. A transitional scenario to DTV should therefore consider converting to digital 4:2:2 the present TV program production to generate an acceptable quality of DTV up-conversions.

By understanding the requirements for new all-digital video and audio production systems, we can develop a new architecture that borrows from the strengths of existing SDTV systems and utilizes new technologies to provide a robust and reliable DTV system for today's environment and for the future.

The broadcast industry is developing suitable technologies to allow a graceful move toward HDTV production and delivery. It is expected that advanced technologies will soon facilitate signal format conversions and make possible the production of complete programs in the compressed domain, thus significantly reducing the operating costs of contemporary HDTV studios. These improvements will impact image acquisition, signal compression, format conversion, storage, and networking capabilities and performance.

The unavoidable mix of 16:9 and up-converted 4:3 materials in the same DTV channel has generated some concern regarding the switching from one aspect ratio to another. Recent experience has showed that such a switch is not distracting for viewers as long as it takes place during an environment change.

12.7.1 Transition in video production and distribution

The present 525-line digital component format quality will continue to satisfy viewers for many years. Line doubling or quadrupling of these component format signals will provide real subjective picture improvements at the receiving end. Consequently, the transition in video production must take into account existing production facilities, be they recent NTSC or digital component studios.

- Minimum scenario. A minimum scenario involves the ATSC encoding of the existing analog program to generate an SDTV-based feed (480 at 30I) for DTV broadcasting, as shown in Fig. 12.42. This DTV program uses a small part of the available channel bandwidth. Optional data services (for home and business applications) can use the remaining channel bandwidth while providing additional revenue. PSIP and data services information are multiplexed with the DTV program before being sent to the digital transmitter through a digital STL.

Note 1: If the broadcast DTV format is not 480 @30I, an up-converter is necessary between the ATSC decoder and the analog MCR switcher.

Note 2: The NTSC encoder is necessary when the existing SDTV format is 4:2:2.

Figure 12.42 Migration of existing analog 4:2:2 or NTSC production facilities to DTV—minimum scenario.

If the broadcaster chooses to generate an HDTV-based feed from the existing analog program, an up-converter must be used. Local branding is inserted in the analog SDTV program before up-conversion. This scenario provides less flexibility than that of existing analog TV facilities and no possibility to broadcast real high-definition programs. Consumers are able to display the received program in a 4:3 or 16:9 aspect ratio.

To take full advantage of the 6-MHz bandwidth, the primary DTV program can be multiplexed with one or several secondary DTV programs created from the existing SDTV analog facilities. In this case, an additional analog SDTV switcher is used to feed one or several ATSC encoders, which send MPEG-2 program streams to the ATSC multiplexer.

■ Pass-through scenario with minimum local insertion. This scenario is most suitable for local stations that want to start with a minimum-cost DTV system. The network feed is decoded to its baseband format, e.g., from 1080 at 30I down to 480 at 30I. Local branding is essential to local stations and can be progressively added through small existing packages offered by manufacturers, such as Ez-HD from Leitch Technology. The DTV implementation provides flexibility for expansion (an HD-DVTR or server can be added) and full reusability in a future studiowide DTV infrastructure, as shown in Fig. 12.43. A frame synchronizer must be used to provide a clean switch between HDTV sources. The HD switcher is controlled by the NTSC MCR and must provide for HD source expansion. If the local station retransmits in a DTV format different from that of the DTV network feed, a format converter is necessary between the ATSC decoder and the HD router.

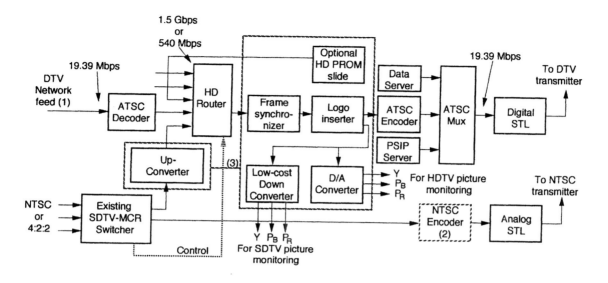

Note 1: The received DTV signal format can be 1080I, 720P or 480P.
Note 2: The NTSC encoder is necessary when the existing SDTV format is 4:2:2.
Note 3: These two system components are available in a package named Ez-HD from Leitch Techology.

Figure 12.43 Migration of existing analog 4:2:2 or NTSC production facilities to DTV—pass-through with minimum local insertion scenario.

System implementation flexibility requires a careful choice of DTV products having a common hardware platform that can be expanded and reused for subsequent improvements and additions during the transition period. For example, a DigiBus* platform can house up to 10 front processing cards and 10 rear I/O cards. It can incorporate HDTV cards such as generator, logo inserter, frame synchronizer, electronic store, digital-to-component analog and component analog-to-digital converters, DAs, a low-cost monitoring down-converter, and fiber-optic I/O cards meeting SMPTE 292M specifications that allow a maximum distance up to 65 km between DTV equipment. This modular platform can also be completed with one or several clusters of SDTV cards such as SDTV MPEG-2 encoder or decoder, VBI processor, audio AC-3 encoder, and MPEG-2 audio encoder. DTV signal interface cards such as ATM, DS-3, DVB-ASI, and SSI are available for signal distribution to affiliates or to the transmitter. The DigiBus platform offers a flexible hardware solution to accommodate a progressive transition to DTV, in HTDV, SDTV, or a mixture of both formats.

Broadcasters may elect to start their DTV broadcasting with HDTV local news insertions into a network feed. Local news is the most watched prime time program and the easiest to produce in an HDTV studio. Several man-

---

*DigiBus is a trademark of Leitch Technology International.

ufacturers have developed low-cost HDTV cameras and compressed DVTRs to provide a cost-effective solution to news production in HDTV formats. In addition, this same equipment can be used for shooting field content, such as news or documentaries. Newly designed HDTV cameras are optimized for a 16:9 aspect ratio. If a 4:3 output signal is simultaneously required, a DSP system derives it from the 16:9 signal. This provides a very good signal conversion quality and a low-cost solution for simulcasting. Figure 12.44 shows a modified pass-through scenario with local production and news material insertions.

■ Pass-through scenario with minimum local insertion—alternative. There is a great temptation to build a complete MPEG-2 studio environment. Several products are available to help in producing a simple MPEG-2 program, such as MPEG-2 encoders, splicers, logo keyers, servers, etc. The general use of the first-pass encoding decisions by subsequent encoders can facilitate the integration of MPEG-2 compressed video production equipment. However, it is premature to design such a complete MPEG-2 system due to the limited choice of production equipment as well as unsolved important compression-related issues. The main issues are:

*Note: The HD DVTR can be a Sony HDCAM format, or a Panasonic HD-D5 or DVC-Pro100. These DVTR's can downconvert HDTV signals to a SDTV format.

Figure 12.44 Migration of existing analog 4:2:2 or NTSC production facilities to DTV—pass-through with minimum local production and news insertions.

Production mixer and effects are unavailable.

There is a plethora of MPEG-2-compliant combinations of profiles and levels, making interoperability more difficult to realize.

A unique tape-recording format is unavailable. HD-Cam, DVC-Pro HD, HD-D5, and digital Betacam use different incompatible compression schemes, and no solution exists at present to ensure direct interfacing without a decompression-compression process.

There is difficulty in editing materials in the compressed domain.

MPEG-2 encoders and production and monitoring equipment are expensive.

It is difficult to achieve the convenience of current analog TV production facilities.

Figure 12.45 shows a pass-through scenario allowing the local insertion of MPEG-2-compressed materials in ATSC feeds.

■ HDTV production island scenario. This scenario is designed around a fully operational HDTV production facility that includes editing and graphics. DTV network feeds are ATSC-decoded, and existing SDTV production is up-converted to HDTV. Figure 12.46 shows a typical HDTV production studio, with all the convenience of current analog TV production facilities. This is

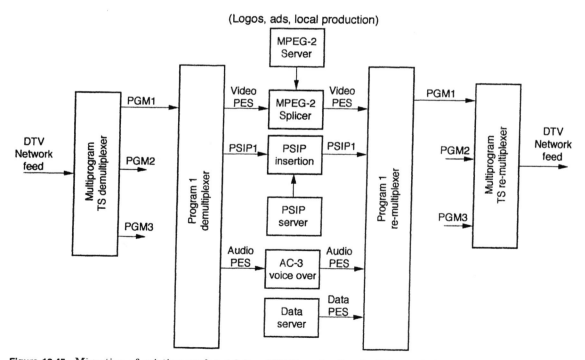

Figure 12.45 Migration of existing analog 4:2:2 or NTSC production facilities to DTV—alternate pass-through scenario with local TV material insertion in the MPEG-2-compressed domain.

Figure 12.46 Migration of existing analog 4:2:2 or NTSC production facilities to DTV—full HDTV production studio scenario.

the most expensive scenario, but it offers the best production quality and flexibility for future business expansion while maintaining the same ease of operations found in current analog SDTV facilities.

12.7.2 Transition in audio production and distribution

The ATSC transmission standard requires that the sound delivered suit everything from the inexpensive monophonic television receiver to the home theater picture and surround sound system. The two-channel stereo and the discrete 5.1 multichannel configuration are the most important. Multiple audio bit streams may be delivered simultaneously for multiple language transmissions or for services for the visually impaired. Other associated services provide for dialogue, commentary, voiceover, and emergency broadcasting. For DTV transmissions, ATSC has selected the AC-3 encoding scheme while the DVB consortium has preferred the MPEG audio layer 2.

Several experiments have shown that multiple successive AC-3 encoding/decoding processes bring significant sound quality degradation to program material due to repetitive signal conversions. To overcome this problem, the ATSC top-down report recommends that a contribution-quality coding scheme be developed to use the existing distribution infrastructure for surround sound audio signals. This recommendation has led to the concept of an intermediate level of compression, or mezzanine compression.

The AES-3 was designed for stereo audio transmission within a production or broadcast facility. Current digital and analog TV broadcast facilities support four analog or two AES-3 digital audio channels per video feed. Typical DVTRs can handle four audio channels only. Surround sound requires six mono channels (5.1 as described in Sec. 8.3.3.3). However, there is a requirement for a separate stereo-compatible mix (left and right); therefore, a minimum of eight audio channels must be considered in designing an audio production system for DTV broadcasting.

12.7.2.1 Requirements for surround sound and stereo mix audio production.
These requirements are as follows:

- High quality in production facilities. Production, postproduction, storage, and content delivery applications require high-quality audio materials that can be processed in either a PCM or a mezzanine compression format.

- Interoperability between the existing mono or stereo service and the new multichannel/multiservice broadcasting.

- Voiceover a mezzanine-compressed data stream. This is an essential requirement for national, regional, and local stations. Local branding activities (commercials, announcements, station IDs, alerts) ensure revenue. AC-3 signals cannot be decoded for local insertion and reencoded. Poor audio quality may result.

- Accurate multichannel audio synchronization. In production and distribution systems, the selected format must ensure that the original audio source materials are image-accurate in order to maintain the original spatial perception of the program. Sample slips in processing steps must be avoided.

- PCM format. In a studio environment, multiple audio signals can be distributed using a 12-channel serial digital interface (18.432 Mbps at 48 kHz) as defined in the SMPTE Standard 324M. This standard provides a reliable method of distributing multiple cophased channels of digital audio around the studio without losing the initial relative sample-phase relationship. For postmix applications, a channel assignment is defined in SMPTE 320M to facilitate interchange. Another transport alternative is the embedding of up to 16 audio channels in serial data streams, complying with SMPTE Standard 259M (for SDTV systems), or up to 24 channels for the SMPTE 299M (for HDTV systems). Both standards are described in Secs. 7.7 and 11.7 of this book. Embedded signals are then routed through existing bit-serial digital distribution and routing equipment. Multiple cascading of audio-video multiplexers should be avoided, as the relative timing of the six or more audio channels cannot be preserved. Audio embedders with four AES inputs are presently available on the market.

New ways of producing higher-quality audio signals and their transportation in a production studio are being developed by AES and IEEE. AES is defining a standard document for a 96-kHz audio sampling and is also considering 192-kHz audio sampling, along with new ways of distributing multiple AES channels on a high-speed interface. The IEEE 1394 standards committee has already started the documentation of a high-capacity and high-speed digital interface.

- Mezzanine compression encoders. Two solutions exist on the market, the enhanced apt-X and the Dolby-E compression schemes. They both compress eight discrete audio channels into a 3.072-Mbps data stream that can be carried by an AES-3 link. Mezzanine compression allows the reuse of an existing digital AES-3 audio infrastructure. Most of the current digital audio equipment can handle 20-bit audio data words. However, a few devices (DVC-Pro DVTRs, T1 PDH networks, satellite, etc.) have a 16-bit audio processing capability, and mezzanine compression schemes must be flexible enough to adjust to the suitable word length. The requirements for such a mezzanine scheme are a light compression ratio to obtain a transparent multigeneration quality performance, a low encoding-decoding processing delay (latency) to preserve audio-video synchronization, switching and editing capability, audio metadata throughput, an auxiliary data channel, time code embedding capability for time stamping, and a good robustness to errors in the transmission path.

  Given broadcasters' needs to produce multiple programs for DTV broadcasting and the use of data asset management to retrieve successive audio material generations, original audio materials will be reused and processed with an increasing number of generations. More than 10 generations are thus needed in a broadcast facility.

- Audio metadata. This information is generated in production and postproduction areas and is distributed with the 5.1 channel compressed signals to the AC-3 data stream multiplexer. It is called consumer metadata and is needed at the receiver end to exploit all capabilities of AC-3 signals (dynamic range control, dialogue normalization, audio mode, etc.) by listeners according to their preference and home installation. Information such as audio mode (5.1, 2.0, 1.0, Lt/Rt), channel assignment, compression type, time stamp, station ID, and location must be provided for processing, transport, distribution, and data asset management and constitute the professional metadata for studio environment use only. A separate metadata path and routing layer are needed for authoring and editing during the production process, as shown in Fig. 12.47. Metadata are created at the point where the surround mix is generated.

- Editing/switching. Uncompressed or compressed audio and video signals must have an exact alignment of frame boundaries to allow editing or

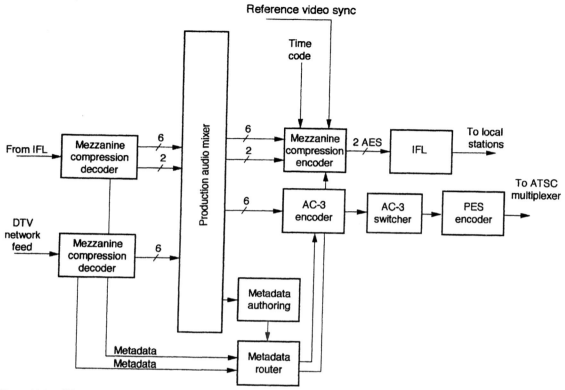

Figure 12.47 Migration to DTV broadcasting—digital audio metadata processing in a production studio.

switching of data streams. Compressed audio data are stored in time blocks. The time block duration limits the editing resolution, which is 80 μs for apt-X and 24 ms for the Dolby-E scheme.

■ Latency issues. Low latency is a requirement for real-time production. Time stamping is recommended at a time where the relative audio video timing is known. Audio-video synchronization issues are discussed in Sec. 12.6.5.

12.7.2.2 Enhanced apt-X compression scheme. The original version of this scheme was developed several years ago to allow the compression by a factor of 4 of 16-bit audio signals. It is described in Sec. 8.8.3.1. Enhancements to the original algorithm have made it possible to create a backward-compatible apt-X scheme, thanks to recent advances in DSP technology and speed.

Apt-X operates in the time domain and is based on a subband predictive coding (also called ADPCM) to reduce data. It uses historical information of the previous 122 samples to accurately predict the next sample value, which is compared with the actual sample when it arrives. This difference (or prediction error) is then encoded with a lower resolution and transmitted to the apt-X decoder. No masking has been used to permanently remove sound information in the original material. With the use of an accurate prediction, a near lossless compression-decompression cycle can be obtained.

Four AES channels can be encoded in a single 20-bit data word and embedded into one AES transport stream. The apt-X scheme features the following:

- A 4:1 audio bit rate reduction.

- High-precision conversion (16/20/24-bit selectable) to suit various transport capabilities.

- Minimal encoding-decoding processing delay (122 samples/48 kHz = 2.5 ms).

- Editing and switching quasi-synchronous with video frames. The very short packet length (80 μs) ensures fast switching with minimum disturbances that the compression algorithm can easily dissimulate. Editing can then take place without audible artifacts at edit points. Synchronous and asynchronous audio-video switching can be used. Low-cost editing tools already exist on the market.*

- Robustness to errors. No audible distortion can be noticed on an apt-X-based compressed program material at a bit error ratio of 10^4. One of the reasons is that the use of adaptive prediction and requantization reduces the significance of the errors by spreading their effects over a number of samples. With a short compressed data time block, apt-X features a high immunity to distribution and production data errors, such as bursts and dropouts.

- Capacity for extra data such as metadata and AES channel status.

- Increased dynamic range.

- Improved THD + noise.

- Virtually transparent performance allowing much more than 10 compression-decompression cycles.

- Low hardware complexity.

- Time code that can be embedded in the compressed data stream for time stamping.

DVTR manufacturers have developed apt-X-based compressors/decompressors to extend audio record/playback capabilities to 8 or 16 channels, such as the Panasonic AU-Z2123 that is combined with the D-5 DVTR. The apt-X scheme has been selected in Europe for DAB distribution to transmitter and for many other contribution links and audio distribution activities (movie com-

---

*A processing card is available from a well-known manufacturer and can be used to replace the incoming compressed audio data with newly compressed audio data.

panies and theaters, surround sound, disk-based storage systems, ISDN, digital audio workstation, etc.). In addition, its concatenation with an auditory masking-based compression scheme does not generate compounded artifacts.

12.7.2.3 Dolby-E compression scheme. Dolby-E and MPEG layers 2 and 3 use a psychoacoustic-based algorithm to remove sound information in the original audio signal and achieve a bit rate reduction. It operates in the frequency domain.

Dolby-E encodes up to eight audio channels plus the necessary metadata and inserts this information into the payload area (20 bits $\times$ 2 = 40 b per audio frame) of a single AES-3 transport stream. The encoded audio (payload) data rate is 1.92 Mbps (20 bits $\times$ 2 $\times$ 48 kHz). The Dolby-E scheme assumes that an error-free channel exists between the encoder and the decoder. Dolby-E frames include error detection information to be used by the decoder that mutes bad audio frames. The Dolby-E scheme features the following:

- A 4:1 audio bit rate reduction.

- Editing and switching synchronous with video frames. However, asynchronous audio-video switching cannot be used.

- Latency of encode-decode cycle of one video frame (33.33 ms).

- Capacity for extra data such as audio metadata.

- Up to 10 encode-decode generations in production applications. This limitation is due to the fact that successive transform coding generations can result in a cumulative noise that exceeds the auditory mask and then becomes audible. Problems in cascading two psychoacoustic algorithms such as Dolby-E and AC-3 may occur.

- Time code can be embedded in the compressed data stream for time stamping.

- Robustness to errors. Decoded data errors below the masking level are not audible. With a long compressed data time block, the Dolby-E scheme does not have a high immunity to distribution and production data errors, such as bursts and dropouts. An additional but expensive error correction system may be needed in some production and distribution areas.

12.7.2.4 DVB multichannel audio. DVB has adopted the MPEG-1 and MPEG-2 audio standards for transmission of multichannel audio signals to the receiver. Forward and backward compatibility with existing systems on the consumer markets and interoperability between other services, such as satellite, cable, terrestrial, and DAB, and computer software have been considered in selecting these compression schemes. DVB is considering the optional use of AC-3 in broadcast bit streams. Draft ETR-154 standard document Rev.3 describes in Annex C the mapping of AC-3 audio data streams in DVB-compliant transport streams. In addition, the embedding of audio-related ancillary data (metadata) into MPEG-2 audio elementary streams is described in Annex

D. AC-3 can be optionally used by the receiver to decode audio data signals having access to information such as the proper dynamic range, the dialogue normalization information, down-mixing levels, audio coding mode, and coarse and fine resolution time code.

12.7.2.5 Transition scenarios. The start of DTV transmission in North America has been made with a mono sound only. But the move to enhanced and high-quality pictures must be accompanied by a similar move to the surround sound system for audio signals. This makes the listener a real participant in the TV program, not just an observer.

The highest possible quality should be maintained in audio program origination, using a single 48-kHz sampling frequency. New production, distribution, and contribution systems must accommodate the 5+1 channel and mezzanine compression format signals of future TV programs. Transmission and delivery systems use AC-3 surround sound in North America and MPEG audio layer 2 in European countries. Network stations are making efforts to implement surround sound, but most of the affiliates have no intention of expanding their existing mono or stereo infrastructure. An audio mix is generated and converted to two discrete audio channels.

In designing digital audio installations, analog audio cables should be avoided. Their impedance is between 40 and 70 Ω and is very different from the 110 Ω requested by the AES-3 standard. This impedance mismatch creates AES signal reflections that degrade the received signal and can cause data stream transmission errors.

▪ Minimum scenario. A minimum scenario involves the existing audio production environment, be it mono or stereo. The audio console output signal feeds the AC-3 encoder that is PES-encoded and then sent to the ATSC multiplexer. Local branding is performed in the analog or digital uncompressed domain. The IEC-958 standard* specifies the formatting of AC-3-encoded data signals into a single AES/EBU data stream. Regular AES/EBU equipment can convey and route AC-3 signals for contribution and final delivery. Delay lines may be necessary to synchronize audio and video signals. Figure 12.48 shows a minimum audio studio implementation for DTV broadcasting. When transporting nonaudio data on AES-3 interfaces, any audio DAC must detect the C-bit information status and mute the converted signals to prevent any operational error, such as a distribution to an audio monitoring system.

▪ Pass-through scenario. This scenario is the most suitable for local stations that want to start with a minimum-cost DTV system. An audio console with 5.1 channel capability is used. The use of analog audio for DTV applications is not recommended, as it cannot handle time stamps that are required for

---

*This standard is included in Annex B of the ATSC A/52 document.

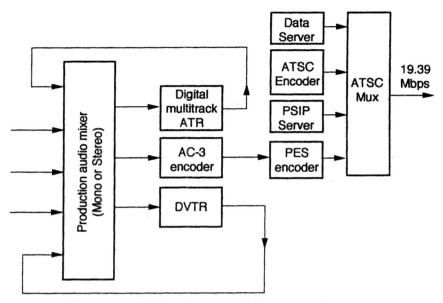

Figure 12.48   Migration to DTV broadcasting—minimum scenario.

maintaining an acceptable sound-picture synchronization. Digital consoles facilitate the work in surround sound in a similar way to stereo. In addition, they have built-in delay lines for audio-video synchronization. Digital audio workstations (DAWs) are available with surround sound processing by means of either dedicated hardware or software.

Audio processors and distribution systems must be adapted to DTV audio requirements. Mezzanine compression encoders are used in areas where bandwidth is limited. As an example, all current DVTRs have a processing capability of four discrete analog or two AES audio channels. The use of a mezzanine compressor-decompressor allows DVTRs to handle eight audio channels (six for surround and two for a compatible mix used in NTSC broadcasting). Figure 12.49 shows a typical block diagram of a such an audio processor with its associated DVTR. These compressed audio signals must be considered as nonaudio data. DVTRs must provide a totally transparent recording; i.e., audio signal adjustments and processors must not be used such as gain, preemphasis, error concealment, and sample rate conversion. SMPTE is developing an engineering guideline to define "characteristics of equipment in order to properly handle non-PCM AES3 streams."

Figure 12.50 shows a pass-through scenario where several options are included for the distribution of the DTV program to other broadcast facilities. Mezzanine compression data feeds can originate from sports events and are contributed by means of satellite or microwave links. As AC-3 DTV network signals cannot be decompressed and recompressed, an AC-3

Figure 12.49 Mezzanine compression encoding-decoding with associated DVTR.

Figure 12.50 Migration to DTV broadcasting—pass-through scenario.

Figure 12.51 Migration to DTV broadcasting—full implementation of a multiple-channel audio production system.

switcher allows the selection of network feed or the local audio production signals to feed the PES encoder and ATSC multiplexer.

▓ Full implementation scenario. Figure 12.51 shows a full audio implementation scenario. A mezzanine compression data encoder has been added to the previous pass-through scenario. It adds the possibility of distributing compressed audio feeds to affiliates and local stations. This IFL distribution can be realized by embedding the mezzanine compression signals into an MPEG-2 audio stream in a similar manner as for the AC-3 signals. A new standard defining the embedding of mezzanine compression scheme (apt-X or Dolby-E) into an MPEG-2 audio stream (with profile and PID to be determined) would thus be necessary for contribution links.

▓ A separate audio metadata path and routing layer must be added for authoring and editing during the production process, as shown in Fig. 12.47. Metadata are created from the production audio mixer information that is sent to an authoring system. This audio mixer must simultaneously process surround sound and stereo mix signals. These two separate audio signals have different processing latencies that must be compensated for in order to ensure synchronization of the surround sound signal with the MPEG-2 video signals for DTV broadcasting and the mono/stereo signal with the analog video signal for NTSC broadcasting.

List of Acronyms and Abbreviations

AAL	ATM adaptation layer
ABSOC	Advanced Broadcasting Systems of Canada
ACATS	Advisory Committee on Advanced Television Service
A/D	Analog-to-digital
ADM	Adaptive delta modulation
ADPCM	Adaptive differential pulse code modulation
ADSL	Asymmetric digital subscriber line
AES	Audio Engineering Society
AFD	Active format descriptor
ANSI	American National Standards Institute
APA	All-point addressable mode
API	Applications programming interface
ASCII	American Standard Code for Information Interchange
ASI	Asynchronous serial interface
ATEL	Advanced television evaluation laboratory
ATM	Asynchronous transfer mode
ATR	Audio tape recorder
ATSC	Advanced Television System Committee
ATTC	Advanced test television center
ATV	Advanced television
AUI	Attachment unit interface
AWG	American wire gauge
BER	Bit error rate
BIOS	Basic input/output system

B-ISDN	Broadband-integrated service digital network
BPM	Biphase mark
BRI	Basic rate interface
BRR	Bit rate reduction
BTA	Broadcasting Technical Association (Japan)
CAPM	Carrierless amplitude phase modulation
CATV	Community antenna television
CBR	Constant bit rate
CCIR	Comité Consultatif International en Radiodiffusion
CCITT	Consultative Committee on International Telegraphy and Telephony
CCS	Closed-captioning signal
CD	Compact disk
CD-I	Compact disk interactive
CD-ROM	Compact disk-read only memory
CDTV	Conventional definition television
CFF	Critical flicker frequency
CIF	Common Intermediate Format (MPEG-1)
CLUT	Color lookup table
Codec	Coder/decoder
COFDM	Coded orthogonal frequency division multiplexing
CPU	Central processing unit
CRC	Cyclic redundancy check
CRT	Cathode-ray tube
CSA	Canadian Standards Association
CSIF	Common Source Intermediate Format (MPEG-1)
CSMA-CD	Carrier sense multiple access with collision detection
CTI	Computer telephony integration
CTV	Conventional TV (NTSC, PAL, and SECAM systems)
CWS	Computer workstation
D/A	Digital-to-analog
DAB	Digital audio broadcasting
DARS	Digital audio reference signal
DAT	Digital audio tape
DAVIC	Digital Audio-Visual Council
DAW	Digital audio workstation
DBS	Direct broadcasting satellite
dc	Direct current
DCT	Discrete cosine transform
DMT	Discrete multitone

DPCM	Differential pulse code modulation
DPI	Dots per inch
DPRAM	Dual-port RAM
DRAM	Dynamic random access memory
DSP	Digital signal processing
DSS	Digital satellite system
DTV	Digital television
DU	Data unit
DV	Digital video (consumer-format VCR)
DVB	Digital video broadcasting—the European version of the ATV project in North America
DVCR	Digital video cassette recorder
DVD	Digital video disk
DVE	Digital video effects
DVI	Digital video interactive
DVTR	Digital videotape recorder
EAV	End of active video
EBU	European Broadcast Union
ECS	Entropy-coded segment
EDTV-II	Enhanced-definition television system
EIA	Electronic Industry Association
EISA	Extended Industry Standard Architecture
ELF	Extremely low frequency
ENG	Electronic news gathering
EOB	End of block
ES	Elementary stream
ETSI	European Telecommunication Standards Institute
FC-AL	Fiber channel-arbitrated loop
FCC	Federal Communications Commission
FDCT	Forward discrete cosine transform
FDDI	Fiber distributed data interface
FDM	Frequency division multiplexing
FEC	Forward error correction
FFT	Fast Fourier transform
FIFO	First in, first out
Gbps	Gigabits per second
GBps	Gigabytes per second
GBR	Green, blue, red
GOP	Group of pictures

GUI	Graphical user interface
HANC	Horizontal ancillary
HAS	Human auditory system
HDSL	High-bit-rate digital subscriber line
HDTV	High-definition television
HFC	Hybrid fiber coaxial
HSV	High-speed video (bus)
HVS	Human visual system
IDCT	Inverse discrete cosine transform
IDTV	Improved-definition television
IEC	International Electrotechnical Commission (part of the ISO)
IMA	Interactive Multimedia Association
I/O	Input/output
IRD	Integrated receiver-decoder
ISA	Industry Standard Architecture
ISDB	Integrated services digital broadcasting
ISDN	Integrated services digital network
ISO	International Standards Organization
ITU	International Telecommunication Union
JBIG	Joint Binary Image (Experts) Group
JPEG	Joint Photographic Experts Group
kbps	kilobits per second
LAN	Local-area network
LDTV	European definition for lower-definition television
LMDS	Local multipoint distribution service
LPF	Low-pass filter
LSB	Least significant bit
MADI	Multichannel audio digital interconnect
MAN	Metropolitan-area network
Mbps	Megabits per second
MBps	Megabytes per second
MCA	Microchannel architecture
MCP	Motion-compensated prediction
MCR	Master control room
MCU	Minimum coded unit
MDA	Monochrome display adapter
MDCT	Modified discrete cosine transform
MFLOPS	Millions of *floating-point* operations per *second*
MHEG	Multimedia and Hypermedia (information coding) Experts Group

MIPS	Millions of instructions per second
MMDS	Microwave multipoint distribution system
MMOS	Multimedia operating system
MNR	Mask-to-noise ratio
Modem	Modulator-demodulator
MPC	Memory and peripheral controller
MPEG	Moving Pictures Experts Group
MPEG-2 TS	MPEG-2 transport stream
MPR	Swedish Board for Measurement and Testing (initials in Swedish)
MSB	Most significant bit
MSL	Maximum signal level
MTBF	Mean time between failures
MUSE	Multiple sub-Nyquist encoding
MVP	Multimedia video processor
NII	National Information Infrastructure
NMS	Network multimedia system
NTSC	National Television System Committee
NVOD	Near video on demand
OSI	Open system interconnection (model—basis of data network standards)
PAL	Phase-alternating line
PAM	Pulse amplitude modulation
PCI	Peripheral component interconnect
PCM	Pulse code modulation
PCMCIA	Personal Computer Memory Card International Association
PCR	Program clock reference
PDS	Processor direct slot
PES	Packetized elementary stream
Pixel	Picture element
PQMF	Pseudo-quadrature mirror filter
PRI	Primary rate interface
PS	Program stream
PS/2	Personal System/2, computers introduced by IBM
PsF	Progressive segmented frame
PSIP	Program and system information protocol
PSTN	Public switched telephone network
PTS	Presentation time stamp
PVC	Permanent virtual circuit
PWM	Pulse width modulation

QAM	Quadrature amplitude modulation
QMF	Quadrature mirror filter
QoS	Quality of service
RAID	Redundant array of inexpensive disks
RAM	Random access memory
RAMDAC	Random access memory digital-to-analog converter
R-DAT	Rotary-digital audio tape
RISC	Reduced-instruction-set computing
RLC	Run length and level coding
RMS	Root mean square
SAV	Start of active video
SCSI	Small computer system interface
SDH	Synchronous data hierarchy
SDIF2	Sony Digital Interface Format
SDTV	Standard definition television
SIF	Source intermediate format (MPEG-1)
SMATV	Satellite master-antenna television
SMPTE	Society of Motion Picture & Television Engineers
SNR	Signal-to-noise ratio
SOL	Standard operating level
SONET	Synchronous optical network
SPA	Significant picture area
SPDIF	Sony Philips Digital Interface Format
SSA	Serial storage architecture
SSI	Synchronous serial interface
STC	System time clock
STL	Studio-to-transmitter link
STM	Synchronous transfer mode
SVC	Switched virtual circuit
SVGA	Super VGA (VESA standard)
TC	Transfer controller
TCO	Swedish Office-Workers' Union (initials in Swedish)
TDM	Time division multiplexing
THD	Total harmonic distortion
TS	Transport stream
TSL	Transmitter-to-studio link
TTL	Transistor-transistor logic
UTP	Unshielded twisted pair
VAFC	VESA—advanced feature connector

VANC	Vertical ancillary
VBI	Vertical blanking interval
VBR	Variable bit rate
VCR	Video cassette recorder
VESA	Video Electronics Standards Association
VGA	Video graphics adapter
VGA	Video graphics array
VITC	Vertical interval time code
VL-bus	VESA local bus
VLC	Variable-length coding
VLF	Very low frequency
VLSI	Very large scale integration
VOD	Video on demand
VRAM	Video random access memory
VSB	Vestigial sideband
VTR	Videotape recorder
WAN	Wide-area network
WORM	Write once, read many
XGA	Extended graphics array

SMPTE Standards

SMPTE 125M-1995 Television—Component Video Signal 4:2:2—Bit-Parallel Digital Interface

SMPTE 170M-1994 Television—Composite Analog Video Signal—NTSC for Studio Applications

SMPTE 207M-1992 Television—Digital Control Interface—Electrical and Mechanical Characteristics

SMPTE 224M Television Digital Component Recording—19-mm Type D-1—Tape Record

SMPTE 225M Television Digital Component Recording—19-mm Type D-1—Magnetic Tape

SMPTE 226M Television Digital Recording—19-mm Tape Cassettes

SMPTE 227M Television Digital Component Recording—19-mm Type D-1—Helical Data and Control Records

SMPTE 228M Television Digital Component Recording—19-mm Type D-1—Time and Control Code and Cue Records

SMPTE 240M-1995 Television—Signal Parameters—1125-Line High-Definition Production Systems

SMPTE 244M-1995 Television—System M/NTSC Composite Video Signals—Bit-Parallel Digital Interface

SMPTE 245M-1993 Television Digital Recording—19-mm Type D-2 Composite Format—Tape Record

SMPTE 246M-1993 Television Digital Recording—19-mm Type D-2 Composite Format—Magnetic Tape

SMPTE 247M-1993 Television Digital Recording—19-mm Type D-2 Composite Format—Helical Data and Control Records

SMPTE 248M-1993 Television Digital Recording—19-mm Type D-2 Composite Format—Cue Record and Time and Control Code Record

SMPTE 259M-1993 Television—10 Bit 4:2:2 Component and 4 f_{sc} NTSC Composite Digital Signals—Serial Digital Interface

SMPTE 260M-1992 Television—Digital Representation and Bit-Parallel Interface—1125/60 High-Definition Production System

SMPTE 263M-1996 Television Digital Recording—$\frac{1}{2}$-in Type D-3 Composite and $\frac{1}{2}$-in Type D-5 Component Formats—Tape Cassette

SMPTE 264M-1993 Television Digital Recording—$\frac{1}{2}$-in Type D-3 Composite Format—525/60

SMPTE 265M-1993 Television Digital Recording—$\frac{1}{2}$-in Type D-3 Composite Format—625/50

SMPTE 266M-1994 Television—4:2:2 Digital Component Systems—Digital Vertical Interval Time Code

SMPTE 267M-1995 Television—Bit-Parallel Digital Interface—Component Video Signal 4:2:2 16 × 9 Aspect Ratio

SMPTE 269M-1994 Television—Fault Reporting in Television Systems

SMPTE 272M-1994 Television—Formatting AES/EBU Audio and Auxiliary Data into Digital Video Ancillary Data Space

SMPTE 273M-1995 Television—Status Monitoring and Diagnostics Protocol

SMPTE 274M-1995 Television—1920 × 1080 Scanning and Interface

SMPTE 276M-1995 Television—Transmission of AES-EBU Digital Audio Signals over Coaxial Cable

SMPTE 279M-1996 Digital Video Recording—$\frac{1}{2}$-in Type D-5 Component Format—525/60 and 625/50

SMPTE 291M-1996 Television—Ancillary Data Packet and Space Formatting

SMPTE 292M-1996 Television—Bit-Serial Digital Interface for High-Definition Television Systems

SMPTE 293M-1996 Television—720 × 483 Active-Line at 59.94-Hz Progressive Scan Production—Digital Representation

SMPTE 294M-1997 Television—720 × 483 Active-Line at 59.94-Hz Progressive Scan Production

SMPTE 295M-1997 Television—1920 × 1080 50-Hz—Scanning and Interface

SMPTE 296M-1997 Television—1280 × 720 Scanning, Analog and Digital Representation and Analog Interface

SMPTE 297M-1997 Television—Serial Digital Fiber Transmission System for ANSI/SMPTE 259M Signals

SMPTE 298M-1997 Television—Universal Labels for Unique Identification of Digital Data

SMPTE 299M-1997 Television—24-Bit Digital Audio Format for HDTV Bit-Serial Interface

SMPTE 302M-1998 Television—Mapping of AES3 Data into MPEG-2 Transport Stream

SMPTE 303M Television—Color Reference Pattern

SMPTE 304M-1998 Television—Broadcast Cameras—Hybrid Electrical and Fiber-Optic Connector

SMPTE 305M-1998 Television—Serial Data Transport Interface

SMPTE 306M-1998 Television Digital Recording—6.35-mm Type D-7 Component Format—Video Compression at 25 Mbps—525/60 and 625/50

SMPTE 307M-1998 Television Digital Recording—6.35-mm Type D-7 Component Format—Tape Cassette

SMPTE 308M-1998 Television—MPEG-2 4:2:2 Profile at High Level

SMPTE 309M Television—Transmission of Date and Time Zone Information in Binary Groups of Time and Control Code

SMPTE 310M-1998 Television—Synchronous Serial Interface for MPEG-2 Digital Transport Streams

SMPTE 311M-1998 Television—Hybrid Electrical and Fiber Optical Camera Cable

SMPTE 312M-1999 Television—Splice Point for MPEG-2 Transport Streams

SMPTE 314M-1999 Television—Data Structure for DV-Based Audio, Data, and Compressed Video—25 and 50 Mbps

SMPTE 315M-1999 Television—Camera Positioning Information Conveyed by Ancillary Data Packets

SMPTE 316M Television Digital Recording—12.65-mm Type D-9 Component Format—Video Compression—525/60 and 625/50

SMPTE 317M Television Digital Recording 12.65-mm Type D-9 Component Format—Tape Cassette

SMPTE 318M Television and Audio—Reference Signals for the Synchronization of 59.94- or 50-Hz Related Video and Audio Systems in Analog and Digital Areas

SMPTE 319M Proposed Standard for Television—Transport of MPEG-2 Recoding Information through 4:2:2 Component Digital Interfaces

SMPTE 320M Television—Channel Assignments and Levels on Multichannel Audio Media

SMPTE 321M Proposed Standard for Television—Data Stream Format for the Exchange of DV-Based Audio, Data and Compressed Video over a Serial Data Transport Interface

SMPTE 322M Proposed Standard for Television—Format for Transmission of DV Compressed Video, Audio and Data over a Serial Data Transport Interface

SMPTE 323M Motion-Picture Film—Channel Assignments and Levels on Multichannel Audio Media

SMPTE 324M Television—12-Channel Serial Interface for Digital Audio and Auxiliary Data

SMPTE 325M Digital Television—Opportunistic Data Broadcast Flow Control

SMPTE 326M Proposed Standard for Television—SDTI Content Package Format (SDTI-CP)

SMPTE 327M Proposed Standard for Television—MPEG-2 Video Recoding Data Set

SMPTE 328M Proposed Standard for Television—MPEG-2 Video Elementary Stream Editing Information

SMPTE 329M Proposed Standard for Television—MPEG-2 Video Recoding Data Set—Compressed Stream Format

SMPTE 330M Proposed Standard for Television—Unique Material Identifier (UMID)

SMPTE 331M Proposed Standard for Television—Element and Metadata Definitions for the SDTI-CP

SMPTE 332M Proposed Standard for Television—Encapsulation of Data Packet Streams over SDTI (SDTI-PF)

SMPTE 333M Proposed Standard for Television—DTV Closed-Caption Server to Encoder Interface

SMPTE Recommended Practices

RP 154-1994 Reference Signals for the Synchronization of 525-Line Video Equipment

RP 155-1997 Audio Levels for Digital Audio Records on Digital Television Tape Recorders

RP 159-1995 Vertical Interval Time Code and Longitudinal Time Code Relationship

RP 160-1997 Three-Channel Parallel Analog Component High-Definition Video Interface

RP 164-1996 Location of Vertical Interval Time Code

RP 165-1994 Error Detection Checkwords and Status Flags for Use in Bit-Serial Digital Interfaces for Television

RP 168-1993 Definition of Vertical Interval Switching Point for Synchronous Video Switching

RP 169-1995 Television, Audio and Film Time and Control Code— Auxiliary Time Address Data in Binary Groups—Dialect Specification of Directory Index Locations

RP 174-1993 Bit-Parallel Digital Interface for 4:4:4:4 Component Video Signal (Single Link)

RP 175-1997 Digital Interface for 4:4:4:4 Component Video Signals (Dual Link)

RP 176-1997 Derivation of Reference Signals for Television Camera Color Evaluation

RP 177-1997 Derivation of Basic Television Color Equations

RP 178-1996 Serial Digital Interface Check Field for 10-Bit 4:2:2 Component and $4f_{SC}$ Composite Digital Signals

RP 184-1996 Measurement of Jitter in Bit-Serial Digital Interfaces

RP 186-1995 Video Index Information Coding for 525- and 625-Line Television Systems

RP 187-1995 Center, Aspect Ratio and Blanking of Video Images

RP 188-1996 Transmission of Time Code and Control Code in the Ancillary Data Space of a Digital Television Data Stream

RP 192-1996 Jitter Measurement Procedures in Bit-Serial Digital Interfaces

RP 196-1997 Transmission of LTC and VITC Data as HANC Packets in Serial Digital Television Interfaces

RP 197-1998 Film-to-Video Transfer List

RP 198-1998 Bit-Serial Digital Checkfield for Use in High-Definition Interfaces

RP 199-1999 Mapping of Pictures in Wide-Screen (16:9) Scanning Structure to Retain Original Aspect Ratio of the Work

RP 200-1999 Relative and Absolute Sound Pressure Levels for Motion-Picture Multichannel Sound Systems

RP 201-1999 Encoding Film Transfer Information Using Vertical Interval Time Code

RP 202 Proposed Recommended Practice—Video Alignment for MPEG-2 Coding

RP 203 Proposed Recommended Practice—Real Time Opportunistic Data Flow Control in an MPEG-2 Transport Emissions Multiplex

RP 204 Proposed Recommended Practice—SDTI-CP MPEG Decoder Templates

RP 205 Proposed Recommended Practice—Application of Unique Material Identifiers in Production and Broadcast Environments

RP 206 Proposed Recommended Practice—Opportunistic Data Flow Control Using Ethernet as a Control Channel in an MPEG-2 Transport Emissions Multiplex

SMPTE Engineering Guidelines

EG 27-1994 Supplemental Information for ANSI/SMPTE 170M and Background on the Development of NTSC Color Standards

EG 32-1996 Emphasis of AES/EBU Audio in Television Systems and Preferred Audio Sampling Rate

EG 33-1998 Jitter Characteristics and Measurements

EG 34-1999 Pathological Conditions in Serial Digital Video Systems

EG 35 Time and Control Code Time Address Clock Precision for Television, Audio and Film

EG 36 Transformations between Television Component Color Signals

AES Standards

AES3-1992 (ANSI S4.40-1992) AES Recommended Practice for Digital Audio Engineering—Serial Transmission Format for Two-Channel Linearly Represented Digital Audio Data (Revision of AES3-1985, ANSI S4.40-1985)

AES5-1984 r1992 (ANSI S4.28-1984) AES Recommended Practice for Professional Digital Audio Applications—Employing Pulse-Code Modulation—Preferred Sampling Frequencies

AES10-1991 (ANSI S4.43-1991) AES Recommended Practice for Digital Audio Engineering—Serial Multichannel Audio Digital Interface (MADI)

AES11-1991 (ANSI S4.44-1991) AES Recommended Practice for Digital Audio Engineering—Synchronization of Digital Audio Equipment in Studio Operations

AES17-1991 (ANSI S4.50-1991) AES Standard Method for Digital Audio Engineering—Measurement of Digital Audio Equipment

AES18-1996 (Revision of AES18-1992) AES Recommended Practice for Digital Audio Engineering—Format for the User Data Channel of the AES Digital Audio Interface

AES Information Documents

AES-2id-1996, AES Information Document for Digital Audio Engineering—Guidelines for the Use of the AES3 Interface

AES-10id-1995, AES Information Document for Digital Audio Engineering—Engineering Guidelines for the Multichannel Audio Digital Interface (MADI) AES10

ITU-R (ex-CCIR) Standards

ITU-R BT.601-5—Studio Encoding Parameters of Digital Television for Standard 4:3 and Wide-Screen 16:9 Aspect Ratios

ITU-R BT.656—Interfaces for Digital Component Video Signals in 525-Line and 625-Line Television Systems Operating at the 4:2:2 Level of Recommendation 601

ITU-R BT.962-2—The Filtering, Sampling and Multiplexing for Digital Encoding of Color Television Signals

ITU-R BT.709-3—Parameter Values for the HDTV Standards for Production and International Program Exchange

ITU-R BT.711-1—Synchronizing Reference Signals for the Component Digital Studio

ITU-R BS.1116—Methods for the Subjective Assessment of Small Impairments in Audio Systems Including Multichannel Sound Systems

ITU-R BT.1119-2—Wide Screen Signalling for Braodcasting—Signalling for Wide Screen and Other Enhanced Television Parameters

ITU-R BT.1120-1 Digital Interfaces for 1125/60/2:1 and 1250/50/2:1 HDTV Studio Signals

ITU-R BT.1126—Data Transmission Protocols and Transmission Control Scheme for Data Broadcasting Systems Using a Data Channel in Satellite Television Broadcasting

ITU-R BT.1361—Worldwide Unified Colorimetry and Related Characteristics of Future Television and Imaging Systems

ITU-R BT.1365—24-Bit Digital Audio Format as Ancillary Data Signals in HDTV Serial Interfaces

ITU-R BT.1376—Serial Digital Fibre Transmission System for Signals Conforming to ITU-R BT.656, ITU-R BT.799, ITU-R BT.1120.

ITU-T (ex-CCIR) Standards

ITU-T G.702—Digital Hierarchy Bit Rates

ITU-T G.703—Physical/Electrical Characteristics of Hierarchical Digital Interfaces

ITU-T G.707—Digital Transmission Systems—Terminal Equipment—General

ITU-T G.804—ATM Cell Mapping into Plesiochronous Digital Hierarchy (PDH)

ITU-T I.363—B-ISDN ATM Adaptation Layer (AAL) Specification

ITU-T J.82—Transport of MPEG-2 Constant Bit Rate Television Signals in B-ISDN

ITU-T J.117—Home Digital Network Interface Specifications

ANSI Standards

ANSI X3T10.1—Serial Storage Architecture

ANSI X3.230-1994—Fiber Channel—Arbitrated Loop (FC-AL)

ANSI X3.131-1994—Small Computer System Interface (SCSI-2)

ANSI T1.E1.4/94—Asymmetric Digital Subscriber Line (ADSL)

ANSI/VITA 5-1994—RACEway Interlink Architecture

ANSI T1.519-1999—Specifications for Transport of Generic Packets over DS Hierarchy

IEEE Standards

IEEE 1394-1995—High Speed Serial Bus

IEEE 802.3—Ethernet

IEEE 802.3ab-1999—Physical Layer Parameters and Specifications for 1000 Mbps Operation over 4-Pair of Category 5 Balanced Copper Cabling, Type 1000BASE-T

IEEE 802.3z-1998—Gigabit Ethernet

ATSC Standards

A/49—Ghost Canceling Reference Signal for NTSC

A/52—ATSC Digital Audio Compression (AC-3)

A/53—ATSC Digital Television Standard

A/54—Guide to the Use of the ATSC Digital Television Standard

A/55—Program Guide for ATSC Digital Television Standard

A/56—System Information for ATSC Digital Television Standard

A/57—Program/Episode/Version Identification

A/58—Harmonization with DVB SI in the Use of the ATSC Digital Television Standard

A/63—Standard for Coding 25/50 Hz Video

A/64—Transmission Measurements and Compliance for Digital Television

A/65—Program and System Information Protocol for Terrestrial Broadcast and Cable

A/70—Conditional Access System for Terrestrial Broadcast

A/80—Modulation and Coding Requirements for Digital TV (DTV) Applications over Satellite

ETSI/DVB Standards

ETS 154 Implementation Guidelines for the Use of MPEG-2 Systems, Video and Audio in Satellite, Cable and Terrestrial Broadcasting Applications

ETS 300 421 DVB Framing Structure, Channel Coding and Modulation for 11/12 GHz Satellite Services

ETS 300 473 DVB—Satellite Master Antenna Television (SMATV)

ETS 300 744 DVB Framing Structure, Channel Coding and Modulation for Digital Terrestrial Television (DVB-T)

ETS 300 813 DVB Interfaces to Plesiochronous Digital Hierarchy (PDH) Networks

ETS 300 814 DVB Interfaces to Synchronous Digital Hierarchy (SDH) Networks

DVB-A010 Interfaces for CATV/SMATV Headends and Similar Professional Equipment (EN-50083-9).

ISO/IEC Standards

ISO/IEC IS 11172 MPEG-1 Systems

ISO/IEC IS 13818 MPEG-2 Systems

Index

ABOUT THE AUTHORS

MICHAEL ROBIN was initially a studio systems project engineer in charge of various national and international studio projects early in his 20-year tenure with the CBC. Initially he participated in the various stages of the CBC Maison de Radio Canada (CBC French Network Headquarters) consolidation project. This was one of the few multistudio projects designed specifically with NTSC color in mind, unlike other organizations where NTSC color was added to an existing inadequate black-and-white television structure. His subsequent projects were standards conversion facilities (NTSC, PAL, SECAM) for international program exchange, film-to-to-tape transfer facilities, automated program delay facilities, as well as CBC London and Paris studio facilities. Later activities include the development of video and audio corporate acceptance test procedures, test equipment for initially analog and later digital composite (4Fsc) and component (4:2:2) digital equipment and systems. In this capacity, he was involved in R&D, technical investigations, and SMPTE standards committee work, and he has trained CBC engineers and technicians in various areas of analog and digital audio and video applications. Together with Michel Poulin, he was heavily involved in the development and implementation of the CBC all 4:2:2/SDI digital Toronto Broadcast Center (CBC English Network Headquarters). He is currently an independent broadcast consultant to various national and international broadcasters and gives analog and digital video and audio courses to broadcast engineers and technicians. In addition, he acts as a digital video technical consultant for the *Broadcast Engineering Journal* and contributes a monthly "Transition to digital" column.

MICHEL POULIN has gained his knowledge in the CBC Engineering laboratories, working for 14 years as a technology evaluation engineer, then as a supervising engineer in charge of the Technology Evaluation Group. He is now with Leitch Technology International, in charge of the R&D laboratories and the Standards Compliance Group. He is an active participant in the SMPTE and IEEE standardization committee activities. He is a Fellow of the SMPTE. He is also co-author of the first edition of *Digital Television Fundamentals*, and lives in Toronto, Ontario, Canada.